# The Professional Housekeeper

Second Edition

Georgina Tucker
&
Madelin Schneider

A CBI Book
Published by Van Nostrand Reinhold Company
New York

A CBI Book
(CBI is an imprint of Van Nostrand Reinhold Company Inc.)


Library of Congress Catalog Card Number
ISBN 0-8436-2252-0



Printed in the United States of America

Published by Van Nostrand Reinhold Company Inc.
115 Fifth Avenue
New York, New York 10003

Van Nostrand Reinhold Company Limited
Molly Millars Lane
Wokingham, Berkshire RG11 2PY, England

Van Nostrand Reinhold
480 La Trobe Street
Melbourne, Victoria 3000, Australia

Macmillan of Canada
Division of Canada Publishing Corporation
164 Commander Boulevard
Agincourt, Ontario M1S 3C7, Canada

16 15 14 13 12 11 10 9 8 7 6 5

**Library of Congress Cataloging in Publication Data**

*This book is dedicated to the great housekeepers of today who have raised their ancient calling to the status of an important new profession . . . and to the Professional Housekeepers of tomorrow, the imaginative young people of today who will be able to see beyond the stuffy, old-fashioned word "Housekeeping" and realize that it can be a noble profession—this business of keeping a beautiful building beautiful. Their innovative contributions to the planning and polishing of accommodations and public places and spaces will add measurably to fine living for the increasing millions of guests and residents whose stay away from home will last a night, a week, a semester, a season or longer.*

# Contents

# *Foreword*

The greatest demand made by the public on the hospitality industry is for proper value received from lodging establishments. The very keystone of this demanded and deserved value is comfortable and clean accommodations. The success of any organization which provides accommodations depends on good housekeeping. Inadequate attention to detail in this critical area is the single largest factor causing a reduction of the perception of value received.

To a lodging establishment dependent on profit, the importance of housekeeping is paramount. Certainly there are many areas of activity which demand attention from managers and owners, but the maintenance and cleanliness of the structure is the factor which must precede other concerns if profit is to continue. No amount of advertising, entertainment, glamour, or friendliness will overcome poor housekeeping in the eyes of the guest. Proof of this is easily seen in examples of hotels and resorts that have received worldwide recognition for decades. Some modern, elegant operations which are not well kept do not gain acceptance from the public. Good housekeeping is the base that is required first and is necessary for success regardless of the quality of other marketing efforts of management.

*The Professional Housekeeper* reviews in depth the myriad ideas, talents, and disciplines needed in this varied and challenging field. In my experience the two most important skills of the professional housekeeper lie in the administration, and motivation of people. A sense of humor is also useful.

The success of any hospitality enterprise is totally dependent on the professional knowledge and the personnel skills of its housekeeper.

Harry Mullikin
Chairman and Chief Executive Officer
Westin Hotels

Past President—1975
American Hotel & Motel Association

# Preface

*The Professional Housekeeper,* 2nd edition, is intended as a resource for professional housekeepers and teachers in the institutional field. It is also designed to instruct students in technical schools and colleges who are seeking career information about this important part of the hospitality and lodging field.

Housekeeping is responsible for providing suitable accommodations for guests throughout the world, and this book stresses the management techniques necessary to operate this department. Newcomers to the lodging industry will find solutions to many problems suggested by veteran supervisors and housekeepers. The solutions may be adjusted and amended.

The greatest advance in housekeeping has been the introduction of new business techniques: the use of computers and the realization of management of the necessity of training in these areas for the benefit of the industry.

The one constant factor continues to be the comfort and enjoyment of the guest. To this end, the book continues to emphasize the number one product and concern of the industry.

In revising the book the main object has been to update products, people motivation methods, and techniques. The new material should prove beneficial to many institutions in this industry.

# *Authors' Note*

## *Equal Rights Versus the English Language*

Despite all the efforts to change the English language in deference to the cause of equal rights for men and women, the authors would like to explain that when the term "he" is used it describes a person of either sex. This is the traditional form of English and is not meant to apply only to the male. In the sentence that follows there is no intention that the titles, or the "he's" and "his's" refer only to the masculine gender. "The director of rooms and his division will take care of the reception area, cashiers, bellmen, parking attendants, doormen—all the sections concerned with a guest, from greeting him to adding up his bill as he leaves." Using "he" is conventional English and helps simplify sentences.

## *The Battle of Maids Versus Room Attendants*

In this increasingly unisex society there has been a recent tendency to change the term "maid" to "room attendant" because men might become maids and would prefer the less feminine appellation. Throughout this book we will use both terms but more often the traditional "maid"—especially to separate the customary duties of the maid and houseman. We prefer to avoid such ludicrous expressions as houseperson and bellperson.

Some hotels call their room attendants housekeepers, some call them roomkeepers. Whatever fits the personnel, the environment, and the clientele is the right name to use. A maid by any other name would still be welcome.

# *Chapter 1*

# *Evolution of the Lodging Industry*

**OVERVIEW** Hotel-keeping is anything but lackluster and monotonous. Older than most fields of work, the hotel industry has a long colorful history beginning with missions, inns, and lodges, and evolving into the hospitality industry we know today with hotels, motels, grand resorts, and finally the formation of the eminently successful chains across the United States. In this chapter you'll learn about the roots of this romantic business of which housekeeping is such an integral part.

## *History*

Perhaps the earliest mention of the hospitality industry and the incident with which most of us are familiar is found in the Christmas story. In it we are told that two famous travelers, Joseph and Mary, could find "no room at the inn," and thus spent the night in a stable. Today the alternatives are hardly as uncomfortable or unsanitary since front-desk clerks at crowded inns can often locate available accommodations by means of electronic equipment—and would certainly have summoned an ambulance for history's most famous mother.

The oldest hotels in existence can be seen at Pompeii in southern Italy, where the eruption of Mt. Vesuvius in 79 A.D. buried the town under hot lava that preserved numerous buildings almost intact. This has enabled us to obtain a glimpse of what "hotel life" was like when Rome ruled the world.

After the fall of Rome, religious orders built hospices as places of shelter and rest for pilgrims and strangers alike. The seven Crusades of 1096 to 1270 saw the initial success and final failure of the Christians of Europe to retake the Holy Land from the Muslims. The Crusades did stimulate the expansion of trade and rise of a middle class that gave impetus to the lodging industry. This was first felt in northern Italy, where the original innkeepers' guild was formed in Florence in 1282.

Because the main purpose of the lodging industry has always been to provide accommodations for travelers who are away from home on business or pleasure, its growth was understandably hampered by the very primitive transportation methods that prevailed prior to the 17th century.

The real beginning of the industry may be traced to 1658, when the first passenger-carrying stagecoach lines began to operate in England. Inns, or taverns as they were customarily called, quickly sprang up along every coach route to provide overnight hospitality to the 17th century forerunner of today's truly mobile society.

Between 1825 and 1850, the railroads replaced the coach routes. The trains sped directly from city to city without making the frequent overnight stops along the way that had supported the network of English country inns, which now fell rapidly into disuse.

Not until the 20th century were the charm and commercial possibilities of these quaint establishments recognized by major British hotel chains. Their properties today include scores of 17th century inns, as well as a few that date back 400 years. All have been beautifully restored, with the addition of only such essential features as electric lights and modern plumbing.

Perhaps the oldest inn still standing in the

United States is a stone house in Guilford, Connecticut, built in 1640. In 1641 the Dutch built a four-story stone tavern at the southern tip of Manhattan to ,accommodate the English who sailed daily between New England and Virginia.

## *Early Inns in the United States*

The two most authentic examples of pre-Revolutionary War inns are Raleigh Tavern and Wetherburn's Tavern in colonial Williamsburg. Capitol and principal city in Virginia from 1699 to 1779, it has been magnificently restored as a result of grants and bequests from John D. Rockefeller, Jr. (Figure 1-1). Raleigh Tavern is an exact reconstruction of an inn built before 1742. Its patrons included Washington, Jefferson and Patrick Henry.

Another outstanding example of an early American inn is New York's Fraunces Tavern near Wall Street. Built in 1791, it is best known as the site of Washington's farewell to his troops in 1783. Today it is a restaurant and museum.

Across the continent, between 1769 and 1823, the Franciscan padres under the leadership of Fathers Junipero Serra and Fermin Lasuen, took the first step toward the modern motel chain by establishing a chain of missions (Figure 1-2). These were each spaced a day's journey apart, along the 600 miles of *El Camino Real* (the King's Highway) that stretched from San Diego to Sonoma, north of San Francisco.

Each mission was designed to be a self-contained community and a resting place for persons of all religious persuasions, whites, and Indians. Housekeeping consisted of sweeping tiled or dirt floors, and replacing straw mats that became vermin-infested.

By the early 1800s, it was apparent that leadership in the hotel field was considered by our status-makers to be still another inalienable right of our manifest destiny. Luxury was piled upon luxury and service upon service. In those days, sheer size was equated with true magnificence.

Europe had its castles, but nothing to compare with Boston's Exchange Coffee House, with 200 rooms on seven floors to make it the largest building in America; and nothing to compare with the City Hotel built in Baltimore by David Barnum in 1826, which hotel historians rank as the first of the *grande luxe* establishments. But neither of these could hold a candle to two of the earliest creations of celebrated hotel architect Isaiah Rogers: Boston's princely Tremont House of 1829, and New York's Astor House of 1834, a veritable palace.

By this time, the hotel craze had spread to

**FIGURE 1-1** *Market Square Tavern, Williamsburg, Va. has been used intermittently as an inn for three centuries, today is part of the Williamsburg Inn. Thomas Jefferson was its most celebrated lodger. Oldest section of the inn contains the Great Room, which is still sheathed in the original pine paneling.*

**FIGURE 1-2** *America's first lodging chain was established as early as 1769 by Father Junipero Serra, who set up missions one day's journey apart along the 600-mile El Camino Real (King's Road) in California. Food and lodging were provided to all who requested it in these 21 self-contained communities. Pictured here is the Mission at San Rafael.*

the point where the rich considered the amenities of hotel living to be in many ways superior to those of home. This attitude gave rise to the resort hotel where the affluent could idle away a week or two, a month, or even an entire season.

## *Resorts for the Rich*

A 300-room resort hotel, the Mountain House, was built in New York's Catskill Mountains as early as 1824. By 1834, Saratoga Springs was earning recognition from New York's bluest blue bloods as Queen of the Spas. By 1864, Saratoga unveiled its breathtaking Grand Union Hotel, with 12 acres of carpeting and a record-making front porch that stretched out for over a quarter of a mile, almost as far as the eye could see.

The Greenbrier, one of the world's truly fabled resorts, can trace its ancestry to pre-Colonial times, when the Indians discovered the curative power of the waters at what is now White Sulphur Springs, West Virginia. By 1778, crude shelters had been built for visitors seeking relief from rheumatic maladies. Early owners Bowyer and Wiley subsequently built cabins and a main tavern. When Bowyer died in 1809, his estate included 11 wash bowls, 33 cots, 23 mattresses, 647 pounds of pillows and bolsters, 48 blankets, 45 counterpanes, 70 sheets, and 1 chamber pot.

The Spring House, now the Greenbrier's trademark, was built in 1830, but it was not until 1858 that the first full-facilities resort structure was opened on the premises (Figure 1-3). Called Old White, it became, briefly, a social mecca for persons of great affluence and influence; then from 1861 to 1865 it served as a hospital for North and South alike—soldiers from both sides are buried on its grounds.

Two subsequent events assured the revival of the resort's prewar eminence: in 1867, Robert E. Lee made White Sulphur Springs his summer home; in 1870, the railroad arrived. Bad management almost led to its demise, which was averted when the property was sold to the Chesapeake & Ohio Railway.

In 1910, the Greenbrier as we know it today was built by the C&O. Through the years additions have included three golf courses, indoor and outdoor swimming pools, mineral baths, tennis club, presidential cottage, and three new wings.

**FIGURE 1-3** *The grandest of the old U.S. spas was in White Sulphur Springs, W. Va.; it was originally called Old White, and today is known as the Greenbrier. Though this picture was taken in the early 1930s, the elegant building has changed little except that it is more carefully maintained. The U.S. Government used it as an army hospital and internment center during World War II. After the war the Chesapeake & Ohio Railroad bought the Greenbrier back from the army. Millions of dollars were invested in the restoration of the hotel and cottages and in landscaping its 7000 acres. Interior decorator Dorothy Draper gained world renown from furnishing its enormous public areas on a grand scale. After 191 years, the Greenbrier continues to draw prestigious guests.*

## *Early Southern Spas*

Other early Southern spas included Florida's Ponce de Leon, built in St. Augustine in 1888. With 208 rooms on 350 acres, the Ponce de Leon is now Flagler College.

Perhaps the Greenbrier's most serious contender for southern spa supremacy is the Boca Raton Hotel and Club, located on 500 acres of the world's most valuable earth, between Palm Beach and Fort Lauderdale. The Boca Raton takes a championship golf course and mile-long private beach for granted, and offers its super-rich clientele such appropriate options as skeet shooting and polo.

The Boca Raton recently added a 246-room, 26-story tower. The original 368-room, 6-story building was designed as a private club in the early 1920s by architect Addison Mizner, whose fees are as legendary as his fame. Mizner was the absolute master of a style described by some as nouveau Spanish rococo. Boca Raton is perhaps his handsomest and most restrained creation. Among its charms for the housekeeping department are quarters that compare favorably with the guest rooms at almost any other hotel.

While resorting touched the tip of grandeur in the southeast and northeast, the midwest made two contributions that are worth mentioning. The Grand Hotel is located on an island in the Straits of Mackinac, a body of water separating the major portion of the state of Michigan from its remote, northernmost Upper Peninsula. Famed for a front porch second in length only to that of Saratoga's Grand Union, the Michigan resort is idyllically located on an island where one still travels from place to place by horse and carriage or by bicycle. As an alternate, a guest may be able to arrange transportation in one of the island's two motor vehicles—an ambulance and a fire engine.

Far to the south of Mackinac stands French Lick Hotel on 1600 glorious acres of southern Indiana countryside. French Lick reached its first peak during the prohibition era, when Chicagoans arrived by the trainload to savor a lifestyle that anticipated Las Vegas by decades. The hotel skidded toward oblivion when the trains stopped running, but is now prosperous as a convention center, having been taken over by Sheraton and restored at a cost of millions. French Lick is as famous as the Greenbrier for its waters—in this case, Pluto Water, which is bottled for world consumption in a building located immediately across the street from the restored spa.

## *Last of the Extravaganzas*

California State Historical Landmark No. 844, the Hotel del Coronado, refers to itself as the last of the extravagantly conceived seaside hotels (Figures 1-4, 1-5, 1-6, and 1-7). (This claim would obviously be hotly contested on the ocean side of Collins Avenue in Miami Beach. There is to be found the world's longest uninterrupted stretch of oceanfront resort hotels, some of the most sumptuous built after World War II.)

The del Coronado was built by a Chicagoan and a Hoosier from Evansville, Indiana, on the then almost deserted Coronado Peninsula. It is now connected by a bridge to downtown San Diego. The front page headline of the San Diego Union for August 14, 1887 announced, "A Hotel That is Surpassed by None in the World; Like Unto an Old Norman Castle." More accurately, the del Coronado is one of the seven wonders of the world of Victorian architecture. Each of its 399 rooms, now complete with private bath and central heating, has been restored. A new 200-room addition and many meeting rooms preserve the remarkable character of this great dowager, while adding modern conveniences.

The first hotel in California to win world renown was, of course, the Palace in San Francisco, which opened in 1875, was destroyed in the 1906 earthquake, and when rebuilt became the most celebrated hotel in the west. Now owned by

**FIGURE 1-4** *One of America's great seaside resorts, Hotel del Coronado was built in 1887 outside San Diego on Coronado Island; it has now been officially designated a California Historical Landmark and is listed in the National Register of Historic Places, so it will not be demolished. Built around a courtyard that originally enclosed monkeys, canaries, and parrots, the hotel has a quiet, beautifully landscaped garden.*

**FIGURE 1-5** *Bridal chamber of del Coronado was modestly decorated with floral wallcoverings and separated from its parlor by tasseled silk draperies. Floral wall-to-wall carpet was a luxurious rarity in the 1880s. Rooms today retain their Victorian flavor with white wicker furniture.*

Japanese and operated by Sheraton, it has long since been overshadowed by such prestigious hostelries as the Fairmont atop Nob Hill. The Palace's present claim to fame is the Garden Court, an historic landmark.

Almost a contemporary of the San Francisco Palace was Denver's Brown Palace, started in 1888 and completed in 1893 at the then princely sum of $2 million. No luxury was omitted from the 10-story hotel that boasted of a 300-mile view of the Rockies. Electric lights, central heating, elevators, ice water at the turn of a faucet, exquisite custom-made carpets, the finest Irish linen and English china, an onyx-laden lobby with a cathedral ceiling that soared to a ninth floor dome—all of these helped to make the Brown Palace a magnetic attraction for virtually every celebrity who ever passed westward. On the list were socialites, political leaders, and business tycoons, not to mention presidents and royalty.

South of Denver, at the foot of Cheyenne Mountain in Colorado Springs, stands the newest of the west's truly legendary resorts (Figures 1-8, 1-9, 1-10, and 1-11). The Broadmoor is an impressive legacy of Spencer Penrose, whose name looms large in the history of Colorado mining. Its amenities include two championship golf courses designed by Robert Trent Jones; indoor and outdoor ice rinks for public use, as well as for hockey matches and major ice shows; an international center for expositions as well as superstar nightclub performers; saunas, pools,

**FIGURE 1-6** *In the 1980s, the hotel remains almost unchanged on the exterior. Its wooden structure must continuously be repainted, its shingled roof constantly repaired. Full-time security staff plus closed circuit TV coverage of areas that might be damaged, help prevent theft and arson. The sprinkler system is comprehensive and often tested.*

**FIGURE 1-7** *The Crown Room, del Coronado's original main dining room, has not one pillar supporting it, nor one nail in its vaulted natural wood ceiling. Exposed wood abounds throughout "Del's" charming old structure, requiring several housemen who spend full-time cleaning and polishing it with lemon oil.*

**FIGURE 1-8** *A truly unlikely development, the famous Colorado Springs resort the Broadmoor was built in 1891 by Count James de Pourtales from Silesia as a replica of a Georgian casino. It was located on the rim of a lake created by the dammed-up waters of Cheyenne Creek, and modeled after Imperial Palace, Potsdam, Germany.*

**FIGURE 1-9** *The original casino burned to the ground in 1897. The following year another building replaced it, however, by the summer of 1916, this second casino was moved a few yards south and the spectacular hotel that is now the Broadmoor was built. The casino pictured here is now the Broadmoor Golf Club.*

polo, skeet, and even the spectacularly located Cheyenne Mountain Zoo.

## *Early Hotel Development*

The first of the American hotel tycoons with a name that became known around the world was Ellsworth Statler. Entering the field in 1901,

**FIGURE 1-10** *"New" Broadmoor held its gala grand opening in 1918. Not just a hotel, this comprehensive resort in the Colorado Rockies has 25 major buildings plus a zoo, a shrine, and a ski lodge. With over 500 guestrooms and 60 suites, the resort employs over 800 people, and still continues to expand.*

**FIGURE 1-11** *This formal and luxurious ballroom corridor at the Broadmoor acquired its new elegance in 1960 after an extensive redecorating program was carried out by W. & J. Sloane Co., New York City. The ceiling of intricate plaster lattice work is original, meticulously applied by Italian workmen brought to the United States in 1918 just for that job.*

before his death in 1928 he built two hotels in Buffalo, plus Statler Hotels in Cleveland, St. Louis, Detroit, Boston, and the 2200-room Pennsylvania Hotel in New York in 1922. In 1954, Hilton Hotels purchased the 10-unit Statler chain (including units under construction in Dallas and Hartford) from Statler's widow for $111 million.

Ernest Henderson, who built Sheraton into the largest hotel chain of all time, was actually an investor rather than a hotelier. At age 44, in 1941, he had become the owner, purely for investment purposes, of four hotels: the Lee, Wayland Manor, Stonehaven, and the Sheraton in Boston. When he died in 1967, the 154-unit Sheraton chain was grossing over $300 million a year. Shortly after Henderson's death, the chain was sold to International Telephone and Telegraph Company (ITT).

Unquestionably the best known name in the history of the hotel industry is Conrad Hilton, whose son Barron Hilton presides over an empire that includes some 205 hotels and inns owned or franchised as Hilton Hotels and Hilton Inns. Hilton International, a 75-unit worldwide complex sold by Hilton to TWA is now a separate corporation.

Hilton owns the most famous hotel in the world, New York's Waldorf-Astoria (Figures 1-12 and 1-13), and at one time owned its most famous competitor, the Plaza (immortalized by the *Eloise at the Plaza* books) (Figures 1-14 and 1-15). Prior to Hilton's purchase the Waldorf had lost money every year of its existence; during the first year under Hilton it showed a profit in excess of $1 million.

Hilton also owns America's largest hotel, the 3000-room Chicago hostelry that was built in 1925 as the Stevens Hotel, and now bears Conrad Hilton's name. The New York Hilton is the world's most lucrative hotel. Costing over $50 million to build, it grosses well over $70 million a year. The sales reported for Hilton's properties totaled $895 million for food and room sales in 1980.

Figures 1-16 through 1-20 are examples of other famous multi-unit luxury hotels.

Howard Johnson is perhaps as well known a name as Conrad Hilton, except that it is primarily associated with restaurants trademarked by their orange tile roofs. Today, the number of motor inns owned by the company (approximately 525) is rapidly catching up with their restaurant count, although room sales total only about two-thirds of what the firm makes in yearly food sales.

**FIGURE 1-12** *Picture taken after it was newly built shows modernistic design motifs in interiors that were characteristic of the time. (Old Waldorf, built in 1893, was one of the first to offer a bath for each guest room.)*

## *Impact of Tourism*

While the heads of Ramada Inns and Marriott Hotels have been publicized in the lodging industry and are well known to that part of the public that reads the financial pages, the name of the most important figure in the history of the business has been almost an unknown to the general public. Kemmons Wilson first amassed a modest fortune in the construction business. In 1952, Wilson decided to put his fortune into the development of a national hotel chain that would provide the kind of clean, comfortable, reasonably priced accommodations that he had not been able to find consistently when vacationing with his own family.

**FIGURE 1-13** *The Waldorf-Astoria Hotel, New York City, as it looks today. It takes up an entire city block; has had no structural changes since it was built in the early thirties, although interior renovation is continual.*

Unaware that TraveLodge had set the same goal some years earlier, Wilson rapidly built six pilot units, five in Memphis and one just across the Mississippi state border (Figures 1-21 and 1-22). He called the new company Holiday Inns, Inc., and vowed that the chain would ultimately include 500 units. Today there are well over 1700 Holiday Inns in 54 countries (and the company has barely begun to invade the international market) grossing about $2 billion per year.

Kemmons Wilson's impact on tourism was perhaps second only to that of Henry Ford, Sr. on the automobile industry. Wilson spawned scores of imitators among motel builders whose efforts made the second-class hotel an economic dinosaur that could no longer compete. First-class hotels that had been cutting corners on quality were forced to return to their original high standards to meet competition.

As hotels began to regain lost ground, the motel industry fought back by moving right into downtown areas, constructing the kind of elegant high-rise structures that Holiday Inn has erected all the way from Philadelphia to Honolulu.

Today, motels have one foot in and one foot out of the hotel business. Except for their locations, most Holiday Inn structures are indistinguishable from luxury hotels (Figure 1-23). Hyatt Corporation's properties range from small hotels to the Regency buildings considered architectural wonders in downtown Atlanta, San Francisco, Los Angeles, Chicago and other cities. Only

**FIGURE 1-14** *The Plaza towered over its neighbors in 1907 when it was opened primarily as a residential hotel. Building on left is Chateau du Blois, 100-room mansion owned by Cornelius Vanderbilt, torn down in 1927, replaced by the famous Bergdorf Goodman store. Pulitzer fountain was constructed in 1912 in front of the Plaza, which has remained virtually the same* (see Figure 1-15) *while New York has grown up around her. Mansions on Fifth Avenue have been replaced by shops and towering office buildings. The Plaza is probably America's only hotel without an identifying sign or a coffee shop. It is now the property of Westin Hotels.*

FIGURE 1-15

**FIGURE 1-17** *Benjamin Franklin Hotel, Seattle, Wash., was the oldest property of Westin Hotels, a leading hotel company with hotels around the world. The company was founded in 1930 by S. W. Thurston, Frank Dupar, and Peter Schmidt, competitive hotel operators who combined their talents to survive the depression.*

*In Seattle's new skyline, the old Ben Franklin, now becomes the Westin headquarters. The new tower addition is the Westin Hotel (formerly Washington Plaza).*

FIGURE 1-16

FIGURE 1-18

recently a small number of companies has expanded their budget and economy motels (see Chapter 2) to fill the needs of a less affluent market.

The big beneficiary of the fierce competition between hotels and motels has been the luxury-loving public. Ironically, Kemmons Wilson's dream may have only magnified the basic problem he set out to eliminate: the high cost of lodgings encountered by the family traveling on a modest budget. At San Francisco's posh Holiday Inns in Chinatown and at Fisherman's Wharf, a room for parents and two children rents for three times the cost of the lowest-priced Holiday Inns. During the season at Palm Beach the same family will pay four times the rate that the Holiday Inn chain originally set out to charge. *Plus ça change, plus c'est la même chose.*

Kemmons Wilson retired, and today Holiday Inns is in the forefront of the casino–gambling business, with two Atlantic City hotel ventures and a merger with casino pioneer, Harrah's. Plans are to expand nationally the regionally strong chain.

Despite these moves into new business, Holiday Inns is aggressively enlarging its worldwide chain with a target addition of 80,000 new rooms to the current 286,000 before 1985, according to the head of its hotel group.

Coinciding with the development of an interstate highway system, the emergence of franchising, and an explosion in family travel, Wilson's dream gained unbelievable momentum (Figure 1-23). But the new Holiday Inns corporation is the product of new managers with a new set of dreams. Wilson retired to luxuriate on his well-deserved laurels in 1980.

**FIGURE 1-20** *The Fairmont Hotel lobby presents plushness reminiscent of luxurious European hotels or movie sets: deep red carpeting, red velvet sofas and chairs, crystal chandeliers, gilded wall decorations, marble columns and stairs. Continuous upgrading is necessary as it has one of San Francisco's highest room rates, more restaurants and lounges than its competition. Rooms are luxurious, although not Victorian, and in good taste thanks to local interior designers Barbara Dorn & Associates.*

**FIGURE 1-19** *San Francisco's Fairmont Hotel's famed carriage entrance with its circular drive provides impressive area for arriving guests. Built in 1907, post-earthquake, it is located atop Nob Hill, center of society and wealth in mid-1880s when men who had made fortunes in gold, lumber, railroading built homes there. The Mark Hopkins Hotel, Pacific Union Club, and Stanford Court Hotel are Fairmont's immediate neighbors. Rich in tradition, hotel keeps up with the times—added 250-room tower in 1961.*

**FIGURE 1-21** *One of America's first budget motels, TraveLodge was built two stories high in the 1940s. This was a breakthrough because prior to World War II motels were a series of room and bath units on a horizontal plan, usually highway-side, called tourist courts. Interiors of TraveLodges contained simple basics, but boasted wall-to-wall carpeting and television—then considered luxury features.*

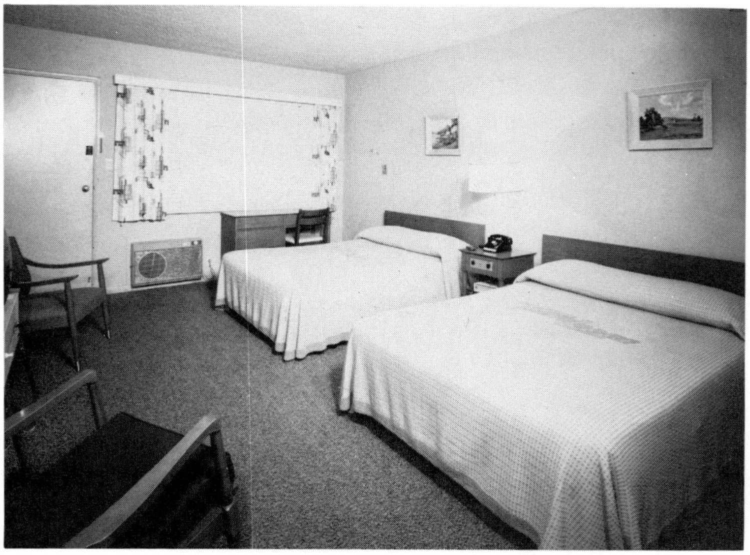

**FIGURE 1-22**

**FIGURE 1-23** *No longer a low sprawling inexpensive motel, the largest lodging chain in the country now has high-rise buildings that resemble hotels more than their original form, the tidy little Holiday Inn. Even the famous oversized Great Sign has changed over the years, no longer dominating the structure.*

**FIGURE 1-24** *Walt Disney World's Contemporary Resort hotel is among the spectacular attractions at new Vacation Kingdom in Central Florida. Built by United States Steel with preassembled room units, there are 1057 guest rooms. Through its huge Grand Canyon Concourse pass silent all-electric monorail trains. Top-of-the-World restaurant–nightclub crowns the hotel. Some 400 rooms are in main building while another 600 are in Bayshore and Lakeside wings.* (Copyright Walt Disney Productions.)

## *Bibliography*

Christensen, Erwin O. *The Index of American Design.* New York: The Macmillan Company, 1959.

Davidson, Marshall B. *The American Heritage History of Notable American Houses.* New York: American Heritage Publishing Company, 1971.

Hilton, Conrad N. *Be My Guest.* Englewood Cliffs: Prentice-Hall, Inc., 1957.

Lundberg, Donald E. *The Hotel and Restaurant Business,* Third Edition. Boston: CBI Publishing Company, Inc., 1979.

Olcott, William. *Greenbrier Heritage.* Arndt, Preston Chapin, Lamb & Keen, Inc., n.d.

## *Hotel Histories*

*The Broadmoor Story,* Helen M. Geiger.

*The Brown Palace Hotel,* Denver, Colorado.

*The Crown City's Brightest Gem: A History of the Hotel del Coronado,* Marcie Buckley, Coronado, Calif.

# Chapter 2

# Types of Lodging Establishments

**OVERVIEW** Before anyone interested in the hotel business makes any decision about the kind of hostelry to join for a career, it is wise to study the various types of hotels, motels, clubs, or residence halls. This chapter will give you some idea of what each type of place is really like.

"Lodging industry" and "housekeeper" are somewhat unfortunate terms in that they fail to indicate either the infinite variety of the industry or the fact that the Housekeeper not only functions as a department head but actually, in many establishments, serves as a vital member of the management team.

As evidence of the latter, witness Margaret McCaffrey Kappa. Mrs. Kappa majored in the Housekeeping arts at Cornell University. Her practical experience was gained at leading New York hotels. Today she is a respected member of the management of one of the world's famed resorts, the Greenbrier, with the title of assistant manager in charge of housekeeping and decorating.

Lodging industry as used in this book applies to all gathering places that provide away-from-home sleeping and/or living accommodations. It includes not only the YMCA and YWCA, small roadside motels of the early fifties, and small-town hotels of even earlier vintage, but the great and near-great resorts and hotels and worldwide empires of the international chains.

It includes cruise ships, which are nothing more than seagoing hotels; boatels, which are marina-based motels; and residential hotels, some of which reserve a certain number of rooms for transients, with the rest converted to co-op apartments (e.g., the beautiful Hampshire House on Central Park South in New York).

Also discussed are retirement homes, which are basically similar to residential hotels except that the latter usually offer European plan only (meals not included in rent), while the retirement home may offer European or American plan but is more apt to let the resident determine which meals are to be included in the flat monthly fee.

The lodging industry also includes boarding schools, which may be high schools or junior colleges but are in most cases residence halls at the four-year college level. The clubs of many types that will be discussed are also part of the industry. Exclusive of hospital rooms and jail cells, the industry covers all other away-from-home accommodations, some of them as exotic as the White House itself.

### *Housekeeping at the White House*

The executive housekeeper of the White House has a unique job, since the facility she cares for is a national shrine containing priceless historical artifacts and visited by more than one million people a year. It is decorated with fabrics, fittings, and furniture that are as difficult to refurbish as rare museum pieces, which is, of course, what they really are.

The White House staff of 18 persons cares for both the public areas and the presidential family's private living quarters on the second and third floors. The housekeeper is probably the only one in the world who reports to a management executive who holds the title of chief usher, and one of the few who requisition repair and replacement of materials from a chief curator. (This, of course, is also the case at such museums as the Smithsonian Institution, the Metropolitan in New York, and other such institutions.)

The housekeeping staff avoids aerosol sprays because of the residue they leave. Lemon oil, which feeds the wood, is used to polish all furniture. Chandeliers are washed by hand with ammonia and water. Marble floors are simply damp-mopped and in some areas of the White House are being covered with wood for aesthetic reasons. In the Red, Blue, and Green Rooms, black rubber runners are put down daily to protect the priceless carpets from excessive wear during sightseeing hours.

Guest rooms are small, simply furnished, and seemingly designed to discourage long stays. The notion is belied, however, by the very ample closets, whose late-20th century functionalism is one of the few departures from design concepts of the previous century. In most respects, however, the earlier ways are carefully adhered to.

The White House's executive housekeeper lives graciously, but hers is a way of life that can obviously be duplicated by very few professional Housekeepers in any one generation. Within the endlessly varied hotel industry, however, there are countless establishments in which people may enjoy a similar ambience and refinement, but there are many others that provide totally opposite lifestyles for those in housekeeping.

**FIGURE 2-1** *Small, elegant and overlooking Central Park, the Pierre Hotel accommodates half transients and half permanent guests. The tenants who live there on a cooperative basis use the hotel's housekeeping staff and so maids (called housekeepers there) are especially well-trained and pleasant.*

## *Hotels of Every Kind*

Hotels come in all sizes and with many kinds of atmospheres. They may be urban, suburban, located at airports, or in remote resort locations. They may be independently owned, part of a major chain, or franchised (an arrangement usually requiring the owner to make a substantial investment and pay a fee for the use of a famous name; for example, Ramada Inns). Type of ownership and the size of the property are prime factors in determining the Housekeeper's authority and responsibility (Figure 2-1). In chain properties, policies tend to be carefully controlled by national headquarters.

Housekeeping policies also vary with the clientele. Airport inns cater primarily to businessmen who fly in one night, out the next, and spend their entire stay closeted in a corporate conclave held in one of the inn's many meeting rooms. Nowadays even the finest resorts solicit business meetings, although they are more likely to attract the board of directors than the field sales force. Basically, however, they cater to the affluent guest who stays from a week to an entire season and expects the ultimate in fine service.

## *What Chains Offer*

Because the major hotel and motel chains advertise widely and effectively, they have built reputations for apparent stability that makes them particularly attractive to many interested in hotel careers (Table 2-1). The table also includes such referral organizations as Best Western and Quality Inns. These are associations of independent hostelries that show considerable variation in architecture as well as in the names of member units. Each member, however, prominently displays the association sign, adheres to uniform quality standards, and can make confirmed reservations at any other unit in the association within a matter of minutes.

In the 1950s and early 1960s the motel and hotel industries were clearly differentiated from one another. Motels were usually located where land was cheap because it was inconvenient to anything except a highway. The guest parked his own car, carried his own bags, provided his own room service, and made his own arrangements for laundry, cleaning, and pressing. Except for Holiday Inns, most of the earlier motels also omitted food and beverage service.

Built and furnished relatively inexpensively, many motels constructed before 1960 have not aged nearly so gracefully as many far older hotels built of stone, steel, marble, the finest hardwoods, and thick plaster walls. On the other hand, the quality of motel architecture, interiors, and services has since soared to the point at which instead of a distinct separation between hotel and motel, there is now a considerable overlapping (Figures 2-2, 2-3, and 2-4). It is still difficult to find a hotel that offers free parking, and just as difficult to find a motel that doesn't. Beyond that, the differences that once distinguished the two have been largely blurred or eliminated.

Marriott Hotels of Washington, D.C., is probably the most prominent proponent of the total merger of the amenities of a hotel with the conveniences of a motel. Marriott's real objective, however, has not been the ultimate glorification of the great American motel, but to compete with Hilton for the high-profit convention, banquet, and other group business trade for which the average motel chain could not bid seriously (Figure 2-5).

## *Budget Motels*

As the pioneers of the motel industry have increasingly tended to join the ranks of hotel so-

**TABLE 2-1** United States' Top Hotel and Motel Chains

| Organization and Headquarters | Room Sales ( $ millions ) 1980 | Number of Properties | Number of Rooms |
|---|---|---|---|
| Holiday Inns, Inc.<br>Memphis, Tenn. | 2500 | 1755 | 303,578 |
| The Sheraton Corp.<br>Boston, Mass. | 1145 | 418 | 108,000 |
| Ramada Inns, Inc.<br>Phoenix, Ariz. | 615.4 | 635 | 94,060 |
| Best Western, Inc.<br>Phoenix, Ariz.<br>(Association of Independents) | 1340 | 2761 | 207,299 |
| Friendship Inns International<br>Salt Lake City, Utah<br>(Association of Independents) | 450 | 1502 | 112,500 |
| Howard Johnson Co.<br>Braintree, Mass. | 470 | 525 | 60,150 |
| Hilton International Co.<br>New York, N.Y. | 441 | 81 | 29,485 |

**Continued**

**TABLE 2-1** Continued

| Organization and Headquarters | Room Sales ($ millions) 1980 | Number of Properties | Number of Rooms |
|---|---|---|---|
| Hilton Hotels Corp.<br>Los Angeles, Calif. | 545 | 209 | 73,494 |
| Intercontinental Hotels Corp.<br>New York, N.Y. | 415 | 81 | 30,182 |
| Westin Hotels<br>Seattle, Wash. | 421.1 | 51 | 28,786 |
| Hyatt Hotels Corp.<br>Rosemont, Ill. | 416 | 57 | 34,000 |
| Marriott Corp.<br>Washington, D.C. | 506.4 | 75 | 33,756 |
| Trusthouse Forte, Inc.<br>El Cajon, Calif. | 321.9 | 517 | 38,461 |
| Days Inns of America, Inc.<br>Atlanta, Ga. | 204 | 315 | 45,000 |
| Quality Inns International<br>Silver Spring, Md.<br>(Association of Independents) | 260 | 346 | 41,415 |
| Americana Hotels<br>Chicago, Ill. | 106.1 | 40 | 12,150 |
| Rodeway Inns of America<br>Dallas, Texas | 133 | 155 | 18,720 |
| Loews Hotels<br>New York, N.Y. | 142.8 | 14 | 6788 |
| Motel 6, Inc.<br>Los Angeles, Calif. | 131.3 | 303 | 32,132 |
| AIRCOA<br>Cleveland, Ohio | 78.7 | 34 | 9700 |
| Amfac, Inc.<br>Honolulu, Hawaii | 101.4 | 97.9 | 8709 |
| Red Carpet Inns<br>Daytona Beach, Fla. | 94.2 | 94 | 11,000 |
| Hyatt International Corp.<br>Chicago, Ill. | 116.1 | 36 | 11,648 |
| Dunfey Hotels<br>Hampton, N.H. | 112 | 22 | 8950 |
| La Quinta Motor Inns, Inc.<br>San Antonio, Texas | 81 | 103 | 12,300 |

**FIGURE 2-2** *Lobbies usually have one dramatic spot to give the motor inn a touch of luxury. Many are franchised; Ramada also has overseas properties.*

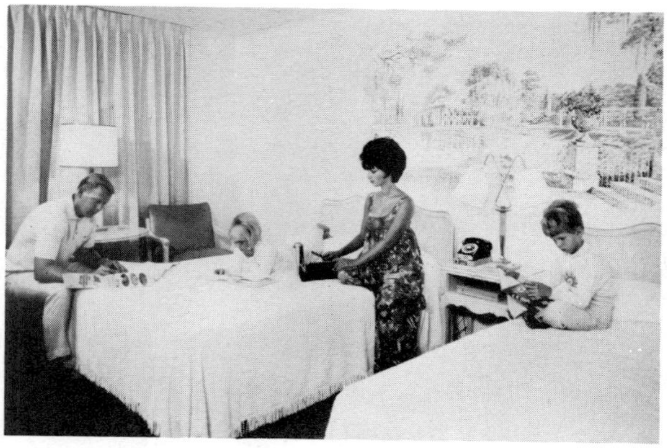

**FIGURE 2-4** *Ramada welcomes families and traveling men with moderately-priced rooms that, while having no-nonsense washable chenille spreads, also include a touch of French provincial to take rooms out of cold motel category.*

ciety, a whole new generation of motel men has moved onto the scene—aware of the large number of travelers who cannot afford deluxe accommodation, or who, regardless of means, resent having to pay for conspicuous grandeur when all they actually want is a clean room with a comfortable bed (Figure 2-6).

Leaders in the movement to return the industry to its original concept are such firms as Imperial 400 and Days Inns in the economy bracket. Priced just under the economy motels are the budget chains, including Scottish Inns of Kingston, Tennessee, which has found the solution to escalating construction costs in prefabricated modular rooms.

The whole case for the so-called budget motel was stated in an advertisement by the oldest and biggest of the budget chains, Motel 6, founded in Santa Barbara, California, in 1962.

"Has The Whole Idea of Motels Gotten Out of Hand?

". . . Motels were invented as an alternative to the bother and expense of a hotel. Someplace simple, where you could drive right up to your door and have a nice functional little room for the night without taking out a bank loan to do it.

"But today many motels are just a little simpler and slightly plainer than Buckingham Palace.

"What happened?

"Runaway luxury.

**FIGURE 2-3** *Ramada Inns are distinguished by white columns and colonial-style entrance, suggesting Southern hospitality.*

**FIGURE 2-5** *The beauty of the multimillion dollar Los Angeles Marriott Hotel is reflected at night by its exterior lighting. A gracefully designed automobile ramp in the foreground is one of two that allow hotel and restaurant guests to arrive at the lobby entrance at the same time convention and banquet guests are using the lower ramp. Located near the International Airport, the 1020-room hotel is the thirty-first and one of the largest of the fast-growing Marriott Hotel chain.*

"That's what happened.

"Somebody started building a motel that was a little fancier than the next guy's. And the next guy got fancier right back to meet the competition.

". . . At Motel 6 our idea of a motel room is more like the original idea. A nice functional, inexpensive little place to put up for the night.

". . . What you get at Motel 6 is what you need.

"A decent place for spending the night.

"Without spending the bankroll."

Although Motel 6 rates are double what they were in 1962—$6.00 per night for one person—they are still far under most motel rates in the country. Even four people can stay overnight, without frills, of course, for about $18.00 plus tax.

Housekeeping is never a placid job, but Motel 6 and similar chains reduce the number of possible frustrations for the department by reducing the number of material objects in rooms, halls, and public places that can be soiled, cracked, torn, broken, stolen, or worn out.

**FIGURE 2-6** *Typical guestroom interior of a current TraveLodge is furnished in cheerful good taste. TraveLodge offers moderate or low-priced rooms to travelers who simply want basics, not frills. Chain also has high-rise hotels but is best known for its smaller Sleepy Bear motels. Newer properties now must have a dining facility and at least 100 rooms. Because Trusthouse Forte, Ltd. bought TraveLodge there are now many elegant establishments such as the Westbury in the chain–all distant cousins to the Trusthouses in Australia and England.*

## *Challenge of Commercial Properties*

Toward the other end of the spectrum, and especially appealing to housekeepers who prefer a fast-paced cosmopolitan atmosphere, are giant commercial properties. These attract businessmen and their families from all over the world. The Sheraton Boston, the Sheraton Center in New York, Palmer House in Chicago, the Rice in Houston, Roosevelt in New Orleans, Olympic in Seattle are all large, well-run hotels in fascinating cities.

Since 2000-room hotels like the Sheraton Center might well have 1000 people checking out and an equal number checking in during a single day, the able professional Housekeepers who head huge departments in these out-size properties are well compensated for their challenging assignments.

## *Working Outside the Central City*

For housekeepers who prefer a suburban environment, opportunities have multiplied greatly in recent years with the development of airport-oriented properties in almost every city.

Other opportunities have been created as the lodging industry has infiltrated the suburbs in the wake of the movement of business and industry away from the central city. Outside every major city, hotels abound both because they provide an isolated place for meetings, and because they are adjacent to airports.

### *Where Elegance is the Standard*

Still another opportunity for the professional Housekeeper is represented by the elegant hostelry that caters not to big business, the beautiful people, or the swinging generation, but to a genteel clientele that values impeccable service, demands extreme privacy, and does not care at all about the cost (Figure 2-7). In this bracket fall New York's Pierre, St. Regis (now a Sheraton property), Carlyle, and the Regency Hotel on Park Avenue (not to be confused with the flamboyant, contemporary Hyatt Regencies—see Figure 2-8). One of the best of the breed, perhaps, is the Ritz-Carlton in Boston; the name alone impresses.

**FIGURE 2-8** *Renowned for the spectacular skylighted atrium faced by all 800 of its rooms, the Regency Hyatt in Atlanta was first of a series of Hyatt Hotels that have used avant garde architecture (by John Portman) to gain attention and attract business. Glass-enclosed elevators that run up and down the central court are part of drama that lobby-sitters can enjoy.*

**FIGURE 2-7** *The Benson Hotel, Portland, Ore., is small and luxurious, old but very much alive. Its original architectural features—ornate ceiling moldings, working marble fireplaces, polished wood paneling—are a challenge to its operators, Westin Hotels, in their efforts to preserve the charm. Although plush drapery, upholstery, and carpet are new wear-resistant synthetics, the hotel's character has been retained. It is equally appreciated by travelers on business or local patrons who cherish its ambience and cuisine.*

Sophisticated travelers like to debate which hotels are the most elegant. Very often the names figuring importantly in such arguments include the Plaza in New York City (a Westin Hotel), the Stanford Court in San Francisco (as well as the Fairmont in that same city), and many of the new Canadian chain, the Four Seasons, which includes such lovely hotels as the Clift in San Francisco, the Ritz in Chicago, and the Four Seasons in Washington, D.C.

As a general rule, the more elegant the hostelry, the greater the demand imposed on Housekeeping. Quebec's famed old Chateau Frontenac, for example, makes minimum use of vinyls and other easy-to-clean synthetics. Rather, it makes maximum use of marble, elaborately carved woodwork, expensive wall hangings that cannot be wiped clean with a damp cloth, and countless dust-catching niches for displaying valuable artifacts.

High ceilings and correspondingly high draperies also add to the Housekeeping burden in palatial old hotels. Those establishments necessarily seek out exceptionally well-qualified housekeepers and compensate them accordingly. An important fringe benefit, incidentally, is the feeling that one's "home" is almost literally a palace.

For those to whom natural outdoor beauty is as important as indoor grandeur, career opportunities abound in giant chain hotels or charming little inns plentiful throughout the Caribbean, Florida, and Hawaii (Figure 2-10). In a different climate zone, Housekeeping managerships can be found in larger ski lodges such as those at Aspen and Vail in Colorado; Sun Valley, Idaho; Stowe, Vermont; on both the California and Nevada sides of Lake Tahoe; and at many other sites.

At Harrah's Hotel (Figure 2-11) on the Nevada side of Lake Tahoe, the executive Housekeeper reports that the exceptional levels of intelligence and education represented on the staff are matched by exceptional turnover. The "ski bums" who sign on for Housekeeping jobs stay just long enough to be able to afford to get back out on the slopes again.

**FIGURE 2-9** *Town and Country Hotel on San Diego's Hotel Circle is one of city's largest hotels, with an enormous convention facility that can handle conventions for thousands. Like many hotels today, it consists of an older section—low-rise—and a tower, the illuminated building shown here.*

**FIGURE 2-10** *Vacation Village, on an island in San Diego bay, has several acres of uniquely landscaped, one-story living units, which the staff reaches on battery-powered golf carts. Waterways abound, but the decor and cleanliness standards for the hotel, constantly upheld, are high. Buildings and Grounds Department here has a challenge seldom encountered.*

The Tahoe economy, as well as that of Reno and Las Vegas, is obviously heavily dependent on gambling. The better hotels in Nevada compete for the well-heeled guest not only with topflight entertainment, but with luxurious accommodations. At Harrah's rooms are so lavishly appointed (each has two bathrooms and computerized self-serve bars) that the maids customarily make up only nine units a day, compared with an average of approximately 15 in the typical commercial hotel.

Some of the most sumptuous rooms in Las Vegas or Reno are to be found at the MGM Grand Hotel. Set on prime acres of the Strip and billed as the World's Largest Resort Hotel, the 26-story MGM Grand is the $106-million bauble of Kirk Kerkorian, who sold off virtually all of Metro-Goldwyn-Mayer's assets, and finds hotel-keeping and gambling concessions more profitable than movie-making.

The Grand Hotel's 2100 deluxe rooms include four-room penthouse suites with full kitchen, dining room, bar, and two bedrooms.

**FIGURE 2-11** *Harrah's Hotel in Reno is a far cry from the brassy, overdecorated caravansaries that dominate Las Vegas and Reno. Although its lobby has the usual complement of gambling facilities, the remainder of the hotel is done in excellent taste—contemporary styling with interesting artifacts. Profit from gaming gives the hotel the extra means to improve its service, furnishings, and salaries. Completed in 1969, Harrah's has 326 rooms and its new sister in Lake Tahoe was completed in 1973 with 250 rooms—small enough properties so it is possible to cater to guests' needs.*

Other rooms feature split-level, gargantuan step-up baths with round tubs, circular beds, and mirrored ceilings and walls.

The hotel has five restaurants, the usual array of swimming pools and tennis courts, health club, shopping arcade, youth center, convention facilities, and (a first) a 2000-seat jai alai fronton. Its 50,000-square foot gambling casino, complete with 1000 slot machines, is almost 50 percent longer than a football field. The Metro Club, with dues of $1000 per year, is the only private-membership gambling casino in America.

The hotel has over 3500 employees and the largest Housekeeping staff in the world. A few quick statistics will indicate why. The Grand has 600 chandeliers weighing two tons; 1.5 million pounds of furniture; 125,000 square yards of carpet; 30 tons of bathtubs; 300 miles of draperies; 4.5 tons of brass stars (they adorn all hotel doors), and 200,000 light bulbs.

For the Housekeeper who has had everything, the gambling casino provides a fascinating challenge. Since it operates 24 hours a day Housekeeping must do its job unobtrusively even though the public areas are always in use, which poses a neat problem for the carpet shampooers. Not to mention the problem of keeping those 200,000 light bulbs relamped!

## *Seaside Retreats*

Seaside retreats run to extremes. Cape Cod is known for its small picturesque cottages. The Gulf coast from New Orleans to Mobile offers little more than conventional motel facilities. On the Georgia coast are such famed hostelries as the Savannah Inn and Country Club offshore at Wilmington Island, and another perfect hideaway, the Cloister on Sea Island just off the coast from Brunswick (Figure 2-12).

At the other extreme is Miami Beach where mile after solid concrete mile of glittering high-rise hotels are packed with northerners fleeing from winter, or with tens of thousands of conventioneers. The competition for professional Housekeepers is keen among the thousands of these hotels, many of them in the *grande luxe* class. Because so many of the Miami Beach hotels were built in the 1920s and 1930s, they are attracting the attention of the National Trust for Historic Preservation and in the interest of maintaining Art Deco, many of these now run-down old hotels will be restored. (See the later chapter on Restoration.)

The beauty of the Hawaiian Islands can be enjoyed at the Kona Hilton (Figure 2-13).

## *Desert or Mountains*

For those who prefer the desert to the seashore, there are Palm Springs, California, and Scottsdale, Arizona. For those who prefer the mountains there is all of Colorado, but especially Colorado Springs and adjoining Manitou Springs. The most magnificent mountain compound of them all, of course, is Grand Teton Lodge in Jackson Hole, Wyoming, owned by Rockresorts. With its faithful following of the

**FIGURE 2-12** *The Cloister Hotel, Sea Island, Georgia, has that irresistible combination of Spanish architecture and subtropical climate that makes a resort everpopular. Remote from the nearest town, the Cloister lives up to its name, separated as it is from Brunswick by the Marshes of Glynn (inspiration for Sidney Lanier's poem) over which a toll causeway has been built. A year-round vacation spot surrounded by lush gardens, it offers all sports including pirate-hunting in Blackbeard's haunts.*

**FIGURE 2-13** *Kona Hilton on the big island of Hawaii has been designed so that all rooms have a bay view. Salt spray and wind make problems for Housekeeping and Maintenance, but most of the time this leeward side of the island is sunny and serene. With its addition, built in 1970, hotel has a total of 452 rooms.*

very very rich, Rockresorts has been able to concentrate completely on perfection of design and service, knowing that its clientele will find its luxurious accommodations affordable regardless of price.

The same is true to a somewhat lesser extent of the great golf resorts: the Broadmoor, Del Monte Lodge at Pebble Beach, California; Pinehurst and Southern Pines in North Carolina; and, of course, Palm Springs.

## *Cruise Ship Housekeeping*

The professional Housekeeper aboard an ocean liner or cruise ship faces some challenges not customary in the industry.

Climate conditions have some dramatic effects as do insect and plant life in the ports. Special sprays should be available for use against bugs in the tropics. As guests may be allergic to many sprays, however, care must be taken to allow time for sprays to dissipate while cabin doors are open and the passengers are absent.

As cruise passengers are a captive group governed by set meal hours and shore excursions, a Housekeeper must be flexible in arranging cleaning schedules.

Passengers tend to be older people (they have both the time and money for this expensive leisure activity) who arise early and use the lounges and public areas a great deal. They expect these areas to be clean and in order early in the day and then maintained that way. Such facilities include card rooms, theaters, pools, decks, game deck, main lounge, show lounge, dining rooms, offices, and public rest rooms. Food and beverages are served in or carried to all of these areas, causing soil and spots on carpets and upholstery, and most cabins are subjected to similar use. Keeping carpets, spreads, and upholstery clean is difficult as cruises are booked back-to-back. This precludes areas being put out of order for general cleaning.

Bed-making in cabins is a special sea-going art. Top sheets are folded around blankets and never tucked in, as passengers are given to afternoon naps while at sea which means that beds often must be made twice a day. Although most ships have a laundry on board, bed linen may be changed only two to three times a week. Use of terrycloth items is heavy, towels being used for showers and at pools, gyms, and saunas. Theft of linen is a minor consideration on ships (but a major one for hotels and motels).

Vacuuming corridors may never be done before 9:30 to 10:00 AM so as not to waken late sleepers. Every cabin must be vacuumed every day, but this is easy as there is no movable furniture. Dining rooms must be vacuumed three times a day.

Windows and portholes are subject to constant sea water and the elements of sun and wind. Thus they must be cleaned frequently.

Table tops are nearly all synthetic, which simplifies that area of cleaning.

Passengers' cabins are about one-third the size of hotel rooms. Few bathrooms have tubs, sitting areas are limited, and wall space is plain with little art work, except in deluxe cabins that may also have draperies.

Ship builders contribute to cleanliness and sanitation if shower floors are slanted toward the drain. This helps to reduce build-up of soap scum in the shower well. Care must be taken to avoid accumulation of mold around showers. Painted surfaces tend to chip and collect moisture and rust, thereby requiring considerable scrubbing and drying out.

Cruise lines are operated by many countries in which sanitation standards differ greatly from those of the passengers. These standards are a main contributing factor toward gaining or losing a full complement of passengers. News travels fast among travel agents and cruise enthusiasts. Poor sanitation or food, or a surly staff can be disaster signs for a line.

## *For Those Who Prefer Permanence*

Ironically, relatively inaccessible locations have proved more of an attraction than a deterrent to the loyal clientele of many of America's leading resorts. The employees tend to react much like the guests. Either they disappear forever after a brief lonely stay, or they adopt a resort like the Greenbrier as their second home; and if they stay long enough, as their real home.

Many of the employees live nearby and almost all management personnel live on the grounds. The Greenbrier is where they eat, sleep, play, entertain, shop, and even raise their children. The sole distraction is the tiny town of White Sulphur Springs—population 2500.

## *Colleges and Universities*

A similar life style is pursued by these professional Housekeepers who establish careers in educational institutions, many of which are located in college towns even smaller than White Sulphur Springs.

Prior to World War II, Housekeeping was largely a student function in residence halls as well as in sorority and fraternity houses. In today's more affluent society, a residence hall maintained by students qualifies as a notable experiment in communal living.

Reports B. H. Bennet, maintenance supervisor at the University of Florida at Gainesville: "We have one residence that is a male co-op. Residents perform all custodian functions, painting and maintenance, within their capabilities. They also elect their residence hall government, screen applicants and handle all disciplinary problems. It has worked out exceptionally well. They are proud of their accomplishments, and frankly their hall has a more consistently clean and better appearance than the majority of the others."

For the most part, however, on-campus Housekeeping is handled today by highly skilled professionals. They not only care for the resident halls and many sorority and fraternity houses, but for faculty clubs, student unions, and campus guest room facilities for visitors.

The leaders in student housing are shown in Table 2-2. Note that the University of Wisconsin houses almost 40,000 students, administering more rooms than many hotel chains. The University of California has a bed capacity of 27,927, more than double that of most chains.

In addition to being in lodging's big league, college Housekeepers enjoy the advantage of an unusually stimulating environment, while at the

**TABLE 2-2** United States Colleges and Universities that House Greatest Numbers of Students in Residence Halls

| University | Sales Volume (millions) as of 1978 (Commercial Equivalent) | Number of Buildings | Number of Beds |
|---|---|---|---|
| University of Wisconsin<br>Madison | 44.5 | — | 35,277 |
| Allen & O'Hara, Inc.<br>Memphis, Tennessee | 35 | 40 | — |
| Michigan State University<br>East Lansing | 24.8 | — | 21,910 |
| California State Universities and Colleges<br>Los Angeles | 22.6 | 19 | 15,904 |
| University of Michigan<br>Ann Arbor | 17.2 | 15 | 11,453 |
| University of Minnesota<br>Minneapolis | 17.1 | 38 | 11,449 |
| Indiana University<br>Bloomington | 16.8 | 1488 | 17,003 |
| Ohio State University<br>Columbus | 16.7 | 608 | 11,466 |

**FIGURE 2-14** *Residence halls in ladies' colleges in the early days were highly dignified and very institutional, as is this dining room in Main Hall at Vassar College, Poughkeepsie, N.Y., pictured in 1878, 17 years after the college opened. The dormitories there and at Bryn Mawr had Warden's Suites where house-mothers lived.*

same time finding respite from several of the profession's more common problems (Figures 2-14 through 2-18). The campus population does not vary from day to day, or even from month to month, the Housekeeper is not plagued by concerns incident to wildly fluctuating occupancy rates that are encountered in other segments.

For the same reason, forecasting, staffing, and scheduling never give cause for panic. Nor is the labor problem often acute, since there is usually a large pool of part-time student workers to

**FIGURE 2-15** *Student Center, Rio Hondo Junior College near Los Angeles, exemplifies the exciting, handsome structures in which Maintenance and Housekeeping staffs work today. In addition to glassed lounge that occupies the entire second floor, center includes offices, meeting rooms, rest rooms, and kitchen. Some of the country's campus unions have banquet halls on a scale with, and as expensively equipped as, any hotel meeting hall.*

**FIGURE 2-16** *Student lounge is dramatically built around a central fireplace, focal point in the room. It gives students a feeling of excitement—something happening in a social center. Furniture is not arranged in a permanent way because it is almost impossible to prevent students from adjusting furniture groupings. Wood floor, vinyl upholstery, and plastic laminate are easy-maintenance elements; light fixture, ceilings, and windows may require special equipment and dedicated employees when cleaning time comes.*

**FIGURE 2-17** *Designed by famed avant garde architect Eero Saarinen, Noyes House, a residence hall at Vassar College, was completed in 1958, yet has a fresh original look. Women residents today clean their own rooms but Maintenance and Housekeeping are important on a campus such as this where over 1000 live.*

**FIGURE 2-18** *Distinctive Yosemite Hall, student residence at "Cal Poly," San Luis Obispo, Calif., was cited by College & University Business Magazine the year it was built, as one of 15 outstanding examples of college building design. Designed by Corwin Booth & Associates, architects, the co-ed hall houses 600 students. Bright poppy-colored curtains are provided for student rooms, a handsome contrast to the grey concrete. Lobby (below) is surrounded by different levels of the building (glass presents primary Housekeeping problem); is furnished with easily refinishable, warm wood furniture; floor is textured cement.*

**FIGURE 2-19** *At the Missouri Athletic Club in St. Louis, members and their guests are greeted by a gracious, quiet, traditional atmosphere in the main lobby. In background is grand staircase leading to the main dining room on second floor. Marble floors and delicate area rugs pose double set of problems in floor care. Club members tend to take care of property better than hotel guests or students in residence halls.*

draw on. With few exceptions, these students are competent and reliable.

Advancement prospects are well above average. Housekeeping supervisors are often promoted to positions as housing managers or residence hall directors. These positions can lead in turn to deanships. There have, as a matter of fact, been instances where the chief housekeeping executive actually rose to the presidency of the university. By no means is campus housekeeping a terminal job.

## *Housekeeping Potential in Clubs*

Clubs represent still another major opportunity for prospective Housekeepers. Like hotels, they occur in all sizes and with many kinds of atmospheres. There are city clubs (often called athletic clubs) (Figures 2-19 and 2-20), country clubs (commonly golf clubs), university, dining and drinking, and fraternal clubs, and retirement residences or communities that are clubs in all but name. There are also numbers of other clubs organized along professional and other lines.

Many clubs, but especially those in the city, have extensive sleeping-room facilities that are available to members, their guests, and to affiliates of out-of-town clubs with which membership privileges are exchanged. Visitors with such exchange privileges constitute an important part of every club's lodging population.

Some also house a sizable population of elderly widowers who endure rather cramped sleeping quarters in order to enjoy the society of a community that they consider to be uniquely their own. Because guest rooms tend to be rather spartan and their inhabitants tend to have rather spartan habits—as opposed, for example, to

**FIGURE 2-20** *Interior view is one of many, newly-decorated suites available to Missouri Athletic Club families and guests.*

conventioneers—the complexity of the Housekeeper's functions insofar as room care is concerned is minimal.

The interior decoration, care, and maintenance of public rooms is of overwhelming importance, however. To a large extent, a club succeeds or fails on the basis of the appeal of its public areas, not only for present but for prospective members.

## *Hospitals*

This important opportunity for prospective Housekeepers has been left unmentioned because it is beyond the scope of this book. Hospital Housekeeping is an art of its own, involving different problems, different systems, and most importantly, the medical environment.

Hotel, college, and club Housekeepers may move comfortably and effectively within all three of these fields, but not into that fourth one represented by hospitals, without specific training in aseptic techniques and hospital systems. It goes without saying that hospital housekeeping can offer immense rewards to those with a genuine sense of dedication to or a fascination with medicine and science.

# Chapter 3

# The Importance of Housekeeping

**OVERVIEW** Management and guests consider the business of keeping a place clean and in good repair to be vitally necessary for a hotel or dormitory to command a fair price and get repeat business. This chapter deals with ways that some individual hoteliers and some of the chains keep their standards high.

In the 1980s both men and women are competing vigorously for the position of professional Housekeeper. The fact that women still dominate should not be interpreted to mean that the role demands characteristics and abilities more likely to be found in the female sex. What it does reflect is that women were fully aware of the scope of this major career opportunity at a time when few men outside the lodging industry had any conception of its potential.

Few titles are more misleading than that of housekeeper. If men thought about it at all, they tended to think of a housekeeper as "one who keeps house." By this same perverse logic, they might just as well have equated one who manages a five-room apartment with one who manages a thousand-room hotel!

Today, male housekeepers are becoming more commonplace, especially when the position is combined with departments of property maintenance or guest services. Hotel experts rate housekeepers as prime candidates for management positions. The Housekeeper, they point out, is the only department head with legitimate access to every area of the institution; the only one who maintains regular liaison with every other department head, and thus obtains a singular overview of the entire operation, including an almost unique insight into the ways that guests of all types react to the facility, its personnel, and its services.

That the Housekeeper relates to the entire macrocosm rather than to an isolated function or area has real significance for management. Here is how Hospitality Management Corp., the Dallas-based firm that manages hotels for others as well as operating its own Registry Hotel and Latch Key Inn chains, describes this in its manual for managers:

"The housekeeper is the eyes and ears of management. Use her; her opinions and suggestions must be seriously considered. Many times the housekeeper will see and hear things that will never reach you. Only through her, if she trusts you, can you obtain pertinent information. Make her welcome—meet regularly."

HMC insists that its hotel managers establish continuing liaison with their Housekeeping departments. The manual recommends that the managers be "present several mornings a week when carts are being loaded," and requires that they personally inspect at least five of their guest rooms each day.

Occasionally, management makes the mistake of overlooking the fact that, like most other businesses, the hotel business has a line of products to sell—and that the quality, value, and price of those products determine to what extent the business will succeed or fail. While this so-called product line includes the lobby, hallways, rest rooms, meeting rooms, restaurant, coffee shop, banquet rooms, and so on, the principal product of every hotel and motel is obviously the guest room.

Guests do indeed form instantaneous impressions when they walk into the lobby of a hotel they have not previously visited. These impressions are enhanced or diminished as the guest moves from the registration desk to the elevator, and then down the corridor toward a room.

When the guest enters his room all prior impressions are immediately supplanted by a virtually total response to the room itself. "The rooms," said the executive Housekeeper of the Las Vegas MGM Grand Hotel, "are the heart of the hotel. Unless their decor is appropriate, the air odor-free and the furnishings spotlessly clean, you have lost the guest as a potential repeat customer." This same housekeeper also attaches great importance to the quality of service, and the appearance and courtesy of all Housekeeping personnel.

## *The Mark of Excellence*

That these remarks reflect the convictions of management has been confirmed repeatedly by the public declarations of many of the most prominent executives in the lodging industry (Figure 3-1). Karl Eitel, president and general manager of the Broadmoor in Colorado Springs, has stated that without any doubt "the mark of excellence of any hotel is quality housekeeping."

His counterpart at the Plaza in New York has pointed out that "national surveys have shown again and again that the number one reason for a customer's returning to a hotel or *not* returning is the cleanliness of the facilities. Therefore, the most grandiose advertising campaign, or the most unique location, or the most unique design and architecture may attract a customer to try a hotel once—but if the cleanliness is not up to snuff, he or she will not return. The best insurance for investment in a hotel is excellent standards of housekeeping and cleanliness."

In addressing the subject of *Housekeeping and the Management Team,* Hanne Dittler, housekeeping expert at Westin Hotels—operators of deluxe facilities from Alaska to Acapulco, including San Francisco's St. Francis, the Houston Oaks, Portland's Hotel Benson, and the Plaza in New York—noted: "Study after study conducted by researchers underscores the critical importance

**FIGURE 3-1** *Valeri Richardson, executive Housekeeper at the Broadmoor, feels that room attendants will be far more motivated to carry out their jobs if they can be certified. She puts so much stress on the importance of housekeeping that she promotes the idea of having maids and housemen earn their credentials.*

of cleanliness as the primary reason for the guest's selection of a hotel.

"Nevertheless," she added, "we often tend to overlook Housekeeping in the perspective of its responsibilities in the area of operating costs. The Housekeeping Department is, in almost every instance, the largest single 'employer' in the hotel. The director of housekeeping carries, literally, an hour-to-hour responsibility for the control of labor costs, perhaps the most critical factor of all the 'bottom line' of the operating statement of the hotel.

"The recruiting, training, supervision and retention of Housekeeping personnel is a never-ending challenge. The director of housekeeping and her supervisory team must seek out potential employees, train them in hard, demanding work, encourage them to take pride in their skills, and provide positive supervision to assure the quality of guest room service that today's customer demands.

"And then," continued Ms. Dittler, "there is the 'fact dollar' responsibility the Housekeeping Department assumes in the management of costly inventories of linens, uniforms, guest room supplies, and equipment. The director of housekeeping carries major management responsibilities in the purchasing, inventorying, issuance, and care of many thousands of dollars worth of hotel property. The director of housekeeping must be constantly alert to new products, new cost-saving techniques, and new requirements such as those posed by the new polyester blends in guest room and Food and Beverage linens, not to mention all the uniforms for which Housekeeping must always assume responsibility.

"But perhaps," Ms. Dittler added, "the most important function discharged by the director of housekeeping is that of inter-departmental support and communications. Since no hotel wants to room a guest in an unmade room, the housekeeping department must provide the front desk with assurance of rooms in order to meet the anticipated arrivals for the day. With the changing patterns of travel, this becomes no small challenge. And as hotels experience more and more large group check-ins and check-outs, the coordination between the Front Desk and the Housekeeping Department becomes critically important.

"The director of housekeeping is very much a part of the operating management team of the hotel. Today's director of housekeeping must be involved in management planning—not just on the day-to-day basis of arrivals and departures and group requirements—but in terms of total knowledge of hotel operations.

"The responsibilities of the director of housekeeping are truly management responsibilities, running the gamut from budgeting to communications, from personnel administration to systems evaluation."

Ms. Dittler's insistence that Housekeeping must be involved in management planning "in terms of total knowledge of hotel operations" reflects the fact that a hotel houses many businesses, and almost all involve Housekeeping (Figure 3.2).

**FIGURE 3-2** *Hanne Dittler, left, is corporate director of property management for Westin Hotels. She confers here with a director of housekeeping on color/furnishings schemes for guestroom renovation—important because it represents heavy dollar investment. Updating helps establish repeat business as guests enjoy staying in latest room design.*

## *A Hotel Is All Things . . .*

Granted that a hotel's main business is room rentals, to the members of the community it may actually represent an altogether different set of symbols—in many cases, incidentally, the same ones that the clubs of the community represent.

The businessmen of a community may think of the city's hotels and clubs in terms of fine restaurants where they entertain customers or dine with fellow executives. Others may think of them in terms of the private dining rooms in which their professional or other associations meet weekly or monthly. Still others as social centers, where they hold their banquets and balls, entertain their friends, or find entertainment for themselves if only in the form of a small orchestra for dancing.

And where is the best place for guests or natives to meet? In the lobby, of course—especially a dramatically designed, superbly upholstered lobby like those now appearing in hotels from coast to coast. Because cocktail lounges are becoming an integral part of so many lobbies, special maids or housemen are called on to assist the food and beverage personnel to keep lobby areas litter-free and sparkling.

## *The Role of Housekeeping*

For all of these subsidiary activities and many more, Housekeeping is responsible for setting the scene and for maintaining the quality of the scenery. This obviously dictates cooperation with the Food and Beverage Department, Engineering, Maintenance, Accounting, Personnel, Sales, and the Front Desk. A close working relationship with the general manager is also essential. Without his enlightened understanding of the many-faceted housekeeping functions and his strong support especially at budget time, the quality of housekeeping cannot help but falter, and with it the quality and reputation of the establishment as a whole (Figure 3-3).

**FIGURE 3-3** *Because executives of well-managed hotels take an active interest in having their properties beautifully maintained, they must have good working relationships with their executive Housekeepers. Here, Henrietta LaBorde, director of Housekeeping for the San Francisco Hilton discusses with Henri J. Lewin, senior vice president of division, guest room redecorating and refurbishing.*

One such model relationship is that of Margaret Kappa, assistant manager in charge of decorating and housekeeping at the Greenbrier resort, and the general manager. The latter speaks of the total housekeeping responsibility in terms of "we," rather than as a function that even the multitalented Mrs. Kappa could not execute with maximum effectiveness if the relationship were otherwise.

"*We,*" says the vice president, "check every guest room and every public space every year, and do what is necessary to bring that room up to perfection. Constant checking is, of course, basic to an outstanding program, and as a matter of fact, I, informally but thoroughly, check every public and private room in the house about six times a year. We have a revolving decorating program whereby we continually replace draperies and upholstery fabrics as necessary, along with the large-expense item of carpeting.

"Speaking for myself, I feel that management must support its housekeeper with proper maintenance crews—including night cleaners, an upholstery shop as well as other craft facilities, not to mention adequate floor supervisors and housemen."

Similar teamwork exists at the Hyatt Regency Embarcadero in San Francisco. Once a week, general manager Peter Goldman, the executive housekeeper, the rooms division manager, and the chief engineer spend at least two hours or more spot-checking 20 of the hotel's guest rooms. In another weekly ritual, Goldman and the engineer check all the public rooms, restaurants, cocktail lounges, the lobby, and hallways. The slightest imperfections are corrected almost immediately.

"Our maintenance standards," says Goldman, "are among the highest in the world. We keep track of repeat customers and their individual preferences, whether it be for special pillows, bedboards or a daily copy of *The London Times.* And when our maids (whom we actually call 'housekeepers') do the nightly turn-downs, they'll even tidy up the room and bath for a second time if there is the slightest sign of disorder."

They also place a mint and a goodnight card on the guest's pillow, along with a breakfast menu in the shape of a large key. (The guest can check off his preferences, and hang it on the outside doorknob where it is picked up by a Room Service person patrolling the corridors in the wee small hours.)

As regional vice president for Hyatt on the West Coast, Peter Goldman is instigating these same niceties and high standards of housekeeping in the nine other Hyatt hotels he oversees.

Another industry leader who esteemed the personal touch was the late I. A. Harned, former general manager of the Cloister on Sea Island, Ga. He said, "The most important thing a maid can do in a resort is to learn and use the names of guests. They feel recognized, and realize that the management knows they are here."

Unlike commercial hotels where the check-out rooms are made up first, the Cloister reverses that policy as a welcome courtesy to guests who are staying on. One of Harned's more interesting observations was, "The impression a guest gets from the hotel comes from the condition of carpeting, draperies and bedspreads—the three things he looks at when entering a room. These quickly say whether there's good housekeeping. Old furniture is more easily forgiven than a worn carpet or a bedspread that is not pressed."

Since World War II, lodging has been one of America's fastest growing industries. Today, top 10 lodging organizations gross far in excess of $8 billion annually. Yet as recently as 1950, almost none of the organizations even existed. Other than the consolidated school and college systems, plus the state and federal institutional complexes and the military, virtually the only members of today's top 100 on the scene in 1950 were Hilton, Sheraton, and TraveLodge.

Since then, three events have combined to make lodging a great industry and Housekeeping a basic career:

1. Hundreds of new companies moved into the lodging market; although most of them failed, the few that did survive joined the ranks of the titans.
2. The complexes they erected were bigger and much more luxurious than anything previously conceived; lodging's new leaders are more likely to put up a 200-room motel than a 20-room unit; or put up a 500-room hotel than one with just 50 rooms.
3. As the new chains became bigger and bigger, the number of companies controlling the industry became fewer and fewer. As long ago as 1979, 10 top hotel or motel companies did 57 percent of the total business.

The classic example, of course, is Holiday Inns, which grosses considerably over $2 billion a year from over 300,000 rooms in a worldwide lodging network that includes over 2000 separate facilities.

Sheraton Corp. total sales are over $900 million, and Hilton around $500 million (Hilton International, with about 80 properties, owned by Trans World Corp. actually grosses more than the domestic giant Hilton).

What does all of this mean to Housekeeping—other than more directorship at headquarters, more executive housekeepers at the unit level, and the evolution of Housekeeping from a miniscule to a major activity in 20 years?

For one thing, it means that the industry has grown faster than the supply of trained executives, supervisors, and workers at all levels. This is a problem that education is only now beginning to catch up with, although no one really believes that the supply will actually catch up with this demand until lodging can attract more competent people. Were it not for the documented (and often undocumented) aliens there would be an even more acute shortage of maids or room attendants.

Also, demands have far outpaced available technology. In thousands of lodging establishments, Housekeeping is still a primitive craft of the type usually passed down from one generation to another with little change. In more enlightened properties, management has sought out directors of Housekeeping and executive Housekeepers possessed of genuine creative talents; and there, the profession has become perhaps the most fascinating of all the hotel arts (Figure 3-4).

As always, however, it is the leaders, the big chains, who are beginning to set up basic techniques, procedures, and standards. Check-lists are being codified into scores of manuals for distribution to thousands of housekeepers, who are charged with the indoctrinating their tens of thousands of employees in basic housekeeping science.

In still other cases, directors of Housekeeping for large chains are using at least a section of their headquarters hotel as an experimental laboratory for refining present procedures, developing new techniques, innovating entirely new systems, and testing every new product that shows promise of advancing the science of professional Housekeeping.

Henri J. Lewin, senior vice president for the Western Division of Hilton Hotels, points proudly to the important experimental work performed by Henrietta LaBorde, director of Housekeeping for the division. He adds that executives from Housekeeping and other departments from company properties throughout America are coming to the San Francisco Hilton to study not only Mrs. LaBorde's system for cleaning and employee control, but her techniques for scheduling and budgeting for room refurbishing. Her regular routines and schedules are adhered to in the Housekeeping Department as rigidly as recipes are followed in the kitchens.

**FIGURE 3-4** *Bill McCreary, far left, resident manager; William G. Quinn, vice president and managing director; and Inge Krieg, director of property management, discuss plans for the renovation of guest rooms in the Century Plaza. The entire 800-room hotel is being redecorated. Ms. Krieg oversees the departments of Housekeeping, Laundry, and even Property Maintenance.*

## *Manual and Quality Audit*

The corporate director of Housekeeping for all Sheraton Hotels and Motor Inns devised a master manual suitable for properties with from 100 to 1900 rooms. The tool for enforcing the requirements set forth in the manual is the Quality Audit. It is a 16-page form listing 282 regulations that must be complied with if the individual unit is to meet the Sheraton standard for housekeeping quality.

The Quality Audit is a controversial document. Many industry figures consider a 282-point program too complex to install and to regulate. It is almost universally agreed, however, that the points are not only valid, but that their number if actually conservative.

To assist the units in implementing the program, Larry Walker, vice president of rooms, and his assistants travel to Sheraton properties in the United States and around the world.

At the local level, each Sheraton unit has a Quality Improvement Team comprised of department heads and their assistants, who check quality control "from the guest's point of view," and meet frequently with the Housekeeper to discuss problems that might have to be overlooked unless interdepartmental cooperation can be obtained.

General managers are particularly welcome at these meetings, as they tend to underestimate the extent to which the executive Housekeeper's job has grown during the past decade. "Housekeepers," says Florence Cunningham, former corporate housekeeper, "have to keep their employees producing 24 hours a day—even when a particularly high-spirited army of conventioneers is determined to turn the hotel into a 'round-the-clock playground. The real point we are trying to drive home to top management is that it is the Housekeeping Department that actually delivers what the Sales Department can only promise!"

The point is very effectively being driven home by Holiday Inns, which has a special facility to be used for communicating such points as the importance of Housekeeping to its innkeepers around the globe. It is Holiday Inn University—a 174,270-sq. ft. educational complex located on an 88-acre campus near Olive Branch, Mississippi, just 15 miles southeast of the chain's Memphis headquarters.

The university, which is supported by a voluntary assessment of 1 percent per room per night, grants certificates of achievement to thousands of Holiday Inn executives. The facility includes six classrooms, four meeting rooms, a 200-seat amphitheater, audiovisual studio, test kitchen, expanded laundry facilities, and a model Holiday Inn with an operational and educational staff of 160. In 1981 it resumed its Housekeeping training after reassessing and reorganizing the entire important program.

## *Management of People*

Nena Holder, Holiday Inn University's first director of Housekeeping, focused her curriculum on the four major areas of responsibility:

1. Management of people, equipment, and supplies
2. Preservation of building finishes, fabrics, and furnishings
3. Controlling costs
4. Keeping records

"At least 60 percent of class time," Mrs. Holder reports, however, "was spent on personal problems since, without exception, every class entering Holiday Inn University makes it clear that the management of people is their No. 1 problem."

Because of the importance of housekeeping at Holiday Inns, inspectors from headquarters make their unscheduled calls on each of the approximately 1743 units in the chain on an average of three times each year. That a clean room and a well-maintained facility are what the lodging business is basically all about is dramatically emphasized by the importance that Holiday's headquarters attaches to these inspection reports. Innkeepers whose units fail to receive high marks for housekeeping are held personally responsible until these deficiencies are corrected.

This policy applies without regard to the profitability of any inn that fails to pass inspection. The fact that it has been put into effect by the world's largest lodging chain indicates that a similar policy will eventually become almost universal. Should that occur, the demand for the most highly educated, skilled, and experienced housekeepers will, quite obviously, far exceed the supply.

Proper housekeeping and good maintenance in university residence halls is especially important because there is little profit involved in those operations. That means that major renovations cannot be undertaken as often as in hotels or motels. Because students are not only usually hard on room furnishings and equipment, and they may not clean their rooms very frequently, Housekeeping personnel need to be doubly thorough when doing inspections and periodic cleanings.

B. H. Bennett, director of housing at the University of Florida, Gainesville, says, "Despite the problems inherent among young people who rarely concentrate on having spotless rooms, Housekeeping's function is extremely important because it's essential that we keep the residence halls attractive enough to make students want to live there."

It also takes a bit of doing, he says, to get to know the residents and their wants. "For example, the more barren the room and the more bulletin board-like the wall, the better the students like it so they can do their own wall treatments and have some individuality."

## *TraveLodge, "Housekeepers Needed"*

"With the increasing labor costs, we are constantly showing the motel operators how to control these costs and still maintain efficiency. We feel that Housekeeping is one of the most important products in the motel industry and believe that every motel chain in existence continues to keep a sharp eye open for people to specialize in the Housekeeping Department. The lack of supply and the increasing demand for experienced personnel is becoming very serious," says Glenn H. Pyetzki, Operations Division of TraveLodge International, Inc.

At Best Western properties a five-star categorization system has been used internationally for decades and works well once established. It provides the guest with a quick indication of the expected level of facilities, services, and cleanliness. It does not include size of property or room rates.

Says the president of its board of directors, Lin Zoder, "We have asked the staff to develop a new quality control plan that goes beyond our present system of measuring housekeeping, maintenance, and the cleanliness of our facilities. It is a difficult job, but we have to learn to measure and improve properties which may be clean, but do not project a quality image. How does your property look to our guests? Does the exterior appearance represent the best in today's lodging market? Are the public areas and guestrooms attractively furnished and professionally decorated? Are the wall coverings, bedspreads, carpets, and drapes tastefully coordinated? Do they add to a look of professionalism?"

An inconsistent image, Zoder maintains, will drag down Best Western, thus member properties are regularly inspected. The passing inspection score for the 1980s has been raised from 750 to 800 points. "This will help," says Zoder, "but it is not the whole answer. We have a reasonable rate structure which appeals to the large central portion of the market. Our market is the folks who will not pay $100 a night for a room, but who do not want budget accommodations either. We already dominate the roadside lodging product line. We'll get even stronger if we continue to offer the finest service at reasonable cost. A perfect example is the Best Western Little America in Salt Lake City, one of the most beautiful motor hotels in the world."

# Chapter 4

# Areas of Housekeeping Responsibility

**OVERVIEW** You can practically design the kind of position you'd like if you want to become a professional Housekeeper. The range of properties for which housekeepers are needed is so wide that one can choose anything from controlling thousands of people and millions of dollars in a huge 2000-room hotel, to simply supervising three maids in a posh little resort lodge. The wide gamut of specialized knowledge required to accomplish a variety of tasks makes professional Housekeeping endlessly stimulating.

The general public tends to think of Housekeeping as the department that makes the bed, empties the ashtray and wastebasket, tidies up the bathroom, and leaves fresh towels. The truth of the matter is, of course, that it takes more manpower to accomplish Housekeeping's almost countless functions than those of any other department in the lodging industry.

Housekeeping's responsibility may begin with fabrics, and progress to the selection and purchase of all furniture and furnishings; it encompasses not only their cleaning, but their maintenance, repair, refurbishment, and eventual replacement (Figure 4-1).

Areas of concern include sleeping rooms, meeting rooms, halls, and elevators, all considered "front-of-the-house." The designation also includes all other public areas from lobbies to washrooms to restaurants and cocktail lounges, to any furnished outdoor areas such as pools and patios (Figures 4-1 through 4-5). Housekeeping also oversees almost the entire "back-of-the-house" (that part of the facility the public seldom sees), from kitchens to locker rooms to employee dining—virtually everything, in fact, except the boiler room and the carpentry and upholstery workshops. Being willing and able, the Housekeeper may even be responsible for these areas!

Almost all large properties, however, have Engineering and Maintenance Departments staffed with engineers, electricians, plumbers, painters, carpenters, and perhaps even a skeleton cleaning staff. A primary responsibility of the Property Maintenance Department is building and grounds care.

The range of skills required of a housekeeper in a major property is so extensive that it can be mastered only after years of on-the-job training. In the future it will undoubtedly be necessary for a person hoping to enter the profession, even as an assistant housekeeper in medium-sized (500 rooms) properties, to have a college degree in administrative housekeeping.

At its most elementary level, Housekeeping requires a thorough knowledge of cleaning supplies, equipment, and techniques. It also requires the ability to hire the right people to use those materials, the skill to train them, and the wisdom to motivate them.

Furthermore, even the seemingly simple cleaning function isn't simple at all. It involves far more than cleaning the sleeping, eating, meeting, and washroom areas. It may mean the grim job of cleaning the grease-coated vents over

**FIGURE 4-1** *The guest rooms are, of course, always Housekeeping's number one responsibility. They may be simple, austere, and rugged or, like this one, delicate and beautifully appointed. Continental Plaza in Chicago has elegant sleeping rooms but most of the fabrics and surfaces are washable.*

kitchen ranges and fry kettles, the delicate job of cleaning switchboard equipment, and, perhaps, the never-ending job of cleaning the thousands of crystals used in a chandelier valued at $1000 or more.

Even where an interior designer is employed, the professional Housekeeper works very closely in the selection, refurbishing, and replacement of furniture, furnishings, and floor and wall coverings, and even such minor artifacts as soap dishes, towel bars, and ash trays, as well as uniforms and linens. In many cases, the Housekeeper assumes full responsibility for all of these functions.

Since Lost and Found and the Sewing Room are usually under the direction of Housekeeping, these are additional distractions that must be coped with expertly. The professional Housekeeper also needs at least a working knowledge of such functions as accounting, sanitation, and

**FIGURE 4-2** *Outdoor gardens, streams, and waterfalls sometimes come under the professional Housekeeper's aegis. This landscape gardener's master-work is at the Four Seasons Sheraton Hotel in Toronto, Ontario. Often a contract is made with a local landscaping firm to take care of plantings and replace when necessary.*

**FIGURE 4-3** *Other outdoor cleaning problems may include alfresco dining areas such as this one at Mauna Kea Beach Hotel on big island of Hawaii. Sumptuous daily buffet luncheons mean stone floor must be swept and washed, furniture continually damp-wiped and waxed, and air kept insect-free during buffet hours.*

**FIGURE 4-4** *Exotic swimming pool at the Marriott in Los Angeles has literally hundreds of pieces of outdoor lounge furniture around it that must be cleaned often to prevent the smog from penetrating and destroying the plastic or metal. In addition to the need for gardeners to keep shrubbery well trimmed and swept so that it does not pose problems with pool filters, the outdoor bar that serves swimmers (under the pagoda-style house) requires constant vigilance to keep plastic glasses cleaned up and sticky beverages off walkways and furniture. Housekeeping may contract out the actual pool care to professionals to assure that chemical content of the water is properly balanced.*

**FIGURE 4-5** *One of the country's most dramatic lobbies is at the Hyatt Regency Embarcadero, San Francisco, where about 20 stories rise around an atrium that contains a 50-foot sculpture, enormous reflecting pools, a stream, countless large light fixtures, and vegetation of all sorts. Trees are a challenge to pick up after but pedestrians off the street cause even more difficulties with litter, wading in the streams, and so forth. Such lobbies, however, with their cocktail lounges and open patio restaurants, generate so much business that the added housekeeping and maintenance is more than justified.*

record-keeping, and special skill in the field of hiring and handling personnel.

### *Discovering Qualified Employees*

Since Housekeeping is the largest department in any lodging facility, the professional Housekeeper, especially in a major property, needs a touch of genius for discovering and keeping highly qualified assistants and supervisors. Given that gift, it becomes possible to recruit, supervise, and motivate the scores of competent maids, housemen, and other semiskilled workers without whom the job becomes intolerable. The greatest gift of all is the ability to keep the turnover rate down to a manageable level, despite the fact that 100 percent turnover in any year is not at all uncommon.

A manageable level is that level at which it remains possible to replace people within a very short time of their departure. Elsewhere in this text the turnover problem is discussed in detail, but it may be stated unequivocally at this point that there is presently no known cure for the situation; it can only be treated.

The hourly rates for housekeeping workers, whether unionized or not, are among the lowest paid by the facility. The work in many cases is very repetitive, and opportunities for advancement are quite limited. Therein, of course, lie both challenge and opportunity for the professional Housekeeper, who at the higher levels of competence, is among the lodging industry's most eagerly sought-after member of the management team, and who is becoming increasingly better compensated.

"Winning Housekeeping" requires both a sound game-plan and outstanding leadership.

Planning must take into account both long-term and short-term objectives. It must be programmed to provide maximum support for management's goals. It must be based on classic principles, and standard procedures, and also on:

1. The uniqueness of the institution
2. The perspectives and policies of the individuals who comprise management
3. The nature of the union contract
4. The character and policies of the other departments with which Housekeeping (in accordance with organizational charts shown elsewhere in this text) must interrelate

## *It Takes Leadership*

Above all, the Housekeeper must talk and act like a leader, and actually be one. It has been repeatedly documented that entry-level Housekeeping workers feel a maximum sense of security when they are not required to make decisions, and have complete confidence in the decision-making abilities of a strong person whose leadership they eagerly accept.

Leadership avoids extreme measures—both the soft extreme of leading by pleading, and the hard extreme of leading by force. The late Vince Lombardi was enshrined in professional football's Hall of Fame because his legendary qualities of leadership welded the Green Bay Packers into the game's most celebrated power. Lombardi summarized his philosophy of leadership at a conference of the American Management Association in these words: "Leaders no longer understand their relationship to those they lead. People shout for independence, but want to be led by a capable leader. They *want* someone to tell them what to do."

Housekeeping arts and sciences are still in their infancy, and offer lifetime opportunities to people-oriented leaders who can motivate, innovate, and use automation. The truly professional Housekeeper will find new ways to increase the productivity and stability of labor, while at the

**FIGURE 4-6** *Lounge areas must be kept spotless even when lavish velours and damasks are used, as at La Quinta Club, Palm Springs. Carpeting has a scrollwork edge that helps disguise soil along bar. Schedules at clubs often involve sizable night shifts to clean up after late parties and prepare for group functions the next day.*

**FIGURE 4-7** *Some lounges, such as this one at Del Monte Hyatt House, present fireplace problems: ordering firewood; keeping it supplied; cleaning out ashes. Light fixtures represent relamping chores; floor with carpet and plank requires two cleaning procedures; twisted iron dividers must be dusted.*

same time finding other ways to cut the amount of labor required to accomplish Housekeeping's mission, with no diminution in the quality of the department's total work output.

The pictures throughout this chapter are testimony to the wide assortment of people and places for which the professional Housekeeper in a large institution may be responsible.

**FIGURE 4-8** *Banquet rooms for thousands may be set up and dismantled by Housekeeping personnel working with Catering Department housemen. All napery (table linen) is handled through Housekeeping, as are fireproofing draperies and polishing the dance floor.*

**FIGURE 4-9** *Meeting rooms at the Sheraton Boston are typical of those that housemen in the Housekeeping Department may have to set up following specific layouts provided by Sales or Catering, covering banquet tables with cloths, and completing other necessary tasks.*

**FIGURE 4-10** *Public rest rooms, very much in Housekeeping's domain, are becoming less unpleasant to clean thanks to new chemical formulas, better tools, super-smooth surfaces.*

## *Contract Cleaning Services*

The professional Housekeeper should confer with management about the advisability of using outside contract services for such jobs as window washing, painting, and so on, in case of a lack of available labor in the area or if management deems it advantageous to limit the number of full-time Housekeeping employees. In any event, it is advisable to become familiar with the various types of services available, and the economics of handling the job internally versus having it done externally.

The best approach is to get bids from three companies that offer contract cleaning. Sometimes they will perform more than one operation. For example, carpet cleaners often repair carpet, lay it, and clean upholstery too; wall washing is sometimes contracted to the same company that washes windows.

Occasionally, contract companies will even supervise an institution's own employees, directing and training them to clean public areas.

Although total Housekeeping programs have been contracted for in hospitals and college residence halls, there are few, if any, instances

**FIGURE 4-11** *Kitchens may be assigned to Housekeeping Departments and specially trained crews will be needed. Chemicals that degrease and sanitize make job easier than in soap-and-water days, but they are also tricky to handle as food must not be contaminated by cleansers. Most kitchen floors have drains so that steam, pressurized sprays, or hoses can be used to wash them down.*

**FIGURE 4-12** *Entryways such as this one at Stanford Court, San Francisco, need to be kept polished and spotless, requiring the handling of such maintenance problems as a running fountain (often filled with coins), stained glass, skylighted dome, and grease spots on brick from autos (other than Rolls Royces, of course).*

where a hotel has had its guest rooms cared for by an outside company. In office buildings that are cleaned routinely at night contract cleaning works well, but management needs immediate control in a hotel or motel and often the contractor's employees will not take orders from the hotel supervisory staff. In a hotel there are many emergencies that arise due to arrival of unexpected guests or large groups of guests, so priorities must change, and only trained hotel supervisors will know when to redirect an employee's activities based on the needs of the moment.

One of the more difficult problems that occurs when a contract company takes on the Housekeeping chores is that if any thefts or security matters come up it is very often hard for hotel management to interrogate or investigate another company's employees.

Although keeping sleeping rooms and baths clean and ready for guests is always the realm of the Housekeeper, the following is a list of areas that may be included in the responsibility if she is qualified:

1. Guest room supply purchases
2. Cleaning supplies: specifying and purchasing
3. Public area: lobby, bars, dining rooms, banquet rooms

**FIGURE 4-13** *The restaurant interiors that demand tedious long hours from the Housekeeping Department usually are the whimsical, charming ones, full of furbelows and velvet such as this Victorian theme. Still popular with diners-out, these restaurants with "atmosphere" are conducive to spending more for dinner, and thus justify the increased payroll.*

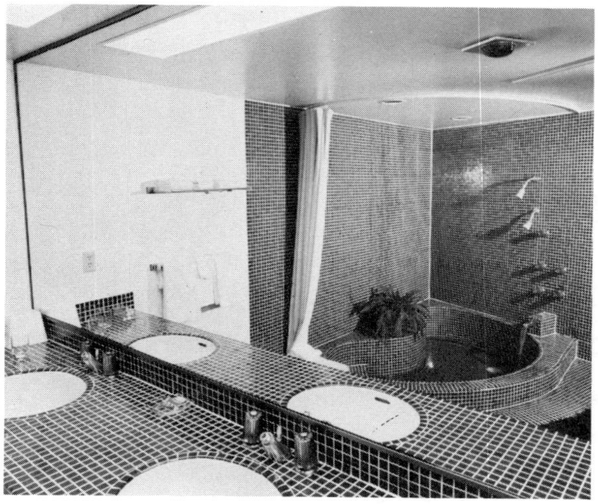

**FIGURE 4-14** *Guest bathrooms may be simply designed or lavish and complex, such as these installed to intrigue guests at the Rodeway Inns in Boise and Salt Lake City. Mirror wall is a plus for guests, but increases maid's cleaning time, as do sunken tubs. Added labor costs are justified because dramatic bathrooms attract customers.*

**FIGURE 4-15** *"Back-of-the-house" means all employee areas that guests or residents do not normally see. Floor care there is as important as in front for morale, efficiency, and for preserving carpet or flooring. Purchasing and care of such costly machines as this pile-brush vacuum are Housekeeping's responsibilities.*

**FIGURE 4-16** *Normally, ice makers are in an upper elevator foyer or back hall, but at Hawaii's Kona Village they are kept in a thatched hut centrally located on the lava rocks, so that all guests can walk to them. It is Housekeeping's responsibility to keep them clean and filled.*

4. Back-of-the-house: halls, kitchens, storerooms
5. Uniforms: issue, purchase, storage, design
6. Linens: guest room, restaurants, banquets
7. Grounds and pools
8. Plants (interior)
9. Shops
10. Offices
11. Garages
12. Mini-Maid service for permanent guests and apartments

# Chapter 5

# What It Takes to Be a Housekeeper

**OVERVIEW** An unusual mixture of qualities, skills, and personality traits are needed to be a professional Housekeeper. Ten important rules for leadership are listed, and educational courses suggested. The description of a housekeeping supervisor for a university is different from that outlined by most hotels, motels, or corporate headquarters.

To be a successful professional Housekeeper—one that management considers a part of the immediate team—one must have leadership qualities, administrative abilities, special personality traits, and an understanding of the techniques and skills required in the modern Housekeeping Department. A genuine interest in people helps immeasurably. Appendix 5-1 provides a typical job description for the position.

Housekeeping and Maintenance Departments have the most employees (and often the least skilled and most disadvantaged) in the organization. Thus there is a continual need to recruit, train, direct, and motivate. About 75 percent of the professional Housekeeper's time is spent with personnel—interpreting policies, explaining assignments, settling problems, and coping with the infinite variety of situations that arise each day.

Despite the need to maintain discipline among employees and keep the organization working through a chain of command, housekeeping executives must display a graceful tolerance of the many "colorful characters" within the group, and let them know that management cares about them as human beings.

Leadership has been described as getting people to do what you want them to do when you want them to do it. This can be achieved better by leading than by driving. There may be occasions, however, when there is no time for subtle psychology; for example, when suddenly and unexpectedly the house is full and the forecast had indicated only two-thirds of the staff would be needed that day. *How do you get an additional 60 rooms made up!* Situations like that don't unnerve experienced executives—they always have a plan.

## 10 Rules for Housekeeping Leadership

The professional Housekeeper must have the administrative ability to:

1. **Utilize manpower effectively.** This involves delegating tasks to others to get the work done properly and on time. To clean 1000 rooms in one day requires that everyone knows where to go and what to do without delay. It is essential to communicate the urgency of having check-out rooms ready for occupancy when incoming guests arrive.
2. **Devise easy methods of reporting work that has been done.** Called feedback, this kind of communication eliminates costly rechecking and retracing of steps.

3. **Develop standard procedures for routine activities.** This enables employees to develop consistent work habits, giving identical services to the guests daily. When routines are standardized employees can substitute for one another easily.
4. **Install inventory controls.** These are essential to any department dealing with costly supplies and expensive equipment.
5. **Motivate all types of labor.** High morale among employees is essential. To keep it high, the Housekeeper and assistants need all the expertise in the field of psychology that they can possibly muster—along with great patience, and a sense of humor at all times (Figure 5-1).
6. **Accept the challenges management proposes.** Perhaps a new uniform that will not show soil is requested, or complete refurbishing of the bridal suite for a sultan arriving next week with his newest wife and complete entourage. Despite such requests, the professional Housekeeper remains unflappable (Figure 5-2).
7. **Involve employees in planning.** Encourage them to use their imaginations to find ways to make their jobs easier. The question must constantly be asked: Is there a better way to do it? It should be addressed as frequently to the employees who carry out the jobs as to the supervisors who direct the maids and housemen.
8. **Increase the educational level of the staff.** This can be done through training and by encouraging employees to attend appropriate adult education classes, and so forth. To take full advantage of new developments the supervisory staff as well as the hourly employees must grow through observation, study, and education.
9. **Set up recruitment programs to find and develop management trainees.** This not only improves the property, but gives the trainees an opportunity to move into better

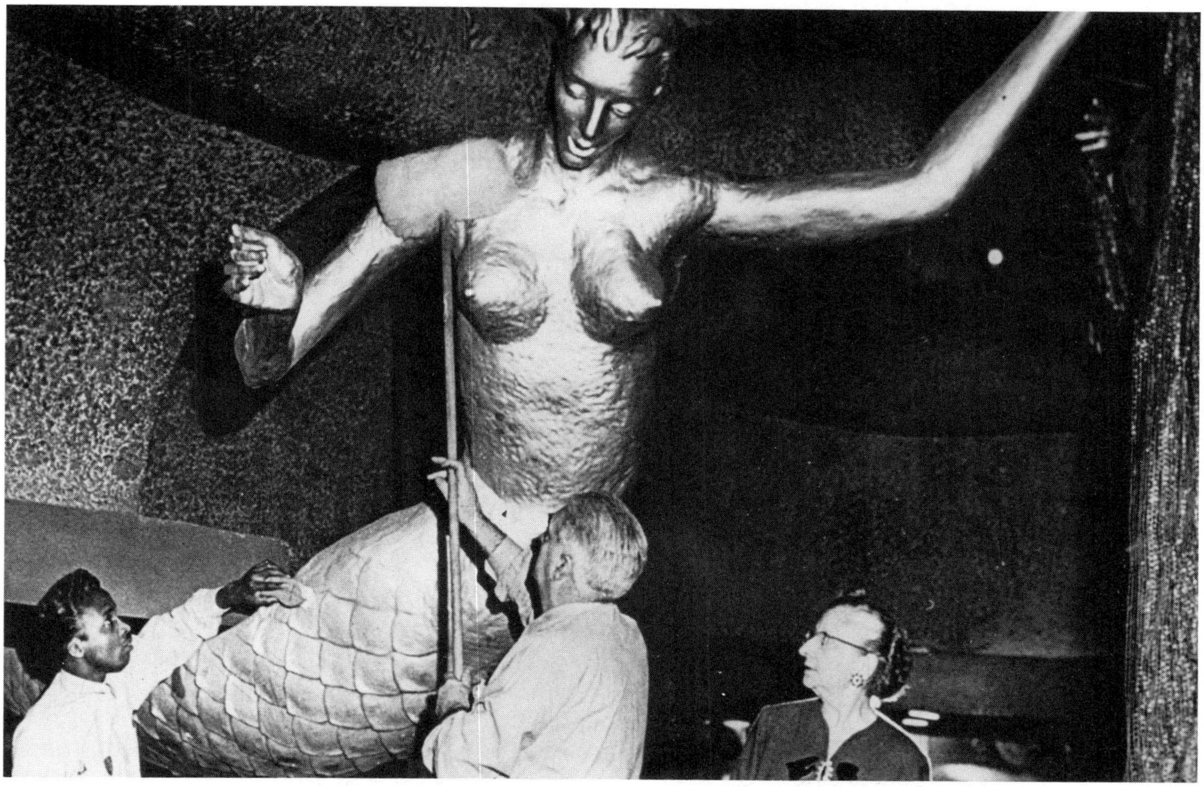

**FIGURE 5-1** *What does it take to be an executive housekeeper? A sense of humor and a lot of ingenuity, since one may have to figure out how to clean a giant mermaid as did this resourceful lady at the Park Sheraton Hotel in New York City. A lamb's wool duster and a short scrub brush for the scales do the job nicely.*

**FIGURE 5-2** *Perhaps the first Director of Guest Services, instead of Executive Housekeeper, Rosalie Hinderleiter for over 30 years was in charge of housekeeping and decorating at the historic old Brown Palace Hotel in Denver. She knew how to furnish a guestroom to give it that "worth more" look so a higher rate would be justified. David L. Owen is now Director of Housekeeping and Design for the hotel with a housekeeper and all crafts in his department.*

positions. It is also essential to train people for that great second level of junior management—the supervisory positions—where there is indeed a dearth.

10. **Cooperate and coordinate with other department managers.** Every well-run institution must be a unified whole, and Housekeeping is simply a part of that whole. The department must serve the good of the entire facility. The professional Housekeeper must therefore understand how each position fits into the total plan of the institution, and must be able to make even the houseman collecting trash realize that his role is indeed vital to the success of the facility. Each employee must understand how his work relates to the entire operation and realize that he is an important member of the team. A good manager also keeps all employees informed on significant aspects of the enterprise, communicating with all levels of workers (Figure 5-3).

## *Inquiry, Research, and Space-Age Innovations*

The professional Housekeeper who is possessed of an inquiring mind will constantly research innovations in equipment and proce-

**FIGURE 5-3** *Ex-Navy Captain Robert J. Martin was director of services for the Los Angeles Marriott Hotel, where he had command of three departments—Housekeeping, Laundry, Pool and Recreation. Now an instructor at the University of Nevada at Las Vegas he is teaching potential Housekeepers.*

dures. Salesmen with new products and developments—from pigeon poison to piano tuning—will be welcome.

This many-faceted manager will readily consider new systems, and try to mechanize manual functions to reduce labor costs in the department. Closed-circuit television, for example, might be used in many more ways to communicate with maids and train them. A manufacturer might be encouraged to develop a machine that would count soap electronically or hand out 30 pass keys. Finding progressive new suppliers—such as specialists who apply new porcelain finish to old bath tubs—is another of the many responsibilities that heighten the professional standing of the Housekeeping Department.

An interest in sociology is a requisite for persons considering housekeeping as a career because it will continually be necessary for them and their supervisory personnel to understand attitudes and behavior patterns of the worker who, in many cases, will be newly-arrived in this country, or perhaps has recently been on welfare. The housekeeper can be of real help in rehabilitating under-privileged people who would like to improve their economic status and enjoy the material and psychological benefits of being part of a respected group.

Working in a lodging facility will often reveal to maids and housemen a totally new way of life. If promotions of entry-level employees are publicized, and workers see that their friends or associates have received increased pay and prestige, department heads may find more employees willing to try a little harder, realizing that indeed "it could happen to me."

**FIGURE 5-4** *Margaret Kappa, director of housekeeping, the Greenbrier, White Sulphur Springs, W. Va., examines rare, white French opaline vase. After 23 years with the resort, Mrs. Kappa contends that, "A qualified woman executive can become a professional Housekeeper without hotel experience. It helps if she knows the industry, but if she is a good business manager she can administer a department without having been in a hotel before."*

## *Personality Traits*

The professional Housekeeper must have self-confidence, knowing that she has the training, ability, and experience to teach and motivate the people working in the department. Teaching, of course, requires dedication and a desire to help others learn—and patience, a quality admired by management because of its importance in Housekeeping (Figure 5-4).

Ambition, too, is important, as it may lead to advancement to other departments, a wider realm of Housekeeping, a more active role in management, or to a bigger housekeeping job in a larger property.

High on the list of necessary traits for success are honesty, loyalty, and sincerity—honesty in relationships with management; loyalty to employees, both within the department and throughout the company, as well as to the institution; and sincerity with the housekeeping staff. The executive's loyalty to the hotel may rub off on the employees, and loyalty in a Housekeeping Department is definitely needed to help reduce turnover.

A sense of humor will help immensely when it comes to granting outrageous requests from other department heads (and especially guests!), as will the poise to meet the unexpected crisis that seems to arise so often. Liberal amounts of tolerance and stability are also necessary. Some days it is even necessary to deal with maids fighting in guest corridors.

Being a persuasive salesman, able to present an idea or explain a need convincingly, will help—whether it is asking engineering to repair a leak or requesting that the laundry stop folding sheets and start washing banquet cloths for a late-scheduled dinner.

Although the professional Housekeeper enjoys the privilege of bringing guests into the hotel dining rooms and using the public facilities, the enlightened ones more frequently take the back elevators, eat in the employee cafeteria, and ask one or more employees to lunch. Just the few such minutes that an executive invests in the employee can pay off handsomely in loyalty and changed attitudes about work. Birthday parties and Christmas get-togethers also help the employee to feel that he really belongs.

Figure 5-5 can help give you an idea of how you might do as a professional Housekeeper.

## One of the "Old World" Housekeepers

Henrietta LaBorde, executive housekeeper, Hilton Hotel's Western Division, is a charming, volatile Frenchwoman who discharges her many duties with fierce energy and just enough drama to get her department *and* other hotel personnel to pay attention. Her control on Housekeeping is somewhat reminiscent of that in an army regiment. Yet she is consumately fair and has won the respect of her employees. She expects her people to be honest with her and to admit their mistakes—for which she does not chastise them as she feels they can grow if they recognize and admit to an error and thus learn from it.

Mrs. LaBorde runs her department by written report. She wants to know what has taken place so her assistants write out everything of importance for the record. Her assistants have the authority and full responsibility for carrying out every function except that of firing employees—they can only suspend until Mrs. LaBorde can confer with the individual.

If, for example, a maid is having trouble with her floor housekeeper, Mrs. LaBorde may try that maid on another floor to see if there was a personality clash. "I operate Housekeeping as if I own the hotel," explains Mrs. LaBorde, who gets little or no interference from her few superiors because she has proved her ability to contribute to maximum prestige and profit for the hotel.

"Housekeeping is *organization*," states Mrs. LaBorde, who points out that thinking ahead is the whole key. "And there is so much necessary paperwork—especially in union towns."

## Educational Requirements

Professional Housekeepers—from the director of housekeeping for a large chain to the housekeeper in a small motel—have a wide range of educational backgrounds. A new director of Housekeeping who had been in another profession during most of her career may have a master's degree in Business Administration; the Housekeeper in a very small motel may not have finished high school. The educational qualifications are related to the size and quality of the institution and the services offered.

It is only in recent decades that Housekeeping has begun to earn recognition as a profession. Thus academic training in vocational schools and at junior and four-year colleges, is not yet universally available. Nor have courses as yet become standardized even among the fewer than 25 four-year colleges whose Home Economics Departments offer a major in Institution Management.

It goes without saying, however, that someone who has successfully completed a two-year or four-year course in Housekeeping at the vocational school or college level starts with a big advantage over the one with no formal education in the area. Part of that advantage lies in the fact that the individual has made the determination over a two-year to four-year period to seek a lifetime career in this profession.

The college graduate with an unrelated de-

CAREER VERSUS JOB QUIZ

Here are some of your most important decisions; answer each question thoughtfully.

| A. Do you want | Don't Care or Don't Know | Yes | No |
|---|---|---|---|
| **1.** A salary between $14,000 and $35,000? | | | |
| **2.** Normal working hours 8 AM–5 PM? | | | |
| **3.** Most week-ends off? | | | |
| **4.** Two weeks paid vacation after one year? | | | |
| **5.** Long-term security, retirement plan, excellent fringe benefits? | | | |
| **6.** Pleasant, even beautiful, surroundings? | | | |
| **7.** A semiannual performance evaluation? | | | |
| **8.** To learn new systems and skills at company expense? | | | |
| **9.** Variety in your daily work? | | | |
| **10.** Opportunity to advance? | | | |
| **11.** Equal job opportunity for men and women? | | | |
| **12.** To be able to find employment in almost every city of the world? | | | |
| **13.** To be proud of your position? | | | |
| **14.** To use your technical knowledge and skills? | | | |
| **15.** To have career power? | | | |

| B. Are You | Don't Know | Yes | No |
|---|---|---|---|
| **1.** People-oriented? | | | |
| **2.** Goal-oriented? | | | |
| **3.** Of potential management caliber? | | | |
| **4.** An organizer? | | | |
| **5.** One who can delegate responsibility? | | | |
| **6.** A good trainer and supervisor? | | | |
| **7.** A good listener? | | | |
| **8.** Capable of spending large sums of money judiciously (and based on a budget)? | | | |
| **9.** A peacemaker? | | | |
| **10.** A good judge of people? | | | |

If you answered Yes to more than eight in A group and more than five in B group with two Don't Know's, then by all means continue reading this book.

By your answers did you discover anything about yourself—your potential talents or unknown goals?

Open your mind to entering the wonderful world of Housekeeping. Know that all hotels/motels, hospitals, convalescent homes, retirement facilities, residence clubs, or college residence halls need housekeepers, supervisors, property managers, environmental service directors. Whatever the title, the work is challenging, creative, interesting, filled with learning opportunities, and continually rewarding.

**FIGURE 5-5**

gree lacks this advantage. Persons who fall into this classification account for one of Housekeeping's highest drop-out rates. One must also recognize, however, that even college graduates with degrees in administrative housekeeping make their contribution to the drop-out rate when they find the actual work to be somewhat less fulfilling or more demanding than it seemed in the classroom. Many of them simply lack the temperament to cope with the labor problem.

## *NEHA Certification Standards*

The National Executive Housekeepers Association (NEHA) has set specific standards for the certification of housekeepers and requires college sponsorship of its courses, however, the number of college-trained housekeepers is barely sufficient to meet the demand for qualified managers. Even certified personnel are hard to find, and graduates of vocational schools are also in short supply. Efforts are being made to set up more courses.

The American Hotel & Motel Association, adult education courses available in the community, summer courses sponsored by the school of Hotel Administration at Cornell University, all help to train for supervisory Housekeeping positions.

Experience indicates that the following are among the more valuable courses for potential executive: (1) English, (2) a foreign language, preferably Spanish, with French and German being second choices on the East Coast, and Japanese and Chinese on the West Coast, (3) psychology, (4) accounting, (5) labor relations, (6) interior design, (7) chemistry, and (8) sociology.

**FIGURE 5-6** *Extensive greens and gardens at Los Angeles' Century Plaza are kept in perfect shape by gardeners who prune one of the larger rubber trees as Inge Krieg checks their progress. She has enormous versatility, moving from laundry to landscaping in a few minutes.*

Most mistakes made by employees can be traced to their not having understood instructions. The importance of being able to communicate clearly, both verbally and in writing, cannot be overestimated.

The opportunity to travel the world is ever-present as the chain hotels always need experts to help train employees and open foreign properties. The importance of the first two subjects on the above list thus becomes obvious. The importance of the others is explained in other chapters.

## *Careers for Both Sexes and All Ages*

Because the housekeeper is in charge of the lodging facility's largest, yet lowest-paid department, with a turnover problem of formidable proportions, it is imperative that the executive have the competence to command respect, the grace and good humor to earn a measure of affection, and the maturity to guide firmly on behalf of management yet understandingly on behalf of the workers. These qualities occur almost equally in men and women, but more often in persons who are over 40 than in persons under 30 years of age. One earns the executive position the same way as for any other management area. It does not come easily or quickly; it is seldom awarded on the basis of anything other than merit.

The professional Housekeeper is almost invariably more familiar with the total lodging facility than any other executive or staff member (Figure 5-6). For this reason, the normal problems incident to the generation gap may be aggravated still further if the Housekeeper must report to a very young resident manager or assistant housing director. To eliminate this problem the Housekeeper, especially the one who demonstrates outstanding ability, increasingly reports directly to the general manager of the facility—a significant sign of the profession's upgraded status in this changing industry.

Good health is perhaps more important to the professional Housekeeper than to most executives. It is extremely important for that person to be seen by employees inspecting some areas of the property on a daily basis. This amount of activity necessitates physical endurance to function in top form. The Housekeeper should formulate a system of physical exercise that contributes to good health. Employees should be encouraged to do the same.

Margaret Kappa, a second-generation assistant manager in charge of housekeeping and decorating at the Greenbrier (with a Cornell degree in hotel management), is fond of quoting her mother's candid analysis of the lofty standards that management sets for its executive housekeepers: "The manager of any property expects the executive housekeeper to be the manager above the lobby floor, expects him or her to be a peacemaker, to provide leadership, to settle disputes, to keep a clean house which is pleasingly decorated, and to maintain meticulous records."

Despite her mother's definition of management's awesome expectations, Mrs. Kappa readily admits to having found real fulfillment in her chosen profession. She especially enjoys the splendid isolation of White Sulphur Springs, having previously been at midtown Manhattan's bustling Waldorf-Astoria and Plaza hotels. To Mrs. Kappa, who raised two sons there, the Greenbrier is literally her home—and not just an ordinary home, for as a large part of the world knows, that hotel is one of the most civilized and celebrated retreats in this or any other country.

## APPENDIX 5-1

### Position Description

**Title of Position:** Housekeeping Supervisor for a university residence hall

**Classification:** Staff Exempt

**Department:** Office of Housing and Food Service Operations

**Function of Position:** Responsible to the manager of housing services for administering and coordinating housekeeping functions in a specified residence hall area.

**Principal Duties and Responsibilities:**

1. Coordinate activities of the assistant housekeeping supervisors to ensure that uniform operational practices are followed and that these practices are in conformity with departmental practices.
2. Formulate plans and establish work schedules for employees to ensure equal distribution of work loads; revise schedules as required.
3. Plan and recommend to the manager of housing services plans for systematic housekeeping. Supervise housekeeping employees as needed.
4. Plan and provide care for equipment, furnishings, and buildings. Report items needing maintenance. Initiate work orders to include repair of windows, doors, plumbing, heating, and electricity, etc.; follow through to completion.
5. Counsel employees on job performance, recommended increases, discipline, etc.
6. Discuss with employees grievance matters, suggestions, and complaints. Recommend appropriate action.
7. Control employee time cards, vacation and sick leave records, etc.
8. Regularly inspect work being performed and ensure that standards are being met and maintained. Conduct on-the-job training for all assigned personnel. Develop training for cleaning all types of floors, proper methods of dusting, sweeping, waxing, buffing, cleaning rooms, washing walls, etc.
9. Responsible for maintaining a perpetual inventory of supplies and equipment, requisitioning supplies and equipment as required, and verifying orders when received. Responsible for linen and bedding inventories.
10. Control and issue supplies and equipment to employees as required. Maintain record of materials used; prepare monthly and periodic reports. Correct waste, misuse, or abuse of supplies or equipment.
11. Inspect student rooms and urge students to maintain acceptable standards of housekeeping; report damage or use of unauthorized items.
12. Coordinate work with the Dean of Students area staff regarding all areas of activity including making special arrangements for activities.
13. Investigate accidents and initiate accident reports. Recommend corrective action as required. Maintain first aid supplies and administer first aid if required. Responsible to ensure safe working conditions in assigned area.
14. Prepare for conferences as required and coordinate these activities with other areas involved.
15. Encourage work simplification, efficiency, and general pride in work performed.
16. Receive, hold, and process monies received as income from refrigerator and bicycle locker rentals, etc.
17. Eat meals for the purpose of evaluating quality and quantity of food products; correct food preparation techniques and overall menu appeal.
18. Perform other duties as assigned.

**Supervision:** Work is performed under general supervision where standard practice requires the use of several procedures and enables the employee to proceed alone. The employee is expected to refer highly questionable cases to the supervisor while supervising assigned support personnel. Supervises up to three assistant and/or associate housekeeping supervisors, one clerical employee, and twenty or fewer technical service employees.

**Minimum Qualifications:** A bachelor's degree or equivalent is required, plus one to two years of effective experience.

*This Position Description (from Pennsylvania State University) explains what the housekeeping supervisor of a large residence hall, or halls, is expected to do. On a very large campus, where 10,000 or more students live, there may be several of these supervisors.*

# Chapter 6

# Staffing the Housekeeping Department

**OVERVIEW** A person who does hiring should be very familiar with the jobs for which he or she is interviewing applicants, and should use professional techniques that will help reveal the potential of each person. Careful hiring is the key to keeping turnover down. Here are pointers on effective and *legal* interviews, the latter being very tricky these days when fair employment practice laws impose many contradictory rules.

Basic to the lodging industry is adequate staff to maintain continuously fine service and a well-groomed environment. The ultimate (although idealistic and unreachable) objective of personnel management is to make the selection process so foolproof that turnover is virtually eliminated and job security is recognized as an established feature. This can only be achieved if management concentrates on creating the right work climate so that employee morale and efficiency are reinforced constantly by proper rapport with co-workers and supervisors.

Because the majority of the applicants interviewed for Housekeeping's entry-level jobs—maids and housemen—are from the less advantaged groups, it is often not only necessary to make good employees of them, but also to help them adjust to different living standards. For many, the job may be their first exposure to middle-class customs. They may have to learn such basics as personal hygiene, proper eating habits, good grooming, and at least enough about the social graces to get along with fellow employees and to be unfailingly courteous to guests.

Many entry-level employees may need help in filling out legal papers or understanding how to register and vote, or in finding where to go to learn English. The executive housekeeper must face the fact that not all employees will succeed or even adjust to the ranks of the regularly employed, but she takes pride in the ones who make it, and keeps trying with the others.

### Plan the Ideal Staff

To develop an ideal staff, the Housekeeper must begin by defining an ideal set of positions to be filled. This makes it necessary to decide what levels of supervision the department should have; to write brief job specifications and detailed job descriptions for each position; and to determine approximately how many people will be in Housekeeping. (See Figures 6-1 and 6-2 for samples of a job specification and job description.)

A job specification is a written itemization of the minimum requirements that must be met by an applicant. These specifications are derived from a job description, which is a detailed report of all functions that must be performed by a maid, houseman, and so on, and the manner in which they must be performed.

JOB SPECIFICATION

**Job Title:** Linen Controller

**Number Employed:** 1

**Hours:** 7:30 am to 3:30 pm; Monday to Friday

**Days Off:** One week after 1 year; two weeks after 3 years

**Salary:** Union scale

**Extras:** Uniform furnished; 1 meal

**Working Conditions:** Pleasant

**Contact:** Employees of Food and Beverage, Banquet, all other Departments that use linens

**Supervisor:** Assistant housekeeper

**Type of Labor:** Unskilled

**Responsibility:** All food and beverage linens be in good condition

**Age Limits:** 20 to 60 years (see Physical Qualifications below)

**Mental Qualifications:** Average, ability to follow instructions, alert

**Personality and Appearance:** Very pleasant, careful

**Experience:** None required

**Education:** Able to read and write, speak, or understand Spanish or whichever second language is widespread in the area.

**Physical Qualifications:** Able to stand light sustained physical activity including frequent use of arms, able to lift moderate weight, good eyesight for examining fabrics for stains, snags, holes etc., and stitching in linens.

**FIGURE 6-1**

## *Astute Hiring Is the Key*

Having set up the positions (see organization charts in Chapter 7), Housekeeping then begins the crucial process of selecting employees.

Interview time can be cut appreciably if applicants are initially screened by members of the Personnel Department to weed out persons who are misfits, illegal aliens, or physically unable to perform the required duties. To illustrate the point, one of San Francisco's newer and better hotels recently screened 4000 applicants to find 300 employees.

Final interviewing may be conducted by the Housekeeper, along with the assistant or resident manager in smaller properties. In larger properties maids and housemen will probably be interviewed by assistant housekeepers, who will narrow the choice to two or three applicants and leave the final decision to the head of the department or executive assistant.

Choosing the wrong applicant because of unscientific selection procedures is the major cause of turnover—that "antipersonnel" monster that devours employees at a yearly rate that is incredibly costly. Studiously avoid the unqualified, the misfit, and the over-qualified.

An unqualified employee cannot learn to do the job at all. Hiring him is comparable to hiring an accountant who cannot add. The misfit has chronic personality problems that fall into the field of psychotherapy, but not Housekeeping. The over-qualified applicant—who might be compared to the CPA hired to be a payroll clerk—should also be avoided since he, too, is a misfit when relegated to an inferior position.

The unqualified employee and the misfit seldom consider resigning, but continue to stay on the payroll. They aggravate such chronic problems as absenteeism, tardiness, and inefficiency—all of which deflate the morale in a department where it is essential that all employees feel the workload is evenly distributed and that all jobs are of equal importance.

## *Seek Sources of Prospective Employees*

The employer who has several applicants to choose from has a better chance of hiring the right person than the employer who has only one and will take any "warm body" because he needs help immediately. Crisis hiring is not one of management's better habits. Too often, these important sources of potential employees are overlooked:

1. Present employees and their friends
2. Private and public employment agencies
3. Schools—vocational, secondary, college
4. Personnel files
5. Ads in newspapers, especially local ones

**JOB DESCRIPTION**
**LINEN CONTROL (Tablecloths and Napkins)**

**Title:** Linen Controller

**Responsible to:** Director of housekeeping

**PURPOSE AND FUNCTION**

**A.** Control and coordinate the receiving of clean tablecloths and napkins and issue those linens requisitioned by the various food and beverage facilities within the hotel. The functions of this service are as follows:

1. Receiving linen several times throughout the day as it is processed by the laundry and delivered.
2. Inspection of linens for burns, tears, fading, etc., maintaining standards outlined to you with samples.
3. Stocking of linens on shelves according to size, color, and use for each food or beverage facility.
4. Issuing linen only for the amount requisitioned that day.
5. Communications with laundry manager for a specially requested quantity, color, or size linen that is in short supply or not available on the shelf when requested.
6. Issuing ONLY the linens designated for each facility's use.
   REMEMBER! Banquet department has priority on linens necessary for its use.

**B.** Remove all linens requiring repairs to the Sewing Room.

**C.** Adequately control, and give proper attention to, the issuance of linen. Failure to keep track of linens results in their misuse, damage, and disappearance, thus wasting time and money.

**FIGURE 6-2**

6. Veterans hospitals, churches, and charitable institutions
7. Boys' farms or homes for the mentally retarded (if trained properly, these people often can do simple, repetitive, routine tasks very well)
8. Military bases—wives of servicemen
9. Union offices

Posting notices in civic or recreation centers and in laundromats or supermarkets in sections of the community with high concentrations of ethnic groups is often effective if young people are needed. Employees in all departments should also be made aware when Housekeeping is looking for employees.

The recruiting program should be continuous. Today's prospects may fill tomorrow's jobs. Anyone who has been interviewed and shows promise should have an application on file, even if his skills are not needed today. (Perhaps he's already employed but is interested in making a change at a later date.)

## *Don't Sugar-Coat the Job*

In talking to applicants one must be specific and accurate in describing the job that's open. The employer who has all of his specifications written out, who is concerned with finding exactly the right person to fill a position, has a much better opportunity of securing a satisfactory employee than the employer whose requirements are vague and unrealistic. Nothing is worse for the employee than to discover that what he was told in an interview about the type of work he would be expected to do was highly inaccurate.

Accurate job specifications and a realistic job description go hand in hand in a good recruitment program. They are invaluable tools for interviewers and applicants alike, describing not merely the daily tasks, but those that must be performed once a month, once a week, every three days.

Turnover in any service industry is high, but in many areas of the hotel business six to ten people are replaced out of every ten hired during the year. The high cost of turnover (which averages over $600 per employee who leaves) is one of the major reasons managers in all types of institutions should constantly be on the lookout for prospective employees, and try to convince the younger generation of the opportunities in the service industries.

Demand for labor with only entry-level skills is increasing, and many industries compete for the limited number of workers available at any one time. All branches of the hospitality and lodging industry must be aggressive in seeking out new employees and eager to improve the working conditions and wages that are offered.

It is particularly important that management strive to upgrade the unit's reputation in the community with the kind of enlightened employee relations program that workers will discuss with their friends. This helps to make the surrounding neighborhood the major source of prospective new employees.

### *Turn Off Turnover*

Whoever does interviewing and makes hiring decisions is performing a very critical service in a turnover-plagued industry. The situation is parallel to attempting to cut the divorce rate by making it more difficult to get married. If people wouldn't hire so casually because of a crisis condition, which tempts them to hire anyone just to get the work done, turnover could be cut noticeably.

The rate would also be lowered if employers were more wary of hiring persons whose applications indicate strong job-hopping tendencies. A little of this may indicate ambition; a lot is a signal, warning of serious instability.

It is doubly unfortunate that the turnover problem affects employees as adversely as it affects the institution. Too often the individual leaves his job before he has had time to demonstrate enough competency to be considered for promotion, or has established a sound enough track record to convince employers that he is qualified for responsibility. This is a situation that all supervisors should be reminded to mention frequently to employees, not solely in the interest of the institution, but perhaps even more in the interests of the employees.

### *Interview Means Show and Tell*

The interview must be conducted so that it develops as much information about the applicant as possible if it is to provide a sound basis for predicting whether he can do the job. The interviewer must try to determine the applicant's needs, both psychological and monetary, to determine if the job will satisfy him, and whether he is a self-starter or at least responsive to ordinary motivation.

The prospective employee must clearly understand the position for which he is being interviewed. In some cases it may be worthwhile to talk with him about the possibilities of promotion, either in Housekeeping or other departments. Obviously, his main interest is in the specific immediate benefits accruing to him, particularly the rate of pay, when payday occurs, and what sort of health, welfare, and fringe benefits are included.

He should also be fully acquainted with such other company benefits as coffee breaks and incentive programs. These factors are, of course, routine in large operations, but conspicuous for their absence in small institutions.

Paid vacations may not be of interest to the person doing the interviewing, but are often uppermost in the mind of the applicant. He must understand exactly what the job entails; what the working hours are; what the pay is; whether the uniforms are furnished; what meals are included; the policy regarding absenteeism both with and without due cause; and all other pertinent facts.

If employees are not allowed to come into the institution and its restaurants as guests while off duty, applicants should know this because many cannot understand the rule. (The basis for this policy is that employees on duty will pay too much attention to the employees off duty, and neglect customers or behave too informally.)

If the person who fills the job must become a union member it is essential that he be told and that all matters pertaining to membership be clearly described, including the date he must join the union, the cost, what his union benefits will be, and how he will be able to withdraw from the union should he leave or be promoted.

With the exception of large cities, particularly New York, San Francisco, and Chicago, employees in most hotels or motels are not unionized. Chains such as Hilton, Sheraton, and Westin have unions for Housekeeping employees. (Carpenters, plumbers, and electricians, of course, belong to other specific unions and come under different jurisdictions.)

One major hotel chain—Marriott—has, however, managed to avoid having any of its service employees become unionized. It boldly fought back the union advances, causing a particularly flagrant stir in Chicago in 1978 when it opened its 1200-room hotel on Michigan Avenue surrounded by pickets. The hotel spokesmen admit that the picketing during its first week cost the hotel $19,000 worth of canceled room bookings.

Only one of Marriott's 55 hotels—the Essex House in New York City—is unionized; that contract was in place when the hotel was purchased.

Union officials themselves admit that it has been difficult to win over Marriott employees because they are paid above union scale and given substantial benefits.

One of the many currently forbidden questions during an interview is how the applicant feels about the union. The law refusing to permit this question is enforced by the National Labor Relations Board.

Important to consider when hiring, especially for brand new institutions, are local labor conditions. Is the population floating or stable? What other employers in the area might be interested in hiring this employee?

These and similar matters may require some research if a new property is located in a geographic area unfamiliar to the employer. Usually, local conditions can be determined by checking with other institutions nearby. Employers in colleges and hotels maintain it is easier to attract and hold employees in smaller towns, even though there may be less of a field from which to choose.

Enlightened management will take all of the following into account: types and frequency of public transportation, parking facilities, whether applicants are mostly from one section of the city. Will employees come primarily from ghettos, and how far away are those areas? Are they served by adequate public transportation? If the employer hires enough people, local transportation companies may be persuaded to modify schedules to accommodate employees, or it may be worthwhile to help arrange carpools or provide a charter bus.

Management of very large institutions may even go so far as to discuss new low-cost housing with the city fathers in order eventually to relocate the labor market within walking distance. Other considerations: what conveniently located child-care facilities are available? What is the unemployment rate in the area? What is the educational level of the unemployed? Is the employer competing with welfare payments?

Nearly all areas in which lodging units are being operated include sizable ethnic groups. Their racial and religious practices must be understood by persons doing the hiring. It is imperative that these practices (for example, ethnic or religious holidays) be respected by management, department heads, and supervisors. The point to be considered is not discrimination but intelligent planning, inasmuch as the lodging industry, with few exceptions, operates on a 24-hour day, 7-day week, 365-day year.

The FEPC regulations make it difficult to learn much about an applicant's home life from a job application, but casual conversation often enables the interviewer to determine whether the prospective employee's personal life is so unsettled that it will probably affect job performance unfavorably. In the case of a prospective maid, for example, does she have a husband and children? Are they healthy? Will there be a babysitting problem? Will there be trouble at home if her work hours do not coincide with her husband's?

Most institutions now advertise themselves as Equal Opportunity Employers. The affirmative action program adopted by many hotels has to do with the equal opportunity of women and minority personnel for positions of leadership at all levels. The hospitality and lodging field has historically been dominated by men, but women are entering the industry through:

1. Hotel and restaurant degrees from major colleges
2. Promotion from within up through the ranks
3. Becoming consultants in the industry
4. Stockholder's voice from ownership
5. Teacher influence

Women are now beginning to make themselves known by taking responsibility for leadership and showing their ability to do so.

By government regulation it is illegal for position titles to indicate the sex of the worker. Therefore the old titles of maid, houseman, bellman, and so on have been deleted in many instances. There is also a good deal of resistance in the industry to the suffix–person, which seems either ludicrous or depersonalizing to many. Because of common usage in our industry, however, certain words have come to indicate a position and the work defined rather than the sex of the worker.

## *Physical Requirements*

All applications must include a health history because much of the work done in the Housekeeping Department requires strenuous physical activity. If the applicant indicates that she is completely healthy, it is well to get into an informal conversation and make remarks that might evoke additional health information. "Wonderful that you never catch a cold, and your stomach never gives you any trouble. You are lucky. What about your period; don't you ever have any discomfort?" One might also discuss how many people have back trouble or arthritis these days, or mention other conditions that could affect performance.

Generally, the applicant will not be able to resist talking about some physical problem if there is one. If the interviewer's comments do not seem to be too direct, but rather in the nature of a friendly chat, health history that was never entered on the application will usually surface. It takes a clever interviewer to ferret out this information.

With safety in mind, one of the very important questions to be probed during an interview is whether the employee has ever been away from work for an extended period of time due to a major accident. If he has, he may be practically nonemployable for Housekeeping labor. If he has been in an auto accident and suffered whiplash, back injuries, or similar disabilities, and was away from work for a year, six months, or even as little as three months, every effort should be made to determine residual after-effects that will become readily apparent whenever physical exertion is involved.

This does not preclude hiring the handicapped for positions whose requirements are unrelated to the infirmity. There are many jobs in housekeeping that handicapped people can do exceedingly well, and for which they should be given serious consideration. Handicapped workers have an established record of outstanding performance.

Personnel will automatically verify the information on a prospective employee's job application immediately after personally interviewing the prospect—assuming their favorable response to the interview. High school or college attendance may have to be checked with school authorities or substantiated by transcripts. Military service may have to be confirmed by discharge papers. Credit reports may also be helpful. If for any reason it is considered necessary (because of gaps in a work history) to check on an applicant's police record, the local police department will probably be very cooperative, as will the FBI.

A telephone call to an applicant's previous employers is the quickest and easiest way to check on his work history. Printed forms are available to help one make certain that all of the key questions are asked of former employers—especially "Would you re-employ?"

Sometimes it may be possible to give certain tests to find out if the employee has the ability to perform elementary but vital functions. Employees in entry-level jobs in Housekeeping have to have enough education to sign their names, and read and write room numbers, floor numbers, and dates. Simple tests will reveal these abilities. Before these are given, however, it must be determined whether or not they are permitted by government and union regulations.

Personnel experts point out that tests for

manual dexterity, mental ability, job aptitude, and job knowledge are most helpful in determining what an applicant *can* do, while personality tests are most helpful in determining what an applicant *will* do. Any worthwhile insight into an applicant's personality must be based on test information *plus* information from previous employers, *plus* information from a personal interview. The best early indicator of the employee's ability to get along and cooperate with other employees is probably his attitude during the interview.

## *Improved Interviewing*

In an employment interview the applicant and the interviewer try to find answers to two questions: (1) will the applicant satisfy the job requirements, and (2) will the job satisfy the applicant's requirements? Therefore there must be an honest exchange of information. Furthermore, when one is interviewing for a supervisory or management position it is often advisable to have more than one person talk with serious candidates.

## *Ten Simple Rules*

1. **Be prepared.** A hurried or harried person often does not take the time to review the requirements of the position. Study the job specs.
2. **Find a private place.** The applicant will feel freer to talk if others are not listening and if it is quiet. Try to use an office with no one else around unless it is essential to have a translator.
3. **Practice.** If possible, the interviewer should be trained and his techniques checked while he conducts practice interviews under the guidance of someone experienced in the procedure.
4. **Study the application.** Take a few moments before asking questions to see if the application form seems to indicate such essentials as good health and a reasonable work record. If the interviewer ignores the application, the applicant may think that just anyone will be hired to fill the job because it is a temporary situation.

   It is essential that one try to hire permanent employees, even for entry-level positions. Some of these people will become the supervisors and managers of the future (Figure 6-3). Others will perform even a menial job zealously, preferring security to even modest advancement.
5. **Get the applicant to talk.** By sure, however, that the questions you ask do not violate the law. The FEPC, the U.S. Fair Employment Practices Committee, state departments of employment, and various other agencies will not permit certain questions to be asked because the answers are not considered pertinent to whether the applicant is qualified.

   Although trained interviewers know very well which queries are forbidden, the uninitiated might consider it quite normal to ask the very questions that are outlawed. The interviewer must not give the idea that he is trying to find out the religion or national origin of the applicant or any member of the applicant's family, or that he is prying into his personal life.
6. **Ask questions that involve explanations.** Questions requiring only a yes or a no do not reveal much about the background of the prospective employee. Also, interviewers should never ask a question that suggests the answer they wish to hear.
7. **Recheck dates on the application with the applicant.** Sometimes in looking at the form, time gaps will be evident and by questioning an interviewer may find out a great deal that was not on the application.
8. **Give complete, honest, and accurate information about the job and the company.** In the case of entry-level positions, the interviewer must give detailed descriptions of the job so that after being hired, the employee cannot say, "I didn't know I had to do that." As an example, when interviewing for a maid, show an applicant without previous experience both a check-out room (one in almost complete disarray) and a fin-

THE GREENBRIER PERSONNEL ACTION REQUEST

NAME ______________________________

DEPARTMENT ______________________ DATE OF REQUEST __________ EFFECTIVE DATE _______

SIGNATURE OF DEPT. HEAD ______________________ SIGNATURE OF MANAGER (For Additions, Rate Increases, New Positions) ______________

FURTHER INFORMATION MAY BE OBTAINED FROM ______________ PHONE EXT. ______________

## PART A - REQUEST FOR PERSON ACTION

**1 REQUISITION FOR EMPLOYEE(S)**

Job Title ______________________ Rate of pay ______ Shift _______ No. Needed ___________

☐ Addition
☐ Replacement for ______________________
☐ For new position described in Part B, below

☐ Permanent
☐ Temporary (How long? ____________________)

☐ Male ☐ Female

Special Requirements:

**2 CHANGE IN AN EMPLOYEE'S STATUS**

| PRESENT STATUS | NEW STATUS (*) |
|---|---|
| Name | Department |
| Job Title | Job Title |
| Wage rate | Wage rate |
| Shift | Shift |

(*) ☐ Addition ☐ Replacement for ______________________

If wage rate change, check one: ☐ Merit ☐ Seniority ☐ Other (Explain)

## PART B - REQUEST FOR POSITION ACTION

Type of action requested: ☐ Establish a new position ☐ Change a present position

| PRESENT | PROPOSED |
|---|---|
| Job title | Job title |
| Starting rate | Starting rate |
| Rate range | Rate range |
| Hours | Hours |
| Days | Days |
| Name of present occupant | ☐ present employee remains ☐ new employee |

Describe proposed position:
☐ Permanent
☐ Temporary (How long? ____________________)

☐ Male
☐ Female

Other qualifications: ______________________

Brief statement of duties and responsibilities:

REMARKS:

*The above form is filled in by the Greenbrier department managers when they want (1) a new employee, or to change the status (salary) of a present employee, or (2) to promote or change an employee's position. The form is sent to the Personnel Department. Figure 6-5, which is the reverse of this form, gives details about the employee's termination so that information for government, insurance companies, credit bureaus, or other employers is easily obtained. One piece of paper front and back takes care of the entire history of an employee and can be kept on file for future reference.*

ished room so she will have some idea of what the job requires. Negative aspects of the position must be explained. It is far better for the interviewer to soft-sell the position rather than to oversell it.

9. **Guard against using technical terms.** There are many common industry expressions (e.g., high-occupancy, back-to-back, etc.) that an applicant will not understand.

10. **Evaluate and rate the applicant.** This is the difficult balancing process that goes on in an interviewer's mind when he tries to weigh and judge the information and the impressions he has accumulated. It is difficult because he is evaluating the attitudes, aptitudes, and abilities that the applicant has developed over an entire lifetime.

    The interviewer's impression is more likely to be based on impressions and intuition than on strictly factual data. There is no way to eliminate this subjectivity, but the interviewer must be reminded constantly to give somewhat more consideration to the provable facts than to ideas that often border on mere suspicion.

## *Eight Pitfalls*

1. Try to avoid feeling that the employee will be just right based on a few outstanding characteristics rather than on the sum of his characteristics.
2. Don't be influenced by neatness, grooming, expensive clothes, and an extroverted personality—none of which have much to do with housekeeping competency.
3. Over-generalizing can be bad news. Interviewers assume too much from a single remark (e.g., an applicant's assurance that he "*really*" wants to work).
4. Avoid hiring the "boomer"—the one who always wants to work in a new property and keeps changing jobs whenever another one opens.
5. Projecting one's own background and social status into the job's requirements is unfair. Which school the applicant attended, or whether he has the proper look is beside the point—job performance is what matters.
6. Confusing strength with weakness, and weakness with strength should be avoided. What is construed by one to be overaggressiveness might be interpreted by another as confidence, ambition, and the quality of leadership—the last two traits being in chronic short supply in Housekeeping and most other departments. These are also the very qualities that make it possible for management to promote from within.
7. Don't be impressed by a smooth talker or, the reverse, assume that silence reflects strength and wisdom. The interviewer should concentrate on what the applicant is saying rather than how he is saying it, and decide whether his personality will fit into the organization.
8. Try not to be tempted by overqualified applicants. Persons with experience and education that far exceed the job's requirements may be unable for some hidden reason to get jobs commensurate with their backgrounds. In any case, if they are not concealing skeletons in the closet, they still tend to become frustrated and dissatisfied with jobs far below the level of their abilities.

## *FEPC Is Listening*

To prevent being accused of unfair employment practices by state and/or federal agencies, *employers must not discriminate.* The ban on discrimination applies to all elements of the employer-employee relationship—hiring, firing, wages, terms, conditions, or privileges of employment. An employer cannot, solely because of race, color, religion, sex, age, or national origin (1) fail or refuse to hire an individual, (2) discharge or otherwise discriminate against him, or (3) limit, segregate, or classify employees in any way that would deprive them of employment opportunities or adversely affect their status as employees.

In an interview one may ask, "Have you ever worked for this company under a different name?" or "Have you ever been convicted of a crime under another name?" but *not* "What was your maiden name?" If the applicant has been convicted of a crime he may be asked, "What crime, committed when and where, and what was the disposition of the case?" One may *not* ask if the applicant has been arrested, only if there has been a conviction.

One may not ask about an applicant's religious affiliation, church, or any holidays observed; one may *not* mention (unless asked) the religious beliefs of anyone in the organization for which the applicant is being interviewed.

There can be one area on the application that shows which days the applicant can work. If he states he cannot work on Saturdays or Sundays, the employer is permitted to ask why. If he says it is because he goes to church, it can be explained that this is something many employees would like to do, but unfortunately most institutions operate on a 7-day week, 24-hour basis, so the employee may have to try to attend off-hour services on some Saturdays or Sundays.

If it is possible to arrange for the employee to be off on specific days in order to attend his church, fine, if it doesn't interfere with normal operations. New employees almost never can be given weekends off in hotels, although in such institutions as residence halls and retirement homes it is usually possible.

Employers may not ask the prospective employee to submit a photograph until after he has been hired. Then if the company uses identification cards the employee's picture may legally be required.

As far as citizenship is concerned, interviewers may ask, "Are you a citizen of the United States, and if not, have you the legal right to remain permanently in the United States, and do you intend to do so?"

A statement should be made by an employer that, if hired, the applicant may be required to submit proof of citizenship. It is *illegal,* however, to ask, "Of what country are you a citizen?" "Are your parents or spouse naturalized or native-born U.S. citizens?" "What date did you or your parents or spouse acquire U.S. citizenship?"

One cannot require the applicant to produce his naturalization papers or first papers. After the person is hired, however, he may then be required to show his visa and number so that it may be attached to the employment application.

If the employee does not fill in his application correctly, that can become the reason for termination at a later date. In fact, one should notify the applicant that any misstatement or omission of material facts in his application may be cause for dismissal.

Any of these regulations is subject to change and probably will be changed in the future due to pressures being put on state and national legislators by certain ethnic groups seeking to open more doors to advancement.

Interviewers should not talk too much. If the interviewer is not interested in an applicant, he should thank the applicant for coming in. The applicant should be disqualified only for lack of merit, necessary skills, or aptitude. He should be told that others are being considered, and that the choice will be made later.

## *Employee Orientation and Rejection*

All job applicants interviewed, whether hired or not, quickly form their own impressions of the institution, and must therefore be handled with tact, politeness, and respect. Word of mouth is a powerful weapon and rejected applicants should feel that they have been treated in a fair, impartial manner so that they will remember the institution in a positive light.

Once the decision to hire is made, the application should be processed and the applicant notified when to report for work. Processing usually also includes filling in a bonding application. (Housekeeping employees are in such close contact with guests' belongings that insurance companies usually insist upon maids and housemen being bonded.)

On the first day of employment two things must be arranged for: orientation and training. Before the employee is exposed to any type of training, he should understand the building in which he will be working. A tour should include enough of the building so the employee will know how to get from his own work area to the locker room, the employee cafeteria, and other departments in which he may be involved.

Employees should, during their orientation, have rules and regulations of the institution explained to them, and also the benefits accruing to them as a result of long-term employment.

## TERMINATION OF EMPLOYEE

__________________________ ________________ ______________________

Name

### I TYPE OF WORKER

Regular worker, last hired ______________________
First day worked

and last worked ____________________________
Last day worked

Number of days in normal work week ____________

Odd job or casual worker, employed for period shown under REMARKS

### II PAYMENTS ON SEPARATION

Received ______________ days paid vacation

### III CAUSE FOR SEPARATION

*The above individual was separated from our employment for the reason set forth below.*

DISCHARGED BUT NOT FOR MISCONDUCT

1 ☐ Lack of work  2 ☐ Retired  3 ☐ Other (see remarks)

DISCHARGED FOR UNSATISFACTORY SERVICES

4 ☐ Violation of company rule pertaining to (a) ☐ Absence (b) ☐ Safety regulation (c) ☐ Neglect of duty
(d) ☐ Intemperance (e) ☐ Company property (f) ☐ Dishonesty
5 ☐ Insubordination (see remarks) (6) ☐ Other (see remarks)

QUIT AS PER STATEMENT OF EMPLOYEE

7 ☐ No reason given
8 ☐ To accept other employment
9 ☐ To seek other employment
10 ☐ Dissatisfied:
(a) with working hours
(b) with working conditions
(c) with pay

11 ☐ Distance of work from residence
12 ☐ Working conditions hazardous
13 ☐ Work detrimental to health
14 ☐ Leaving locality
15 ☐ Sickness
16 ☐ To retire
17 ☐ Refused to transfer to other work

18 ☐ To enter self-employment
19 ☐ Personal business or affairs
20 ☐ To marry
21 ☐ Pregnancy
22 ☐ Family duties
23 ☐ To attend school
24 ☐ Other (see remarks)

Employee's Signature ____________________________

### IV RATING BY DEPARTMENT HEAD

| | Excellent | Very Good | Good | Fair | Poor | |
|---|---|---|---|---|---|---|
| Ability | | | | | | |
| Efficiency | | | | | | |
| Conduct | | | | | | |
| Attendance | | | | | | |

Would you re-employ? ________________  ☐ Replacement desired  ☐ Position abolished

Remarks:

DEPARTMENT ____________________________________ Date _______________
SIGNATURE OF
DEPARTMENT HEAD ______________________________________________
FURTHER INFORMATION MAY BE OBTAINED FROM __________ PHONE EXT. _____

**FIGURE 6-4**

Initially, the new employee may be placed in a work area with an experienced employee who will see that he gets wherever he needs to go. In many places, the system is called big sister or big brother, or the one-for-one (one new employee with one old employee). In a medium-sized or small institution, a buddy system may be almost impossible, and the department head will probably be the only person available to take the employee around.

### *Performance Records*

Keeping meticulous employee performance records will always prove worthwhile when it becomes necessary to justify one's position on terminations, promotions, or reassignments (Figure 6-4).

In a small property, it may work a hardship on the housekeeper who keeps his or her own payroll and records to have to fill in slips on misconduct, attendance, and all the other information that pertains to the employee. It pays off, however, because the institution's unemployment payments to the state are determined by the number of cases that may be charged against the institution. To avoid payment for people who have been fired for misconduct, poor work performance, or other legitimate reasons, Housekeeping's performance records are the most important substantiating evidence that can be submitted to an unemployment board.

Misunderstandings about unemployment insurance should be cleared up early. The employee should understand that he will not be able to collect unemployment insurance if he was terminated for any activities considered misconduct. Only a reduction in staff or dismissal without legitimate cause will enable the employee to collect.

### *Additional Advice*

Discussing the pressurized position of the Housekeeping employee, Valeri Richardson, the Broadmoor's executive housekeeper says, "Not only is the maid or housekeeping supervisor fair game for guests, bellmen, management, and other maintenance personnel; but he or she is often deprived of dignity and self-esteem that might otherwise compensate for a difficult and poorly paid profession." As a means of relieving the stress and increasing stature of roomkeepers and housekeeping attendants, Ms. Richardson says, "Licensing did remarkable things for the nursing profession. Not only did it ensure professionalism, but it gave dignity and helped motivate people who had been looked down on and poorly paid for a highly skilled job. With the labor market what it is today, maybe the only way out of our current difficulties (shortages of personnel, turnover, and lack of motivation) is to develop some sort of certification for our industry."

Richardson comes close to this in her own operations with a regimented training system and formalized feedback sessions and evaluations. A certificate, she maintains, can be shown to others as proof of training and time invested in this profession. "The restlessness, the 'I don't care' attitude and turnover in employees is due to a lack of dignity in this business which is totally unfair . . . this is a hard job; not everyone can measure up; it takes knowledge, physical strength, and coolness with all kinds of people even at the entry level."

According to Richardson, the best way to make someone more serious and satisfied with what they do is to make it more competitive. Inventing titles and trying to give more status by simply changing the name of the job from maid to room attendant hasn't worked for her. "We tried calling the maids 'room attendants' for 30 days, and either they would become confused about their positions or would say point blank that they were maids and wanted to be known as such."

Richardson makes sure to tell all prospective employees that being a maid is no easy job. "If you don't intend to work hard, learn and act with a cool head, or if you're looking for a free ride, then just walk on down the road," is the advice she gives most prospects.

# Chapter 7

# Who Does What in Housekeeping?

**OVERVIEW** Organization charts are helpful, but one involving Housekeeping takes many different forms depending on the range of the department's responsibilities. Staffing and scheduling are the most important functions of the professional Housekeeper as they involve the largest expenditure of people and thus of money. Here you'll see how different hotels handle these critical elements, including team housekeeping, a significant new trend.

Although an organization chart can be regarded as a demoralizing document in that it graphically shows the hierarchy of an institution and emphasizes the idea that there are those few chiefs on the top and many lowly Indians beneath, it is nonetheless a very helpful tool. It shows lines of authority and how responsibility is distributed, and it helps applicants and employees interested in moving up (or crosswise) to see who their supervisors are, just where they fit in, and where they might go in case of promotion. Charts should be kept in a desk if one wants to avoid reminding those with entry-level jobs of their relative positions, but there should be one handy for easy reference. At least one well-known housekeeper has been known to invert the chart, putting the guest on top so that maids see themselves in closest relationship to the guest.

There is no single ideal way to organize an entire institution or a Housekeeping Department. How the chain of command works depends on the nature of the institution, whether there are Maintenance or Engineering Departments, and whether either of those is under the jurisdiction of the professional Housekeeper.

Many kinds of organization charts are shown in this chapter. It would be impossible to try to fit every operation into one of these. The professional Housekeeper will work with management to ensure that no area is forgotten, and provision is made for adding personnel as the plant expands and duties increase. This requires understanding of company organization.

## *Hierarchy of a Company*

In a corporation, the highest man on the totem pole may be the chairman of the board or the president. In some cases the chief executive officer may actually carry both titles. Immediately below him there may be an executive vice-president, followed by a series of other vice-presidents in charge of everything from sales to guest services. In any one unit of a chain, a vice-president and/or managing director will be in charge, followed by a general manager (in a large hotel) or a manager (Figure 7-1). Often under the general manager is a resident manager, executive assistant manager, or a senior assistant manager. Those three titles are often interchangeable, depending on the size of the property. Assistant managers on each shift are sometimes also called reception hosts or duty managers. They usually greet and show VIP guests to their rooms.

In some large properties there will be a director of rooms. His division will have charge of the reception area, cashier area, bellmen, parking,

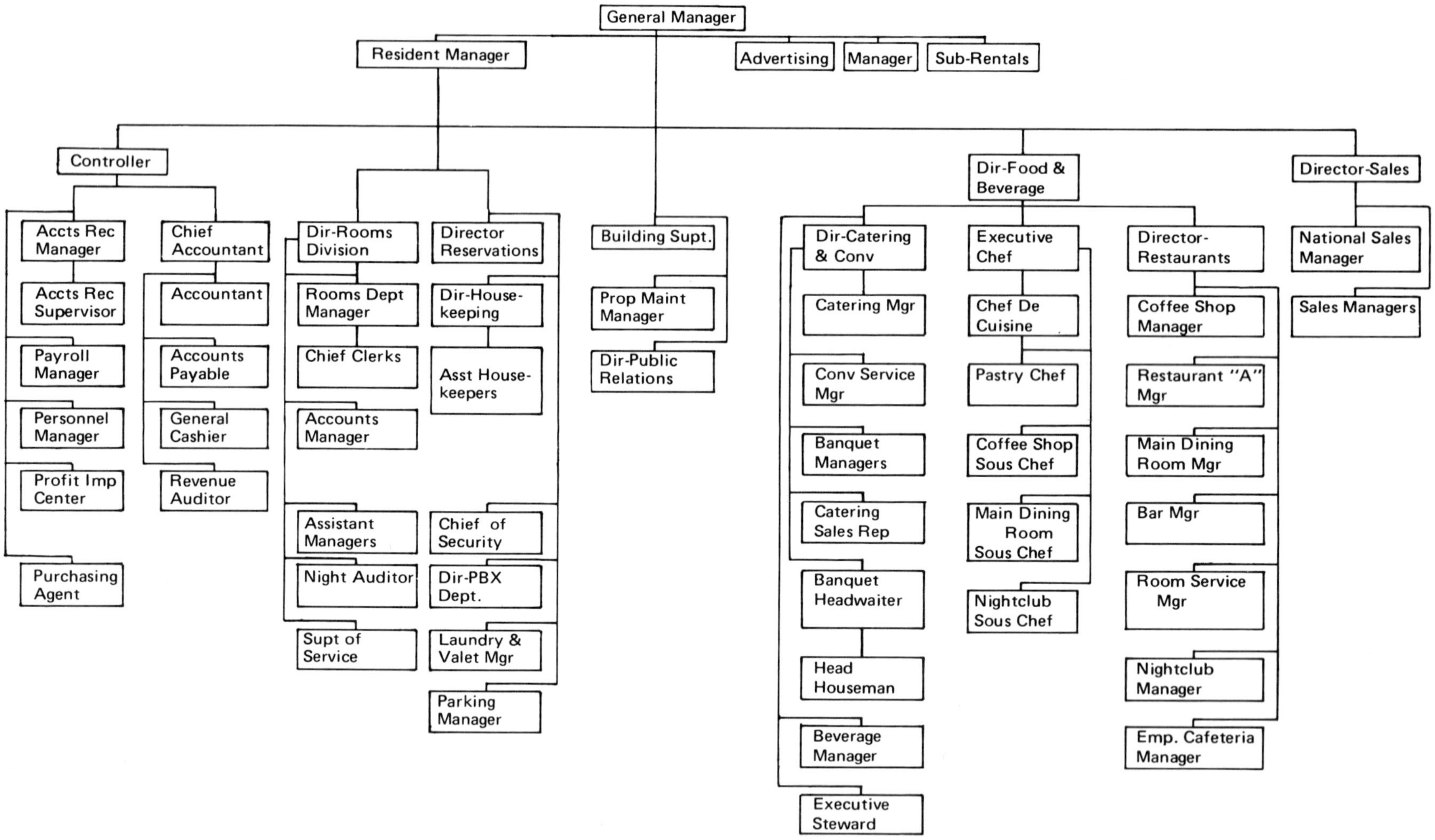


**FIGURE 7-1** *Organization Chart for Large Hotel (500–1500 Rooms)*

doormen—all of the sections concerned with a guest, from greeting him to adding up his bill as he leaves. Under the director of rooms, there may be a junior assistant manager, rooms manager, chief clerk, front office clerk, reception host or hostess, mail and information clerk, cashiers, credit manager, front office secretary, key clerks, and others.

The Housekeeping Department also has a long series of titles (Figures 7-2, 7-3, and 7-4). The person in charge may be director of housekeeping, assistant manager in charge of guest services, manager of environmental services, manager of environmental facilities, executive housekeeper, and so on. But to the guest it is *housekeeper.*

Next in command are usually the executive assistant housekeeper and a series of assistants, some of whom specialize in handling guest floors, some of whom may be responsible for training or scheduling, but all of whom are also relief housekeepers. There are also inspectresses and floor supervisors (or floor housekeepers) and reporting to them are all maids—training, bath, and apprentice maids. Powder room maids who work in the public areas usually work under an assistant housekeeper. The person filling the position of maid can also be called room attendant, room-keeper, or even housekeeper.

Often there is a clerk-typist who takes care of Lost and Found and the Housekeeping Department payroll, and also functions as a secretary. The manager of the Sewing Room may have under her the drapery makers and menders. In most facilities she will also supervise the seamstresses.

A head houseman (or house attendant) who reports directly to the executive or assistant housekeeper will be in charge of housemen. He may have charge of carpet repair, wall washers, window washers, drapery men, carpet shampooing, heavy vacuuming for general cleaning, chandeliers, electric light changers (re-lamping personnel), bathmen if they are used, furniture men, floor men, lobby housemen, linen delivery men, linen sorters, linen truckers, linen runners, and warehousemen. This is a responsible position.

A handyman who does such chores as replacing damaged furnishings in guest rooms or public areas (whatever small jobs the union will permit) is one of the highest paid of the union employees and is usually resourceful, hardworking, and able to perform a wide variety of tasks. He most often reports directly to the housekeeper or assistant housekeeper.

Banquet stewards, kitchen stewards, and various members of the Engineering and Building Maintenance Departments all do jobs similar to those performed by Housekeeping's housemen and handymen. The area of authority for all of these should be firmly established so union lines or department duties do not get tangled to cause problems.

Once all of Housekeeping's cleaning and maintenance functions (both hourly and supervisory) are titled and described and the employees are hired, the next problem is to determine how many people are needed, where, and when—in other words, staffing and scheduling. Before those activities can be carried out, however, the department head and those who assist must have some idea of how many rooms will be occupied to determine what proportion of the total work force will be needed on duty.

In a small motel, housekeeping duties are usually performed by one of the operating team, and maintenance and repair functions are the responsibility of the other. Motels of from 12 to 40 units require additional help on at least a part-time basis—for example, maids, cleaning women, a houseman, or handyman. When units number over 40, a full-time housekeeper and staff are needed. Whether the housekeeper also supervises the maintenance crew, yardman, gardener, and laundry workers depends on the staff organization plan. The housekeeper's responsibilities are further influenced by the size and complexity of the overall facilities. If maintenance and utilities are to be the responsibility of a full-time engineer or superintendent, he and his staff will cooperate with the Housekeeping Department.

## *Forecasting: Educated Guessing*

In a college or university residence hall (Figure 7-5) the population is stable, so the entire matter of who does what and when can easily be

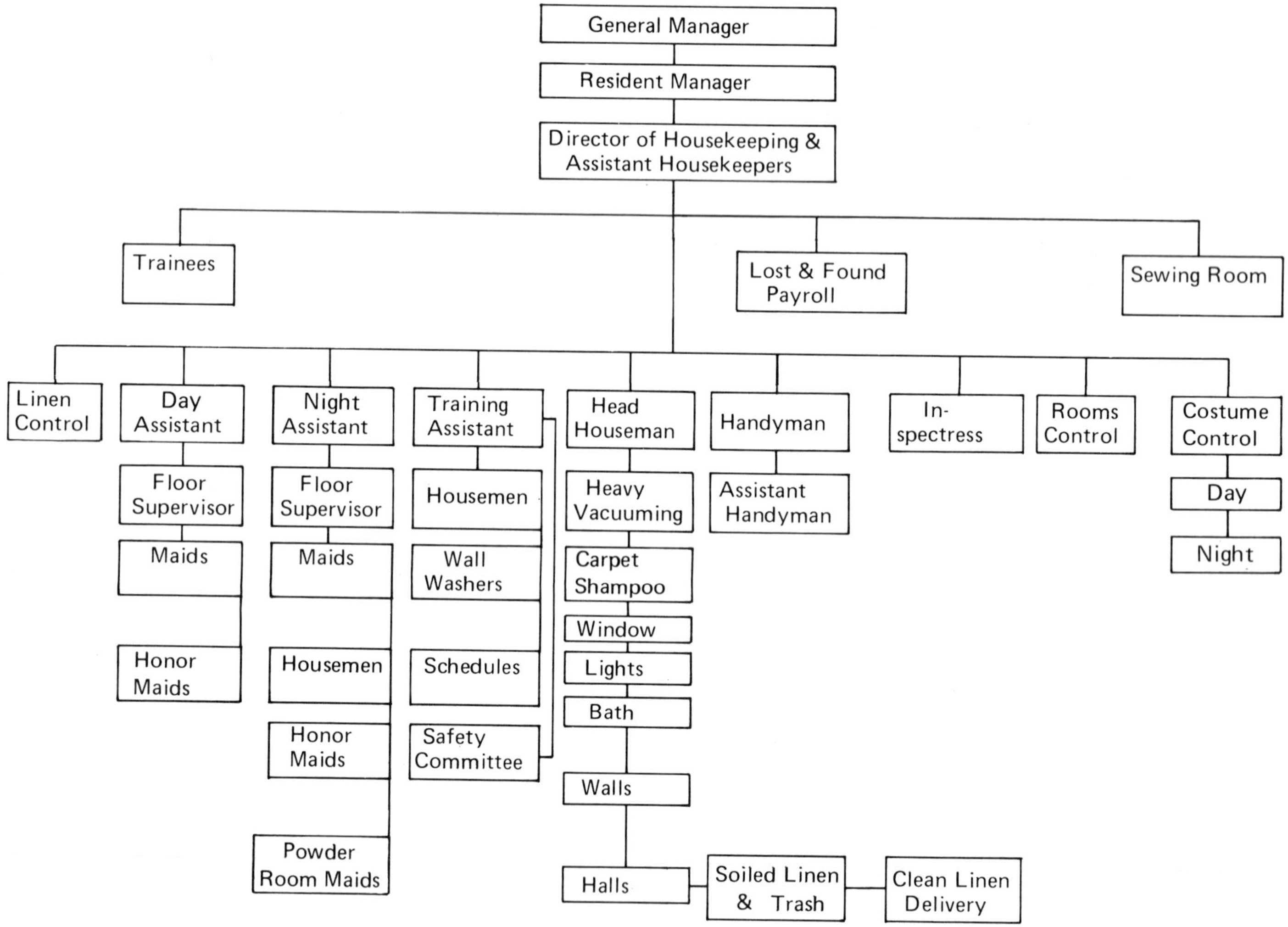


**FIGURE 7-2** *Organization Chart of the Housekeeping Department in a Large Property*

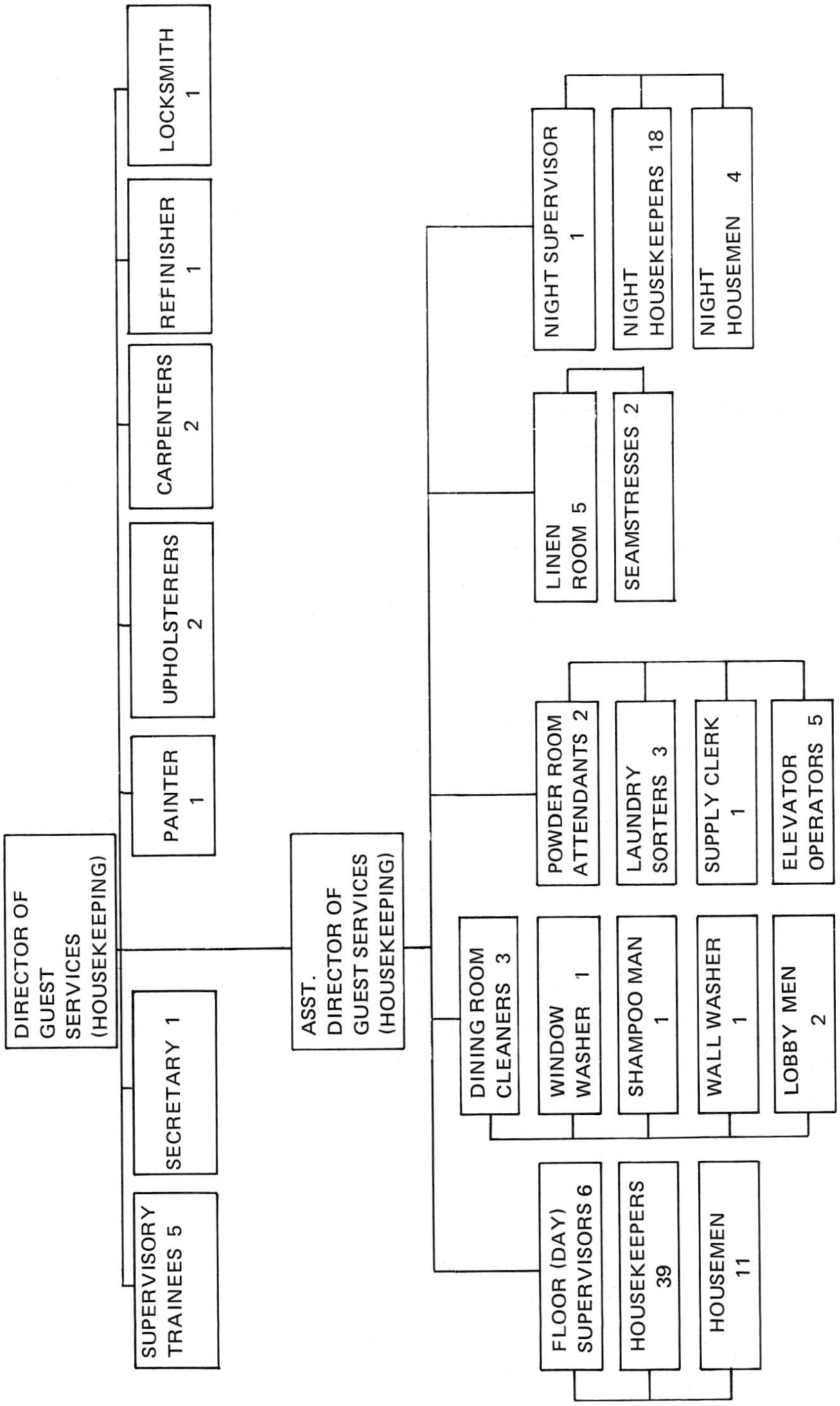


**FIGURE 7-3** *Housekeeping Department—600 Room Hotel*

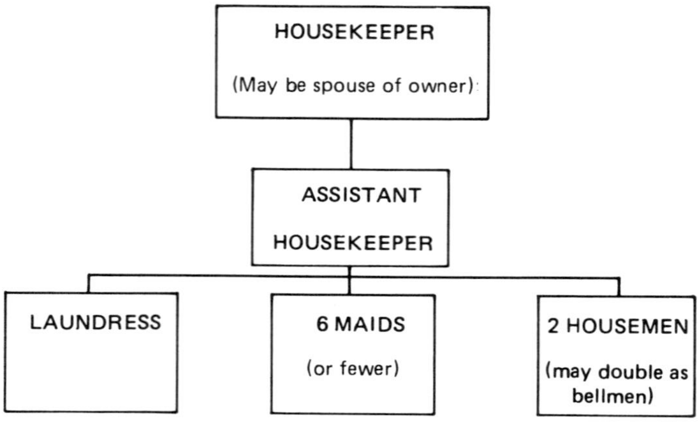


**FIGURE 7-4** *Typical Organization Chart for 85 Rooms or Less*

set down and followed on a routine basis. In a hotel, motel, or club where occupancy fluctuates, the staff must be flexible and the schedule easily changeable.

Before hotel management will know how many maids are needed next week, or even tomorrow morning, it is necessary to do some crystal ball-gazing to determine just how many guests will be there. This magic soothsaying called forecasting is usually performed by the assistant manager who, with the help of the Reservation Department, predicts how many rooms will be occupied that night, that month, that week, and in the case of making out a budget, next year.

The approximate occupancy rate is figured on: (1) number of reservations, (2) last year's occupancy, (3) conventions or special events in town and extent to which other hotels may send their overflow, and weather conditions—will travelers be stranded downtown in a snowstorm; will fog prevent planes taking off or landing?

In many hotels the forecast is sent to the Housekeeping Department each day between 12 noon and 2 pm so that there is time to figure out exactly how many maids and housemen will be needed the next day before they leave at 4 pm. Most Housekeeping Departments will have received sufficient advance notice to make it possible to give employees 7 to 14 days notice of anticipated schedules.

Many hotels schedule employees 15 days in advance for the morning and afternoon shifts and the night crew. All Housekeeping employees must be instructed to check posted schedules every day, both in the morning and at night, since the house count may change, or the sudden booking of a large function (banquet, party, or meeting) may require an increase in staffing.

In the Sheraton Hotel Corporation's Quality Audit—a lengthy list of items to be checked in the Housekeeping Department to be sure it is functioning at 100 percent efficiency—there is a section headed Staff Control. The following quote from it will give the reader a fair idea of how involved the professional Housekeeper must become in the hotel's activity to be able to staff and schedule properly:

1. Does the housekeeper receive a written weekly rooms forecast?
   Is the forecast received by the housekeeper on Thursday? On Friday?
2. Does the housekeeper receive a revised three-day rooms forecast? In writing?
3. Does the housekeeper receive a forecast of group and convention bookings?
4. Does the housekeeper receive a weekly function sheet?
5. Does the housekeeper receive the revenue or split-key machine slips each morning from the Front Office?
6. Does the housekeeper receive a list of occupied room numbers each morning from the Front Office?
7. Does the housekeeper receive a list of expected check-out rooms each morning from the Front Office?

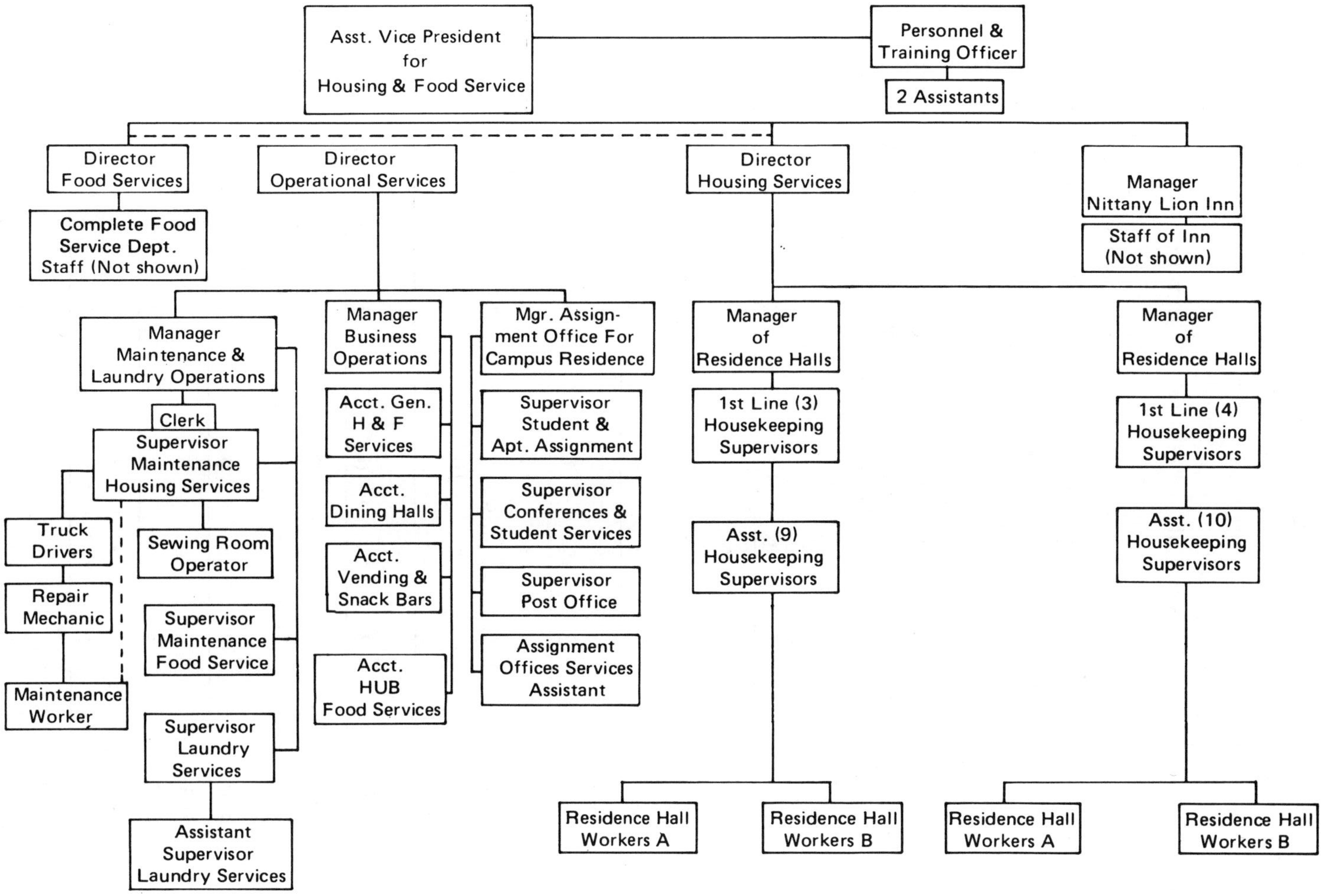


**FIGURE 7-5** *Organization Chart for Pennsylvania State University's Housing & Foodservice Department*

## *Astute Staffing Saves Labor Costs*

The whole business of staffing is based on the number of man-hours needed to accomplish the work. Those who make out charts and work schedules for Housekeeping Departments need only consult the man-hour tables to figure the number of employees needed when they know the occupancy expected and all scheduled events. The Staffing Table for Maids (Table 7-1) in a 750-room hotel shows not only how many are needed but also what the costs will be for that labor.

Each maid is expected to do a certain number of rooms each day—usually 14 to 16. That number is based on the following arithmetic: each room takes 30 minutes to clean; $14 \times 30 = 420$ minutes or 7 hours. Each maid works an 8-hour

**TABLE 7-1** Staffing Table for Maids—Housekeeping Department

MAID VARIABLE COST AT VARIOUS LEVELS OF OCCUPANCY
(Wage scales are approximate—vary in different communities)

| Per-cent of Occ. | No. of Occu-pied Rooms | Supervisors (25.50) AM | Supervisors (25.50) PM | Maids (22.89) AM | Maids (22.89) PM | ** | 5 PM Maids (22.89) | Powder Rooms* AM | Powder Rooms* PM | TOTAL Empl. | (34.33) OT AM | (34.33) OT PM | CASH TOTALS ($) |
|---|---|---|---|---|---|---|---|---|---|---|---|---|---|
| 100.0 | 750 | 5 | 2 | 52 | | | | | | | | | 2067.96 |
| 96.3 | 720 | 5 | 2 | 50 | 12 | 1 | 1 | 3 | 3 | 82 | 6 | 1 | 2022.18 |
| 94.4 | 705 | | | 48 | | | | | | 79 | 4 | | 1906.69 |
| 92.6 | 690 | 5 | 2 | 47 | 12 | 1 | 1 | 3 | 3 | 78 | 2 | 1 | 1840.63 |
| 90.8 | 675 | | | 45 | | | | | | 76 | | | 1666.35 |
| 88.9 | 660 | 5 | 2 | 44 | 12 | 1 | 1 | 3 | 3 | 75 | | | 1644.96 |
| 87.1 | 645 | | | 43 | | | | | | 71 | | | 1597.68 |
| 85.2 | 630 | 4 | 2 | 42 | 11 | 1 | 1 | 3 | 3 | 70 | | | 1549.00 |
| 83.4 | 615 | | | 41 | 11 | | | | | 69 | | | 1526.40 |
| 81.5 | 600 | 4 | 2 | 40 | 11 | 1 | 1 | 3 | 3 | 68 | | | 1507.17 |
| 79.7 | 585 | | | 39 | 10 | | | | | 66 | | | 1457.73 |
| 77.8 | 570 | 4 | 2 | 38 | | 1 | 1 | 3 | 3 | 65 | | | 1434.84 |
| 76.0 | 555 | | | 37 | | | | | | 64 | | | 1411.95 |
| 74.1 | 540 | 4 | 2 | 36 | 10 | 1 | 1 | 3 | 3 | 63 | | | 1389.06 |
| 72.3 | 525 | | | 35 | 10 | | | | | 62 | | | 1366.17 |
| 70.4 | 510 | 4 | 2 | 34 | | 1 | 1 | 3 | 3 | 61 | | | 1343.28 |
| 68.5 | 495 | | | 33 | | | | | | 59 | | | 1321.89 |
| 66.7 | 480 | 4 | 2 | 32 | 9 | 1 | 1 | 3 | 3 | 58 | | | 1253.47 |
| 64.9 | 465 | | | 31 | 8 | | | | | 56 | | | 1228.83 |
| 63.0 | 450 | 4 | 2 | 30 | 8 | 1 | 1 | 3 | 3 | 55 | | | 1205.94 |
| 61.1 | 435 | | | 29 | 8 | | | | | 52 | | | 1183.05 |
| 59.3 | 420 | 3 | 1 | 28 | 8 | 1 | 1 | 3 | 3 | 51 | | | 1109.16 |
| 57.4 | 405 | | | 27 | | | | | | 50 | | | 1092.27 |
| 55.6 | 395 | 3 | 1 | 26 | 8 | 1 | 1 | 3 | 3 | 49 | | | 1063.38 |
| 53.7 | 380 | 3 | 1 | 25 | 7 | 1 | 1 | 3 | 3 | 47 | | | 1017.60 |

* May be reduced whenever Banquet & Meeting areas are not busy
** Mezzanine Maid

day with 30 minutes for lunch, and a 15-minute break afternoon and morning.

Different properties have different expectations in this regard, however. A small, very elegant hotel such as the Bel-Air in Los Angeles, where each guest room is a suite and has complicated individual furnishings, expects its maids to do only 9 rooms a day, whereas some large commercial hotels with small rooms and different standards may require as many as 21 rooms a day—meaning that the maid spends only about 20 minutes per room. These two examples are extremes, and most hotels expect maids to do 14, 15, or 16 rooms.

So that the exact assignments for maids can be made each night, the night auditor at the Front Desk makes out a list of vacant rooms, of early check-outs, and of expected new arrivals with reservations, including MIPs and VIPs, and sends it to Housekeeping by 7 am.

Maids usually arrive by 8 am, get their keys, check the schedule board, and go to their usual floors. If they are expected to work in other areas after they have cleaned all the rooms in their regular section, their floor supervisors so inform them.

If occupancy is expected to be low (50 percent or under) it is advisable to understaff and leave rooms unmade until the next day, since there will be so few new guests requiring clean rooms. Overstaffing can be quite costly, as some unions insist that maids must be paid for a full day, regardless of the fact that there may be far fewer rooms to clean than the usual quota. This rule eventually may be dispensed with in some areas as jobs become more scarce.

Many directors of Housekeeping find their employees do want to work more or less than the standard 40-hour week, and accommodate them if union regulations do not forbid it.

Again, Sheraton Hotel's Quality Audit questionnaire asks some very pertinent questions relative to staffing:

1. Does housekeeper have a copy of the latest hotel Staffing Guide on file in the office?
2. Does the number of people currently on the Housekeeping payroll agree with the Staffing Guide?
3. Is the Housekeeping payroll in excess of the Staffing Guide's allowances?
4. Is the Housekeeping payroll below the Staffing Guide's minimum allowable staff?

Understaffing can save the institution considerable payroll costs IF the maids on duty are anxious to do extra rooms for extra money—and in most places they are very pleased to get the additional dollars. This extra payment, even if given out in the form of cash, must be recorded in the payroll records and withholding tax figured on it. Some institutions pay time-and-a-half for doing rooms above the quota. This is not only expensive but may encourage maids to rush through their quota and do a sloppy job in order to have enough of the normal work day left to earn time-and-a-half pay for equally inferior work in over-quota rooms.

Understaffing is also recommended on days before holidays for which maids receive double pay—otherwise they may have very little work to do in return for the extravagant holiday wage. With occupancy at its lowest on holidays (except at resorts) there will obviously be few check-outs the following day, so understaffing then is also recommended.

A "general clean" is a thorough cleaning of guest room and bath to be done periodically, anywhere from once a week to once a month. In most hotels, each maid is expected to do one of these a day. During slow days, however, it is possible that a maid may do this major cleaning in two or three rooms.

At the Fairmont Hotel in San Francisco, each room receives a major cleaning once a month by one of five teams, each consisting of a man and a woman. He vacuums, turns the beds, and positions furniture—all the heavy work, while the woman does other cleaning jobs.

## *Special Assignments*

In addition to the normal cleaning duties, assistant housekeepers and the inspectors may assign several cleaning projects in areas where increased attention is necessary. Special assignments have been designated by a number of

Hyatt hotels as "the perfect room" and the "daily special." The perfect room approach assigns special attention to bright work, cleaning under the tracks of sliding doors, bleaching bathroom grout, cleaning bathroom ceilings, and the other details of heavy cleaning that are often missed. Room cleaners complete one perfect room each day, and thus over a short period of time each room will have been given special attention.

The daily special concept also follows the objective of additional attention to room cleaning, but assigns one of these types of duties for each room on a given day. Inspection plays an important part in assuring that special attention is, in fact, being given.

It is always advisable to have phone numbers of employees, and to warn them in advance of possible emergency calls. For example, one major hotel in a large city had anticipated a large Friday night check-out and very small check-in, so the housekeeping staff for the next morning had been cut to a minimum. Unexpectedly the weather caused two 747 jets to be grounded and 450 people were given complimentary rooms by the airlines, all in that one hotel.

At 4 am the duty assistant manager called the executive Housekeeper to tell her of the situation—actually a friendly and professional gesture despite the fact that it seemed an outrageous hour! This groggy but resourceful lady figured out that her maids would never be available to handle all that, even if she roused them at dawn, so she called every major hotel in town to ask if any maids would work that Saturday morning—for cash, time-and-a-half—and 20 were found to assist her own staff in quickly cleaning the 450 unexpectedly occupied rooms.

## *Resort Scheduling*

One problem unique to resorts is that of lay-off and recall scheduling. During the off-season there is sufficient work for only a skeleton crew. This makes scheduling very difficult; it could be done on the basis of seniority but in order to keep one's staff intact it is far better to share the work. In one typical off-season system, employees work two weeks, and take two weeks off. This and similar solutions are understood and accepted by all employees—including those of long standing—who are eligible, of course, to collect unemployment insurance during lay-off periods.

In popular tour-group and convention areas (e.g., Hawaii), it is sometimes necessary to assign guests to unmade rooms, because when groups are back-to-back (arriving and leaving almost simultaneously in large numbers) it is not always possible for the maids to have all of the rooms ready. When large groups attending business meetings check out, they often leave on the same plane on which their replacements come in. If a plane arrives at 12:30 and leaves at 1:30 pm, there is not much time for maids to make up all of the check-out rooms. Occasionally, luggage will be sent to rooms but arriving guests are encouraged to enjoy the resort's facilities until rooms are ready.

Harrah's Hotel on Lake Tahoe is a sumptuous establishment with 250 luxurious rooms. Gambling is the *raison d'etre* for the hotel so it is run quite differently from conventional commercial properties. More money is spent on services for guests to make them want to stay. Maids have only nine rooms a day to make up because the rooms are large and lavish. For example, there are pushbutton bars (so-called beverage units) in the rooms that are tied to the computer so that charges are recorded in the Front Office. Each room has two bathrooms (one with a shower and one with a tub), each with a tiny black-and-white television set. The executive housekeeper reports that despite the higher wages paid the employees in the gambling hotels and the beautiful setting, there is still a problem finding maids and housemen who will be steady workers. Staffing and scheduling for room cleaning in these hotels is no different in commercial properties except that the maids work from 9 am to 5 pm because people sleep later. A few are scheduled to come in at 10 am. In some of the very large Las Vegas hotels there are even later shifts because there are larger numbers of guests who sleep late.

An entirely different kind of resort is Disney World in Orlando, Florida, where the Contemporary Hotel and the Polynesian are run primarily for guests and their families who are going to want to arise as early as possible to spend their days at the park.

Maids at the Contemporary Hotel have a heavy chore in cleaning the guest rooms because the check-out time is 12 noon and check-in time is 3 pm. There has been 100 percent occupancy since the hotel opened in 1971. Each room averages 3.8 people per room generally, and 4 persons during holiday seasons, so there are more beds to make, more general disorder to repair, and more linen to handle than in a business travelers' hotel or in a less family-oriented resort property. Maids handle over 25,000 pieces of soiled linen daily, and of course that much clean linen daily to take care of more than 1000 always-occupied rooms.

### *Labor Market Affects Schedules*

At the Houston Oaks hotel maids do 14 rooms, one of which must have general cleaning. They do "check-outs" until 11:00 am, then those with "early make-up" signs on doorknobs, unless a guest has specifically requested early service.

At Houston Oaks "general cleaning" means the maid cleans everything not attended to on regular cleaning. She even moves the furniture, (against union rules in many other areas), turns the mattresses, spot-washes the walls, vacuums high places, and so on. Housemen are difficult, often impossible, to find in the city of Houston; maids, too, are scarce as the unemployment level is less than 5 percent.

### *Staffing Problems Reversed*

One of the larger hotels in Los Angeles experimented with an unorthodox system for organizing its Housekeeping Department by setting up mini-hotels. Each senior housekeeper (floor supervisor) had four floors under her jurisdiction, just as if she had a complete small hotel. She worked 10 hours a day, 4 days a week (her pay, with overtime, was equal to that for regular-time, 5 days a week), being on duty from 7 am until 5 pm. During those hours she was expected to inspect 120 rooms (a mammoth order) and supervise the maids. After they had done their 14 rooms, the maids were permitted to leave, even if they had finished their work in under eight hours, although they were required to work a minimum of six hours. Housekeepers checked them out from 3 pm till 5 pm.

These senior housekeepers had to be particularly careful to inspect rooms to be sure that maids had not rushed through their chores and done an inadequate job in their haste to be finished in six hours. (In the beginning of an "honor section," a lick-and-a-promise routine may not be intentional but it can become a habit if supervisors do not check carefully.)

For that reason, this mini-hotel organization did not work out satisfactorily despite the fact that each housekeeper had a sense of importance, having complete control over his or her own section. The plan could work well with the right supervisors.

In areas where part-time employees are permitted there is a system that does work well. All persons scheduled for a day are assigned four or six hours work depending on occupancy. When the duties are completed they are dismissed and credited with the hours worked.

Within the budget framework there must be strict attention to labor costs. No over-staffing can be allowed in most types of operations if the narrow profit margin is to be maintained.

### *The Marriott Way*

The management of one Marriott Hotel devised an effective organization plan for handling housekeeping and service functions. At the new 1020-room Los Angeles facility the Housekeeping, Laundry, and Pool and Recreation are all under one director of services (Figure 7-6), each department with its own manager.

Housekeeping is comprised of five sections: (1) guest rooms, (2) linen room, (3) evening crew, (4) special housemen, and (5) midnight crew. Sections are set up similarly to those found in any traditionally-organized Housekeeping Department.

There are two supervisors under the housekeeping manager each of whom has seven floor

**FIGURE 7-6** *At the Los Angeles Marriott Hotel, director of services talks with two of his housekeeping supervisors. They are checking the permanent rotating schedule for Housekeeping employees who always know in advance when they will be called on to work, although low occupancy may cause some to be given additional days off. Scheduling is done with teams that rotate their days off. During the seven-week cycle each team has two long weekends off.*

supervisors. The floor supervisors have teams of up to six section housekeepers (not maids) depending on occupancy, and one houseman. This gives one a better span of control than the traditional organization: the ideal involves between three and seven people reporting to a supervisor. Having fewer is not efficient, and having more means one may lose control.

At the Brown Palace in Denver, David Owen, director of guest services, explains the management structure in Housekeeping. "We have an Executive Housekeeper, and she has three assistants and two supervisors. One assistant is in charge of the main building; one in charge of the Tower building; one is responsible for public areas and soiled linen control. One supervisor is in charge of the evening shift; the other controls all housemen."

## *Scheduling a la San Francisco Hilton*

Another means of scheduling is used by the San Francisco Hilton Hotel. At 8 AM a house count is done by the maids who check rooms to see whether doors are double-locked indicating that they are occupied, whether there is a Do Not Disturb sign hung on the doorknob, or if the room has been vacated. (All rooms with airline personnel are clearly designated and not to be cleaned until vacated or until the Please Clean sign is put on the doorknob. Maids are instructed never to knock on those doors, as crew members who fly at night may be sleeping.)

In addition, the Front Office notifies Housekeeping daily which rooms are check-outs, which are VIPs, and which are complimentary ("comps"). Meeting rooms slated to be used that day are also designated by the Front Office, and are scheduled to be cleaned and set up according to instructions from the Sales Department.

By 8:30 AM, when the floor housekeepers meet in the assistant housekeeper's office to get their assignments (Figure 7-7), it is possible to predict how many rooms need cleaning and which ones each maid will have to do. The number of maids scheduled for daily duty is sufficient for each to accomplish a reasonable amount of general cleaning (Figure 7-8).

Usually the hotel predicts its occupancies fairly accurately, but occasionally Housekeeping has more workers than work because of an un-

**FIGURE 7-7** *At the San Francisco Hilton floor housekeepers gather each morning at 8:30 while assistant housekeeper gives them 10-minute briefing on the day's activity—how many of the hotel's 1800 rooms must be made up immediately, what VIP guests are expected, what conventions are scheduled.*

**FIGURE 7-8** *Floor housekeepers meet with maids each morning in linen rooms on individual floors of San Francisco Hilton, give them keys to guest rooms, tell them which rooms to make up. Each housekeeper has two floors with a total of 136 rooms, except in tower section, where they have 10 to 15 floors each with 17 rooms.*

expectedly low house (Figure 7-9). At such times, the San Francisco Hilton does as many hotels do: Housekeeping decides on the basis of seniority who should stay to do the required work.

Often, however, the San Francisco Hilton housekeepers will ask if anyone would really like to take that day off. As attendance records are very important, it is then noted on the employee's records that "As per your request and a low house, you did not work on X date." Then if a maid complains of "always being laid off," there is some proof that the complaint is not true. It is also recorded in the Payroll Department, with an L, R, or A, whether an employee was laid off, requested to take day off, or absent.

## *Group Cleaning Concept*

Group work in the Housekeeping Department is gaining favor in many institutions. Areas in which group activity seems to fit a work pattern and personnel management include room cleaning, general cleaning, large area vacuuming, chandelier cleaning, silver cleaning, table setting, and so on. These are all areas in which

**FIGURE 7-9** *Scheduling housemen and maids in public areas, property maintenance manager takes into consideration what events are on the docket, how busy the hotel will be, and what periodic maintenance jobs can be done along with the routine assignments.*

three or more employees may work simultaneously.

Group activity is not a new idea in many professions. Teams of doctors often cooperate in surgery, dentistry, and law. Each team member is an expert in a special segment of a problem. This concept has gained many supporters in the hospitality lodging field, with advantages for both employer and employee to be considered.

Time, an always important element, can be cut in half or more by having many persons working on projects such as chandeliers or large area-vacuuming. This means less nonrentable down-time in public rooms. Usually, the presence of a fellow-worker will accelerate the work. Good supervision will lessen the danger of too much talking to slow the project.

Group cleaning fosters a feeling of fellowship and unity that is a very important factor in employee relations. It generates pride in performance and a feeling of partnership directly traceable to morale and profit. A shared feeling of a job well done can be an affirmative factor in overcoming resistance often encountered among employees. Personnel directors tell us this is one of the major deterrents to excessive turnover.

The methods of setting up a group cleaning program are determined by the area or project. In public area cleaning the team could be made up of experts:

1. High dusting—lighting fixtures, paneling, etc.
2. Vacuuming and setting furniture in place
3. Low dusting (tables, chairs, bric-a-brac, bar, etc.)
4. Plants (water, trim, replace, and wash)
5. Polish metal, scrub counters

If each employee is of equal efficiency and training the tasks can be rotated weekly or by area, bearing in mind that employees like variety and challenges in daily routine. In planning guest room group cleaning the number of persons should not exceed three or four. They could be assigned a schedule such as:

1. Remove soiled linen, make beds
2. Replace guest room and bath supplies
3. Dust furniture and dressers, hang up clothes, straighten furniture
4. Vacuum
5. Clean bath

The entire process should be accomplished in 10 minutes (40 minutes total for all four workers per occupied room) with the tasks rotated so that each employee is adept at each one.

If the room is completed in 10 minutes, guests waiting to check in will have less time to wait in the lobby. It would also be possible to use a team of three and complete the room in 12 minutes.

The team should be scheduled for additional work in a nearby area to avoid losing time going to other areas and using the elevator. Supervision is also important to lessen the time lost because of group discussion in a room.

This same concept can be used in what some institutions call general clean. This usually entails the additional work of:

1. Turning mattresses
2. Cleaning vents and high dusting
3. Changing curtains and washing windows
4. Spotting and repairing carpet
5. Additional bathroom cleaning such as tile grout, light fixtures, and polishing metal
6. Exchanging blankets, spreads, and pillows for freshly cleaned items

If these tasks are all scheduled for one room in each section per day, with half an hour allowed per room, each member of the team can probably handle one extra room daily. Specialists can move about an institution performing these extras on a strict schedule. On completion, every area should be thoroughly inspected to be sure that all items have indeed been cleaned.

What are the advantages of this system over the conventional one room cleaned by one attendant?

The team creates a strong feeling among employees of working together toward a common goal. Also, one attendant responsible for one room is often a frightening concept to new employees. Using a team means there is always a leader to consult, someone near to help if needed, an opportunity to relate to fellow-workers. In short, a team can significantly raise morale.

Speed in completing assignments has been mentioned as a benefit, and team competition can also be a factor that results in better cleaning. Better service to the guest—the Number 1 rule in Housekeeping—may be the ultimate result.

For institutions with short hour shifts, group cleaning could be an excellent solution. In hotels with heavy check-out schedules it may seem a slower process, as four employees working singly would do four rooms in approximately 20 minutes per person, while four working together will do two to three rooms in 20 minutes or nine rooms in one hour. Thus the final outcome should be discussed with management before a decision is reached.

For large suites team cleaning is a marvelous solution. The suite can be made ready in 20 minutes by four people, where one working alone would take an hour or more. Another consideration regarding suites is that the maid may be afraid of working in such a large area alone, whereas group cleaning would offer a comfortable situation for everybody.

Certain jobs can be performed by handicapped persons, so the team concept would make it possible to employ these often underutilized persons who want work and can be loyal, productive employees.

Disadvantages must also be considered. The first hurdle may be the union contract. If this document specifies the number of units allowed per employee there would need to be negotiations to do trial runs to set new quotas. Careful scheduling must be maintained so that there would be no increase in the number of employees. This is a delicate management decision to be studied from all angles.

## *Team Cleaning for Guest Rooms*

The whole hotel may not have to be cleaned on the team concept. For example:

1. Select persons who might be compatible, perhaps three maids and a houseman. One or two older experienced people should work with younger ones.
2. Assign area to be serviced, say, 45 to 60 rooms.
3. Have supplies and equipment ready or easily available.
4. Rotate tasks to be performed: strip and remake bed, clean bathroom, dust, hang up clothes, straighten furniture, empty wastebasket, vacuum (do check-out rooms first; early maid-service requests second; stay-overs third).
5. Take breaks and meals at same time.
6. Carts are loaded by the team to be ready for next day.
7. Make it known at the outset that there should be as little talking as possible, with voices well modulated so as not to disturb guests.

Quotas are (suggested) as follows:

60 rooms team of four to seven minutes per room; 45 rooms team of three to nine minutes per room

If teams finish early they can go home early with no penalty. This concept is particularly successful in convention hotels where many rooms must be readied quickly to accommodate large numbers of guests checking in at one time.

## *Absenteeism*

This is, of course, one of Housekeeping's major problems. That is why the San Francisco Hilton keeps a detailed file on every employee. A memo from the doctor is required when a person is out because of sickness. The Housekeeper makes a practice of discussing each absence with the individual before he or she resumes work. Thus if questioned by the unions about an employee's attendance record, the housekeeper will be able to handle the matter intelligently.

No-shows among maids and housemen in any institution can render a schedule almost useless. Having those openings does give those who want to do extra rooms or work late an opportunity, but if the house is full it is not always possible to find sufficient help. No-show problems are especially acute on weekends when employees are tempted to take the day off because their families or friends are not working. Also, because the head of the department is often absent on Saturdays and Sundays, employees may feel their own absence will cause less concern on those days. Thus it is essential on weekends to have a strong assistant on duty. It might also be advisable to overstaff, since even if absenteeism is zero, the extra personnel can be gainfully employed on general cleaning.

Maids in large cities are notorious for being absent and no amount of discussion or warnings will impress them. Many a maid has been known to pay for a doctor's appointment for the main purpose of saying she was sick last Sunday, and asking for a written note verifying that she was too ill that day to go to work.

## *The Importance of Timing*

It is not just the number of rooms that must be cleaned each day and the number of people who show up to clean them that affects staffing and scheduling of maids. Just *when* the maids can get into the rooms to clean them makes a great deal of difference in whether their quotas can be completed. If guests have double-bolted their doors or have put Do Not Disturb signs on the knobs, maids must wait. This occasionally occurs when groups of businessmen conduct meetings until very late hours. Thus it is sometimes better, if the hotel guest list has many late risers, to stagger the morning shifts and have some maids arrive at 9 to 9:30.

Holiday football games and parades also affect entrance time for maids. On New Year's Day at the Century Plaza in Los Angeles, guests glued to television sets all asked maids to clean their rooms between the Rose Bowl parade and the football game.

Hospitality Motor Inns, an 11-unit chain based in Cleveland, has its housekeepers check maids' progress during lunch hour. If a maid is not keeping pace with her room quota, or is perhaps ahead of schedule, she is immediately questioned so that housekeepers can learn what is causing a delay in making up rooms, or why she is not devoting sufficient time to each room.

Staffing and scheduling apply to more than daily guest room cleaning, of course (Figures 7-10, 7-11, and 7-12). Periodic jobs must be scheduled too, and it is well worthwhile to get them done without having to hire extra personnel. This can be accomplished with advance planning, a knowledge of employees' capabilities and ambitions, awareness of when other departments may be able to lend qualified employees, and a training outline ready when a worker is idle.

Public areas, where most clean-up and maintenance work is done by the late-night shift when guests or residents are not present, also must be serviced in accordance with daily work schedules and plans for periodic maintenance. All floors in lobbies and restaurants, for example, must be vacuumed or shampooed with damp foam daily, while metal thresholds are polished perhaps only weekly. Just when those tasks must be performed has to be determined. Permanent records must be kept of just when each job was last done and when it is due to be done again.

## *Keeping Track of Charges*

Because Housekeeping is a support department for many sections throughout the profit-making institution, it is necessary for the professional Housekeeper to know just which charges for housekeeping labor and materials can be allocated to other departments. Thus keeping careful records of schedules for all employees in order to know what kind of work they did and in which department becomes necessary for the Labor Analysis Sheet.

Housekeeping's main activities are charged to the Rooms Department, of course. Housekeeping also handles uniform control, involving every department in the hotel. Its Sewing Room mends napery for Food and Beverage; its housemen may shampoo banquet room carpets, or do similar tasks.

While no actual revenue or increased productivity is shown in Housekeeping's records as a result of this extra activity, the professional Housekeeper can counterbalance this extra workload by obtaining reciprocal assistance from other departments.

(Change in occupancy will affect only categories 7 & 9)

**WORK SCHEDULE SHEET**

DEPT.: Housekeeping OCCUPATION: Housemen

Page ____ of ____
Date ____

| OPERATION | RATE | HOURS | MON. | TUES. | WED. | THURS. | FRI. | SAT. | SUN. | TOTAL |
|---|---|---|---|---|---|---|---|---|---|---|
| 1. Handyman | | | 1 | 1 | 1 | 1 | 1 | | | |
| 2. Handyman | | | 1 | 1 | 1 | 1 | 1 | | | |
| 3. Hallman | | | | 1 | 1 | 1 | 1 | 1 | | |
| 4. Mezzanine Houseman | | | 1 | 1 | 1 | 1 | 1 | | | |
| 5. Linen Delivery | | | 2 | 2 | 2 | 2 | 2 | 1 | | |
| 6. Lights | | | | 1 | 1 | 1 | 1 | 1 | | |
| 7. Lanai Housemen | | | 4 | 4 | 4 | 4 | 4 | 3 | 4 | Depends on occupancy |
| 8. Penthouse Houseman | | | 1 | 1 | 1 | 1 | 1 | 1 | 1 | |
| 9. Linen and Trash Housemen | | | 2 | 2 | 2 | 2 | 2 | 2 | 2 | Depends on occupancy |
| 10. Heavy Vacuum Men | | | 2 | 2 | 2 | 2 | 2 | | | |
| 11. Wall Spotter | | | | 1 | 1 | 1 | 1 | 1 | | |
| 12. Window Washer | | | 1 | 1 | 1 | 1 | 1 | | | |
| 13. Linen Room Man | | | 1 | 1 | 1 | | 1 | 1 | 1 | |
| 14. P.M. Houseman 3:30 to 11:30 | | | 2 | 2 | 2 | 2 | 2 | 2 | 2 | |
| 15. | | | | | | | | | | |
| TOTAL MAN DAYS | | | | | | | | | | |
| TOTAL LABOR COST | | | | | | | | | | |

**FIGURE 7-10** *Staffing Table for Housemen*

(These do not vary with occupancy)

**WORK SCHEDULE SHEET**

DEPT.: Housekeeping OCCUPATION: Miscellaneous

Page ____ of ____
Date ____

| OPERATION | RATE | HOURS | MON. | TUES. | WED. | THURS. | FRI. | SAT. | SUN. | TOTAL |
|---|---|---|---|---|---|---|---|---|---|---|
| 1. Supervision* | | | 9 | 11 | 11 | 11 | 11 | 8 | 7 | |
| 2. Seamstress | | | 1 | 2 | 2 | 2 | 2 | 1 | | |
| 3. Linen Room | | | 1 | 1 | 1 | 1 | 1 | | | |
| 4. Linen Control | | | 1 | 1 | 1 | 1 | 1 | | | |
| 5. Costume Control | | | 1 | 1 | 1 | 1 | 1 | | | |
| 6. Costume Control | | | 1 | ** | ** | 1 | 1 | 1 | 1 | |
| 7. | | | | | | | | | | |
| 8. | | | | | | | | | | |
| 9. | | | | | | | | | | |
| 10. | | | | | | | | | | |
| 11. | | | | | | | | | | |
| 12. | | | | | | | | | | |
| 13. | | | | | | | | | | |
| 14. *Includes all management people executive housekeeper, | | | | | | | | | | |
| 15. assistant housekeepers, inspectresses | | | | | | | | | | |
| 16. | | | | | | | | | | |
| 17. **Days off Relieved by a supervision person | | | | | | | | | | |
| TOTAL MAN DAYS | | | | | | | | | | |
| TOTAL LABOR COST | | | | | | | | | | |

**FIGURE 7-11** *Staffing Table for Miscellaneous Jobs*

## FIGURE 7-12 *Scheduled Work Assignment*

BUILDING Rortrenft

AREA Pollock

RESIDENCE HALL WORKER

WORKER NO. 1 CLASS B

Effective Date Fall

1st thru 3rd Floors

| | | FIRST WEEK | | | | | | | SECOND WEEK | | | | | | |
|---|---|---|---|---|---|---|---|---|---|---|---|---|---|---|---|
| Perform Operations Indicated by (X) on Day Shown | | M | T | W | T | F | S | S | M | T | W | T | F | S | S |
| Daily Make Ready | | 26 | 26 | 26 | 26 | 26 | | | 26 | 26 | 26 | 26 | 26 | | |
| Clean Rooms 101 to 107 Inclusive (7 Rooms) | | 140 | | | | | | | | | | | | | |
| " " 108 to 114 " (7 Rooms) | | | 140 | | | | | | | | | | | | |
| " " 115 to 117 " (3 Rooms) 30/wk | | | | 60 | | | | | | | | | | | |
| " " 118 to 120 & 201 to 204 " (7 Rooms) | | | | | 140 | | | | | | | | | | |
| " " 205 to 210 " (6 Rooms) | | | | | | 120 | | | | | | | | | |
| " " 211 to 217 " (7 Rooms) | | | | | | | | | 140 | | | | | | |
| " " 218 to 220 & 301 to 304 " (7 Rooms) | | | | | | | | | | 140 | | | | | |
| " " 305 to 307 " (3 Rooms) 30/wk) | | | | | | | | | | | 60 | | | | |
| " " 308 to 314 " (7 Rooms) | | | | | | | | | | | | 140 | | | |
| " " 315 to 320 " (6 Rooms) | | | | | | | | | | | | | 120 | | |
| Clean Bath | Heavy | | 117 | | | | | | Balance same as 1st week | | | | | | |
| " " | Heavy | | | 117 | | | | | | | | | | | |
| " " | Heavy | | | | 117 | | | | | | | | | | |
| " " | Regular (2) | | 146 | 146 | 146 | | | | | | | | | | |
| " " | Regular (3) | 219 | | | | 219 | | | | | | | | | |
| " " | Regular | | | | | | | | | | | | | | |
| Clean Hallways | Heavy (3) | | | | | 63 | | | | | | | | | |
| | Regular (3) | 30 | | 30 | | | | | | | | | | | |
| | | | | | | | | | | | | | | | |
| Clean study 1st floor only | Heavy | | | 45 | | | | | | | | | | | |
| | Regular | 14 | 14 | | 14 | 14 | | | | | | | | | |
| | | | | | | | | | | | | | | | |
| Clean laundry 1st & 3rd (incl. vacuum floor and remove dryer lint) | Heavy (2) | | | 42 | | | | | | | | | | | |
| | Regular (2) | 20 | 20 | | 20 | 20 | | | | | | | | | |
| | | | | | | | | | | | | | | | |
| Clean work-room 2nd floor | | 11 | | | | 11 | | | | | | | | | |
| | | | | | | | | | | | | | | | |
| | | | | | | | | | | | | | | | |
| Other Non-Routine Duties as Directed | | X | X | X | X | X | | | | | | | | | |
| | | 460 | 463 | 466 | 463 | 473 | | | 460 | 463 | 466 | 463 | 473 | | |

*Schedule for Residence Hall Worker "B" (called a "maid" in hotels but union on campus has ruled against terms such as "porter," "custodian," "janitor" and "maid.") This schedule which is the type in use at Pennsylvania State University includes minutes spent on all aspects of the job. Work is repeated in 2-week cycles so that employees know schedules ahead of time and can depend on weekends off—quite different from hotels where weekend work is part of life.*

During the slow season in the Banquet Department (few business meetings are held in summer), housemen or stewards may be loaned to Housekeeping for certain annual or semiannual tasks that can be performed with minimum training anytime during the year. Doorknobs can be cleaned by almost anyone, but they require a special kind of cleaner, a template to protect the wood on the doors, and elbow grease for polishing. The on-loan worker will do nothing but doorknobs for several weeks, or he may rotate that job daily with washing vinyl wall covering in corridors, or cleaning banisters on back stairs.

Sheraton's Quality Audit emphasizes the need for allocations to other departments:

1. Is the housekeeper showing on the daily Personnel Report all hours of housekeeping labor assigned to other hotel departments?
2. Does the housekeeper know what dollar allocations are made monthly from her payroll to other hotel departments?
3. Are all allocations from Housekeeping payroll to other departments being made on an equitable basis?
4. Is ALL labor used by Housekeeping for project work being charged to projects regardless of budget?

Another unusual task that can be handled by an inexperienced person is changing worn decals in guest rooms. The worker takes a box of the decals needed for the phone, the bathroom, the door, closets, and so forth, and replaces the ones that seem soiled or dog-eared. Or one person might do nothing but deliver current phone books to guest rooms.

Planning ahead for intelligent exchange of help with other departments will not only enable Housekeeping to get the odd jobs done in every room and on every floor, but will increase employee productivity throughout the institution.

Automatic electronic staffing and scheduling are being done by computers that are increasingly more indispensable to the hotel industry. Holiday Inn was the first chain to use the computer for this purpose, although there is no one system that is in use throughout the chain. Hyatt Hotels were the first to be totally computerized, but every year more improvements will be developed in computer capabilities, and many Housekeeping Departments will be able to amass, store, and retrieve the information needed to make scheduling and planning easier. No amount of automation, however, can really accomplish the business of cleaning guest rooms. It still takes people who care and are willing to expend physical energy.

# Chapter 8

# Cleaning the Guest Room and Bath

**OVERVIEW** A clean guest room and bath is the primary product that lodging (and residence halls too, although to a lesser degree as the students usually clean their own rooms) has to sell. This chapter offers suggestions and even specific rules on how to get the basic cleaning work done. The routines and duties of maids and an inspectress (or supervisor) are reviewed.

We first consider the techniques of how to clean the guest room, although the main purpose of this book is not to provide a how-to-do-it guide—that sort of thing is available in many publications and pamphlets. Rather, we explain what is involved in all the tasks of the Housekeeping Department, and reveal the scope of the Department's province.

Certain rules must be adhered to and followed by maids when they clean rooms in any establishment. Among these are the following: (1) maids and housemen should use only service elevators; (2) no employee is allowed to open a guest room door for anyone; Linen Room or Front Desk should be notified and bellman will bring a key to the guest who requires entry to his room, or in a residence hall the manager will have to be called; (3) no one working in rooms is allowed to use guest telephones; however, in a few small operations it is necessary to call the Linen Room or other Housekeeping headquarters to report when a room is ready, but employees usually have other means of doing this, and most institutions have intercom telephones in the back halls or beepers for the employees to use when contacting their departments; (4) no employee is allowed to use a guest bathroom—rest rooms are generally provided for them throughout the institution; (5) where and when employees can smoke or eat is a matter for the individual property to decide; (6) linen or towels must never be used for cleaning and dusting.

## *Just How Are Rooms Cleaned?*

Systems and techniques differ in almost every hotel, motel, or residence hall, although some chains, such as Holiday Inn company-owned operations may have real uniformity in their procedures.

In most places, maids first go to locker rooms and change into uniforms, then go to Housekeeping headquarters, usually the Linen Room, sign in, and get their floor assignments. They pick up their rags, supply baskets, room status sheets, and pencils. They then report to their floors and sign in with the supervisor who gives out room assignments and keys.

This procedure is advisable if there is a big enough sign-in area for all the maids (40 or 50 in the biggest properties) to congregate in the morning and afternoon. Such an arrangement tends to give them a certain esprit de corps—a sense of belonging to a group and being part of the hotel organization. Checking in and out at a floor linen room where they see only a few other maids and their supervisor, deprives them of the feeling that part of the success of the establishment is heavily dependent upon their effective performance and that of their co-workers.

Every effort is made NOT to disturb the guests—even if they haven't had the foresight to put signs on the doorknobs or double-lock their doors. As long as there are check-out or preregistered rooms to work in (or rooms for which early make-up has been requested), maids are usually instructed not to knock on doors, thus avoiding danger of awakening guests. Obviously they never knock when the door is double-locked or Do Not Disturb sign is displayed. If by 2:30 PM the room is still inaccessible, Housekeeping is notified by the supervisor and the room is checked by phone or in person by the assistant manager.

Even though many hotels are trying to eliminate maids' knocking on doors, almost universally the following is done: maids check room occupancies and report on them to the supervisor. Different designations (an accepted language) include:

1. Bed used (double-locked door, Do Not Disturb sign, or occupied room)
2. C/O means check-out (bed used but no luggage left)
3. Vacant indicates room is clean and there is no luggage.

Carts stored in the Linen Room or housekeeper's room on floors are to be filled with linens and supplies as specified on a chart or list posted nearby. Supplies may include toilet tissues, facial tissues, ash trays, matches, Room Service menus, directories, printed stationery, drawer liners, bath linens, bed linens, guest soaps, shower caps, various other giveaways, and similar items. Carts are always positioned in front of the guest room being worked—never taken inside the room.

After knocking on guest room doors maids should always announce "Housekeeping," and ask if the guest wants service then or at what time later. The time requested is marked on a room report (Figures 8-1 through 8-12).

Unless windows are sealed, for example, in airport hotels, maids usually air out rooms as they

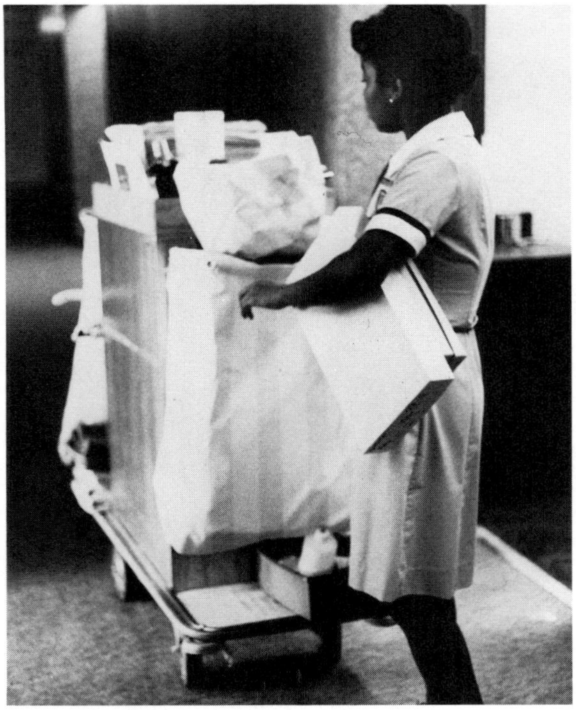

**FIGURE 8-1** *First she removes trash, putting it in refuse bag on a cart that is positioned in front of guest room door.*

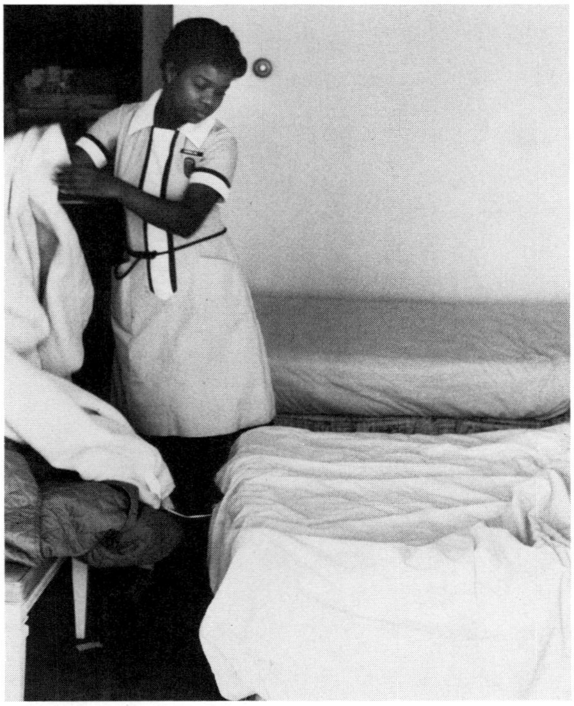

**FIGURE 8-2** *Next she strips bed, keeping blanket, pillow, and spread off the floor, and puts soiled linen in a bag on cart.*

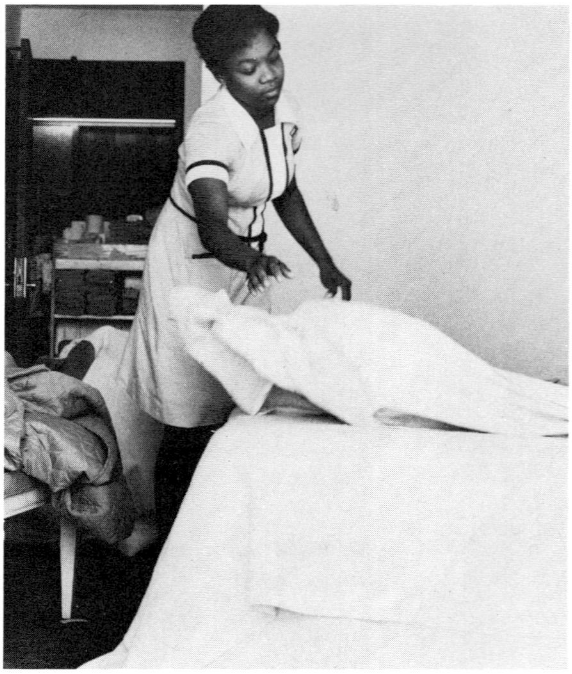

**FIGURE 8-3** *She makes the bed from one side, moving around bed only once, thus saving steps and energy. Learning to fling open sheets from the side speeds up bed-making.*

**FIGURE 8-5** *Carts sometimes become rather unsightly as trash and laundry bags fill up. Housemen continually patrol hallways with rolling bins, collecting soiled linens and trash. Because maids' carts are eyesores in guest corridors, housekeepers constantly explore ways to eliminate them.*

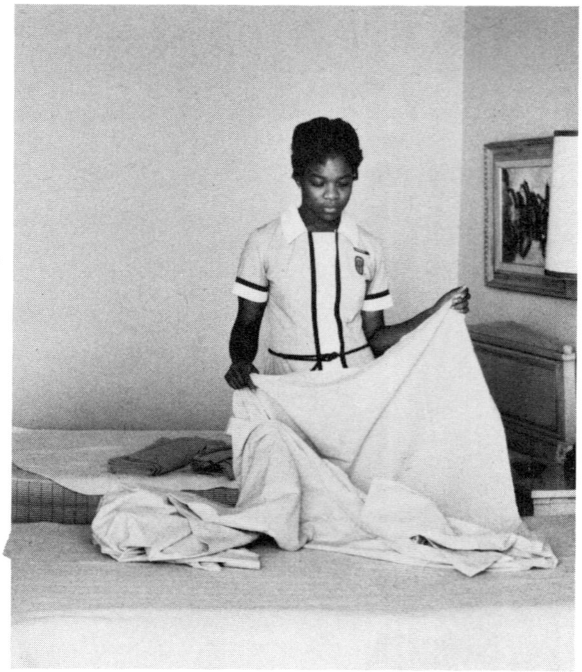

**FIGURE 8-4** *She puts a crinkle sheet over the electric blanket to keep it clean. Note that bedding is not tucked in until every layer is smoothly in its place.*

**FIGURE 8-6** *Special mops with retractable covers permit maids to clean toilets without putting their hands in bowl. Cleaner-disinfectant is left in bowl while the maid finishes other chores.*

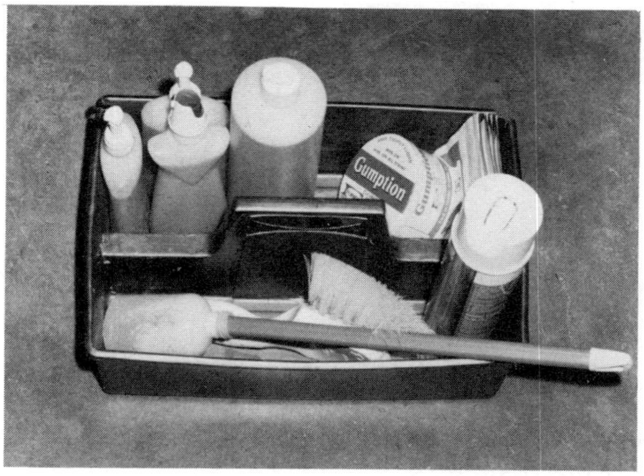


**FIGURE 8-7** *So that maids can easily carry cleaning supplies into guest rooms and bathrooms, their cleaning tools and chemicals are put into a compact caddy that is filled in the Linen Room; then they need not take time to refill containers before starting to clean rooms.*

**FIGURE 8-8** *Removable plastic drain closure is cleaned daily. Note caddy for carrying tools and cleaning supplies. Trend now is to use all-purpose cleaners, thus cutting down on supplies to be carried. Double toilet tissue rolls assure that guests will not run out. Also, they are less apt to be taken by guests than an extra roll stored in cabinet or displayed in the bathroom.*

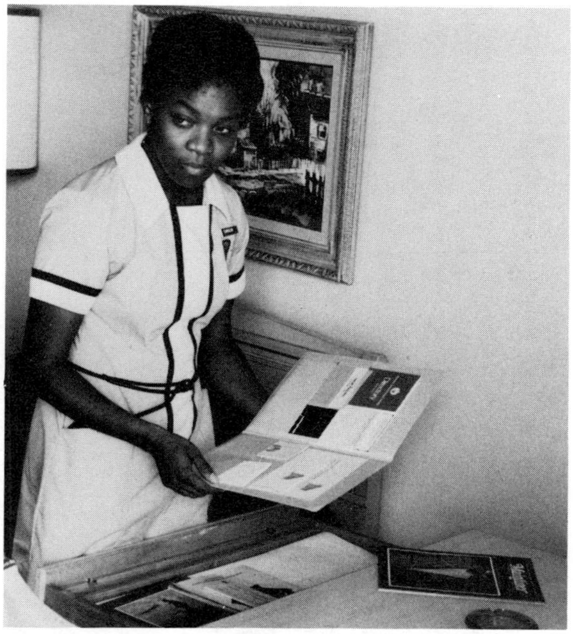

**FIGURE 8-9** *Maid checks stationery packet and other items kept in desk drawers in guest rooms. If paper is missing from packet, the entire folder is replaced. This saves time while cleaning rooms. Folders are filled by employees in the Linen Room periodically as schedules permit.*

begin to work in them. In some hotels they are instructed to hang up guests' clothes, but this rarely occurs in a permanent residence and certainly not in student residence halls.

Making beds first is recommended so that the room will appear neat if guest enters. How they are made depends on the Housekeeping director's preferences, but whenever a maid makes a bed she does change the linens. Most beds in hotels today are made from one side, which saves walking and speeds up the process. Often, posters on suggested procedures are mounted in back halls or Linen Rooms. TraveLodge's manual states that an experienced maid, who uses a practiced technique, should be able to make a bed properly in two and one-half minutes. For a detailed description of how to make a bed see the chapter on procedure manuals.

After the bed is made, the bathrooms are usually cleaned next. Tub, shower curtains, walls, basin and counter, toilet, and floor are usually cleaned in that order, especially as so many operations are now using an all-purpose cleaner that is dumped into the bathroom's plastic wastebasket, which is then used as a bucket. (See Chapter

**FIGURE 8-10** *Drawers are wiped out or paper lining is turned or changed as needed. Although this is usually part of general cleaning done intermittently in most hotels, at Century Plaza Hotel it is part of daily routine.*

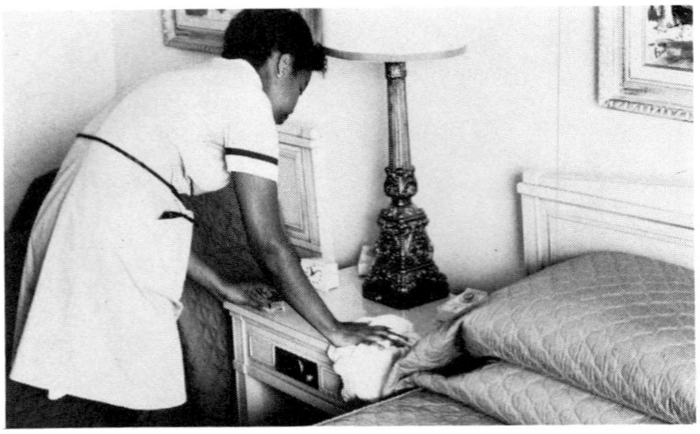

**FIGURE 8-11** *After beds are made (extra-long spreads can be tucked in between stacked bed pillows) and bathroom is cleaned, the maid dusts, as illustrated below, using cloth slightly dampened with lemon oil. Blanket controls and electric alarm clock cords are bound together at plug to discourage guest pilferage.*

18 on Bathrooms and Rest Rooms.) The same solution applied with a sponge is used to clean bathroom areas (disinfectants are often used in the toilets).

If plumbing seems to be clogged or sluggish, maids should report the condition to the Linen Room or their floor supervisor immediately.

How ashtrays and glasses (if not disposable) are washed varies. The former are usually emptied into the toilet rather than into the trash, because they pose a fire hazard. Glasses are sometimes sent to the main kitchen to be washed in the dishwasher, but this requires considerable handling and transporting. Some operations have small automatic glasswashers on the floors, while in others, maids are asked to disinfect glasses in the bathroom basin, rinse, and dry thoroughly. Clear disposable plastic glasses are becoming popular even in luxurious properties, as using them saves labor appreciably.

Heat lamps are cleaned as shown in Table 8-1.

Guest rooms are dusted with a cloth dampened with all-purpose cleaner or, if furniture is wood instead of plastic laminate, with furniture polish. As the last step, rooms are vacuumed.

Guest room and bathroom supplies are replaced as needed, but such matters as whether guests should have freshly wrapped soap every

**TABLE 8-1** Cleaning Heat Lamps

| Supplies | Chemical and Handling | Procedures |
|---|---|---|
| Cup measure, bucket, gloves, sponge, cloth, or brush | Dilute 2 C degreasing cleaner in 1 qt. water | Daily: use gloves: (1) unplug or turn off lamp and allow to cool, (2) soak sponge, cloth, or stiff brush in solution, (3) wipe off lamps, (4) rinse with water, (5) allow to dry before turning on |

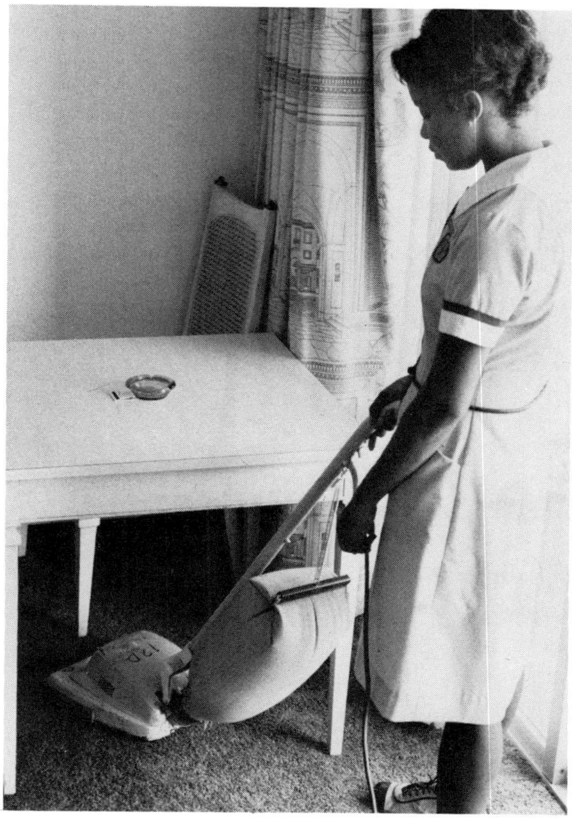

**FIGURE 8-12** *Last step in guest room cleaning is to vacuum the floor. Maid carries upright vacuum on cart and covers only accessible areas, as heavy vacuum crew will be scheduled to move furniture and clean the entire floor.*

day depend on budget and level of luxury to be maintained (Figure 8-13).

Make-up rooms do not receive quite the attention that check-out (c/o) rooms do. For example, locating lost and found items is a very important function for maids doing c/o rooms. Any belongings that are found, regardless of value, are given immediately to a supervisor who takes them to Housekeeping headquarters for safekeeping (Figure 8-14).

If liquor, beer, or wine is found in c/o rooms, it is also to be given to supervisors, and not left on carts or in linen rooms. Disposition will depend on company policy.

In residence halls, regulations containing alcoholic beverages vary from one institution to another. Policies will be clarified by individual managers.

In c/o rooms, clean bath rugs are normally put in bathrooms; bed pads are changed, and possibly bedspreads, if they are washable; drawers are wiped out or paper linings changed; furniture is arranged in accordance with original placements; air freshener may be used if needed; all lights and air conditioners are turned off; and thermostat is set at prescribed temperature.

## *What a Floor Housekeeper or Inspectress Does*

The preceding section has outlined in general what maids do anywhere from 9 to 16 times a day as they clean rooms. To give readers a clear idea of what a floor housekeeper actually does during her day, a complete *Standard Operating Procedure* written by Henrietta LaBorde, executive housekeeper, Western Division, Hilton Hotels Corp., is included here.

Although the schedule varies from one property to another depending on the hotel's physical structure and number of rooms, this description reveals graphically what responsibilities are carried by a floor housekeeper, (sometimes

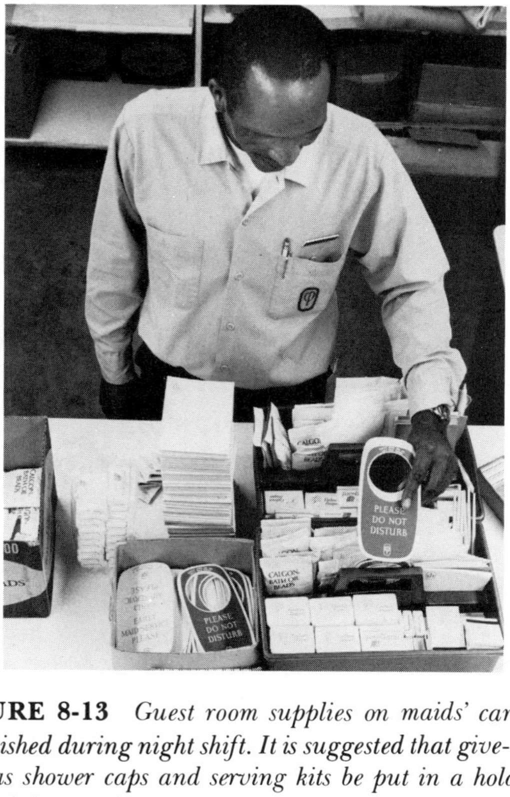


**FIGURE 8-13** *Guest room supplies on maids' carts are replenished during night shift. It is suggested that give-aways such as shower caps and serving kits be put in a holder on lower shelves to discourage guests from collecting an inordinate number of souvenirs.*

**FIGURE 8-14** *Maids take room disorder for granted, are quick to spot items guests left behind, and give them to supervisors.*

called an inspectress or floor supervisor), how many people she must relate to, and an itemization of what must be observed and attended to during the day.

Bear in mind that this particular system is not used in every hotel and that personnel in other Housekeeping Departments may be organized in an entirely different way, with entirely different responsibilities.

## *Standard Operating Procedure for Floor Housekeepers at Hilton Hotels, Western Division*

1. **Distribute Pass Keys to Maids**
   - A. Pick up your own pass key and maids' pass keys from executive housekeeper's office before going to your floor. You are responsible for any pass key while it is in your possession. Do not unlock doors for anyone; courteously refer guests to assistant manager.
   - B. Give each of your maids her pass key and have her sign key slip when she reports for work. (Housemen do not have pass keys.)

     Check appearance of each maid when you give pass key to her, making sure her uniform is clean, shoes polished, stockings without runs, hair neat, little or no makeup or jewelry on. Check for body odor, it might be offensive to a guest.
2. **Schedule for Housemen**

   While maids are taking room counts, assign houseman his work for the day.
   - 1st- Vacuum guest elevator foyer daily.
   - 2nd- Clean sand jars on guest foyer daily. (Refill sand jars once a week, on day before days off.)
   - 3rd- Vacuum out-of-order rooms going in-order that day.
   - 4th- Vacuum "Special Arrivals."
   - 5th- Vacuum standard set-up of eight rooms a day; guestrooms and parlors four days a week.
   - 6th- Wash service landings once a week, sweep every day.

7th- Day after days off, clean all glass in guest corridors, glass covering indicator lights, and face of clock in guest foyer; polish all brass such as fire extinguishers, etc.

Note: You will help the assistant executive housekeeper in preparing out-of-order rooms going in-order that day or you may assign those rooms to maids, when possible.

**3. Prepare "Housekeeper's Room Report"**

A. As soon as all your maids have handed in their "Supplementary Housekeeper's Room Reports" to you, immediately prepare your "Housekeeper's Room Report;" indicate by cross mark (×) in proper column and opposite room number on your report all information shown on maid's report.

B. Make out your report in triplicate; then take the three copies at once to:

(1) Room clerk in Housekeeping office.

(2) Executive Housekeeper's office.

(3) Keep one copy until 4:30 pm.

**4. Check Progress of Maids in Their Work**

A. Check each maid's truck to see that it is properly equipped according to standard equipment listed in "Manual for Maid Training," and adequately supplied with linen and items listed on "Standard Guestroom Requirements Card." If not properly equipped or supplied, have the maid equip and supply it properly (or as nearly as possible, as sometimes laundry is delayed).

B. Inspect sink closets, fire closets, exit landings and stairways, employees' lavatories, and service landing to see that they are clean and free of rubbish, and misplaced pieces of furniture; that lights on stairways, exits, and service landings are burning and of correct wattage as indicated on wattage list posted in maid's closet; and that service landing contains only standard equipment. Check rubbish and trash can area. If not clean, notify houseman. Every day a different maid cleans the maids' toilet, schedule to be posted in your linen room.

**5. Inspect Guestrooms**

A. Look over your "Housekeeper's Room Report" to find out which rooms are listed as "sleep-outs" and O.K. Inspect these rooms first so they will be ready to sell (although obviously not until the sleep-out becomes a c/o).

B. Inspect a check-out room as soon as possible after it has been put in order. A check-out room should not be out of order over an hour after a guest departs.

C. At least every hour during your shift, pick up list of check-out rooms to be made up. Turn in O.K. rooms as soon as possible (Figure 8-15).

D. Following is a list of items to inspect in an O.K. room (one that had not been occupied previous night):

(1) Inspect bathroom (Figure 8-16).

a. FLOOR: See that it is clean and that there are no broken or loose tiles.

b. WALLS: See that they are clean and that there are no cracks or broken or loose tiles, and that ceiling is clean and has no cracks. Make sure top of tiling is clean.

c. NICKEL FIXTURES: Check to see that they are polished.

**FIGURE 8-15** *Floor supervisor stops at maid's cart to see security sheet that lists rooms maid has cleaned. Supervisor then calls Rooms Controls person in Linen Room to report rooms ready to rent; room numbers are immediately relayed to Front Desk.*

d. FAUCETS AND SHOWER HEAD: See whether they leak. (Be sure that shower head regulator is open so shower can drain.)

**FIGURE 8-16** *In bathroom inspectress pays special attention to whether all guest's give-aways are in order, whether proper number of bath linens are in place, as pictured below, and whether fixtures are clean and shining.*

e. SOAP RECESS: Grab handle part of soap recess in bathtub to see that it is not broken.
f. FIXTURES: Check toilet bowl, tub or shower, and washbasin to be sure they are clean, not cracked, and that drains are working.
g. WATER PRESSURE: Check water pressure by turning on faucets and flushing toilet. (Also check ice water faucet at this time.)
h. TOILET SEAT: Check to see that it is thoroughly clean with sterilized band on it.
i. BATH CABINET: See that it is clean, has no broken shelves, and that door works.
j. SUPPLIES: Check supplies such as towels, soap, etc., to see that bathroom is correctly supplied according to "Standard Guest Room Requirements Card."

(2) Inspect guest room:
a. Check carpet to see that it has been properly vacuumed, has no tears or spots. Be sure to look along baseboard and underneath beds, dresser, desk, and chairs for dust and dirt.
b. Move out furniture from wall when inspecting underneath. Backs of beds, dresser and desk are also to be inspected to see if they have been dusted by houseman (Figure 8-17).
c. Inspect windows, sills, and sashes to see that

they are clean and undamaged; check draperies to see that they are clean, linings not torn, and drapery tiebacks attached on two separate hooks.

d. Inspect furniture, being sure there are no stains on it.

E. In a check-out room:

(1) Inspect bathroom for same things as in an O.K. room.

(2) Inspect beds to see that linen has been changed and beds properly made. Lift spread and blankets to determine this.

(3) Inspect carpet in bedroom for same things as in an O.K. room.

(4) Inspect windows and draperies for same things as in an O.K. room.

(5) Inspect walls and ceiling in bedroom to see if they need washing or have cracks. Pictures on walls should be checked at the same time to see that they are clean, undamaged, and hung straight.

(6) Check all lamps in bedroom to see that they are dusted, in working order, their shades in good condition, and that bulbs are of correct wattage as indicated on wattage list posted in maid's closets (Figure 8-18).

Note: Lamps on dresser, desks, beds, etc., are checked when that piece of furniture is inspected by you.

(7) Inspect clothes closet to see that walls are clean and free of baggage marks, shelf and floor are clean (run hand over shelf for dust); clothes pole is dusted; pants hanger attached to wall or door and not bent out of shape; closet has standard set-up

**FIGURE 8-17** *Men who do heavy vacuuming cover each room every three weeks at the Century Plaza Hotel. They move all furniture so that hard-to-reach spots can be vacuumed, especially baseboards, edges of carpet, and under beds, areas most often neglected.*

**FIGURE 8-18** *Even a small spot on the lampshade is noticed by keen-eyed inspectress. After discovering it, above, she makes a note on the Repair Sheet that a handyman should bring a clean shade to room and return spotted one to Linen Room.*

of hangers and laundry bag (Figure 8-19).
Note: Twin bedroom has extra baggage rack in closet.

(8) Inspect bureau to see that mirror is polished. Open drawers to see that they have been washed, a laundry bag is in bottom drawer, and that "Do Not Disturb" card and "Special Maid Service" care are in top drawer.
Note: Metal fixtures on all furniture such as drawer handles, knobs, etc., should also be checked as you come to them. (This includes door knobs and chains.)

(9) Inspect night table to see that it has a memo pad, a breakfast menu, and reminder card to order breakfast the night before. Check telephone to see that it is working by lifting receiver; inspect mouthpiece for cleanliness. Check cord to see that it is not frayed; check controls (1) and (2) on radio to see that they are working, and radio compartment to see that it is locked; phone book (central section) with cover on night stand; yellow classified book in dresser drawer.

**FIGURE 8-19** *Closer inspection covers items shown below to make sure that there are enough hangers; laundry bags are in place; mirror is clean. Each item is checked off on the Please Do form that is retained in files. (See later chapter for form.)*

(10) Inspect desk to see that it is correctly supplied according to "Standard Guestroom Requirements Card," and that pen works. Note: At least one ash tray with matches on desk or dresser.

(11) Inspect waste basket beside desk to see that it has been properly cleaned.

(12) Inspect all chairs and baggage rack to see that material is not worn, ripped, or burned, springs not broken, wood free of splinters, and legs and arms not broken or cracked.

(13) Inspect servidor to see that it is clean, contains one hanger, and that locks work.

(14) Check ventilator over entrance door to see that it opens and closes easily, and that rate card is in holder (Figure 8-20).

(15) Inspect thumb bolts on connecting doors to see that they are safely locked.
Note: Keep a sharp lookout while inspecting rooms for articles left by guests that maid overlooked. If

you find an article, attach paper with room number on it. Immediately take it to Lost and Found.

(16) Inspect all door frames as you come to them to see that they are clean.

F. In a reoccupied room—Inspect it the same way you do a regular check-out room except that the guest's belongings are not disturbed during inspection.
Note: If guest is in room when you go in to inspect it, explain to guest that you were not able to inspect the room before his arrival but can inspect it now with his permission.

G. In a parlor—Inspect it the way you do a regular check-out room except that there is no bathroom or bed in this type of room.
Note: Be sure to lift up all loose cushions in room to see that they are clean and to see if any small articles have slipped down between side of cushion and chair or underneath cushion (Figure 8-21).

H. In an occupied room—Inspect room to see that it is generally neat in appearance and that bathroom and desk are supplied according to standard; guest's belongings are not disturbed, of course, during inspection.
Note: As far as possible you should inspect all occupied rooms each day.

**6. Prepare Trouble Report**

A. During inspection of rooms, make notes in your memo book of each defect or unsatisfactory condition which you find (Figure 8-22). Your notes will be of two kinds:

(1) Items for which you will need to make out a trouble report.

(2) Items you can turn over immediately to your maid or houseman for prompt handling.

**FIGURE 8-20** *Vents are frequently ignored in many hotels because they are difficult to reach, and require patience and the use of a crevice tool. Vents shown in picture at left are vacuumed regularly; removed annually and washed. Only tall men need apply for this job.*

B. Prepare trouble reports as time permits during your shift as follows:

(1) Make out a separate trouble report for each craft (carpenters, electricians, etc.), listing all items needing attention.
Note: Any furniture, plumbing, etc., reported to you by

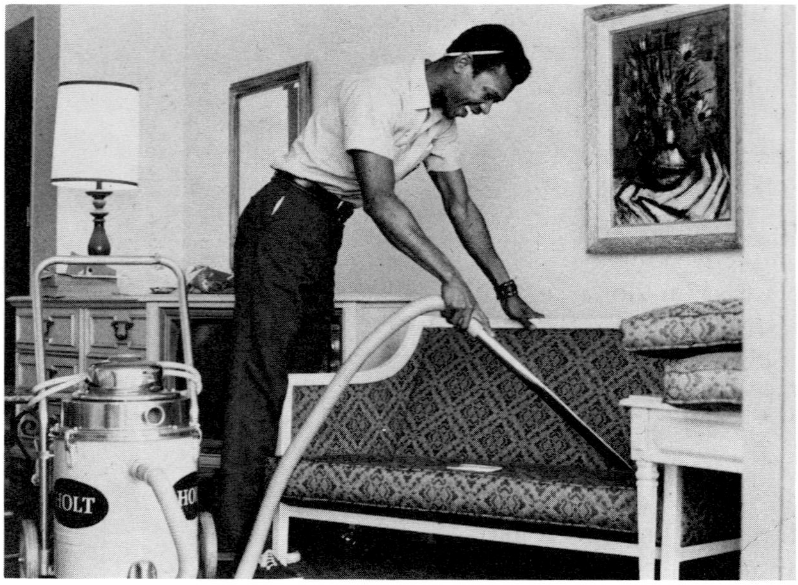


**FIGURE 8-21** *Upholstered furniture is thoroughly vacuumed and cushions are removed so that soil or dropped objects can be collected. Sofas can produce welcome bonus during thorough cleaning, as change frequently falls out of guests' pockets.*

**FIGURE 8-22** *Inspectress is responsible for 60 rooms daily. She checks all VIP rooms, "pre-reg's," c/o's, and other rooms at random. Maids never know which of their rooms will be examined. Working like a white-gloved military inspector, inspectress makes certain no item is dusty, even the tops of picture frames.*

maid or housemen as needing repairs or cleaning is put on a trouble report.

(2) If you find or learn of a defect or condition that must be taken care of at once, report it by "Telautograph" to proper department.
Note: Be sure and follow up your Telautograph report to see that it is given immediate attention; if necessary, telephone executive housekeeper's office to see that it is being done.
Note: When a maid reports to you about guest damage to furniture, or you find such damage during your inspection of rooms, immediately report this by phone to executive Housekeeper.

(3) If, on inspection, you find a defect or condition that had been previously reported on a trouble report and not attended to, telephone department to which you sent report; find out if report was received; if so, find out how soon the condition can be cared for.

(4) Trouble report is prepared in triplicate; copies to be time-stamped and distributed as follows:
a. One to department that will correct the trouble.
b. Second to executive Housekeeper.
c. Third copy to be held till job is done, then taken to executive Housekeeper's office.

C. If you find a room service table in the hall on your round of inspection, notify Room Service by phone to remove it. (Also trays. Ed.)

D. Always be alert for suspicious persons on the floors, including employees who have no obvious business there. Report presence of such persons immediately to assistant manager.

**7. Report Sickness of a Guest**
When a maid reports sickness of a guest to you, immediately notify executive Housekeeper.

**8. Report Loss of Linen or Bedding**
When a maid reports loss of any linen or bedding to you, immediately notify executive Housekeeper's office by phone.

**9. Replace Stained, Torn, Burned Bedding or Linen**
A. When maids bring you blankets, bedspreads, etc., that have been burned, torn, or stained, give them clean fresh ones from your supply closet in exchange; send stained or torn bedding down to Linen Room.
B. When you need supply of clean blankets, go to the Linen Room on the Service Floor and take an ample supply to keep in your closets.

**10. Order Supplies**
When you bring down your cart for the next day's bathroom glasses, attach your requisition for supplies for the following day. The night Linen Room attendant will fill this order.
Note: The removal of soap from occupied guestrooms before it is almost used up is controlled by you through checking the number of cakes distributed daily to a maid in relation to number of rooms occupied in her section.
Note: Old razor blades and used soap are turned in to you daily by your maids. Keep these articles in closet during week, on Friday send to the Linen Room by the houseman.

11. **Check Rooms with "Do Not Disturb" Signs**
At 2:30 pm check rooms with "Do Not Disturb" cards on doors by first finding out from the maid if room has had room service; if not, telephone room and ask occupant how soon he wants room made up. If no answer to phone call, immediately notify Assistant Manager.

12. **Check Maids' Room Slips**
Around 3:30 pm check each of your maids' room slips to find out which rooms they have not made up. List all rooms that will not be made up by 4:00 pm, and turn in to executive Housekeeper's office when you go off duty; also list on blackboard in Linen Room for the night maids.

13. **Cart and Closet Inspection at End of Shift**
   A. At end of shift, inspect maids' carts and maids' closets.
   B. Keep in mind that the company purchases only the best equipment, so it is important that workers under you know how to use equipment properly, and how to take care of it.

14. **Inspect Guest Elevator Foyer**
Just before you leave your floor, inspect guest elevator foyer to see that it is clean and neat looking. Check sand jars to see that they are clean.

15. **Collect Pass Keys**
At 3:50 pm collect all pass keys, take to executive Housekeeper's office, place in cabinet.

16. **Training Maid or Houseman**
When training a maid or houseman, use training material and procedure set up in "Manual for Maid Training" or "Manual for Floor Houseman Training."

Although from the foregoing detailed instructions it seems that nothing should go unattended to in a well-staffed hotel, it is still possible that the inspector will not notice neglected duties because there simply is not time on all days to check every detail in a full quota of rooms.

Many maids build a reputation for doing a good job and supervisors do not bother to check up on them, preferring to spend their time inspecting those whose performance is less praiseworthy or consistent. Even that A-plus maid may be interrupted in her work, however, and forget to carry on where she left off, or because she was hurried by a guest, she may neglect proper cleaning procedures.

Louise Wheeler, executive Housekeeper of the Contemporary Hotel at Walt Disney World feels that inspecting rooms cleaned by experienced maids is degrading and demoralizing. "It's no wonder they aren't motivated; in effect we're saying, 'We don't trust you . . . we have to check.' " Until a new maid has proved her ability and sense of responsibility, however, Mrs. Wheeler does believe in room inspection.

Another reason an unclean or damaged room might accidentally be rented is because an inspectress has noted something on a weekend that needs repair or redoing, but knowing that there is no help available on a Sunday, does not bother to report it. If she has reported the same damaged item before, and found after several weeks it still has not been repaired or properly cleaned, she may simply stop trying to get action since it seems nobody cares. In the chapter on Controls we will consider ways to double-check these kinds of problems.

At the Greenbrier Hotel, supervisors and inspectors apparently are very observing and always alert because Mrs. Kappa has abandoned all checklists—"Our system of inspection is 'Follow your right hand around the room and train your eyes to cover ceiling to floor as you check.' "

# Chapter 9

# Cleaning the Front- and Back-of-the-House

**OVERVIEW** Guests will judge the level of cleanliness by the hotel's public areas. Housekeeping may be responsible for lobbies and the back of the house, kitchens, exterior areas, meeting rooms, and pest and waste removal. Here we take up each area, how to care for it, and each type of professional contractor that might be involved.

## *Front-of-the-House*

The public areas—foyers, lobby, corridors, and dining areas—of any institution are always "on stage," being scrutinized by residents or visitors. They must look their best at all times lest onlookers infer from one visit, when the lobby was littered or a chair back loose, that the whole establishment suffers from inadequate housekeeping and maintenance. This is especially true for hotels, as a large percentage of people who use them are not overnight guests, but *are* prospective patrons, or could be boosters, if after attending a meeting or dinner at the hotel, they feel confident that the establishment is immaculate from top to bottom and front to back.

As the staffing charts in Chapter 7 indicate, there are many employees responsible for cleaning, maintaining, and repairing both the front- and back-of-the-house. Each area presents certain problems when it comes to scheduling workers and selecting tools and cleaning supplies for specific furnishings and equipment. Usually the larger areas involve housemen rather than women because of the heavier equipment used and the frequent need for climbing ladders and lifting heavy objects.

### *Lobbies*

The lobbies are the areas that require constant vigil, not only because heavy public traffic marks its passage by trails of footprints, cigarette stubs, and chewing gum wrappers, but also because the appearance of the lobby is all-important if an institution is to have a favorable impact on new arrivals. Snow and rain tracked in can create serious housekeeping problems. Matting or carpets put down at entrances can be helpful, but if crowds come and go frequently, housemen should still be assigned to see that entryways are kept mopped throughout the day, and that entrance carpets are properly covered and protected. Runners and mats may have to be mopped or changed frequently during wet weather.

Cleaning out sand urns and emptying ash trays is another continual job for the housemen or maids on duty in lobby areas (Figure 9-1). That same employee might also be charged with the job of keeping the rug clean—an on-going process accomplished with a long-handled whisk-broom and dust pan.

Except in emergencies, however, these two activities are the only major cleaning operations that are done during the day in the lobby. Cleaning, maintenance, and repair usually go on late at night when few people are around to get in the way or to be inconvenienced.

Nightly lobby jobs to be scheduled for the night housemen and/or maids are: vacuuming and shampooing carpets; sweeping, buffing, or mopping floors (detailed in a later chapter); dusting furniture; emptying and wiping out ashtrays and wastebaskets; damp-wiping all plastic lami-

**FIGURE 9-1** *Parlor maid removes ashes from sand urns with slotted spoon and conceals debris in attractive, lightweight basket. Wearing a crisp uniform, she may cover all public areas during the day to keep urns clean, or to tidy powder rooms.*

nate surfaces; removing finger marks or spots on walls and woodwork; polishing the drinking fountains; spot-cleaning mail chutes, glass, and metal; removing all foot and hand marks from doors; cleaning all brass door knobs or push handles, and so on. Wastebaskets should be washed inside and out, and all glass doors washed and polished.

Jobs to be done weekly in public areas such as the lobbies and elevator foyers include:

1. Wash house telephones and phone booths
2. Polish wooden furniture with lemon oil
3. Dust window ledges
4. Vacuum louvered doors
5. Polish sides of drinking fountains
6. Dust ceiling vents, using 20-ft. extension
7. Vacuum baseboards, using crevice tool

Jobs done monthly in public areas at the front of the house:

1. Vacuum upholstered furniture and draperies (if dust build-up occurs sooner, vacuum as needed)
2. Wax lamp bases and any decorative items so they will collect less dust
3. Wash and polish all sides and drawers of metal desks, bookcases, chairs, filing cabinets, etc. Do on a skip-clean basis (a few every day) so that all are polished at least once a month
4. Polish all wooden desks, cabinets, and furniture on a skip-clean basis
5. Wash and polish all clear glass except windows—spot-clean daily
6. Dust jambs and door closers

## *Elevators and Escalators*

Like lobbies, the elevators are areas that are used continuously. If they are automatic, and most are these days, hotel employees have little control over what outsiders or guests will do to them. Although walls are often marred by luggage, the damage done by vandals is much more permanent, so walls in elevators require care nightly unless there is a carpet on sides and ceiling—a non-vulnerable material useful as a cover for elevator walls.

Elevator floor carpets, however, are perhaps the single most difficult area to keep clean, particularly because they sustain such concentrated wear with people pivoting on them all day. Wisest solution is to have several extra sets that can be removed and washed—not shampooed in place, but thoroughly washed so that grit and dirt are cleaned out and the carpet has a chance to dry completely before being returned to the elevator.

Escalators in a busy convention hotel should be cleaned nightly (Figures 9-2 and 9-3). Tar and gum on the step-off plates need to be removed and the metal cleaned, railings dusted, and glass shields polished.

## *Restaurants*

Areas used for foodservice in the front of the house often display the most obvious need for immediate maintenance and housekeeping

(Figures 9-4 and 9-5). Waitresses, waiters, and busboys do, of course, clean off tables, change cloths, and keep the dining operations presentable during meal times. Some enlightened food-service managers even have busboys or service personnel make the rounds of the carpeted restaurants with a small broom and dustpan, or one of the new Japanese carpet sweepers that pick up debris almost by the magnetism of static electricity. But for the most part, dining area personnel are too rushed to monitor the carpet.

Picking up spilled bits of food immediately before they are stepped on or permeate the carpet pile will help immeasurably to preserve carpeting and make the vacuuming or shampooing job easier and more effective later. When spills of liquid or salad dressing occur it is extremely important to make a quick effort to dilute or dissolve the spot. If one or two busboys in the dining area are assigned floor care as a specific part of their duties, perhaps the life of the carpet will be extended.

Spotting kits are available from many manufacturers of carpet cleaning equipment. If these are kept handy, employees are much more apt to administer first aid to the carpet when it is on the verge of sustaining a permanent stain. Later that afternoon or evening, complete stain removal can be scheduled.

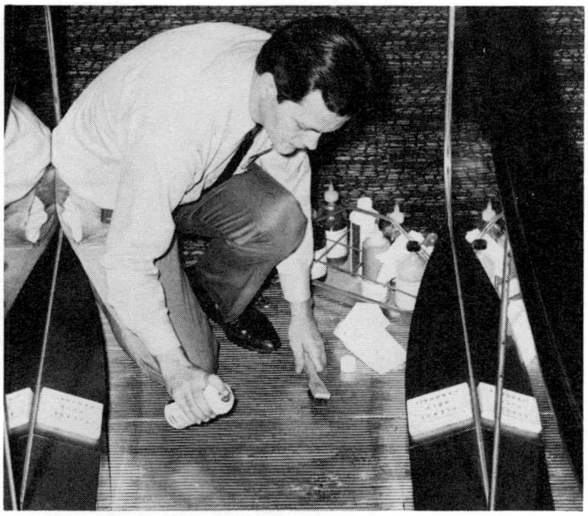

**FIGURE 9-3** *Chemicals that work like magic, plus a brush and lots of elbow grease, are required to remove tar and gum from escalator step-off plate; later, houseman polishes entire metal area—done nightly if escalator gets heavy use.*

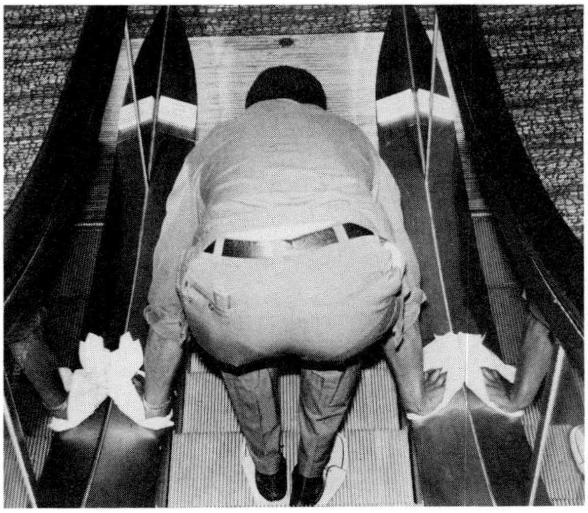

**FIGURE 9-2** *Cleaning escalator at 3 AM is a nightly procedure in hotels where heavy meeting business brings in crowds. To dust sides and rails houseman simply crouches, holds cloths, lets escalator move him down, and thus covers areas effortlessly.*

In many dining areas maids rather than housemen can do an adequate clean-up job, especially if two of them work together.

Jobs that should be done nightly when cleaning a dining area:

1. Pull out tables and chairs.
2. Remove crumbs on seats with a whisk-broom (primitive as it may seem, the time-tested and trusty whisk-broom is really the only tool that can quickly and effectively remove crumbs and dust and bits of food from a closed-back dining chair—especially one with arms that trap crumbs). Crevice tools on vacuums take considerably longer to do the same job.
3. Pick up large pieces of debris around each table before vacuuming to save wear and tear on vacuums.
4. Vacuum around each table.
5. Clean out tufted upholstery. On booth backs, especially those made of fabric, a piggy-back vacuum with a crevice tool is most effective.
6. Wipe window ledges or other horizontal surfaces, table posts, legs, and metal floor vents with a cloth dampened with all-purpose cleaner.

**FIGURE 9-4** *To clean restaurant, certain equipment should be carried on maid's or houseman's cart: trash bag, vacuum, cleaning chemicals in dispensers, spray-cans and squirt-bottles, gloves, and feather duster, and treated dust mop for small tiled areas or for woodwork.*

7. Wash vinyl booths or seats, bar stools, and bar front with appropriate cleaner.
8. Wipe hostess' or maitre d'hotel's desk.
9. Wash their phones.
10. Dust and polish metal chairs (chrome really shows fingerprints).
11. Polish oiled wood chairs a few dozen each night, so that within a week all in the restaurant can be completed.

**FIGURE 9-5** *Safe in foodservice areas when dispensing an insecticide, the steam-activated unit carries sprays up to 20 ft. so large areas can be covered in seconds. The unit can also deodorize or disinfect. Credit: Bishop Wisecarver Co.*

12. Polish foot rail and metal trim on bar.
13. Dust wood on chairs using chemically-treated cloth.
14. Spot-clean walls and if there is a counter, the front of it should be wiped off. (One assumes counter tops are cleaned by waitresses.)

The same piggy-back vacuum with crevice tool can be used to clean any stairs within the dining area and accessories, decorations, or light fixtures. Sometimes the lights on tables need to be washed or dusted. Many candle lamps are designed with glass chimneys that can be put into the automatic dishwasher.

## *Meeting and Banquet Areas*

Not only is Housekeeping often responsible for cleaning these areas, very often it is also that department's task to set up the rooms for meetings as specified by the hotel Sales Department or the university's Business Department.

Today, more and more business meetings with rather elaborate meals are being held on campus all year. During the summer few of them are related to academic activities; very often entire residence halls are turned over to meetings, seminars, or symposiums, some of which last one, two, or three weeks. Participants use the sleeping rooms and take all their meals in the residence hall dining areas.

With the large residence halls often kept reasonably full during the once slow summer periods, it is easier to keep year-round help, and to increase income for the Housing and Foodservice Departments so they can improve facilities or build new ones. It may also become possible to pay a better wage to all of the employees—especially the Housekeeping cleaners who are, unfortunately, lowest on the wage scale on campus, just as in hotels and retirement residences.

## *Setting Up for Meetings*

There is usually considerable flexibility as to how the facility can be set up. The customer who rented the room for his group will instruct the Sales or Food Department as to how many seats will be needed, whether it should be set up schoolroom style, semicircular, with or without tables, and so forth (Figure 9-6). Almost always a stage or raised area is involved, which means that housemen must take portable equipment out of storage and set it up according to a simple diagram. If the stage is an unsightly old structure, pleated skirting must be put around it, perhaps carpet tacked on top.

Tablecloths usually go on all of the tables because most folding banquet tables used for seminars, or the shallow ones reserved for single-side seating only, are unattractive or damaged or both, and covering them with a tablecloth or a removable colored felt cover is often the best way to disguise their condition.

There are on the market today, several folding tables that are very acceptable looking, not easily damaged, and that can be used "as is," thus saving the added labor and laundry expense of using tablecloths. They are relatively expensive and lack the elements of warmth and color that tablecloths bring to the usually somber meeting room. Corners of the tablecloths, incidentally, should be mitred (as on a bed) to keep them neat.

For draping or decorating speakers' tables as well as display and buffet tables, easy-to-attach skirting is available. This snap-drape has strips of grippers that fit onto plastic clips that can be put along the edges of the tables. This is a vast improvement over the pins and tacks customarily used.

A mobile drapery caddy has hangers arranged on a rack—each hanger bar can carry up to 18 ft. of snap-drape skirting. Since it is made

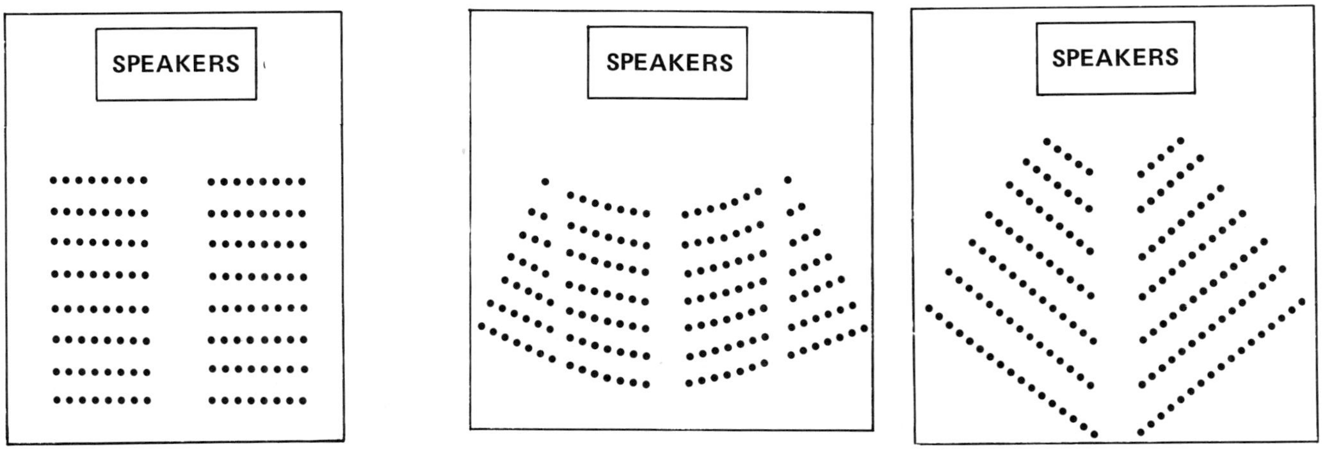


**FIGURE 9-6** *Seating Arrangements Starting with center aisle at front, arrange chairs in eight rows with eight chairs to a row. Total: 128.*

*In a square room, auditorium layout is best; start with center aisle and set up seven rows with seven seats across each side; then outside section has one seat at front, adds one seat as it goes back seven rows. Total: 154 seats.*

*In a rectangular room, seating can be arranged as above. To achieve this effect, starting at center aisle at front, angle chairs, starting with five and adding two chairs on each row running back as far as necessary.*

of polyester and fiberglass there is none of the old problem of wrinkled draping around tables.

### *Only the Most Fit Chairs Survive*

The stacking, folding, or ganging chairs that are used in busy meeting rooms are generally transported on dollies from storage. Since storage for these chairs takes so much room, it is becoming customary simply to leave the stacks in whatever sections of a ballroom or meeting room are not in use—thus saving labor because eventually many of them will be needed in one of those areas anyway.

Function chairs probably sustain more wear than any other furnishings in an institution because they are carelessly tossed about during set-up and dismantling when housemen display more speed than caution. Most are indeed designed to "take it," but the professional Housekeeper will keep a watchful eye on the chairs, noting places that sustain most wear, or whether a heavier vinyl or a different finish should be used, so that when it comes time to reorder a choice can be made intelligently. The matter of selection of function chairs is explained in Chapter 13.

Fire regulations in some states require that chairs set up theater-style to accommodate 300 or more people must be bonded together in groups of not fewer than three. This is so that people scrambling to leave a crowded meeting room in case of fire will not knock over chairs and cause tripping and obstructions. This requirement makes it advisable to purchase function chairs that have a ganging device.

Chairs have to be set up so that they measure 33 inches back to back, thus making the space between rows wide enough to get through easily. The aisles that lead to and from the stage should be a minimum of 44 inches wide, increased by 1½ inches for every five feet of travel to the farthest exit. As these regulations vary, it is important to check with local fire departments on the municipal code that is in effect within specific areas.

Curtains in places of assembly must be properly treated for flame resistance. In some states, they must be processed and tagged by state-licensed applicators unless the drapery fabrics are inherently flame-resistant (e.g., fiberglass).

Audiovisual equipment is especially important to those renting the facilities. Large screens, sound systems with microphones, lights that can be controlled, blackboards, recorders, gavels, perhaps even a projection room above—all are requisites today.

Details such as ashtrays, ice-water pitchers, glasses, and even pencils and note pads all become part of the responsibility of those who set up and dismantle meeting rooms.

Where these rooms have windows, it is especially important to keep them clean and sparkling. People sitting for hours (not always entertained or engrossed) may be unexpectedly observant of the condition of the glass.

### *After the Ball Is Over*

In some institutions banquet rooms and ballrooms require the largest, latest night clean-up crew. The areas may be cleaned by banquet housemen (employees of the Food and Beverage Department) or by Housekeeping staff. After banquet stewards remove all dishes and linens comes the job of sweeping up large pieces of debris with a long-handled whisk-broom (sometimes actually a toy broom) and breaking down the room, that is, stacking or folding the chairs, folding the tables and putting them on dollies to be stored elsewhere. Folding doors, airwalls, or room dividers might further complicate the job if the room will have to be ready the following day in another configuration.

Last comes the carpet-care crew with its "space vac" (a large vacuum with extra-wide 30-inch head). Perhaps a thorough vacuuming will take care of the floor, but more likely a damp-foam shampoo will be needed. That may be followed by a pile lifter that will keep the carpeting from looking dull and matted.

### *Doing Your Thing on the Chandelier*

A number of jobs must be performed periodically in an institution's public area. One that requires an inordinate amount of labor, planning, equipment, and supplies is the cleaning and relamping of chandeliers and light fixtures throughout (Figure 9-7). It represents a great investment in labor for it usually involves gathering together a large inventory of bulbs and setting up a high scaffold or long extension ladder (Figure 9-8).

When a chandelier has dozens of little lights it is more economical to change all of them at once rather than to set up cumbersome rigging every night to change one or two burned-out bulbs. (Used bulbs that are not burned out can sometimes be sold very reasonably to employees.)

Washing or cleaning any large crystal fixture in a high ceiling will present a problem, of course, but today many chandeliers are designed so that they can be lowered and raised on pulleys for cleaning and relamping. Cleaning a conventional chandelier in a conventional manner is often an all-night job, requiring a person with infinite patience to sustain him during the seven or eight hours spent at ceiling level with a large supply of soft rags and ammonia.

One enterprising, inventive maintenance man at the Century Plaza Hotel has devised a most effective means of cleaning a complex crystal fixture in a matter of less than an hour by using a fine spray (from the upholstery shampoo machine) with ammonia (Figure 9-9). The drippings are caught in a huge plastic structure like an inverted umbrella, suspended beneath the chandelier.

The Portman style of architecture such as in the Hyatt Regency in San Francisco and Chicago and Westin's Peachtree Plaza in Atlanta, has posed countless new challenges to Housekeeping and Maintenance directors. Plants of enormous scale, balconies covered with ivy, reflecting pools, canopies, giant decorative fabric panels, transparent elevator shafts—each of these exciting visual elements represents a total cleaning program—or means awarding a contract to an outside firm to care for them.

At the Bonaventure in Los Angeles, wind currents coming into the lobby caused unexpected problems in dining areas, in elevator shafts, with ashes in ashtrays, and the like. The benefits of such dramatic architecture (attracting crowds, especially) are offset by the many prob-

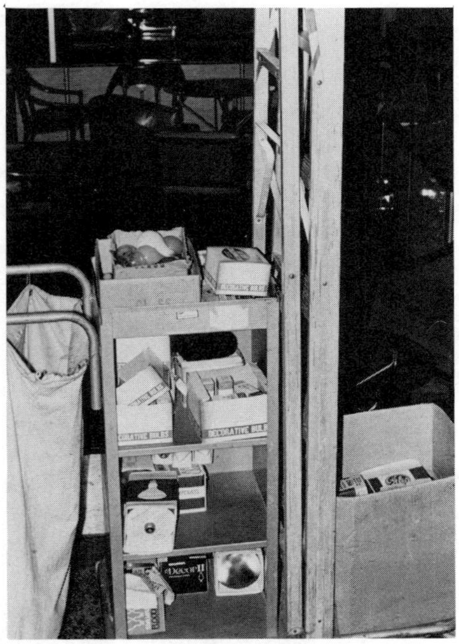

**FIGURE 9-7** *Relamping cart includes a wide assortment of bulbs, a trash holder for burned-out lamps (called burnouts), and a ladder that takes housemen high enough for most relamping jobs except those in lobby and ballrooms, where a special rig is needed.*

**FIGURE 9-8** *This special rig is devised to keep ladder upright while the houseman climbs thirty feet to start relamping lobby chandelier. While he is up there all bulbs are changed; taking this equipment out is something that cannot be done frequently for a few little bulbs in such a hard-to-reach location.*

**FIGURE 9-9** *Cleaning crystal chandeliers at Century Plaza Hotel is done by applying detergent solution with a sprayer nozzle connected to small compressor usually used for upholstery cleaning. Solution drips into plastic "basin" suspended by strings tied to stanchions.*

lems to be solved. Perhaps this is why hotel owners and operators resisted, for so many years, departing from traditional styles.

In the Bonaventure's fine restaurant, Beaudry's, chain-mail curtains hang around the circular booths. They must be kept clean and dust-free, as must all the plain shiny metal surfaces. This is handled by the assistant director of Housekeeping for Public Areas, who uses a vacuum cleaner attachment and periodic washing to remove grease.

## *Back-of-the-House*

In most large hotels there is almost as much back-of-the-house as there is public area. A rule of thumb indicates that there is an employee for every guest room. If the property earns much of its gross revenue from conventions, meetings, banquets, and group business in general, it will require considerably more to take care of all the extra activities. Where there are hundreds of employees behind the scenes, it is reasonable to assume that the areas in which they work, eat, rest, and go to the bathroom are all going to have to be cleaned. Thus the business of maintaining service corridors and elevators, employee washrooms and lockers, kitchens, storage areas, loading docks, and supervisors' offices becomes almost as time-consuming as cleaning the front of the house (Figure 9-10).

Unfortunately, management in too many institutions fails to allot sufficient budget to keep the back of the house clean. It is understandable that many employees become resentful, at least subconsciously, if their own areas are not cared for at the same time that they are expected to keep the lobbies, guest rooms, cocktail lounges, and even the elevators spotless. This kind of short-sighted management not only ruins employee morale with its double standard of cleanliness, but may also find itself the target of a zealous health department inspector. The well-managed establishments of the lodging industry keep their service areas in proper condition all

**FIGURE 9-10** *Back-of-the-house cleaning frequently involves floor care. Back corridors of a large hotel could add up to many miles, but heavy vacuums with wide heads can clean rapidly. Flat, tightly woven carpets make cleaning faster and quieter.*

the time so that guests or sanitation department personnel can be received there comfortably at any time.

To indicate the kind of carefully planned cleaning that should go on in the back halls, the following instruction sheet tells what is to be done in the service corridors:

### *Cleaning and Maintenance of Service Corridor, to be Performed in Order Listed Below*

Baseboards, walking surface, doors, door jambs, and elevator runners must be maintained nightly; handrails dusted nightly, washed as needed.

Spots, smears, splashes, hand marks, including dust build-up on elevator door jambs and elevator doors, must be removed nightly.

Air supply vents, light fixture globes, sprinkler fire-heads must be kept free and clear of dust and lint build-up.

Service elevator runners cleaned nightly, vacuumed and washed as needed.

**Proper Mixtures for Mopping Water:**
40 parts water to 1 part disinfectant cleaner

**Equipment Needed:**

| | |
|---|---|
| 2 buckets with casters | 1 straw broom |
| 2 wringers | 1 push-broom |
| 2 mops | 1 squirt-bottle |
| 1 putty knife | 1 scrub-brush |
| 1 gal. disinfectant cleaner | Assorted rags |

Every employee will be responsible for his own equipment and for keeping it neat and orderly. Return equipment to its proper place when leaving shift. Inform foreman if stand-by employee failed to leave your equipment clean and orderly after using it on your days off.

## *Buildings and Grounds*

To give the reader a complete check-list of all the possible areas and items that might have to be cleaned front- and back-of-the-house would be almost impossible, but we have tried to cover most of the common cleaning chores (Tables 9-1 and 9-2). If Housekeeping has responsibility for buildings and grounds, as it often does in a small property of 100 rooms or less, then such activities as the following must be scheduled and some sort of instructions written:

### *Special Detail = 11 pm-2 am*

(Housemen in uniform with properly shined shoes and name tag)

1. Dry-pick walkways and outdoor stairways, planters and landscaped areas.
2. Sweep walkways (sometimes outdoor vacuums are used in wide areas).
3. Sweep garages and parking areas.
4. Clean pool area—mop, sweep, or vacuum, and rearrange furniture.
5. Clean rest rooms.
6. Clean all cabana areas.
7. Clean storage and closets and custodian's closet.
8. Dust handrails along all walkways.

Landscaping and gardening are generally under the care of qualified professionals. It is enough for Housekeeping to see that litter is removed; exterior entrances and walkways are swept; outdoor facilities properly cared for; reflecting pools, fountains, and so forth are clean and working; all metal trim polished; and outside surfaces washed and/or painted frequently enough to preserve them and keep them attractive. The professional Housekeeper should not also have to be up on all the technical information required to revise soil content or understand the characteristics of all the vegetation on the grounds!

## *Kitchen Cleaning*

Although it may seem a far cry from the Housekeeping realm with its concern for fabrics and mending, making a bed, and reupholstering a velvet chair for the penthouse suite, competent

**TABLE 9-1.** Back-of-the House Cleaning

| Area/Equipment to be Cleaned | Cleaning Utensils | Cleaner/Sanitizer Dilution | Method |
|---|---|---|---|
| Trash receptacles, compactors | Long-handled brush, 1-cup measure, bucket | Institutional all-purpose cleaner and institutional bleach Dilute: 1 C all-purpose cleaner plus 1 C institutional bleach in 1½ gal. water | Daily: (1) wet long-handled brush with solution, (2) scrub with brush, (3) empty solution, (4) air-dry |
| Trash storage areas | Long-handled brush, bucket, 1-cup measure | Institutional degreasing cleaner and institutional bleach Dilute: 1 C degreasing cleaner plus 1 C institutional bleach in 2 gal. water | Daily: (1) to clean and deodorize, scrub area with solution, (2) air-dry |
| Ice-making machine | Bucket, sponge, measuring cup | Institutional all-purpose cleaner and institutional bleach Dilute: ½ C all-purpose cleaner plus 1 C institutional bleach in 2 gal. hot water | Weekly: (1) disconnect ice-maker; (2) fill bucket with solution of hot water, all-purpose cleaner, and institutional bleach, (3) pour solution into ice-maker and allow ice to melt; (4) wipe walls down with solution using sponge or cloth; (5) allow to dry and rinse thoroughly with warm water; (6) air-dry before reconnecting |

professional Housekeepers sometimes find their departments responsible for cleaning foodservice areas (Tables 9-3 and 9-4).

Of course, this requires optimum coordination with the executive chef or foodservice manager so that the level of cleaning and sanitation comes up to the standards of both department managers. They must be in solid agreement on what is to be done, how, when, and very important, by whom. It should be carefully spelled out just which areas the cleaners are to tackle and which to leave alone. Probably the insides of all equipment, insides of storage areas, and surfaces where any food containers are left would be off-limits to the night cleaners. But these are all matters for the managers involved to agree on.

The National Sanitation Foundation in Ann Arbor, Michigan, can provide detailed information on how to clean kitchens, foodservice areas, and equipment in case the professional Housekeeper is suddenly confronted with this responsibility, or is interested in enlarging her department. Seminars put on by NSF are conducted many times each year in all parts of the country.

**TABLE 9-2.** Exterior Cleaning

| Area/ Equipment to be Cleaned | Cleaning Utensils | Cleaner/Sanitizer Dilution | Method |
|---|---|---|---|
| Parking lots, sidewalks | Long-handled brush or broom, hose | Institutional degreasing cleaner Use full-strength | As needed: (1) pour solution on heavy grease and oil stains, (2) spread with long-handled brush or broom, (3) allow to stand 15 minutes, (4) hose down with water |

**TABLE 9-3.** Dining Area Cleaning

| Equipment to be Cleaned | Cleaning Utensils | Cleaner/Sanitizer Dilution | Method |
|---|---|---|---|
| Counter-tops, tables, booths, chairs | 1-quart bottle with trigger sprayer, 1-cup measure, sponge or cloth | Spray cleaner or institutional all-purpose cleaner<br>Dilute: ¼ C cleaner in 3¾ C water, fill quart-bottle with this solution | Daily: Spray and wipe with sponge or cloth; for all-day touch-up spray and wipe with institutional spray |

**TABLE 9-4.** Kitchen Cleaning

| Area/Equipment to be Cleaned | Cleaning Utensils | Cleaner/Sanitizer Dilution | Method |
|---|---|---|---|
| Floors<br>Heavy grease build-up | Long-handled scrub-brush, mop, bucket with mop press, tablespoon, 1-cup measure | Institutional degreasing cleaner and institutional bleach<br>Dilute: 2 C degreasing cleaner plus 2 C bleach in 3 gal. hot water | Daily: (1) scrub with brush until all soil is loose, (2) when dry-mopping use 2 T institutional bleach per gallon hot water and a clean mop and bucket, (3) air-dry |
| Floors<br>Light grease build-up | Long-handled scrub-brush, mop, bucket with mop press, 1-cup measure, tablespoon | Institutional all-purpose cleaner and institutional bleach<br>Dilute: 2 C all-purpose cleaner plus 2 C bleach in 3 gal. hot water | Daily: (1) scrub with brush until all soil is loose, (2) when dry-mopping, use 2 T institutional bleach per gallon hot water and a clean mop and bucket, (3) air-dry |
| Drains, grease traps | 1-cup measure | Institutional degreasing cleaner<br>Use full-strength | As needed: pour 1 C degreasing cleaner into drain every day until drain is free-flowing; to prevent bad odors use ½ C per week |
| Windows, glass doors, sneeze shields, mirrors | Hose, sponge, squeegee, cloth or paper towels, spray-bottle | Institutional window cleaner<br>Use full-strength | Daily: *Outside* (1) hose down, (2) scrub with sponge soaked in cleaner, (3) dry with squeegee *Inside* (1) wipe with sponge soaked in cleaner, (2) use squeegee, lint-free cloth, or paper towels to dry. *Small areas* (1) spray cleaner on 1-foot area, (2) wipe with clean dry paper towel or lint-free cloth |
| Stainless steel surfaces | Trigger sprayer, soft clean cloth | Institutional stainless steel cleaner<br>Use full-strength | Every other day: (1) spray stainless steel cleaner on surface, (2) wipe off (with the grain) with clean cloth.<br>Alternate days: wipe stainless steel with grain with clean soft cloth only. Initially if stainless steel has heavy build-up of film (oil, grease, |

**(Continued)**

**TABLE 9-4.** Kitchen Cleaning (Continued)

| Area/Equipment to be Cleaned | Cleaning Utensils | Cleaner/Sanitizer Dilution | Method |
|---|---|---|---|
| | | | grime, old polishes), remove with full-strength institutional degreasing cleaner using gloves |
| Grills | 1-qt. container with trigger sprayer or 1-gal. bottle with trigger sprayer and tube extender, rubber gloves, stiff brush, grill, pad, grill scraper, sponge | Institutional degreasing cleaner<br>Dilute: 2 C in 2 C water | Daily: *At night* (1) remove all food particles from grill with scraper, (2) turn off grill, (3) allow to cool for approximately 40 min or to approx. 140° F, (4) spray degreasing solution on grill, let stand overnight *In the morning* (1) spray on degreasing solution and brush, (2) wipe soil off with grill pad, (3) rinse and wipe grill with wet sponge and tap water |
| Pots, pans, utensils, trays | Brush with firm bristles, 1-cup measure, scraper | Institutional panwashing detergent and institutional bleach<br>Dilute: see Method | *3-Compartment sink:* (1) scrape loose soil and prerinse in sink #1, (2) wash out sink #1, fill with hot water and 2 C panwashing detergent, (3) scrub pots, pans, etc., with brush with firm bristles, (4) rinse in sink #2 with running warm water, (5) sanitize in sink #3 using ⅓ C institutional bleach per 5 gal. water; dip pots, pans, etc., at least 1 min., (6) air-dry *2-compartment sink:* (1) scrape loose soil and prerinse in sink #1, (2) wash out sink #1, fill with hot water and 2 C panwashing detergent, (3) scrub with brush with firm bristles, (4) rinse with running warm water in sink #2; stack pots, pans, etc., (5) empty and rinse sink #1 or #2 thoroughly; fill with water and ⅓ C institutional bleach per 5 gal. water; dip pots, pans, etc., at least 1 min., (6) air-dry |
| Soft ice cream–milk shake machines, milk dispensers | Stainless steel container for storage of milk shake mix, long-handled brush, bucket, 1-cup measure, tablespoon | Institutional all-purpose cleaner and institutional bleach<br>Dilute: see Method | At night before closing: (1) sanitize clean container with solution of 1 T institutional bleach per gal. water at least 1 min., (2) drain freezer contents into storage container, (3) turn off power to freezer but not to |

**(Continued)**

**TABLE 9-4.** (Continued)

| Area/Equipment to be Cleaned | Cleaning Utensils | Cleaner/Sanitizer Dilution | Method |
|---|---|---|---|
| | | | compressor, (4) rinse mix tanks and cylinder with cold water for at least 1 min., drain (5) close mix tank valve; fill mix tank with a solution of 2 C all-purpose cleaner and 1 C institutional bleach in 2 gal. hot water, (6) brush down interior walls, (7) drain solution into mixing cylinder, (8) turn on dasher switch, run for at least 30 sec., (9) drain solution into bucket, flush complete unit with solution of 2 T institutional bleach in 2 gal. water for at least 1 min., (10) remove mix tank valve, shaft seals, dasher, and bearings, scrub parts in bucket of all-purpose cleaner solution, (11) reassemble machine, (12) sanitize parts with 1 T institutional bleach in 1 gal. water at least 1 min., (13) air-dry |
| Steam tables, make tables, salad bars, food preparation areas | Damp cloth, gloves, tablespoon, bucket | Institutional degreasing cleaner or institutional all-purpose cleaner and institutional bleach<br>Dilute: see Method | For heavy grease build-up: (1) use gloves, (2) pour full-strength degreasing cleaner on damp cloth, (3) wipe down entire food contact surface, (4) rinse thoroughly with solution of 1 T institutional bleach per gal. water, (5) air-dry<br>For low grease build-up: (1) pour full-strength all-purpose cleaner on damp cloth, (2) wipe down entire food contact surface, (3) rinse thoroughly with solution 1 T institutional bleach in 1 gal. water, (4) air-dry |
| Drink dispensers | Tablespoon, 1-cup measure, bucket | Institutional all-purpose cleaner and institutional bleach<br>Dilute: 1 C all-purpose cleaner in 1 gal. water, dilute 1 T institutional bleach in 1 gal. water | Daily: (1) remove contents to refrigerated storage, (2) fill dispenser with water and drain, (3) fill with solution 1 C all-purpose cleaner per gallon water, (4) recirculate 2 min., (5) flush with clear water, (6) fill with 1 T institutional bleach per gallon water, (7) recirculate |

**(Continued)**

| Area/Equipment to be Cleaned | Cleaning Utensils | Cleaner/Sanitizer Dilution | Method |
|---|---|---|---|
| | | | 1 min., (8) flush with water, drain, (9) air-dry |
| Coffee urns | Long-handled brush, tablespoon | Institutional bleach<br>Dilute: 1 C per gal. water | Daily: (1) fill empty urn with institutional bleach solution, (2) scrub with brush, (3) empty urn, rinse with clear water, (4) drain |
| Toasters | Sponge, gloves | Institutional degreasing cleaner<br>Use full-strength | Daily: (1) always unplug electrical appliances before cleaning, (2) use gloves, (3) pour full-strength degreasing cleaner on damp sponge, (4) wipe down toaster, (5) wipe with clean damp sponge |
| Food preparation equipment, blenders, mixers, slicers, choppers, peelers, dough machines | Sponge or cloth, tablespoon, bucket | Institutional all-purpose cleaner and institutional bleach<br>Use all-purpose cleaner full-strength and 1 T institutional bleach per gal. water | Daily: (1) always unplug electrical appliances before cleaning, (2) pour full-strength all-purpose cleaner on damp sponge or cloth, (3) wipe down equipment, (4) rinse with clear water, (5) wipe, submerge or rinse equipment with institutional bleach solution for at least 1 min., (6) air-dry |
| Broiler chains | Brush with stiff bristles, cooking oil, soaking container | Institutional degreasing cleaner<br>Dilute: 2 C per 1 gal. water | Daily: (1) remove gross food particles with brush, (2) immerse chains overnight in degreasing cleaner solution, (3) rinse chains in clear water in morning, (4) lubricate chains with cooking oil (Note: under ordinary conditions, solution can be reused for approximately 2 weeks) |
| Mops, rags, brushes | Bucket, 1-cup measure | Institutional panwashing detergent and institutional bleach<br>Dilute: 1 C panwashing detergent in 2 gal. water, dilute 1½ C institutional bleach in 2 gal. water | Daily: (1) wash in detergent solution, (2) rinse, (3) soak 5 min. in institutional bleach and water, (4) rinse in clear water, (5) air-dry |
| Refrigerator-freezer | Bucket with mop press, cotton mop, sponge, wire brush | Institutional all-purpose cleaner and institutional bleach<br>Dilute: 2 C all-purpose cleaner and 2 C institutional bleach in 3 gal. water | Every other day: (1) remove all food particles from floors, racks, troughs, (2) using a sponge, wipe all food contact surfaces with solution of all-purpose cleaner and institutional bleach; let stand for at |

**(Continued)**

**TABLE 9-4.** (Continued)

| Area/Equipment to be Cleaned | Cleaning Utensils | Cleaner/Sanitizer Dilution | Method |
|---|---|---|---|
| | | | least 1 min., mop floors with same solution, (3) mop down drainage trough with solution, let stand 5 min., (4) rinse thoroughly with clear water, (5) air-dry all surfaces by leaving door open, do not let freezer temperature exceed 45°F |
| Ventilator filters | 1-cup measure, soaking container, bucket | Institutional degreasing cleaner<br>Dilute: 2 C per 1 gal. water | As needed: (1) soak filters overnight in solution, (2) rinse in clear water, (3) air-dry |
| Deep-fat fryers | Gloves, goggles, vinegar or pickle juice, paper towels | Institutional deep-fat fryer cleaner<br>Dilute: 1 packet per 2 gal. water | Weekly: (1) drain fryer, (2) fill with warm or hot water, (3) leave baskets in unit, (4) add 1 packet per 2 gal. water, more cleaner if fryer is heavily soiled, (5) bring to slow boil, (6) for maintenance cleaning, boil 15 min., for heavily soiled units, boil 30 min., (7) drain solution, (8) to neutralize again fill completely with water and add 2 C vinegar, (9) heat gently to 150°F 5 min., (10) drain and dry unit to prevent grease splattering |
| Ovens—inside<br>Stack, broiler, convection, warming | Gloves, trigger sprayer with 3-ft. extension tube, sponge | Institutional oven cleaner<br>Use full-strength | Weekly: use gloves, (1) Preheat oven to 200° F, (2) turn oven off, (3) spray foam uniformly over areas to be cleaned, starting in back of oven, working forward, (4) Wait 10 to 15 min. depending on how heavily soiled oven is, (5) wipe oven clean with damp sponge, rinsing sponge frequently to remove all dirt and chemicals |
| Ovens—outside | | Institutional degreasing cleaner, institutional stainless steel cleaner<br>Use both full-strength | Weekly: Use gloves, (1) wipe surface with degreasing cleaner on damp sponge, (2) rinse off with water, (3) spray stainless steel cleaner on surface, (4) wipe off (with the grain) with clean cloth |

There are many properties that have old kitchens, which though kept clean, look gloomy or at best colorless. Rockresorts, however, is one hotel chain that makes a point of showing off its shining kitchens. Operating prestigious resorts, it has long had the policy of kitchen tours for guests. Tour time may be 2 or 3 pm, after the luncheon buffet and before dinner preparation. At that time, guests are taken on a full tour so they can see the immaculate working and storage areas. This gives the kitchen staff a tremendous morale boost.

Probably nowhere in a hotel or restaurant is there a greater potential source of fire than in the cooking areas of the kitchen. The exhaust flues in the hoods above the main bank of equipment are the critical danger areas, so removing any build-up of grease and dirt and *keeping* it out is the number one consideration (Figure 9-11). It is equally important to make sure the exhaust system is in perfect mechanical condition. If the system is particularly troublesome outside specialists should be called in to put it safely in order.

In modern kitchen installations the ventilating systems have removable filters that collect grease and dirt (Figure 9-12). They must be cleaned regularly. Disposable filters are also available for certain systems.

**FIGURE 9-11** *Lifting fork facilitates removal of grease filters for cleaning, without need for ladders or climbing on kitchen equipment. Fork of chrome-plated steel is 6 ft. in length. Credit: Flame Guard*

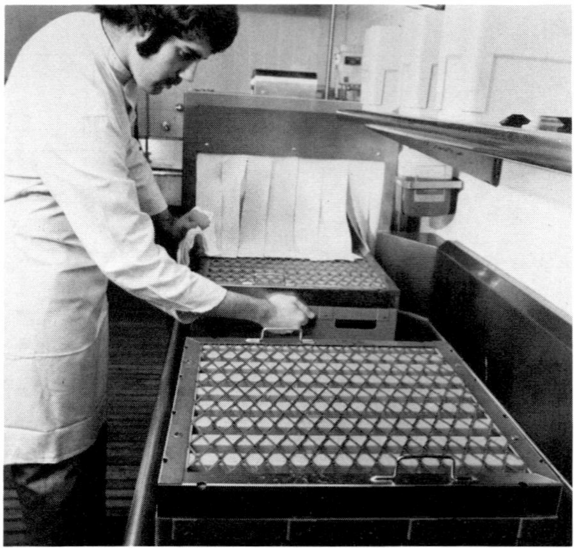

**FIGURE 9-12** *When grease filters need to be cleaned, they are removed from hood and run through dishwasher; the filter's Teflon-coated baffles are easily washed. Credit: Flame Guard*

A second danger area in a kitchen is the floor (Figures 9-13 and 9-14). If it is slippery from food spills and grease, it may cause employees to fall. As spills can occur accidentally and cause a slippery condition within minutes after a floor has been thoroughly cleaned, it is necessary to put some sort of non-skid covering over the area immediately in front of the battery of cooking equipment.

Thick, open-weave rubber matting is often used, but duckboards are most popular as a safety floor covering in kitchens (Figure 9-15). The traditional duckboards are wooden slats wired together so they are flexible, and can be easily rolled up and removed from kitchen areas for cleaning. They are scrubbed and hosed outside near a drain, while the kitchen floor, usually terra cotta tile, is being scrubbed (Figure 9-16). This may be performed by a man with (1) a mop and two pails; (2) a scrub-brush and two pails; (3) a mop and one pail followed by a wet/dry pick-up with high-pressure hot water spray units, or (4) with built-in live steam cleaning equipment.

Where any sort of chemical cleaner is used on kitchen floors, it is usually a detergent sanitizer (also called a germicidal detergent or a disinfec-

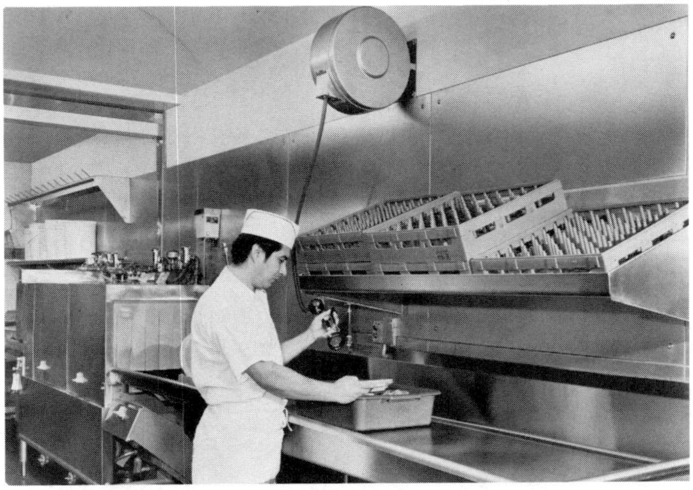

**FIGURE 9-13** *This retractable hose reel and spray, useful for a general washdown of kitchen equipment and floor, has 35 ft. of high-pressure hose to permit continuous use of 180°F water. Spray valve has a squeeze-lever dish guard and insulated handle that remains cool. Credit: Fisher Mfg. Co.*

tant cleaner). The exact type to be used depends on the local water, types of soil, whether deodorizers are needed (Table 9-5). These technical problems should be resolved by an expert, because careless use or improper application of strong chemicals can cause serious problems.

There are sanitation professionals today who are qualified to provide a complete consulting service on all aspects of cleaning and sanitizing the back of the house. They will set up a proper program and train the personnel if their particular line of products is to be used. The companies

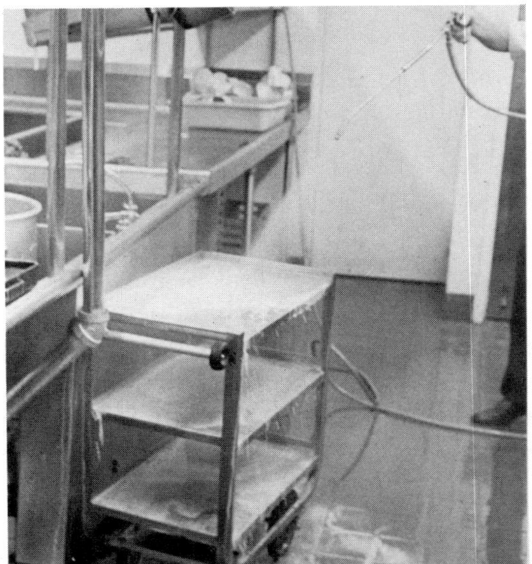

**FIGURE 9-14** *Equipment such as stainless steel cart can be quickly cleaned, sanitized, and rinsed while floor is being washed. Spray nozzle options make different coverage possible—from straight jet for behind ranges to 40°-wide spray for wall washing.*

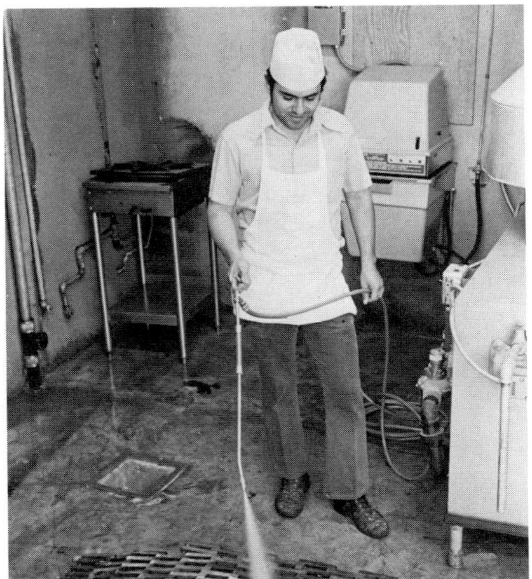

**FIGURE 9-15** *High-pressure spray-cleaning equipment is effective for cleaning duckboards or other nonskid mats in kitchen areas. Power unit is on wall behind operator. Hand control regulates amount of cleaning solutions and water spray. Credit: Sage System*

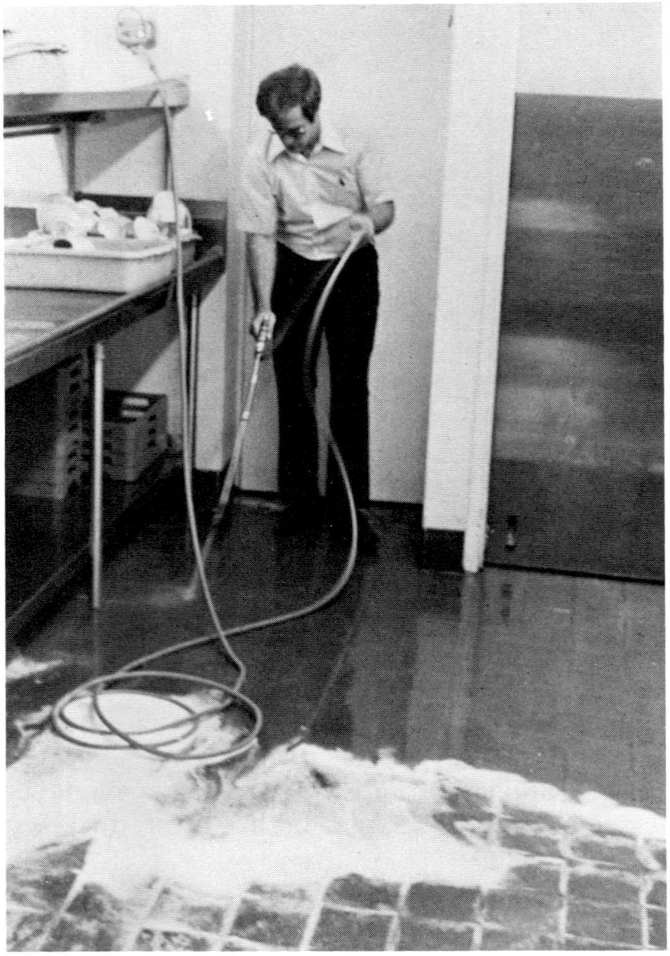

**FIGURE 9-16** *Remote stations (small unit on wall at left of operator) permits use in many areas in and around kitchen up to 300 ft. from central unit. Quarry tile floors are easily washed with pressure spray that has a scouring action.*

that produce cleaning compounds are so competitive and have developed such high standards and fine products that subscribing to their services is hardly risky or extravagant. These companies have done more to upgrade the roles of dishwasher, janitor, kitchen steward, and porter than any other force in the industry.

## *Walls and Ceilings*

Walls and ceilings in some kitchens will need to be washed as often as weekly or as seldom as semiannually, depending on kinds of cooking equipment used, ventilating systems, and types of food prepared. Because grease and steam from the cooking and dishwashing areas combine with normal dust on the walls, it is necessary in almost every case to clean these walls monthly.

On lower areas of the walls, splash marks may have occurred from spills or from sloppy floor mopping. Walls in modern kitchens today are often of tile or stainless steel that can be washed far more easily than painted or vinyl-covered surfaces. Stainless steel requires a special type of cleaner because unfortunately, it does not live up to its name. Although it may not be possible *permanently* to stain the soft lustrous metal, it can become discolored from constant spillage and it does show every fingerprint, every spot of grease. Mild acidic products can renew its finish, but must be applied with great care as they can quickly etch or discolor surfaces if used improperly.

Light fixtures in kitchens require far more frequent attention than those in other parts of the institution because grease and dust will combine to form a rapid build-up of soil; then steam

**TABLE 9-5** Chemical Cleaners for Foodservice Areas

| Product | Benefits | Disadvantages |
|---|---|---|
| All-purpose cleaner | Less inventory needed; housemen need fewer supplies | Must be diluted differently for each surface |
| Cleansers | Will remove heavy soil, grease, and rust with minimum of effort | Scratch stainless steel and fine porcelain |
| Degreasers | Heavy alkali content; cut grease quickly | Highly irritating to skin |
| Deodorizers | Effective where lingering unpleasant smell cannot be removed even after cleaning | Leave film and residue and odor unless used correctly |
| Dishwasher detergents | Specific formulas available to suit water conditions at each institution | Require careful measuring and following of instructions or may leave dishes with film, spots, grease, etc. |
| Disinfectants (quats) quaternary ammonium compounds | Long residual bacteriostatic effect; odorless, nontoxic | Do not control TB; effectiveness varies depending on hardness of water; may not combine with detergents |
| Drain cleaners | Provide fast action in emergency when no plumber is available | May contaminate foods; irritating to skin; may damage plumbing or make repairs difficult |
| Hand-dishwashing detergents | Perform in both hard or soft water | To be effective on grease may be hard on hands; gloves needed |
| Insecticides | May be applied to affected area quickly; insects react immediately | Residue may contaminate foods; insects build up immunity |
| Metal cleaners and polishers | Remove discoloration and soil easily if used regularly | Leave white film unless used correctly |

will combine with this and cause streaking. If this sort of build-up is not frequently removed, then the stronger cleaners that dull the metal and make it even more likely to collect soil will have to be used.

See Table 8-1 for instructions on cleaning heat lamps.

## *War on Pests*

If the foodservice workers keep food containers and trash cans covered it will help to discourage flies, roaches, and rats—the three biggest threats to kitchen areas. Even if all precautions are taken by cleaners and kitchen personnel, it will doubtless still be necessary to use some kind of pest control.

### *Insects and Rodents*

If the situation is serious—and even one rat can indicate a nest with 20 more nearby—a professional exterminator should be called in. Although his services may not seem needed after the initial extermination, it still is important to continue a control program.

If employees have an indifferent attitude about flies, roaches, or rodents it might be a good idea to show them a film about where these enemies come from and how they spread diseases. Employees then may apply insecticides with renewed interest.

If an institution has had any instances of infestation, it is wise seriously to consider setting up a contract with an exterminating company that will give regular weekly, monthly, or quarterly service.

According to Alan Caruba, an expert consultant on the business of pest extermination, "Since World War II, the pest control industry has become increasingly sophisticated and a wide array of materials and techniques now exist to respond to pest-control problems."

It is still very much in the Housekeeper's interest to know how to select a good pest-control operator (PCO) and determine what must be done to stem the invasion. With the knowledge that this is an on-going expense, getting your money's worth begins with the selection of a good pest-control firm. While there is a growing number of regional and national firms, most are small independent outfits.

At the local level, selection should begin by asking a friend. Executive Housekeepers tend to keep in touch with one another, so find out what company others use and if they are satisfied with the service. Ask for references. Any PCO doing his job will be able to name several hotels or residence halls.

It would be a serious error to make your decision based on price alone. The pest-control industry is a highly competitive one according to Caruba, and is likely to become more so in the coming years. What you will be paying for are service and results.

Chemicals and other items PCOs use represent only about 5% of their operating expenses. Trained personnel is the key element. The PCO is selling a detailed knowledge of where the pests are, how they gain entry, and how their access to your operation can be eliminated and controlled. The wise Housekeeper will not hesitate to spend a few dollars more for a PCO who will schedule regular inspections and provide 24-hour service.

Although much pest control has to do with good sanitation practices in the kitchens, pantries, and storage areas, there are some older buildings that no matter how well cleaned and maintained they are, will inadvertently become hosts to roaches, silverfish, moths, mice, and rats.

Whenever you are storing dry goods—paper or linens—the glue on the boxes can attract roaches no matter how clean your premises. So you see, even the innocent are invaded by the omnipresent roach. Housekeepers should work as partners with their PCO, as he cannot achieve effective control if the hotel personnel defeat his best efforts. Cooperation is essential.

Consider the places where insect pests can set up shop; under counters, false bottoms of booths, in and behind drawers, in the hollow posts of tables, in cabinets and behind them, in the support legs of stoves and refrigerators, around sinks and toilets, and all the cracks and crevices providing access to the void areas behind walls. Heating and air-conditioning ducts, wall and ceiling exhaust fans, and the entire electrical wiring system are also suspect places.

Be prepared to ask questions. Everyone knows that pesticides and rodenticides are potent items. The pest-control industry is answerable to the Federal Insecticide, Fungicide and Rodenticide Act, which, among its many provisions, requires all PCOs and their staffs to secure certification from their individual states, according to expert Caruba. Therefore PCOs must ensure that their technicians receive training in the proper use of their materials.

There are three basic chemical groups from which pesticides are manufactured. The most famous, DDT, belongs to the group known as chlorinated hydrocarbons, and these chemicals are generally banned from use with the exception of chlordane that is used to exterminate termites.

The second and most popular group is the organic phosphates. These are invariably applied with a fine oil spray or a fogging process called micron-generation that converts a liquid chemical insecticide into a fog of fine particles less than one micron in diameter. Precautions must be taken to make sure the fog doesn't contaminate areas for food preparation, storage, or service.

Certain strains of cockroach and other pests are not affected by these insecticides quite the way they were several insect-generations ago.

Carbamates, the newest chemical group, appear to hold the most promise. They have the added safety factor of usually being wettable powders that mix with water, and thus when applied, are less likely to damage woodwork and other structural elements. Moreover, they do not become volatile and carry on air currents. There is the added advantage of little or no odor.

Residual is the key word, along with safety, because a pesticide that does not go on killing long after it is applied isn't worth much.

All pesticides have special qualities that either enhance or limit their appropriate use in a foodservice environment. One needs to learn the advantage of dust over sprays, fogging over crack-and-crevice treatment, baits and traps over glue boards.

Technical data in booklets and brochures are available if you want to do your homework on the subject, and it is recommended that you do. Write to The National Pest Control Assn., 8150 Leesburg Pike, Suite 1100, Vienna, Va. 22180.

What you don't want to do is hire a pest-control firm and walk away. You don't want to insist on instant proof that the cockroaches and other pests are dying on the spot.

You do want to cooperate with your PCO, based on a regular schedule of service. Cooperation also means being responsive to his recommendations concerning the areas in and around your establishment that invite or encourage infestations of insect or rodent pests.

This can involve relatively minor maintenance such as sealing holes in the building, screening windows, caulking crevices and cracks, and many other sanitation measures without which the best pest-control techniques are rendered ineffective.

Having provided cooperation asked of you, look at the results. Give it time, but if you still find roaches and other pests in your establishment, after conferring with your PCO be prepared to question the pesticides he's using, whether he is doing so with sufficient regularity, and what alternatives exist.

Remember, roaches and other pests multiply prodigiously, and failure after a reasonable period of time to control them is your legitimate excuse to determine whether or not you are getting your money's worth.

When a property starts any kind of remodeling or additions, one will notice insects appearing. For instance, one hotel had a distressing problem with bugs and insects because the streets in front of the hotel were torn up. This dislodged insects and sent them scurrying for some place that was warm; they chose the hotel basement.

One of the circumstances that usually leads to the Housekeeping Department's having to locate a rather innocent insecticide occurs when bugs discover the outdoor dining and service areas where bits of food inevitably are spilled. Then the problem is to get a fast, effective means of eliminating the pests without any danger of contaminating food.

Electric lanterns that are large enough to cast considerable attractive-to-bugs light at night are a most appropriate solution because insects are cremated instantly within a hot chamber as soon as they fly into the element. There is not even a refuse holder to empty.

A variation of this safe convenient process is also possible in the daytime, and it eliminates the dangers of sprays in the hands of irresponsible employees and around food.

An electric fixture for flying-insect control can be wall-mounted in the areas where insects, especially flies, congregate. It has a light source within the heating chamber that attracts insects, which are destroyed on contact with an electrified grid, and fall into a clean-out tray. The unit can be moved from place to place and there is no installation cost. (Photographs of this unit are in Chapter 27 on Health and Safety.)

## *Other Unwelcome Pests*

Termites, beetles, and fungus are decidedly enemies that should be dealt with by professionals. To help you justify the expense of summoning that service, you might do some checking on certain areas that are especially vulnerable:

1. Gate posts, fence tie-ins, etc.
2. Meter boxes
3. Bathroom inspection doors
4. Ventilation (adequate or insufficient)
5. Planters and posts
6. Windows and sills
7. Door sills, casings, and jambs
8. Frame siding
9. Shingle roofs; eaves and facia boards
10. Faulty grades, cracks in foundation
11. Balconies and landings
12. Abutments, columns, pilasters, etc.
13. Loose stucco

The most frequently encountered problem areas are:

1. *Earth-wood contact.* Wherever wood (a gate post for example) rests on the ground, it provides a means for subterranean termites to come up between the post and the building.
2. *Faulty grade.* If the exterior earth grade is less than two inches below the foundation of the structure the mudsill is too readily accessible to subterranean termites.
3. *Wood siding* or vent frames and earth contact.
4. *Faulty grade* obstructing ventilation.
5. *Meter box.* Important inspection point will frequently disclose the existence of drywood termite infestation.
6. *Faulty grade and insufficient ventilation.* Check especially where a planter is above the foundation, or the soil level in the planter has cut off ventilation.
7. *Another planter situation.* Earth contact and faulty grade at the posts in the patio planter.
8. *Outside post.* If one penetrates below the concrete slab into earth and supports wood attached to building termite infestation moves up inside the post, across the roof, and into attic eaves.

In tropical locations regular termite inspections and preventive treatment are essential.

## *Waste Handling, Whither?*

Investments in waste-handling systems and equipment are often sizeable and require careful study and ultimate approval by management. The professional Housekeeper should be aware of the many new ways to handle waste, and be ready to suggest some of the better solutions when the time comes to confer with other department managers about new systems for the entire facility.

Terminology is confusing but these are definitions established by health organizations.

1. *Solid wastes.* Any item to be disposed of that is not normally disposed of through the plumbing system.
2. *Trash.* A mixture of paper, cardboard, cartons, plastics, disposable "linen," wooden boxes, dunnage, furniture, cans, bottles, glass, and all general wastes.
3. *Garbage.* Wastes from preparation, cooking and serving of food, generally wet waste.
4. *Foodservice waste.* Includes garbage and all kitchen, scrapping, paper, plastic, wax containers, solid and liquid food. Does not include cartons and crates; these are part of trash.
5. *Refuse.* General term for a mixture of trash and foodservice wastes.
6. *Swill.* Liquid waste consisting principally of food preparation wastes.
7. *Street refuse.* Sweepings, dirt, leaves, salt to speed snow melting, contents of litter baskets.
8. *Construction wastes.* Building materials–metal partitions, lighting fixtures, fluorescent tubes, sheet-rock, masonry.
9. *Residue.* Result of incineration, including ashes, cans, bottles, clinkers.
10. *Fly ash.* Solids of ash, charred paper, cinders, dust, or other partially incinerated matter.

When institution management analyzes the composition of its wastes, it is then in a better position to investigate the best equipment or system to cope with that special mix. Whether bags, carts, conveyors, gravity or pneumatic chutes, compactors, shredders, incinerators, food grinders, or pulpers should be used is probably a decision that should be made by a professional engineering service. Meanwhile, until the ideal system is devised, the germicidal cleaners, deodorizers, and insecticides on the market can keep trash areas clean and odor-free.

Pressurized portable spray equipment that can quickly cover refuse areas and loading docks will clean and disinfect with very little labor. If an area has been neglected and a build-up of soil has occurred, it may be necessary to hire a firm to come in and thoroughly steam-clean and sanitize so that employees can then maintain refuse areas with regular procedures.

# Chapter 10

# Hard Floors: Kinds, Covers, Care

**OVERVIEW** More man-hours go into floor care than into any other single pursuit in the Housekeeping Department, so planning appropriate flooring installations and how to clean them with different types of equipment and chemicals is vitally important. How to schedule the labor is explained here, and a detailed procedure for mopping is included. The format can be followed for explaining and teaching other floor-cleaning techniques as well as many different tasks.

No other departmental function is of greater concern to the professional Housekeeper than floor care. Nor does any other function offer as bewildering an array of options. The variety of choices is almost literally endless and constantly changing. The subject cannot be covered adequately in one or a dozen books, much less in one or several chapters. To keep abreast of the kaleidoscopic changes in this field requires continual involvement on the housekeeper's part.

In a new structure, the flooring may be cement, concrete, or wood. Where the area will be used as a loading dock or for warehousing, cement will suffice. Wood floors are occasionally left uncovered, but only when the quality and character of the planking or parquetry are compatible with both physical and aesthetic considerations. In most cases, however, all or part of the raw flooring will be carpeted (a complex subject covered in the next chapter) and/or surfaced with coverings as plebian as linoleum, as patrician as marble, or as popular as vinyl.

For practical purposes, it is merely necessary to note the availability of such exotic floor coverings as marble, brick, stone, slate, ceramic, and so forth. All of these uncommon materials are lacking in the most important characteristic of the hard-floor–covering industry—resilience. Resilient floor coverings are of many different types, which in turn cover wide spectrums of quality and price. They are available in every conceivable color, as well as in a seemingly endless array of textures and patterns. They come in 9-inch or 12-inch squares (commonly called tiles), or in rolls that may be 6, 9, or 12 feet wide and 90 feet long, referred to as sheet goods.

### Continuous Floors

Resilient floor coverings are also available in a semiliquid form that may be premixed and purchased in cans, or may be formulated on the site. The seamless, one-piece poured floor is most impressive when it covers a large, relatively uncluttered, well-lighted public room.

### Epoxy

Epoxy is a continuous flooring material that has been in use in institutions for over 20 years, but is usually confined to areas such as loading and warehousing, storage, and other areas of heavy-duty traffic. In their early formulation, epoxy floors were usually pigmented a solid color and could be aesthetically incompatible with other areas of the same building.

Recently, breakthroughs in product formulations have seen the development of transparent, low-odor epoxies. When combined with decorative quartz granules or flat, multicolored chips, they form a seam-free, highly decorative floor. Generally, they are stable, resist cracking, will not spall (lose chips out their surface), and are dustless.

With a compressive strength of over 8000 psi, they can provide excellent wear or abrasion resistance (usually at least two or three times that of concrete), as well as high chemical resistance. In addition, they lack the seams found in sheet floor or composition tile.

This new breed of epoxy flooring opens up a wide area of applications in institutions. With proper planning and application, a seam-free, slip-resistant, and disinfectant- detergent-resistant surface can be designed for hallways, lobbies, reception areas, walkways, and locker and shower areas.

Another maintenance advantage is that an epoxy floor has the ability to form a cove up the wall surface, thereby eliminating the seam between the floor and wall.

### *Terrazzo*

Terrazzo is classified somewhere between resilient and hard. It is a continuous floor, laid by a contractor, comprised of 70 percent exposed marble chip and 30 percent portland cement (Figure 10-1). If sealed properly (and devoid of cracks) it will stand up well under continuous exposure to moisture and heavy foot-traffic.

It is subject to damage by alkaline or acid cleaners, will stain easily, and unless suitable wax is used, may become somewhat yellowed. Despite these limitations, terrazzo is handsome (available in several marble chip colors), relatively inexpensive, and if properly maintained, *very* long-lasting.

## *Non-Continuous Flooring*

### *Wood and Cork*

Wood and cork are the most resilient of all hard floor coverings, but together with linoleum they offer the least resistance to impregnation by water and other liquids. This severely limits their applications, especially for on-grade, or below-grade installation. Raw wood deteriorates rapidly, and must be kept coated with nonporous sealers at all times.

Some simulated wood flooring products on the market today give what advertising calls "the unmistakable look of wood," in either parquet or plank form. This resilient elastomer material has a texture, color, and grain that is very convincing as a wood surface, yet it is resistant to stains and cigarette burns, and has a durable topcoat polymer to reduce need for waxing. Although it is not recommended for busy restaurant areas, it is an easy-care flooring that, unlike real wood, tends to dampen sound vibrations.

Resin-reinforced waxed cork tile and vinyl-coated cork extend the practical applications for cork, but do little to cut the cost or difficulty of maintenance. The high cost of the product is justifiable only where there are compelling aesthetic considerations.

### *Linoleum and Asphalt Tile*

Linoleum offers good durability and fair resilience, but so little resistance to water, even in the form of high humidity, that its use would seem warranted only when its installed price is relatively low.

Asphalt tile, despite its continued low price, has lost much of the popularity it once enjoyed. When it is cold asphalt tile becomes brittle enough to crack, while it becomes so soft in the presence of extreme heat, grease, solvents, and so on, that permanent indentations present a serious aesthetic problem. The product also tends to lose its sheen and pattern more quickly than competitively priced products.

### *Rubber Tiles*

Rubber tile has excellent cosmetic qualities and outstanding resilience. The product is rugged, waterproof, and easy to maintain. When dry, it offers almost complete safety from slipping.

New breakthroughs in flame-retardant flooring include both real and synthetic rubber tiles,

**FIGURE 10-1** *Terrazzo floors are most often found in spacious lobbies, as they last for decades and are easy to maintain if proper systems are followed. Terrazzo is a marble aggregate mixed with portland cement poured over mortar base into which a grid of metal dividing strips are set. The strips control cracking due to heat or cold.*
*Credit: United California Bank*

**FIGURE 10-2** *Although this is a real marble floor in the coffee shop of Stanford Court Hotel, San Francisco, vinyl asbestos floor tiles today are made to resemble this type of marble square with its diamond-shaped insets.*

one style with raised circular discs for better traction, and one with convex discs. Both are nonstaining, nondiscoloring, and fully homogeneous throughout. The tiles are 24-inch squares, and permit quick installation with few joints—important for cleanliness. Circular discs, both concave and convex, allow for easy cleaning and for easy movement of wheeled equipment. Ideal for ramps, entrances, back elevators, and hallways, these floor coverings have proved to be an ideal substitute for terrazzo or finished concrete floors. They meet the ASTM-E84 Spread Rating of 25 or less.

## Vinyl Asbestos

A close relative of asphalt tile, and with many of the same disadvantages, vinyl asbestos tile has nevertheless increased in popularity so greatly in recent years that it now outsells all competitors (Figure 10-2). It offers little resilience, marks easily when subjected to heavy loads, and has almost no sound-deadening qualities. On the other hand, it keeps its good looks with just normal maintenance. It offers strong resistance to oil, grease, acids, and alkalis, and may therefore be used on-grade or below-grade, outside, in the kitchen, and virtually anywhere else. Priced somewhat above asphalt tile, vinyl asbestos ranks as a real bargain.

### *Solid and Sheet Vinyl*

Solid vinyl plastic tile made of 100 percent homogeneous vinyl is the top-rated resilient tile on almost every count, and is priced accordingly. It has a high-gloss finish that is easy to maintain and almost impervious to attack. Highly resilient, its static load limit is 200 pounds per square inch.

Sheet vinyl, which has a price range as wide as its quality range, is rapidly moving up in popularity, largely because it ranks high in sanitation, ease of maintenance, and installation. In evaluating sheet vinyl, it is important to determine whether the pattern is inlaid or surface printed, and the depth to which it penetrates. The type of backing should also be investigated. Some types of backing make it possible to install sheet vinyl over an existing flooring material in poor condition. Other backings provide no defense against markings, thus requiring complete removal and thorough sanding of a rough existing surface before installation.

One vinyl sheeting specially formulated of PVC (polyvinyl chloride) resins is made to withstand wear, abrasion, sun, and water, so is usable outdoors on balconies, patios, and rooftops. It has elasticity, is resilient, and is highly sound-absorbent. Easy to maintain, it should attain considerable popularity if threatened shortages of PVC do not limit its production and escalate its price.

### *Quarry Tile*

Although not a resilient floor, quarry tile is becoming increasingly popular. It is frequently found in kitchens because, having been fired at high temperatures (as is ceramic tile), it will withstand endless high-pressure or hot steam cleanings, and food spills will not affect its surface. In public areas—lobbies, foyers, and restaurants—it is used in its usual terra cotta color or a gold ochre tone. If sealed and buffed or polished frequently, it can be extremely attractive (Figure 10-3).

**FIGURE 10-3** *To prevent slipping on kitchen floor these treads are available, precut to 5 ½-in. squares for application on 6-in. quarry tile. Treads provide sure footing despite oil or other spills, helping to meet requirements set by the Occupational Safety and Health Act (OSHA).*

### *Cleaning Hard Surfaces and Resilient Floors*

Not many years ago, hard surface and resilient floors were swept, scrubbed, rinsed, dried, and waxed. Then perhaps they were polished with a revolutionary new labor-saving buffing machine. Daily sweeping that raised the dust, or dust mopping, which moved the dust around, were traditional, usually followed by damp-mopping. Then a few wet-moppings would suffice until it was time to strip off all the wax and start over. Soap and carnauba wax were the standard supplies; mops, the primary tools.

Chemists and engineers, urged on by the makers of maintenance products, stepped in to improve and simplify floor care and ease the drudgery. Result: incredibly versatile machines abound, miraculous chemicals proliferate. At the same time, however, the abundance of systems devised has so complicated the whole business of floor care that a neophyte in the housekeeping field is almost invariably puzzled, not just by the array of supplies and machinery, but by the basic disagreements concerning floor cleaning that exist even among the so-called experts.

During the 1970s there was a gradual change in the whole approach to floor care; the scrub-and-wax supporters were challenged by the proponents of spray-buffing with polymer finishes. Makers of vinyl floors and installers of terrazzo further puzzled us by suggesting that no protective finishes were needed. We were caught up in the confusion of change.

If floors and corridors in office buildings were the only areas to be concerned about, it might be possible to be married forever to the spray-buff theory. But in institutions with foodservice, maid service, bathrooms, locker rooms, and loading docks, there is no simple solution. In the next decade, chemistry and technology may give us the permanent floor finish.

Despite such futuristic machinery as the electrostatic precipitator to attract dust, and magnetic forces or sonic waves to clean the decks, one thing we can be sure of is that dust, dirt, and human carelessness will continue to make floor care a major industry.

**TIPS ON FLOOR CARE**

For waxing, use 16-oz. wet-mop, preferably nylon or rayon, not the kind with rope-type strands.
New mops should be soaked *to remove sizing before being used to apply wax.*
When applying finishes, two thin coats are better than one heavy one.
When waxing, avoid walls, cabinets, and fixed objects. Stay 6 to 12 in. away to prevent problem of wax or polish build-up that necessitates floor stripping.
Most buffable floor coatings should not be polished until wax is thoroughly set—wait length of time indicated on instructions before buffing.
Sometimes it is advisable to dry-mop floor after buffing to remove soil loosened by buffing.
Ridged thresholds of doorways and elevators, particularly sliding doors, accumulate dirt; vacuum attachments are best means of picking up dirt from grooves.

According to Ray Pedderson, author of *SPECS: The Comprehensive Foodservice Purchasing and Specification Manual,* CBI Publishing Company, the following describes the chemistry involved with products suitable for cleaning hard surface floors:

"Pick materials that sequester and rinse well but do not corrode adjacent painted or metal surfaces. A mild abrasive to aid in physical action is acceptable but the product should not produce heavy suds or foam. Wax strippers should dissolve and chelate (pick up soil and salts and hold onto them for rinsing without returning them to the soiled surface) but not mar the floor. Floor polishes should have 12% to 16% solids depending on the amount of traffic."

## *From Soap to Buffs*

It may be helpful to see how the current collection of cleaners evolved after chemists analyzed the shortcomings of soap.

The first answer was detergent, which was initially a soap or trisodium phosphate. Today, there are endless varieties of detergents that contain such different components as soaps, alkalines, or acids, abrasives, solvents, corrosion inhibitors, water-softening agents, diluents, and so forth.

Synthetic organic detergents (syndets) do not precipitate in hard water as soap does, forming insoluble lime soap curd; unlike soap, they are not difficult to rinse off. For these two reasons they have found wide use in institutional housekeeping. They can be formulated for any requirement—they may be oil-soluble, water-soluble, high- or low-foaming, good or poor wetting agents, or whatever is needed.

Syndets break down into anionics and cationics, the former being used for cleaning because they have good emulsifying and degreasing characteristics. Cationics are more often used for deodorizing and as germicides. Products containing both of these syndets are true all-purpose cleaners and sanitizers.

To boost cleaning efficiency, alkaline salts similar to those found in soaps were added to improve syndet action in hard water. But because water hardness is so variable around the country, it is impossible to formulate cleaning agents to satisfy all needs. Certain areas require special products that can be determined only through consultation with the local water department.

A high-alkaline cleaner is needed for some metals and for the kind of heavy cleaning that must be done in kitchens, where degreasing is important.

Across the chemical way are the acid cleaners that remove lime encrustations in washrooms, rust stains, and tarnish on copper and brass. The strongest—muriatic acid—is used for urinals; the mildest—citric acid—for cleaning bronze.

Acid cleaners dissolve the soil. To dissolve grease and heavy oils, a solvent with its acid or alkaline builder may be necessary in addition to the syndet. These are flammable, but highly effective. If a wetting agent is used the solvent cleaners may be rinsed with water, thus permitting complete removal of oily soil as well as cleaner. These are now blended and so safe that they are called safety solvents.

The syndets have been altered further with finely ground abrasives so that they become powdered cleansers for use on very heavily soiled basins and floors.

Simply to clean a surface effectively did not satisfy the ambitious chemists—control of bacteria had to be mastered too. Now there are not only sanitizers, there are disinfectants that destroy infectious organisms, including most but not all pathogenic bacteria and fungi. (See chart in Chapter 19 on chemical cleaners.)

What are the different types of chemicals that are harnessed to fight the bacteria found on floors in institutions? Two most frequently encountered are:

1. Chlorinated phenols are generally combined with soaps or syndets to produce germicides or detergent germicides. They kill TB germs, and are strong, effective cleaners.
2. Quaternary ammonium compounds (quats) are used where food or beverages are prepared or served. They are odorless, nontoxic, and nonstaining, but sometimes do not work well in hard water.

Both of these chemical biocides are commonly used on hospital floors, but seldom for housekeeping in a hotel or residence hall. They figure importantly in cleaning locker rooms, washrooms, bathrooms, trash areas, and all food-service areas and adjacent corridors.

## *Polishes, Waxes, Finishes, Sealers*

Floor seals are of two types—*penetrating* and *surface.* Penetrating seals are used on wood or concrete floors and add to hardness and durability as they permeate the pores of the materials. Surface seals are thinner and form a clear film that protects the surface.

Concrete requires a special type of sealer because of its high alkaline content. The process can be complicated; concrete retains its finish properly only if cured for several months.

Quarry tile and terrazzo require a type of sealer that must be rubbed in, because the pores are not as open and thirsty as wood and concrete pores. Terrazzo tends to discolor unless a colorless seal is used.

Floor polishes include both the traditional wax and the new polymer coatings, or finishes, that are highly resistant to scuffs and scratches and are rebuffable so that marks can be removed without removing polish. They resist detergents, so may be subjected to daily scrubbing, yet are removable by suitable stripping agents when necessary. (Stripping agents are alkaline blends.) These polishes will act as sealers for some of the asphalt or vinyl tiles. As they are water-based they can be used successfully on terrazzo without discoloring it.

New polishes, according to Edwin B. Feldman, in his book, *Housekeeping Handbook For Institutions, Business & Industry*, "threaten another major revolution in floor polish technology. Water emulsion pastes and polishes can be applied over any floor surface without stripping old wax or polish . . . simply by burnishing the paste into the floor with a polishing brush. Moreover, they contain small quantities of solvent and cleaner components which will remove surface soils and rubber heel marks as the polish is applied. Buckets and mops are not necessary—only a polishing machine which cleans the floor somewhat as it spreads the polish, buffs the polish to a high gloss, and then maintains the gloss thereafter by regular buffing and occasional reapplication of polish."

Manufacturers of some resilient floors—vinyls, vinyl asbestos, and asphalt tile—claim that no seals or finishes are needed on those floors. They need only regular cleaning supplemented by a steel wool buffing with a floor machine, which is said to make the flooring self-protective by closing the pores to penetration of soils and chemicals.

According to Keith A. Fitch, a floor-care expert of Service Engineering Associates, Atlanta, the combination of pedestrian traffic and grit will seriously affect the appearance and life of these resilient floor coverings, including terrazzo, unless they receive proper care.

**FIGURE 10-4** *Disc floor machine can hold an aerosol spray-can, thanks to a special shock-absorbing, spring-loaded mount. Spray-cleaner covers floor just ahead of buffer. Control cable permits the operator to regulate spray. Rubber bumpers protect walls and furniture.*

In terrazzo, he maintains, the original sealer cannot protect against spills, soil, and traffic indefinitely, and he points out that no owner of a prestigious building expects terrazzo floors to perform without regular recoating.

Fitch says of synthetic finishes: "The evolution in synthetic formulations which brought about the polyacrylics, the co-polymers, and more recently the metalized finishes has made possible the development of products which combine advantages of stable color, good anti-slip properties, toughness, and resistance to powdering and easy repairability. The metalized finishes have given us great resistance to detergents.

"Initially, the synthetic finishes were widely heralded as requiring no buffing or other care except for regular cleaning and occasional renewal. This approach, however, was usually less than completely satisfactory except under very light traffic and soiling. These finishes have been modified in some cases to provide enough elasticity to respond to dry buffing. The later development of spray buffing turned out to be a much more valuable technique (Figure 10-4).

"It is performed by the spraying of a suitable mixture of polymer solids with emulsifying and cleaning agents onto the previously cleaned floor in areas of wear or black marking, and the drying and blending of that spray by means of a suitable pad under a floor machine. When the operation is properly carried out by trained personnel, it is not unusual that heavily trafficked areas can be maintained for periods of one to three years and even longer without floor stripping and refinishing.

"Very porous flooring," says Fitch, "such as asphalt, some of the less expensive grades of vinyl-asbestos, and terrazzo, frequently require several coats of synthetic finish before development of acceptable gloss."

Although it is difficult to predict what costs might be in an institution where floors are rarely unobstructed and where soils vary considerably, Fitch has developed a cost comparison between waxing and scrubbing methods and synthetic finishes with spray-buffing. It is persuasive enough

**TABLE 10-1** Typical Total Time Requirements for Floor Protection Systems
Average Minutes per Working Day Charged to each Procedure

| | Time in Minutes for each System (per 1000 sq. ft.) | | |
|---|---|---|---|
| **Procedure** | **Wax** | **Polymer** | **Terrazzo Seal** |
| Dust-mop | 7 | 7 | 7 |
| Damp-mop | 4 | 4 | 4 |
| Dry-buff | 9 | — | 3 |
| Spray-buff | — | 5 | — |
| Scrub and patch or refinish | 2 | 0.5 | 1 |
| Strip and refinish | 2 | 1 | 1 |
| Totals | 24 | 17.5 | 16 |

to make one a believer in the new polymer system with the metalized finishes.

In this table (10-1) the frequencies and time allowances assumed for each procedure involved in floor maintenance have been translated into average minutes per working day. It is assumed, for example, that dust-mopping will be done daily, so that number of minutes per thousand square feet is charged. Damp-mopping is assumed to be done weekly. (In institutions with foodservice facilities, damp-mopping may be needed more often.)

## *Floor Machines*

Just as the chemicals that are available today will confound the uninitiated because of their diversity, conflicting claims, and confusing nomenclature, so the choice of expensive floor machines will challenge even the experts. Almost as soon as advancements in machinery reach the market, imitations and variations appear to be followed by machines claiming to offer still greater versatility, productivity, and unique features.

There is little question, however, that even a $500 machine soon pays for itself in labor savings. Of every dollar spent for floor care, 85¢ goes for labor and 15¢ for equipment and chemicals, so the importance of high-speed equipment to cut labor costs cannot be overestimated.

A few examples cited by C. Barker Harrison, an authority on maintenance procedures, point up the labor-saving potential offered by high-speed machines:

1. One man using a push broom, can sweep approximately 6000 square feet per hour. A 36-inch power sweeper enables one man to sweep 27,000 square feet per hour.
2. It takes one man 15 minutes to polish 1000 square feet of unobstructed floor area with a 15-inch single-disc floor machine. A man with a 22-inch machine can polish the same area in just 7 minutes.
3. It would take a man 14 minutes to first dry-vacuum a 1000-square foot floor area, and another 16 minutes to polish that area with a 15-inch single-disc floor machine—total time, 30 minutes. An automatic scrubber/polisher/vacuum, which polishes and dry-vacuums in a single operation, enables one man to do the same job in 4 minutes.
4. Using standard mopping equipment (mop, buckets, wringer), one man can wet-mop and rinse 1000 square feet of floor in approximately 45 minutes. One man using a 24-inch automatic scrubber, can do the job in about 6 minutes. This includes time necessary for touch-up work with a damp mop to pick up patches of solution left by the machine at corners and turns.
5. A man using an ordinary wet pick-up vacuum cleaner equipped with a 14-inch wide squeegee floor tool can pick up cleaning solution at the rate of 20 minutes for 1000 square feet. The same vac equipped with a 20-inch squeegee will do the job in 14 minutes. Using a 26-inch floor tool the job takes 11 minutes.

The reader should be familiar with the general types of mechanical floor-cleaning equipment that offer such time-saving in institutions today. Dr. Harrison's booklet, "Building Maintenance Equipment Selector Guide," available from floor machine dealers, covers the subject in detail. The following summarizes salient features of some of the more popular machines.

*Single-disc floor machines* are highly versatile. They are used for buffing, scrubbing, polishing, stripping, grinding, sanding, and troweling fresh concrete. Most models have handles adjustable to different heights and dual-control safety switches (operate with either hand) that shut off when the operator releases his grip. Machines that require lubrication should be avoided, as this important bit of maintenance may be overlooked; parts should be sealed for life. Be sure that the machine is balanced properly; sometimes a heavy handle will interfere with maneuverability and cause operator fatigue.

Spray attachments are available for these machines so that cleaners, waxes, or polishes can be sprayed onto the floor just ahead of machines.

*The wet/dry vacuum with an internal filter* has a by-pass motor sealed off from the fan system used to pick up dust or moisture-laden air. Thus no water can get into the motor. It has a tank to hold from 1 to 16 gallons of recovered liquid picked up after scrubbing. It can also be used as a conventional vacuum, although the filtering capacity of fine dry material is limited due to the small internal filter bag. Any 8-gallon mop bucket or 30- to 55-gallon drum can be converted to wet or dry vacuum by adding a power head and drum adapter. These vacuum-top units also have by-pass motors. Their filters will limit capacity to pick up dry materials, but some have cloth filter bags to fit inside buckets or drums.

**TABLE 10-2** Single-Disc Floor Machine Job/Time Requirements: Scrubbing

| Brush Size (in.) | Time Req. to Scrub 1000 Sq. Ft. (min.) | Labor Cost per 1000 Sq. Ft. Hourly Wages ($) | | |
|---|---|---|---|---|
| | | 3.10 | 4.00 | 4.50 |
| 13 | 32 | 1.65 | 2.13 | 2.40 |
| 15 | 27 | 1.39 | 1.80 | 2.03 |
| 17 | 23 | 1.18 | 1.53 | 1.72 |
| 20 | 16 | 0.83 | 1.06 | 1.20 |
| 22 | 12 | 0.62 | 0.80 | 0.90 |

*The automatic scrubber/polisher/vacuums* will perform three operations simultaneously: (1) lay cleaning solution, (2) scrub, and (3) vacuum up dirty solution. Converted for dry cleaning, the machine can buff the floor and at the same time vacuum up all dust, grit, and dirt loosened by dry-buffing. These machines do *not* strip floors in one single pass because stripping is only accomplished by allowing chemicals to remain on the floor for a few minutes to loosen the finish before it is vacuumed into the machine. Stripping often requires double or triple scrubbing with a scrubber pad.

The automatic scrubber/polisher/vacuum does what the two machines—wet/dry vacuum and single-disc machine—will do. Under some circumstances, using the automatic instead of the other two machines cuts floor care time in half.

*Power sweepers* are available in two types. The high-powered litter vacuum has no brush, and picks up solely by suction. The other type of sweeper has a power brush with a vacuum and dust-control system. It loosens caked-on dirt and can generally pick up heavier debris than the vacuum-only type. It also controls dust problems better. It is available powered by battery, electricity, gas, or propane. Battery models have the advantages of being quiet, fume-free, and needing no connections.

These power sweepers, automatic scrubbers/polisher/vacuums, wet/dry vacuums, and single-disc machines are the major types of floor cleaning units now available. There are, however, many models made by many manufacturers, and with such numerous variations that a detailed study should be made before a purchase is made. A list of major manufacturers can be found in the appendix.

Many machines that are used on hard or resilient floors are also usable on dense low-pile carpets, as are some cleaning pads and even mops.

Selecting one the right size, weight, and horsepower is extremely important; enormous savings may result from investing just a few hundred dollars more in a slightly larger or sturdier machine than would meet minimum requirements. Table 10-2 shows time and labor for wet-scrubbing with single-disc floor machines. Stripping with the same machines would at least double these time requirements.

**TABLE 10-3** Single-Disc Floor Machine Job/Time Requirements: Polishing

| Size Brush (in.) | Time Req. to Polish 1000 Sq. Ft. (min.) | Labor Cost per 1000 Sq. Ft. Hourly Wages ($) | | |
|---|---|---|---|---|
| | | 3.10 | 4.00 | 4.50 |
| 13 | 19.0 | 0.98 | 1.26 | 1.43 |
| 15 | 16.0 | 0.82 | 1.06 | 1.20 |
| 17 | 13.5 | 0.70 | 0.92 | 1.01 |
| 20 | 9.5 | 0.49 | 0.63 | 0.71 |
| 22 | 7.0 | 0.36 | 0.46 | 0.52 |

Here is how to use Table 10-2 to determine what it actually costs to use various sizes of machines. Assume 5000 square feet of floor must be wet-scrubbed once each month, and dry-buffed four times each week, at $4.00 per hour for labor.

Using a 15-inch machine, for example, labor costs $1.80 per 1000 square feet, or $9.00 for the 5000 square feet to be scrubbed once every month. Total annual cost for wet-scrubbing will be 12 times $9.00 or $108.

Refer to polishing time Table 10-3 above. Dry-buffing at $4.00 per hour for labor using the 15-inch machine, costs $1.06 per 1000 square feet, or $5.30 per 5000 square feet. Since the floor must be buffed four times a week, multiply $5.30 by four to get weekly buffing cost—$21.20. Multiply this times 52 for annual buffing cost: $1102.40.

The total annual cost in this example for wet-scrubbing once each month and dry-buffing four times each week with a 15-inch machine, and at the rate of $4.00 per hour labor would be $1210.40.

Using a 22-inch machine, the cost would drop to $586.40 per year, making a savings of $624.00 in labor costs in one year. The difference between the cost of a 15-inch and a 22-inch single-disc floor machine is about $125.00.

## *Floor Machine Weight/Brush-Size Relationship*

To a point, the more weight applied on a brush of a single-disc floor machine, the more efficient it will be for both scrubbing and polishing. If the machine is too light, maximum efficiency is not possible when polishing, burnishing, or scrubbing. Too much weight concentrated on too small a brush area will flatten the bristles, reduce efficiency, and make the machine difficult to operate. The proper relationship of brush size to machine weight and brush speed is essential. As brush diameter is increased (up to 20 inches), the total weight of the machine should be increased. A 90-pound floor machine with a 15-inch diameter brush applies six pounds for each inch of brush diameter. (Simply divide the weight by the diameter.) A 20-inch machine, also weighing 90 pounds, applies only 4.5 pounds per inch of brush diameter. This seemingly small difference can really affect performance.

A floor machine with a 17-inch brush is a popular size. A recent review of 46 different models of 17-inch machines showed weights ranging from a low of 71 pounds to a high of 136 pounds. This is a range of 4.2 pounds per inch of brush diameter to 8 pounds per inch. The average weight was 99 pounds, or 5.8 pounds per inch of the 17-inch brush.

Machines with brush spread of from 15 inches to 17 inches should have a weight per diameter inch of brush of no less than 6 pounds. Larger machines may have slightly less weight per inch for the sake of ease of operation and machine handling, but even these should have 5 pounds per inch.

## *About Horsepower*

It stands to reason that a larger heavier machine will need more horsepower than the lighter one. The type of floor and use to which the machine will be put also make considerable difference. Concrete requires more power than a smooth resilient floor, and spray-buffing puts a heavier load on the motor. The following table (10-4) shows minimum horsepower requirements for machines weighing from less than 95 lbs. to 140 lbs., and indicates correct brush size in relation to weight and power.

Experienced housekeepers warn: don't buy a machine that purports to do too many things, for it may not do any of them well and may pose

**TABLE 10-4** Brush Size, Machine Weight, and Horsepower

| Brush Size (in.) | Machine Weight (lb.) | Minimum Horsepower |
|---|---|---|
| 15 | Up to 95 | ½ |
| 16 | 110 | ¾ |
| 17 | 110 | ¾ |
| 18 | 115 | ¾ – 1 |
| 19 | 120 | 1 |
| 20 | 120 | 1 |
| 22 | 120 | 1 |
| 24 | 140 | 1½ |

serious repair problems as it has too many parts that might break down. Another familiar admonition: no one machine can be all things to all Housekeepers. This is not to say that machines designed to spray and buff or scrub and polish in a single operation are not effective. Their increasing popularity is sufficient evidence to the contrary.

Cleaning the floors in a hotel can seldom be performed on an absolutely regular schedule. Productivity charts that tell how many thousands of square feet a floor maintenance man can do in eight hours are based on cleaning factories or office buildings late at night when the area is unobstructed, no one walks on the freshly cleaned floor, and the worker is uninterrupted. But in a hotel, emergencies arise, and although group meetings are scheduled in advance, they will still interfere with floor maintenance schedules.

## *Floor-Care Schedules*

Perhaps cleaning certain back corridors and behind meeting and banquet rooms will have to be done during a brief three-hour break in the afternoon between luncheon and evening events, instead of during the usual eight-hour late shift. When the floor man finally does gain access, he may find so much excessive soil from the event just held that another type of cleaner may be needed.

Floor cleaners on two different shifts may

TUE. 3rd Shift - 11 to 7 WAXER #1

| AREA | M | T | W | T | F | S | S | Dust Mop | Damp Mop | Wax | Buff | Strip | Base-boards | Corner | Hours | Min. |
|---|---|---|---|---|---|---|---|---|---|---|---|---|---|---|---|---|
| Plaza Service Corridor | | ✓ | | | | | | ✓ | | | | | | | | 25 |
| A&B Elevator Lobby 1 & 2 | | ✓ | | | | | | ✓ | ✓ | ✓ | ✓ | | | | 1 | |
| Cantina 2 Men 3:30 to 5:00 | | ✓ | | | | | | ✓ | ✓ | ✓ | ✓ | | | | 2 | |
| Lobby Service Area | | ✓ | | | | | | ✓ | ✓ | ✓ | ✓ | | | | 1 | 10 |
| Main Entrance Marble | | ✓ | | | | | | ✓ | ✓ | ✓ | ✓ | | | | 1 | |
| Main Cashier Booth | | ✓ | | | | | | ✓ | ✓ | | | | | | | 10 |
| Dressing Room Corridor | | | | | | | | | | | | | | | | |
| Sound Staff Projection | | | | | | | | | | | | | | | | |
| Westside Dressing Room | | | | | | | | | | | | | | | | |
| Westside Dance Floor | | | | | | | | | | | | | | | | |
| Granada Grill Foyer | | ✓ | | | | | | ✓ | ✓ | ✓ | ✓ | | | | 1 | 10 |

**FIGURE 10-5** *Waxer*

Schedule for hotel Floor Waxer No. 1 on the third shift, 11 pm to 7 am, shows exactly where he is to do just what, and even specifies amount of time to be allotted to each area.

have to take care of the public rooms. Usually the shift from 3 pm to 11 pm is comprised of only one or two cleaners who do very little work in the back *or* front of the house because during that time public areas, meeting, and banquet space are subject to heaviest usage (Figures 10-5 and 10-6).

Schedules for cleaning floors in a college or university residence hall can be more rigidly adhered to. The number of residents does not change, the activity (or lack of it) can be determined in advance, and there is less worker turnover so productivity may be more predictable. Here we have included a schedule for both hotel and college floor cleaners.

## *Floor-Cleaning Techniques*

### *Wood Floors*

**Initial Care.** New wood floors must be properly sanded, vacuumed, filled, sealed with three coats of sealer and buffed between applications. Then solvent-base wax or polish is applied. (A water-based wax can be used *only if the floor is properly sealed,* but solvent-base wax is preferable.)

**Daily.** Buff with heavy powered machine to bring up gloss. Apply second coat of wax and buff again.
Use wax-treated dust mop, which should be changed often.
Damp-mop (only if necessary) with just a slight amount of moisture.

**Periodic.** Buff floor with No. 0 steel wool or nylon pads.
Spot-clean with damp mop and dilute solution of detergent.
Rebuff; apply light coat of wax if needed.

**Alternate.** Buffing can be eliminated if glossy solvent-soluble polymers are used. Spot-cleaning can be done with liquid solvent or paste wax, with fine steel wool on power-buffer.

**Caution.** Do not use excessive amounts of water.

### *Linoleum*

**Initial Care.** Surface coating has been applied at factory and should be preserved as long as possible.
Do not scrub for four or five days after installation of linoleum.
Lightly clean with damp mop and properly diluted detergent solution.
Apply one or two coats of water-emulsion polish.
Avoid using excess water or wax that might get in seams.
Sweep with non-oily treated mops or dust-cloths.
Scrub thoroughly with neutral soap or syndet (diluted according to manufacturer's instructions). Apply properly to floor without flooding. Leave on floor for a few minutes, but not long enough to dry. Fine steel wool or mild synthetic pads may be used under floor machine to hasten cleaning. Take up solution with wet-vacuum or mop. Rinse, but do not flood floor. Take up rinse water. Allow to dry, and apply polish. Both water- or solvent-based waxes or coatings are made for linoleum. Apply a *thin* uniform coat of wax or polish; allow to dry. Buff if necessary with soft brush. Apply second or third coats if needed.

**Daily.** Sweep with chemically treated, non-oily mops.
Damp-mop. Do not use excessive water.
Buff lightly.

**Periodic.** Scrub lightly following procedure as outlined above under Initial Care, but use an even more diluted solution and gentle action so as not to remove polishes.
Wax with water emulsion or solvent-based wax, or apply polymeric coating. Strip (when build-up of polish appears—once or twice annually) with solution recommended for linoleum, following manufacturer's instructions.

**Alternate.** Sweep, dry-mop, or vacuum.
Use buffer and synthetic coating that cleans, polishes, and coats.

**Caution.** Avoid ammonia, alkaline, or acid cleaners, as they discolor floor.
Do not flood floor or overclean, as some linoleum wears out fairly rapidly.

**BUILDING** Hartrenft & Mifflin

**AREA** Bullock

**WORKER NO.** 2 **CLASS** A **Effective Date** Fall 1974

| 1 worker serves 2 bldgs. | | FIRST WEEK | | | | | | | | SECOND WEEK | | | | | | |
|---|---|---|---|---|---|---|---|---|---|---|---|---|---|---|---|---|
| Perform Operations Indicated by (X) on Day Shown | | M | T | W | T | F | S | S | | M | T | W | T | F | S | S |
| Daily Make Ready | | 26 | 26 | 26 | 26 | 26 | | | | | | | | | | |
| Clean Rooms to Inclusive ( Rooms) | | | | | | | | | | | | | | | | |
| " " to " ( Rooms) | | | | | | | | | | | | | | | | |
| " " to " ( Rooms) | | | | | | | | | | | | | | | | |
| " " to " ( Rooms) | | | | | | | | | | | | | | | | |
| " " to " ( Rooms) | | | | | | | | | | | | | | | | |
| " " to " ( Rooms) | | | | | | | | | | | | | | | | |
| " " to " ( Rooms) | | | | | | | | | | | | | | | | |
| " " to " ( Rooms) | | | | | | | | | | | | | | | | |
| " " to " ( Rooms) | | | | | | | | | | | | | | | | |
| " " to " ( Rooms) | | | | | | | | | | | | | | | | |
| Clean Bath Extra travel time between bldgs. | Heavy | 25 | 25 | 25 | 25 | 25 | | | | | | | | | | |
| " " | Heavy | | | | | | | | | | | | | | | |
| " " | Heavy | | | | | | | | | | | | | | | |
| FOLLOWING OPERATIONS APPLY TO BOTH BLDG. | Regular | | | | | | | | | | | | | | | |
| Pick up ground floor vacuum runners, rug & elev. | Regular | 50 | 50 | 50 | 50 | 50 | | | | | | | | | | |
| Damp-dry wipe elevator & walls | Regular | 32 | 32 | 32 | 32 | 32 | | | | | | | | | | |
| Clean entryways (3) each bldg. | Heavy | | | 90 | | | | | | | | | | | | |
| | Regular | 42 | 42 | | 42 | 42 | | | | | | | | | | |
| Dry mop ground floor corridor, rec.room, lobby & lounge | | 46 | 46 | 46 | 46 | 46 | | | | | | | | | | |
| Wet mop & buff lobby area | | 64 | 64 | 64 | 64 | 64 | | | | | | | | | | |
| Wet mop & buff other ground floor area | | | 104 | | | 104 | | | | | | | | | | |
| Sweep patio & covered court | | 36 | | 36 | | 36 | | | | | | | | | | |
| FOLLOWING OPERATIONS APPLY TO MIFFLIN ONLY: | | | | | | | | | | | | | | | | |
| Dry mop Hskp, office & 2 locker lounge rooms | | | 17 | | | | | | | | | | | | | |
| Wet Mop " " | | | 27 | | | | | | | | | | | | | |
| Relamp | | | | | | | | | | | | | | | | |
| Relamp ground floors and basement areas as necessary | | x | x | x | x | x | | | | | | | | | | |
| Other Non Routine Duties as Directed | | x | x | x | x | x | | | | | | | | | | |
| Hours available for Household Work | | 321 | 433 | 362 | 285 | 425 | | | | | | | | | | |
| | | 2.5 | | 2.0 | 3.0 | 1.0 | | | | | | | | | | |

**FIGURE 10-6** *Scheduled Work Assignment: Residence Hall Worker*

Schedule for residence hall worker at Pennsylvania State University is assigned by minutes, starting out realistically with Daily Make-Ready time. Director of Housing, Donald T. Arndt has determined that 26 minutes will be used up just in getting machines and supplies ready, in having a few words with the supervisor, in walking to the first point of work, etc.

## *Asphalt Tile*

**Initial Care.** Do not use water for from 5 to 10 days after floor is laid. (Cement must be allowed to harden.)
Sweep frequently with treated mop or broom.
Damp-mop only if it is urgently needed, and with the mop only barely moist.
Thoroughly sweep, vacuum, or dry-mop floor. Use syndet type cleaner and warm water according to instructions. Wet floor with solution but do not flood; let it stand a few minutes; use machine with scrub brush, steel wool, or nylon pad to remove soil and factory finish. Remove dirty solution with mop or wet-vacuum. Rinse thoroughly using lukewarm water; rinse again, keeping water clean. Allow to dry thoroughly. Apply water-based buffable or non-buffable polish. Brush on buffer should be 50/50 mixture of tampico and palmyra fibers. Apply second coat.

**Daily.** Sweep or dry-mop.
Damp-mop with dilute neutral cleaner.
Buff to remove marks if floor was finished with buffable polish.

**Periodic.** Touch up with water-based polish. Strip, rinse, and apply polish as above.

**Caution.** Avoid excessive use of water, especially hot water.
Use *only* water-based wax and polish; non-alkaline cleaners.
Protect against indentation.
Avoid extremes of hot and cold.

## *Rubber Tile*

**Initial Care.** Wait until new floor has been down four or five days so adhesive can set. Sweep with broom, untreated mop, or damp-sweeping tool.
Damp-mop.
Scrub thoroughly using free rinsing syndet designed for rubber tile, with warm (100° to 120° F) water; dilute according to manufacturer's instructions. Be sure to use proper cleaner as rubber tile is sensitive to, and may deteriorate from, soaps containing copper and manganese. Cleaning solution should remain on floor three to five minutes, but not long enough to dry. Scrub with stiff brush and power machine; fine steel wool or mild synthetic pads may be used under brush. Take up solution; rinse with clear cool water. Take up rinse water and allow to dry. Water-emulsion finish (wax and polymer coating) *only* should be applied thinly and evenly. Less polish is needed on rubber tile as it is smoother and less porous than most other floor coverings, which also causes wax or coating to take longer to dry. Buffing is optional.

**Daily.** Sweep or dry-mop with treated mop that leaves no oil.
Damp-mop with cool water and diluted syndet. Do not flood floor. Recoat with polish where it is simply worn off.
Scrub lightly if floor is soiled or has black marks, No. 00 or 000 steel wool can be used for stubborn marks.
Spot-clean at points of heavy traffic.

**Alternate.** Spray-buff with polymer finish frequently to help eliminate scrubbing. This should not be done if there is a heavy build-up of old wax.

**Periodic.** Strip floor when build-up of wax or polish is obvious, when dirt is imbedded, or when color pigment comes loose on surface and appears to bleed. Use a de-waxer that is safe for rubber floors and follow procedure as for initial cleaning.

**Caution.** Avoid solvents and oils, only dust-mops, and any cleaner that has alkaline content. Coarse steel wool or synthetic pads, ozone, and sunlight can cause cracking and fading of rubber tile, so provide protection from open doors or windows.

## *Vinyl Asbestos*

**Initial Care.** Do not scrub for about five days after installation so that adhesive can dry. Lightly clean and polish with water emulsion wax or coating.

Sweep floor frequently or use treated mops or dust-cloths.
After waiting necessary period of time, scrub floor and rinse, using same technique as outlined in initial cleaning for asphalt tile. (Vinyl asbestos will probably dry faster than asphalt.) Buff if higher gloss is desired, but vinyl tile usually requires less.

**Daily.** Sweep or dry-mop with non-oily treated mop.
Damp-mop with cool water and dilute syndet solution; do not flood. Mop as often as necessary.
Buff to remove marks and improve gloss.
After damp-mopping, polish only where needed, for example, where film has worn off in spots.
Scrub lightly to remove heavier soil build-up, but avoid removing polish; No. 00 steel wool can be used for stubborn marks.

**Alternate.** Spray emulsion polish on and buff (some power machines can do both in one operation); use fine steel wool pads to loosen stubborn dirt.

**Periodic.** Strip floor when build-up of wax becomes obvious (from one to six times yearly). Use safe wax stripper diluted as recommended on instructions.
Let solution remain on floor three to five minutes, not long enough to dry.
Scrub with stiff brush under power machine, or use mild synthetic or steel wool pads; take up solution with wet-vacuum or mop; rinse thoroughly; repeat operation if necessary.
Apply two or three coats of polish (e.g., water-emulsion wax, or resinous or polymer coating).

**Alternate.** Older floors may need to be sealed to fill pores, and make surface easier to clean.

**Caution.** Keep sufficient wax or polish on floor. Avoid abrasive cleaners on unwaxed floors.

## *Epoxy Resin*

Although these can be cared for just as terrazzo floors, they *do not require sealing* and waxes or polishes are rarely needed—only sweeping, damp mopping, and scrubbing.

## *Terrazzo*

**Initial Care.** Seal thoroughly. Porous cement matrix requires proper internal sealing with non-yellowing seal.
For first few months mop daily to prevent mineral salts from dulling finish; when cleaned thoroughly, additional sealings may be necessary.

**Daily.** Sweep with push-broom (avoid oily sweeping compounds); vacuum, or dust-mop with non-oily treated mop.
Buff with synthetic or fiber pad (avoid steel wool).

**Periodic.** Wet-mop or scrub with machine, but wet floor with clean, warm water first; use mild detergent (sodium metasilicate is safe when used ½ oz. to 1 gal.); rinse freely and finish with well-wrung mop or wet-vacuum.
Seal with compounds recommended by terrazzo contractors.

**Alternate.** Spray-buff with synthetic finish formulated for terrazzo. (If maintained properly with this technique, floor will not require waxing for years and stripping can be put off three years or more.)
If terrazzo is badly worn, it should be ground down to expose a new surface and sealing processes should be repeated.

**Caution.** Do not use steel wool as bits may break off and discolor floor.
Use scouring powders only when absolutely necessary—then sparingly.
Do not use alkaline cleaners—they may eat into pores and cause spalling (marble pieces chipping out of cement matrix).
Even detergent disinfectants with pH as low as 6.5 will damage terrazzo unless it has been thoroughly and properly sealed.
Select appropriate casters and glides to protect floor.

## *Quarry Tile and Ceramic*

**Initial Care.** Scrub with non-alkaline synthetic detergent, scrubbing and rinsing well. Allow to dry.
Seal floor so that grouting will be protected by penetrating seal.

**Daily.** Sweep with brush or dust-mop.
Damp-mop.
Wet-mop with syndet or use floor machine; take up solution with mop or wet-vacuum; rinse floor; dry floor, since residue from water or detergent may leave marks.
Where quarry tile is in the front of the house, coat or wax floor with a water emulsion polish (waxes with a nonslip ingredient often make hard smooth surfaces less slippery than before application). Bathroom tiles rarely need wax or polish except to protect grout.

**Caution.** Grouts are available that are either acid-resistant or alkaline-resistant, but cleaners to which the grout is *not* resistant can loosen grout.
Abrasive cleaners will scratch surfaces.
Soap tends to build up slippery film.

## *How to Give a Lesson*

Although space does not permit printing step-by-step instructions for all floor-care procedures, the following example indicates the many steps involved in teaching even such an elementary task as mopping and scrubbing. This outline will also provide the reader with a lesson in How to Give a Lesson. The structure of this outline can be used as a format for other training programs. The following is from the maintenance program at Pennsylvania State University.

## *Training Topic—Mopping*

This training topic is to be presented in three sessions. The information to be covered in each session is listed below.

1. First session: Introduction (45 min. to 1 hr.)
   - A. Objectives
   - B. Standards
   - C. Types and uses of equipment
   - D. Types and uses of products
   - E. Preparation of area
   - F. Types of mopping
     - (1) damp
     - (2) wet
     - (3) scrubbing
   - G. Work procedure
   - H. Completion
   - I. Cleaning and storage of equipment
2. Second session: Damp-mopping (30 min. to 1 hr.)
   - A. Objectives
   - B. Standards
   - C. Equipment
   - D. Products and supplies
   - E. Preparation
   - F. Work procedure
   - G. Completion
   - H. Cleaning and storage of equipment
3. Third session: Wet-mopping or scrubbing (30 min. to 1 hr.)
   - A. Objectives
   - B. Standards
   - C. Equipment
   - D. Products and supplies
   - E. Preparation
   - F. Work procedure
   - G. Completion
   - H. Cleaning and storage of equipment

What follows is a detailed instruction for the first session only. Readers can follow the same format when making up specific instructions for second and third sessions.
First session: Introduction (45 min. to 1 hr.)

A. Objectives
   1. Maintain constant standard of sanitation; clean, well-kept floor.
   2. Protect the floor and floor finish.
   3. Prolong interval between major refinishing.
   4. Learn proper techniques and work methods.
   5. Learn use of equipment and products.

B. Standards
   1. Finished floor will be uniformly clean and will show no apparent dirt or mopping streaks.

2. There will be no mop strings on floor.
3. All black marks will be removed.
4. There will be no splash marks on furniture or baseboards.
5. No water puddles will be left on floor and no water allowed to seep under low or fixed furniture.

C. Types and uses of equipment (Figures 10-7 and 10-8)
1. This equipment common to all types of mopping.
2. Dry-mop—24 in. or 36 in., depending on the size of the work area. Used to remove dry, loose soil from floor before mopping.
3. Tank-type vacuum cleaner with hose. Used to remove dust and particles from dry mop as required so that dry mop will continue to gather and hold dry soil. If dry mop is not cleaned frequently enough, loose soil will pass under and remain on floor.
4. Pail(s)—10, 15, or 20 qt. Size will depend on area to be mopped. For lightly soiled areas, only one pail will be needed. For mopping of heavily soiled areas and for scrubbing operations, two or more will be needed; one for cleaning solution and one for rinse water.
5. Mop wringer, one per pail, used to wring excess water from mop after immersion in bucket.
6. Wet-mop, complete. Mop heads are stocked in 12-, 16- and 24-oz. sizes. The 24-oz. size is most commonly used, with the smaller size being used for smaller areas or, if necessary, to reduce the fatigue factor. Handles are available in 54- and 60-in. lengths. The handle selected should be of the length closest to

**FIGURE 10-8** *A sponge mop, swivel-headed tool can be used to rinse tile coving. Cleaning solution is first applied to coving with triangular-shaped brush in foreground. For areas with heavy encrustations, rust and scale remover are needed.*

**FIGURE 10-7** *Ceramic tile floor in employees' washroom is mopped by hand because space is too small for scrubbing machine. All-purpose detergent cleaner is used diluted to proper strength. Mop wringer in bucket is activated by lever so that worker does not need to handle mop.*

reaching from the floor to eye level of the employee using it.

D. Types and uses of products
   1. The only cleaning agent needed for most mopping jobs is neutral detergent. This is received in 30-gal. drums and is then placed in gallon jugs equipped with a plunger-type dispenser. One of these jugs of detergent should be taken to the work area. In later sessions, cover the amounts of detergent to be used in the mopping solution.
   2. Other special products to be covered at time needed.

E. Preparation of area
   1. Take equipment necessary from storage location to the work area.
   2. Move furniture as instructed (items to be covered at time of actual assignment) to clean maximum of work area.
   3. Place signs or barricades, if appropriate, to warn people of possible hazard due to slippery floor.

F. Types of mopping
   1. Damp-mopping
      (a) Used in routine daily cleaning to remove light soil.
      (b) Minimum of water applied to floor, mop wrung out thoroughly.
      (c) Usually done with very mild detergent solution.
   2. Wet-mopping
      (a) Used in periodic thorough cleaning or when needed in heavily soiled areas.
      (b) Mop wrung out lightly. More water applied to floor.
      (c) Usually done with stronger detergent solution, requiring rinsing with plain water.
      (d) Often followed with fresh coat of wax.
   3. Scrubbing
      (a) Usually a wet-mopping procedure in combination with machine-scrubbing.
      (b) Mopping involves putting down detergent solution to be machine-scrubbed, and then picking up residue with mop and rinsing floor.
      (c) Partial removal of floor finish in preparation for refinishing.

G. Work procedure (Figures 10-9 and 10-10)
   1. In a cleared work area, begin with dry mop at one end of the work area in a corner. With one edge of dry mop against wall, push mop ahead of you the full length of the work area to the opposite end. Reverse direction; position mop head so that edge overlaps first stroke by about 2 in. and clean second stroke to the end of work area at which you began. Continue in this manner until entire cleared work area has been cleaned, stopping as required to clean mop head with vacuum cleaner.
   2. Fill one or two pails (depending on type of mopping) ¾ full of lukewarm water. *Never use hot water* as it will remove finish from the floor, loosen the cement, and may damage the floor covering.

**FIGURE 10-9** *Powered with six 6-volt batteries (here being filled with distilled water), this scrubber/vacuum has 15-gal. capacity for foam cleaner laid down by machine. Two nylon brushes, set on springs to maintain contact with uneven surfaces, can scrub or strip. Vacuum picks up water in 20-gal. tank; squeegee finishes job. Machine can be used indoors or outdoors.*

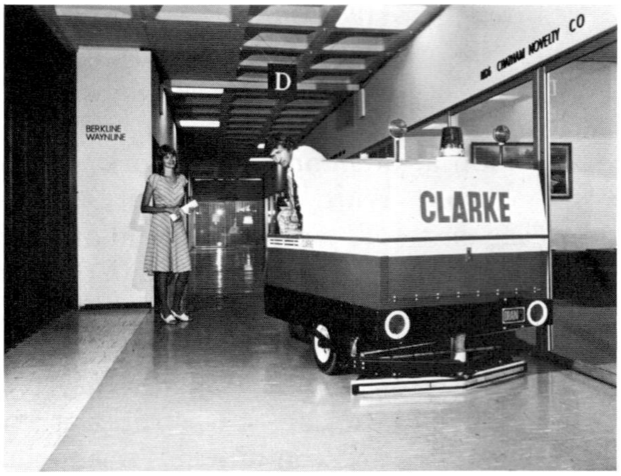


**FIGURE 10-10** *A ride-on scrubber has been designed specifically for cleaning large areas of concrete and terrazzo floors. Under normal unobstructed conditions, it can clean 62,500 square feet per hour, dispensing solution, scrubbing, squeegeeing, and vacuuming up used solution in one pass. Cleaning solution is blasted on floor from 150-gallon storage tank at 500 psi. Scrubbing is accomplished by a 5 ft. wide heavy-duty nylon-bristled brush. Free-floating squeegee and high-speed turbine vacuum leave floor dry enough to walk on safely. Single foot pedal sends machine forward or in reverse. Automotive steering permits turning in radius of only 72 inches.*

3. Take mop (12-, 16-, or 24-oz.) to be used, and pail of water which has had the proper amount of detergent added, to the end of work area where dry mopping started.

   NOTE: The following instruction in paragraphs (a) to (d) should be presented as explanation and demonstration by instructor, then repeated with trainee performing the procedure.

   (a) Immerse mop head in detergent solution, wring out as needed for type of mopping to be done, and place mop head on floor.
   (b) Stand, facing end wall, about 3 ft from end wall and about 3 ft. from sidewall.
   (c) Place mop head to side, just touching baseboard of sidewall. Make first stroke along baseboard to corner and then along baseboard of end wall across in front of you to a comfortable distance on the opposite side from sidewall.
   (d) Work mop from side to side in a slight figure-eight motion. Keep mop handle at about 15-degree angle with mop head passing about 1 ft. in front of toes. Let each stroke reach a comfortable distance to each side, without bending the back, and each stroke overlap the border stroke made along the sidewall. Step slowly backward while mopping in this manner until reaching the end of the border stroke.

   NOTE: Repeat above with trainee performing each step.

   (e) Make another border stroke, and continue mopping as above until reaching the end of the work area. Immerse and wring mop as necessary while working. Change water as needed. (Show employee.) Clean water is the secret of cleaning.
   (f) Reverse direction at end of work area, and mop adjoining section of floor toward place of beginning. This section of mopping should overlap first section by about 1 foot.
   (g) Continue as above until area is completed. This method of mopping is the most efficient and least fatiguing.
   (h) Rinse if required, by repeating the above, using clear water.
   (i) Move furniture to area cleaned and mop balance of area to be done.

H. Completion

1. Waxing of clean dry floor may be required.
2. Reposition furniture to normal locations.
3. Remove signs or barricades.
4. Empty pails in nearest janitor's closet.
5. Take equipment back to storage location.

I. Cleaning and storage of equipment
   1. Dry-mop: vacuum head to remove any remaining soil. Stand or hang in designated location.
   2. Vacuum cleaner: wipe off machine body and cord with damp cloth. Check dirt bag; empty or replace if more than half-full.
   3. Pails: rinse out with hot water, wipe out with cloth, wipe off outside including wringer and casters.
   4. Rinse mop head thoroughly with hot water in sink, wring out thoroughly, shake to loosen fibers, and hang to dry.

## *Bibliography*


Berkeley, Bernard. Floors and Floor Maintenance. *Cornell Hotel and Restaurant Administration Quarterly,* n. d.

Feldman, Edwin B. *Housekeeping Handbook for Institutions, Business & Industry.* New York: Frederick Fell, Inc., 1969.

Fitch, Keith A. The Best Possible Maintenance System for Resilient and Terrazzo Flooring. *The Executive Housekeeper,* January 1973.

Harrison, C. Barker. *Building Maintenance Equipment Selector.* (Distributed through floor machine dealers.)

# *Chapter 11*

# *Carpet Selection*

**OVERVIEW** Enlightened housekeepers should know enough about carpet to be sure the appropriate one (with proper construction, weight, fiber, and so forth) is selected for each different area in the establishment. There are so many considerations and specifications to learn that we have devoted an entire chapter to the selection of carpeting to enhance the readers' understanding. If the wrong carpet is installed over the wrong padding, no amount of cleaning can keep the product attractive and protect the investment.

Until the last few years, carpet was about the only product that had decreased in price in the previous decades. It is still less costly than many other floor coverings and is a far better product, thanks to improved methods of manufacturing and dyeing. Most carpeting was woven on looms until the early 1960s, but today the bulk of it is tufted on high-speed machinery so it is possible to produce less expensive goods that are often thicker and wear better. Thanks to synthetic fibers scientifically engineered for the purpose, today's carpets can be made to last longer and clean more easily. Test-tube fibers also provide a wide variety of either brilliant or muted colors, none of which fade.

Carpet *must* be good today, for it is used in the most improbable, most demanding places—busy back corridors, outside by the swimming pool, even in kitchens. With carpet covering more and more areas throughout the institution, it becomes a major investment and protecting the expenditure is the responsibility of the professional Housekeeper.

Because Housekeeping is involved daily with the carpeting (trying to keep it clean, in a state of repair, and properly replaced as it wears), it is important that the Housekeeper participate in its selection. At the first evidence that new carpeting should be purchased, the Housekeeper is wise to start investigating the latest new products (every few months a new improvement is incorporated in some producer's carpet). The Housekeeper should prepare to make recommendations as to types of carpet and underlays considered best for the particular area in question.

What's the best carpet? It depends on the budget, the use, the place, the care it will or will not be given, and many other factors. Some interior designers may feel a specific carpet is appropriate for the area because of its handsome appearance, but the professional Housekeeper must look beneath the splendid surface and ask: How, and of what, is it made? Will it perform reliably in a particular installation?

The professional Housekeeper knows that any good contract carpet will keep its looks and wear well *if* it is properly maintained. But few institutions can afford to have a carpet maintenance man standing by, poised to wipe up spills in the restaurant, or waiting in the lobby to vacuum behind every guest who tracks in soil. So the carpet must be specified with more than appearance in mind.

Most people today mistakenly think that the choice rests on whether one buys wool, nylon, acrylic, and so on, but fiber is only one of several very important considerations. Just what *is* involved in making selections, writing specifications and purchasing carpeting? Here are 12 important points:

1. Construction—is it tufted, woven, knitted, needle-punched, or flocked?
2. What is the density of the carpet—how close together are the yarns? What ply are they? What is the weight?
3. What is the pile height and character—shag, plush, frieze, twist—and is it cut or loop?
4. What fiber is it? What are the characteristics of those fibers? Are they resistant to mildew, flames; do they absorb soil? Do they resist abrasion?
5. How resilient is the pile? Does it mat or spring back after being walked on? Is it comfortable?
6. How is the carpet dyed and will the color fade?
7. Is the carpet an appropriate style (color, pattern, texture) for the setting?
8. What are the primary and secondary backings?
9. What type of padding will be used, and how will it be anchored? Does that padding mildew, resist bacteria, help the carpet to wear better?
10. Are the carpet and padding being bought from a reputable dealer? Are they made by a manufacturer experienced in producing contract carpets? Does the fiber producer guarantee the fiber?
11. Who will install it? When? How long will it take? Where will it be stored before installation?
12. Will the supplier discuss maintenance techniques with the executive Housekeeper, and train the employees who will be cleaning it?

Let us now examine each of these 12 considerations in detail.

## *Construction*

Carpet construction involves surface (or pile) yarns and how they are anchored, as well as the yarns in the backing that hold woven goods together. Between 85 percent and 90 percent of commercial carpet yardage is tufted, primarily because it is less expensive than most woven goods. Where a distinct pattern is desired, with many more colors than a printed carpet could display, and where long wear is important enough to warrant a larger investment, demand is increasingly for the relatively expensive Axminster weaves—notably, of course, in areas where decor is of prime importance.

Knitted carpets still find some enthusiastic acceptance. The designers of the interiors for Howard Johnson Motor Lodges, for example, have used knitted carpets since their first installations, and continue to specify them.

Other types of constructions are used for specific situations—needle-punched carpet, for example, for outdoor applications.

The forms that various carpets come in are: (1) *broadloom*—rolls 12 ft. or 15 ft. wide; (2) *rolls*—27 in. wide; (3) *squares* (tiles)—usually 18 in. square with semirigid backing for easy replacement; (4) *area rugs*—in standard or custom sizes, from 3 ft. by 5 ft. to 10 ft. by 12 ft. or larger.

Now to examine carpet constructions briefly:

**Tufted.** The tufting process is accomplished when thousands of needles, threaded with pile yarns and extending 12 or 15 ft. across the machine, are forced through a backing material to form loops or tufts (Figure 11-1 and 11-2). Backing may be woven fabric such as jute, and the loops anchored under the jute backing with a hot layer of latex compound. The pile of tufted carpet can be high or low, cut or sculptured. Its main limitation is that intricate, precise, multicolored patterns cannot be produced, except by printing, on the tufted carpet.

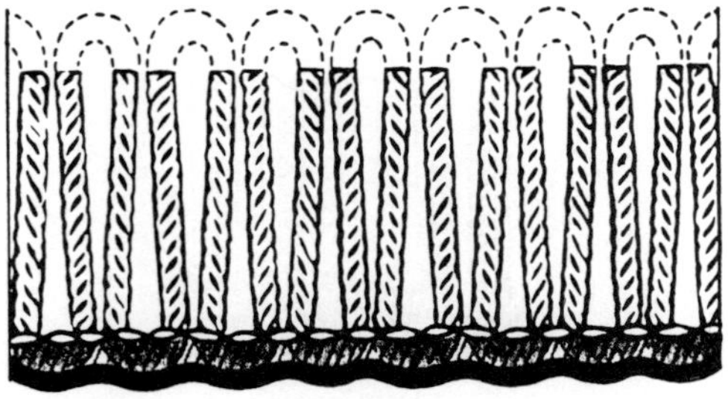

**FIGURE 11-1** *Tufting—About 90 percent of all carpets are tufted.*

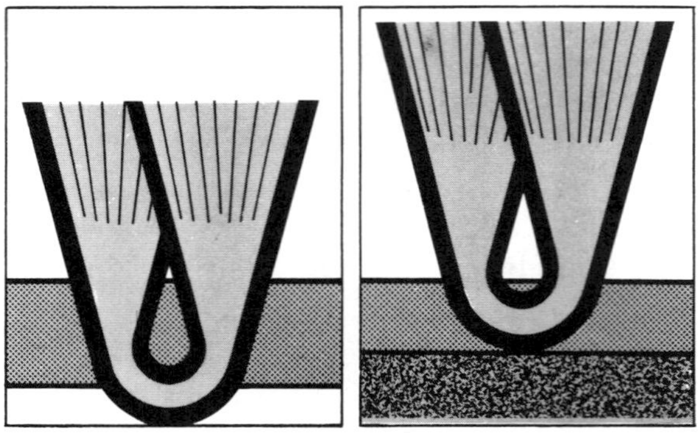

**FIGURE 11-2** *A new tufting technique is fusion bonding, which puts 90 percent of the yarn on the surface. Its heavy vinyl backing also serves as an impermeable layer to stop liquid spills. This construction is said to be more stable because the tuft is locked into the vinyl backing.*

**Woven.** These are constructed on three types of looms—Velvet, Wilton, and Axminster (Figure 11-3). They always have yarns in the backing called warp and weft. Warp yarns run lengthwise, weft (or fill) run crosswise.

The Velvet produces a pile called, naturally enough, velvet, or more recently, plush, which is somewhat longer pile. Although usually only one color and weight of yarn is used on a Velvet loom, variations are possible with the use of multicolored yarns. Note in Figure 11-3 that the tops of velvet pile can be left looped or sheared to make cut pile.

The Wilton carpet is made on a Jacquard loom that produces patterns with up to five colors of yarn. The loom employs punched cards that control the yarns and determine the pattern.

The Axminster loom can accommodate any number of colored yarns and produce precise, complicated patterns. The pile height is always even and the back is heavily ribbed.

All three types of woven carpets are available in 27-in. widths or broadloom rolls.

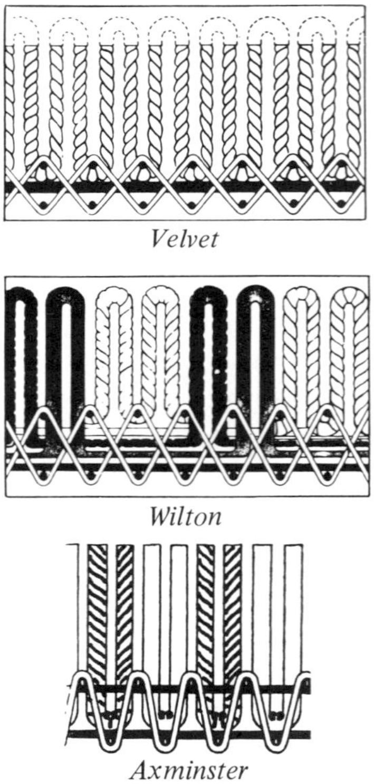

*Velvet*

*Wilton*

*Axminster*

**FIGURE 11-3** *3 Types of Woven Carpets*

**Knitted.** Pile yarns and backing yarns are looped together and secured by a latex backing (Figure 11-4). Usually knitted carpets are produced in solid colors or tweedy effects, although a variety of patterns and textures is possible.

**Needle-punched.** These carpets are made by a process that compacts and mechanically joins fibers to form a felt-like surface. Needles literally punch the fiber into a foam or latex backing. Usually these are used for outdoor installations.

**Flocked.** Synthetic fibers are electronically deposited on soft adhesive that is applied to a woven jute backing. The adhesive is hardened after fibers are firmly embedded. The surface is a single, level-cut pile.

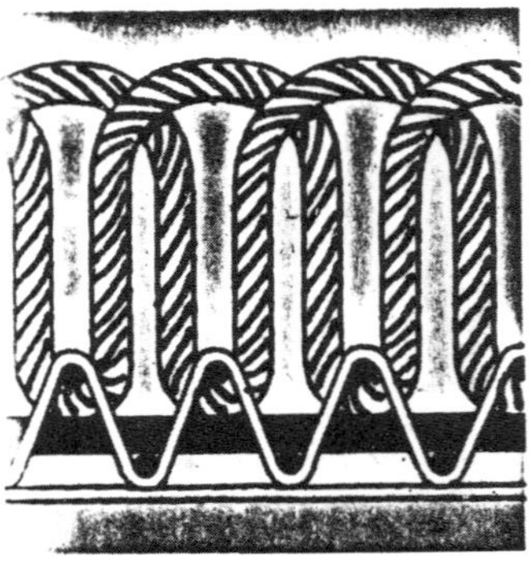

**FIGURE 11-4** *Knitted—A Rare and Sturdy Construction.*

## *Pile Density and Weight*

Any of the carpet constructions described can be made to resist compression despite traffic, and resist abrasion despite soil being ground in and repeated vacuuming, shampooing, and other procedures, if there are sufficient pile yarns in the carpet.

How well any carpet performs is partly a matter of the density of the pile. In tufted carpets, this is expressed by the gauge. A medium-weight carpet would have a gauge of $^{1}/_{8}$, a heavy-duty carpet, $^{5}/_{64}$. Gauge is based on the number of needles (tufts) per width-wise inch expressed in fractions (Figure 11-5). The first half of the fraction indicates number of inches; the second half, the number of the tufts. Thus $^{1}/_{8}$-gauge means there are eight tufts across the inch, while $^{5}/_{64}$-gauge means there are 64 tufts in five inches.

For woven goods, the term pitch indicates the number of warp (lengthwise) yarns in a 27-in. width of carpet (Figure 11-6). A standard pitch for a Wilton is 256, for Axminster 189, and for Velvet 216. The number of tufts per lengthwise inch is expressed in rows for Axminsters, wires for Wiltons and Velvets; this may vary from 4 in an inexpensive carpet to 13 in a densely woven luxury carpet (Figure 11-7).

Yarn weight (face weight) is the number of ounces of yarn per square yard. Maximum is 60 ounces. Usually contract carpet ranges from 20 to 48 ounces. Minimum weights per yard for heavy and medium-traffic areas are given in Tables 11-1 and 11-2, from Bernard Berkeley's *The Selection and Maintenance of Commercial Carpet.*

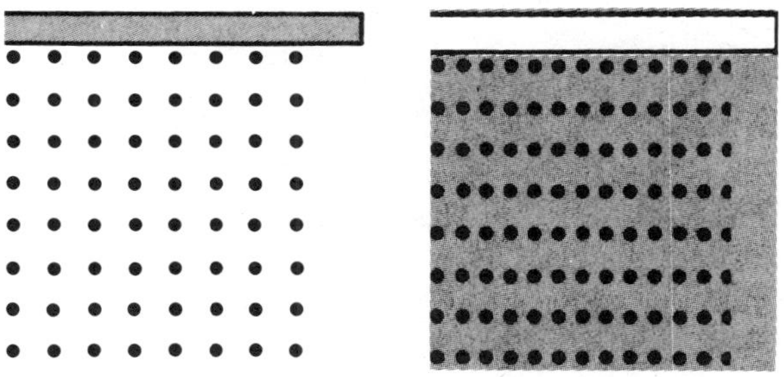

*Tufts per square inch 8 (64 per sq. in.)*

*5/64 (102.4 per sq. in.)*

**FIGURE 11-5** *Methods Used to Measure Carpet Performance.*

**FIGURE 11-6** *Pitch Indicates the Number of Warp Yarns in a 27-In. Width of Carpet.*

## *Pile Height and Style*

The two tables include a reference to average pile height. Variations seem slight (the extremes go from 0.175 to 0.310) but they do account for a notable difference in the total weights.

The length, thickness, and ply of pile yarn will all affect weight of carpet and ultimately help determine its wear. The nature of the pile, however, does have a great deal to do with both the longevity of the carpet, the way it looks, and its performance. On stairs, a longer cut pile is sometimes better than low loop, because there is less "grin," or parting of the yarns, on the stair edge.

A yarn that has been twisted has a different texture than the fluffier one, but they are interesting when used in combination; a high cut-pile yarn is nice next to a low-loop (Figure 11-8). Thus the multilevel, multiyarn surface is appealing in carpets today. It also helps to conceal soil and stains.

Shag carpet (Figure 11-9) has been described as having a "tumbled appearance created in use by the random layover of pile yarn in all directions as distinguished from the normal upright

**FIGURE 11-7** *Rows Per Inch Help Measure Carpet Quality.*

**TABLE 11-1** Minimum Specification Requirements Based on Pile Density for Wool or Acrylic Carpets

| | Average Heavy Traffic | | Average Medium Traffic | |
|---|---|---|---|---|
| | Minimum Weight per Square Yard (ounces) | Average Pile Height (inches) | Minimum Weight per Square Yard (ounces) | Average Pile Height (inches) |
| Axminster carpet | 36 | 0.200–0.310 | 28 | 0.200–0.310 |
| Knitted carpet | 42 | 0.250–0.300 | 36 | 0.200–0.250 |
| Tufted carpet | 42 | 0.250–0.300 | 36 | 0.200–0.250 |
| Velvet carpet: | | | | |
| Woven through the back | 42 | 0.200–0.250 | 32 | 0.175–0.230 |
| Not woven through the back | 36 | 0.200–0.250 | 28 | 0.175–0.230 |
| Twist | — | — | 42 | — |
| Wilton carpet | 42 | 0.200–0.250 | 34 | 0.200–0.250 |

**TABLE 11-2** Minimum Specification Requirements Based on Pile Density for Nylon or Polypropylene Carpets

| | Average Heavy Traffic | | Average Medium Traffic | |
|---|---|---|---|---|
| | Minimum Weight per Square Yard (ounces) | Average Pile Height (inches) | Minimum Weight per Square Yard (ounces) | Average Pile Height (inches) |
| Tufted carpet | 28 | 0.190–0.290 | 22 | 0.190–0.290 |
| Velvet carpet | 28 | 0.210–0.290 | 22 | 0.210–0.290 |
| Woven through the back | | | | |

**FIGURE 11-8** *Nylon—a low loop with pleasant bulk and texture—has low smoke generation and a flame-spread index of zero. Community fire departments should always be contacted for local regulations concerning flammability of carpets and fabrics used in public areas. Credit: DuPont; E.T. Barwick Industries*

**FIGURE 11-9** *Shag (long-cut pile) tufted carpet has jute backing. A blend of four different nylon yarns in four colors, carpet has soil-hiding qualities. This shag is modified, does not have an extra-long pile that mats down easily. Credit: Lees Carpets*

position of the pile in conventional carpeting."[1] To some, shag carpet is a sign of luxury (if the pile is thick as well as long); to others it is simply an inexpensive carpet that contractors put in tract houses.

During the great shag controversy of the early 1970s there was an effort made to upgrade institutions' interiors with shag, but housekeepers and managers alike rebelled. It was too hard to vacuum (creating a drag on the machine); it rarely was vacuumed thoroughly enough; and in guest rooms, pins and staples lurked in the long pile yarns waiting for an unwary bare foot. The fact that shags hide soil is true; the problem is that they hide too much. So the popular pile height for hotel rooms is decreasing into something called "plush," a moderately long pile yarn that is far denser than most shags (Figures 11-10 and 11-11).

The lower the pile height, of course, the denser the weave or tufting should be, for nothing makes a carpet seem sleazier than an exposed jute backing peeking through the pile yarns. Lower pile is far easier to maintain even though it is less glamorous, but with the exciting colors available in carpets today, no dense, level-loop pile need be a bore (Figures 11-12, 11-13, 11-14).

## *Fibers*

Any list of carpet fibers will almost always start with wool as that is the standard against which many of the synthetic fibers are judged. Soft and resilient as it may be, wool has several drawbacks, the most important of which is scarcity. If wool were still available in unlimited quantities at reasonable prices, however, there might have been less impetus for the chemical producers to engineer synthetic fibers especially for carpets. Now in their third generation, some of the test tube fibers are an improvement on wool, since they are mothproof, exceptionally strong, and stain-resistant. Perhaps the contract carpet world is better off as a result.

1. Bernard Berkeley, *The Selection and Maintenance of Commercial Carpet* published by The Cornell Hotel & Restaurant Administration Quarterly, Copyright American Hotel & Motel Assn.

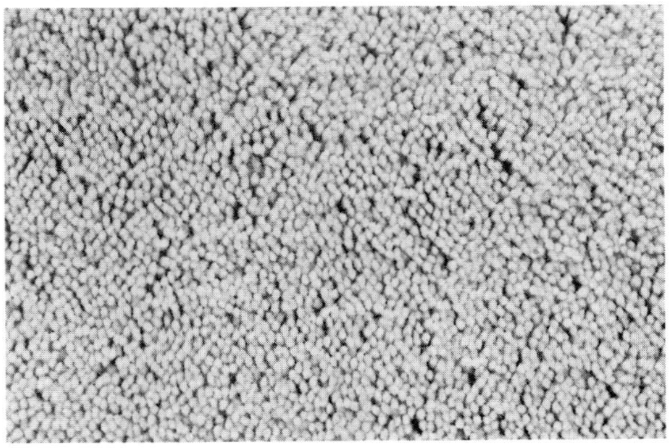

**FIGURE 11-10** *Plush, cut-pile, tufted carpet of Anso-X nylon has permanent static control built into fiber; is of hefty 50-oz. construction. Thick pile such as this resists compression when walked on. Credit: Lehigh Carpets; Allied Chemical Corp.*

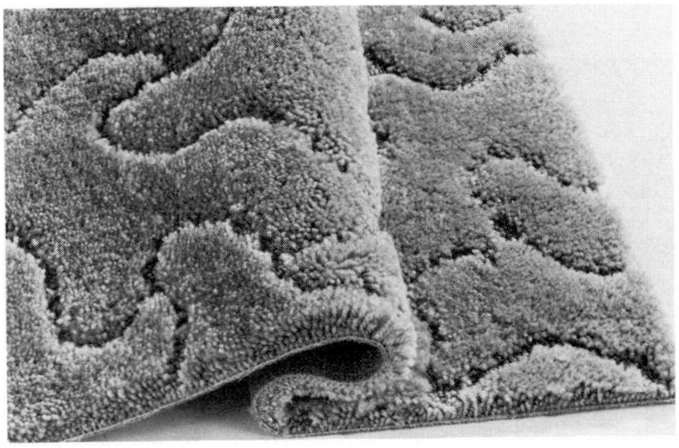

**FIGURE 11-11** *Thick, carved plush effect is made with curving lines formed by low-loop pile yarns surrounded by high cut-pile yarns. Credit: Lees Carpets*

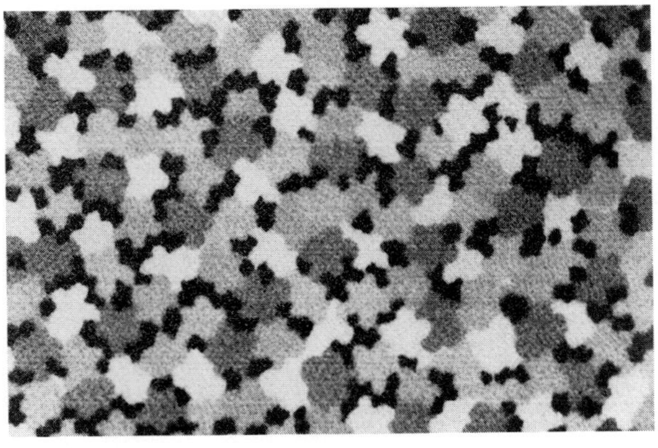

**FIGURE 11-12** *Needlebond carpet is made by punching fibers through high-density foam backing. Pattern is printed on nylon fibers that withstand heavy traffic, even the wear that comes from golf spikes. Credit: Ozite; Allied Chemical*

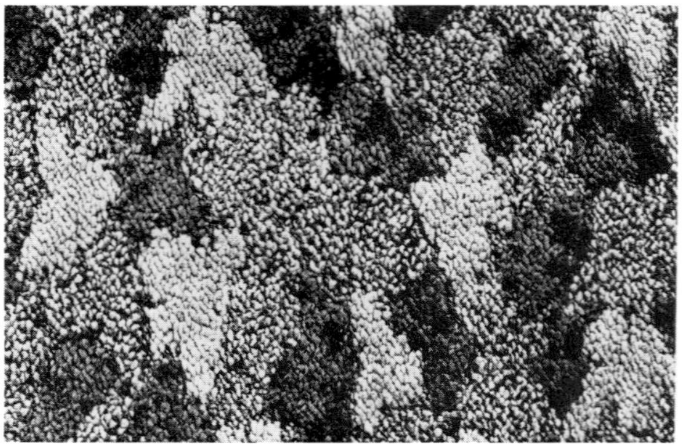

**FIGURE 11-13** *Multilevel loop construction with four colors and textures produces islands of color that will also hide soil and wear. Credit: Burlington House Carpets*

**FIGURE 11-14A** *Plump loop construction with ribbed effect provides tufted carpet of nylon filament available in 11 multi-color combinations. Credit: Stratton Industries; Allied Chemical Corp.*

**FIGURE 11-14B** *Nubby, nylon carpet tufted, loop pile, 1/8 gauge density, for heavy traffic; 12 ft. width on non-woven, synthetic, primary backing with foam secondary backing. Credit: Laurelcrest; Allied Chemical Corp.*

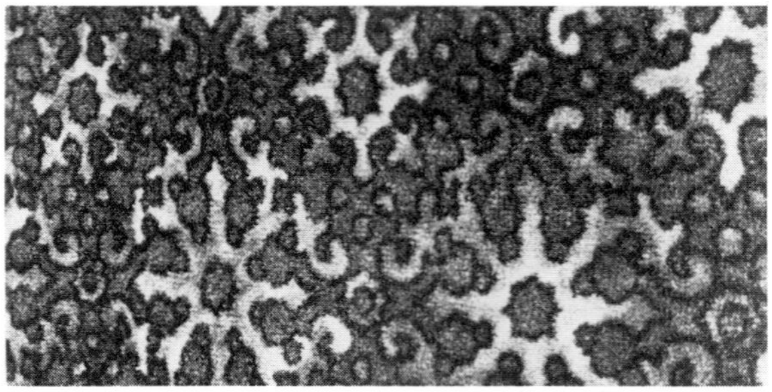

**FIGURE 11-14C** *Printed cut-pile plush of two-ply nylon yarns has a Spanish medallion design in three colors; is being offered in six color combinations, conceals spots. Credit: Monticello Carpet Mills; Burlington Industries*

**FIGURE 11-14D** *Printing the design on carpet after it has been constructed is only way tufted carpet can have distinct pattern; be less costly than woven Wiltons, Axminsters. Credit: Lees Carpets; Burlington Industries*

**FIGURE 11-14E** *"Cigarette-burn-motif" is an ideal pattern for public spaces where burns cause visible damage. This camouflage design was created for an Anso nylon carpet by Hans Koch. Custom carpet designs are frequently used for very large areas in institutions.*

**FIGURE 11-14F** *A distinct pattern best suited for areas where food spills do not occur; this printed, tufted-level, loop carpet of nylon offers a distinct advantage where there is heavy traffic. Credit: Majestic Carpets; Allied Chemical Corp.*

**Example of Carpet Specifications**

| | |
|---|---|
| Type | Velvet weave (woven-through-the-back) |
| Width | 12 ft. |
| Yarn | 100% wool, 3-ply |
| Yarn weight per sq. yd. | 43.92 oz./sq. yd. |
| Gauge or pitch | 216 |
| Pile height | 250 in. uncut |
| Stitches or wires per inch | 8 |
| Total carpet weight per sq. yd. | 77.24 oz. |
| Primary back | Jute |
| Secondary back | Jute |
| Latex per sq. yd. | 8 oz. |
| Brand name | |
| Manufacturer | |

(Name of manufacturer whose carpet was used for above specification and/or names of other manufacturers whose carpet would be acceptable if made to specification, in case several bids are required.)

**FIGURE 11-15**

There are four major types of synthetic fibers used in carpets: nylon, acrylics and modacrylics, polyesters, and olefins (polypropylenes). Although the reader may not recognize some of those highly technical sounding generic names, they do nonetheless represent such nationally advertised, well known products as Kodel, Acrilan, Anso Nylon, Antron, Astroturf, Herculon, Orlon, and so forth.

Table 11-3 and Figure 11-15 give the relative characteristics of the various fibers, but with no subjective evaluation as to appearance or best applications for institutions.

**Nylon.** This is by far the most widely used of the synthetic fibers. It is known primarily for its long-wearing qualities. Experts agree that perhaps, if they had to choose a best carpet it would be a combination of wool and nylon and, indeed, many of the best performing, most beautiful carpets today are just that, made of 70 percent wool and 30 percent nylon. Nylon has gone to charm school since its introduction as a carpet fiber in the 1950s, and has lost its brassy luster, its too-smooth feel, and its tendency to shed and show soil. These defects have long since been corrected. Nylon filaments have been engineered, revised, and improved considerably. Manufacturers have even been able to rid the fiber of any tendency to build up static electricity, and some have been developed specifically as flame-retardant carpet fibers.

**Acrylics and Modacrylics.** These most closely resemble wool in texture and feel. When solution-dyed, the fibers retain color so well that they are even used outdoors (with proper backing) or in a very sunny location inside.

**Polyester.** These fibers resist everything including flame; their long staple gives surface yarn extended life, and they are easy to maintain because of soil-release factors.

**Olefin (Polypropylene).** This relative newcomer to the synthetic world is strong, nonabsorbent, and stain-resistant. Spilled liquids actually do not penetrate the fibers. Olefin carpets are suitable for both indoor and outdoor use; they are the most color-fast of all carpets tested on the standard Fade-o-Meter test. Because of olefin's low specific gravity, more fibers are packed in per ounce, thus allowing greater surface coverage.

**TABLE 11-3.** Comparative Behavior of Carpet Fibers

| Fiber | Appearance | Resilience | Texture Retention | Soil Resistance | Cleanability | Spot Resistance | Fuzzing, Pilling | Static | Dyeing, Colors |
|---|---|---|---|---|---|---|---|---|---|
| Acrylics | Warm, soft, fluffy | Good | Good | Good | Good | Very Good | Some when new | Only in low humidity | Good |
| Modacrylic | Soft, luxurious | Adequate | Fibers usually blended—not a factor | Good | Good | Very Good | Fibers usually blended—not a factor | Resists static | Excellent |
| Nylon | Dull to lustrous | Very Good | Good | Good | Good | Good | Some with staple fibers | Only in low humidity | Good variety of color |
| Olefin | Subdued luster, waxy feel | Reduced | Adequate | Good | Very Good | Excellent | Not a factor | Rarely a problem | Solution- and piece-dyed |
| Polyester | Soft, luxurious—close to wool | Good | Good | Good | Good | Very Good | Some when new | Only in low humidity | Good |
| Rayon | High luster, soft feel | Low | Reduced | Fair | Reduced | Low | Minimum | Minimum | Excellent |
| Wool | Soft, warm, luxurious | Good | Good | Good | Good | May Stain | Some when new | Only in low humidity | Good |

## *Resilience*

How a carpet feels to walk on and how it springs back after crowds tread over it, whether it becomes permanently crushed if furniture rests on it, all enter into carpet selection. In a sleeping room—in hotel or residence hall—warmth and comfort for residents or guests with bare feet or light slippers, are important. Achieving softness and resilience comes in part from using a spongy backing or pad, but the pile yarns for guest rooms should have a more inviting, softer feel than those in busy foyers, lobbies, or corridors.

## *Dyeing*

Color can be introduced at different stages of manufacture in both tufted and woven goods, whether synthetic or wool. The definitions that follow explain some of the different dyeing techniques.

If colors must last in strong sunlight or heavy wear, solution-dyeing is, of course, the most likely to succeed. If more than 600 sq. yd. of carpet in the same color in one area are needed, beware of piece-dyeing because there may be a slight variation in color between different batches.

**Solution-dyeing.** Synthetic fibers spun from a colored solution. When processed under these conditions the filament is completely impregnated with the pigment.

**Stock-dyeing.** Fibers dyed before spinning.

**Yarn- (or Skein-) Dyeing.** Yarn dyed before tufting or weaving.

**Piece-dyeing.** Fabric dyed in vats in one piece after tufting; at this time, a maximum length of 600 yards can be handled in one batch.

**Cross-dyeing.** Multicolored effects produced in one fabric containing fibers or different dye affinities.

The following apply to terms connected with carpet colors:

**Bleeding.** Loss of color when wet due to improper dyeing or from poor dyestuffs; fabrics that bleed will stain fabrics that come in contact with them.

**Crocking.** Color rubbing off as a result of improper dye penetration or fixation.

**Fade-o-Meter.** Standard laboratory device for testing a fabric's fastness to sunlight.

**Greige Goods.** Goods just off the tufting machine or loom waiting to be dyed or finished, usually oyster-colored.

**Mordant.** Chemical used in some textile fibers to increase their affinity for dyes.

## *Aesthetics—Color, Pattern, Texture*

In a spacious lobby or enormous ballroom, a carpet with a large, bold pattern is often needed, and, conversely, keeping the idea of scale in mind, small rooms should not have big busy patterns (Figures 11-16 and 11-17). If people need to be stimulated, the carpet should have warm colors, including some red; if the purpose is to

**FIGURE 11-17** *Grand ballroom of Montreal's Hotel Bonaventure has carpet printed with design in large scale; pictured here, another printed carpet, with large-scale diamond shapes printed over small motifs. Ballroom covers 14,440 sq. ft., accommodates 2500 people. Gold-orange carpet is appropriate for warm wood paneling and contemporary crystal bead chandelier. Credit: Allied Chemical Corp.*

**FIGURE 11-16** *The Chase Park Plaza Hotel, St. Louis, has custom-designed Wilton woven wool carpeting that was basis for redecorating in several areas. Here the Chase Club foyer has a bright active pattern that is focal point of simple contemporary area. Credit: Downs Carpet Co.*

relax people and introduce a quiet air, cool blues and greens are the answer, usually in darker shades.

In dining areas, carpets should not have distinct, precise design motifs; rather they should have vague, somewhat mottled patterns that will not show stains left from spills that could not be wiped up soon enough. Although a certain amount of multilevel pile is attractive and disguises soil, it does complicate the cleaning of carpets in food areas (Figures 11-18 through 11-25).

In heavy traffic lanes or in busy elevators or foyers, the hardiest of all carpets should be specified, low-level loop of nylon or olefin, but here again, not with a distinct pattern or a solid color; any type of shading, speckling, or irregular design motif will disguise wear.

The newer small geometric patterns are of course highly acceptable, and in fact aesthetically desirable *if* there is any real assurance that a relentless maintenance program will be adhered to.

## *Backings*

As was described in consideration number one, ***Construction,*** a tufted carpet has a primary backing into which needles stitch the yarns. That backing may be jute (sisal fiber) or polypropylene in either a punched or woven fabric. A thick layer of latex (a milky adhesive) is applied to the underside of the back, and then a *secondary* backing, usually jute, is bonded to the prime backing for additional strength (Figure 11-26). The primary backing also gives the carpet dimensional stability so that it does not stretch from foot traffic or from carts being pulled across it. A foam back, from ⅛- to ¼-in. thick, may be added to the secondary backing to serve as a self-cushion and eliminate need for separate padding.

Woven and knitted carpets very frequently have a jute or scrim (mesh) backing laminated to the regular woven yarn backing for extra stability and strength.

Vinyl secondary backing sets up a moisture barrier and makes the carpet almost impervious to deterioration. Rubber backing in a waffle design provides grip so that no padding is necessary. A high-density formed latex and rubber material can be attached successfully to the polypropylene primary backing, eliminating the pad and providing a bouncy surface.

## *Padding*

Selection of the proper underlay padding is becoming as complex as choosing the right carpet. In high-traffic areas many carpets are simply glued directly to the floor. Some experts feel that

**FIGURE 11-18** *Ornate Spanish design is printed on low-level loop nylon carpet in an appropriate practical design for a restaurant. Note that carpet is glued down on front and sides of cigar counter in foyer. Often carpet is used up the fronts of bars too, as it conceals scuff marks. Credit: Allied Chemical Corp.*

**FIGURE 11-19** *Carpet in coffee shop of Hotel Bonaventure, Montreal, is a geometric print in level-loop, thickly tufted nylon. "Although it has worn exceptionally well," reports the manager, "the precise design reveals stains that we cannot remove, even with steam-cleaning." Counter seat supports without bases greatly simplify vacuuming and daily foam shampooing. Credit: Allied Chemical Corp.*

carpet installed that way will wear 25 percent longer because friction and air pockets are eliminated (Figure 11-27). New adhesives do not ruin floors but can be scraped up easily if the carpet is removed (Figure 11-28).

Where any type of comfort and luxury is sought, a resilient padding is indicated. If properly mated to the carpet and its backing, a padding can prolong the carpet's life by softening the intensity of abrasive wear. Too much soft padding in itself can cause unnecessary wear on a carpet because of buckling and rippling, especially if carts—such as those carrying room service through guest corridors—go over it frequently.

Paddings or cushionings also provide extra insulation against extremes of cold and heat and add an extra sound-absorbing layer.

Regarding the debatable subject of carpet cushion, we interviewed a knowledgeable hotel executive Arthur Gimson, general manager of the Northfield Hilton outside Detroit, and asked him "What is the principal reason for installing separate cushion beneath carpets in a hotel?"

"Creature comfort," he answered with conviction.

**FIGURE 11-20** *Carpet squares can be removed from the floor of this private dining room to reveal a dance floor of parqueted vinyl tile. Carpet tiles can quickly be replaced. Made of pig's hair, they wear extremely well and resist cigarette burns, which, if they do leave a mark, can easily be sanded off. Credit: Heugatile Corp.*

**FIGURE 11-21** *Characteristic of a new trend toward geometric patterns, the misty effect from this printed cut-pile carpet helps camouflage soil better than distinct motifs. A cut-pile print of tufted nylon, it provides built-in static controls. Backed by Allied Chemical's full five-year wear and lifetime antistatic warranties, it has reduced soiling characteristics so it is a good choice for a restaurant installation. Credit: Gulistan/Allied Chemical*

Gimson's response proved to be an ample summary, as a tour of the 189-unit hotel disclosed. Although not as large as many other Hilton-operated hotels, the initial impression is that it is no less elegant.

The lobby is spacious, well lighted, and handsomely decorated with plush carpeting that provides the setting and the tenor for the rest of the hotel.

If creature comfort is the foremost reason for carpet cushioning, what ranks second?

"Warmth," he declared without hesitation.

"Isn't warmth the same thing as creature comfort?"

"Not necessarily," said Gimson. "Most hotels are constructed using concrete slabs for floors. The type of cushion may vary by geographical location, depending upon weather conditions.

"Here in Michigan we need a cushion that feels soft underfoot, but still gives us the type of insulation that keeps the building warm." In these days of high energy cost, Gimson considers that factor a major plus in the operation of the Northfield Hilton, which relies on electricity for approximately 90 percent of its heating needs.

"Creature comfort and warmth . . . anything else?"

"Sound absorption."

That, too, is obvious in the lobby, where bellpersons have to push heavily-laden baggage carts.

"What about wear and tear, doesn't a heavy cart like that cause a maintenance problem?"

"Absolutely," Gimson replied, "but the problem is greatly reduced if you match the right type

**FIGURE 11-22** *Helping to prevent wear in areas where pivoting and friction occur, sinewy curves and twirls are tufted of 100 percent nylon on a background of natural Belgian linen. This area rug forms a focal point in a luxurious lobby. Credit: Regal Rugs Co.*

**FIGURE 11-23** *Light, sunny, sculptured, high-low, loop-and-cut pile carpet in guest rooms at Harrah's Lake Tahoe Hotel is also used on bench base at foot of bed and on desk bases. Because rooms in this resort area hotel each have two bathrooms to be cleaned and extra furniture (such as computerized beverage unit behind maid), maids are only expected to clean 12 rooms per day.*

**FIGURE 11-24** *Where area rugs are combined with hard surface floors, as in the Missouri Athletic Club lobby, cleaning chores are doubled as two different types of equipment are needed. If elegance and tradition are revered, as in this institution with its favorite room where gentlemen meet to converse over cigars, maintenance budget is adjusted accordingly.*

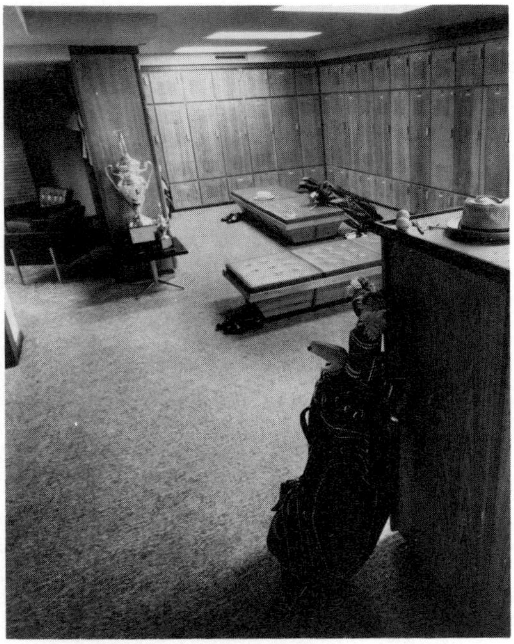

**FIGURE 11-25** *Nylon carpets are used throughout the PGA National Golf Club in Palm Beach Garden, Fla., including men's and women's locker rooms. The nylon pile, engineered to resist effects of soiling, wear, and spike-shoe traffic, provides luxurious comfort underfoot. Credit: Cabin Crafts; Monsanto's Cado Nylon*

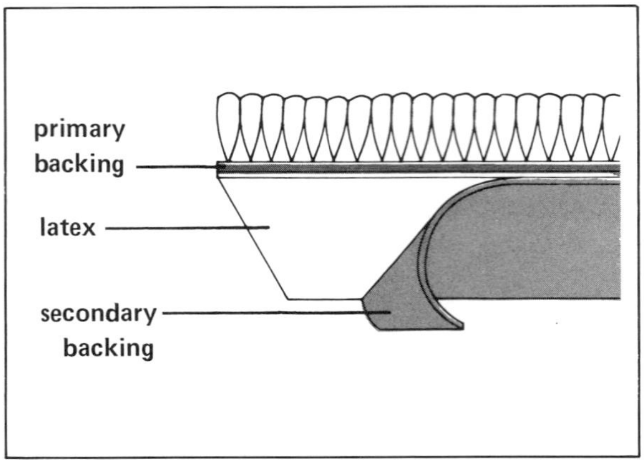


**FIGURE 11-26** *Backing for Carpets*

of cushion to the carpet. Heavy-traffic areas, such as the lobby and hallways, require different cushion than the restaurant and lounge. If you put down too heavy a pad in areas where some of the traffic involves heavy equipment the carpet tends to wave and buckle. That leads to split seams and expensive repairs. On the other hand, if the cushion is too light, it has no feel."

About the financial aspect of cushion, Gimson's carpet file indicated that cushion costs are about 10 percent of the total cost of carpet and installation and that 900 square yards of carpet, with cushion, had been laid down in the ballroom. How do you dance on a carpet? "Easy . . . we put a portable floor down *over* the carpet."

Although new paddings are appearing very often now, the conventional ones include:

**Felted Hair.** This is the most economical. It has a waffle design to provide a skidproof surface and improve resiliency. It is sometimes reinforced with a jute backing or with a burlap center liner. Sizing (adhesive) is used on occasion to strengthen the bond between the fibers and the burlap core. Many manufacturers also sterilize and mothproof hair padding.

Felted padding of hair or hair and jute may mat down in time or may develop mildew, especially if the fibers become wet, as during cleaning. When properly cleaned, sterilized, and treated, however, hair padding is suitable for reasonably dry floors at all grade levels, and also on conventional radiant-heated floors.

**Rubberized Fibers.** Some padding made of jute or hair is coated with rubber on one or both

**FIGURE 11-27** *Double jute-backed carpet is being glued directly to floor. Many experts say carpet wears about 25 percent longer, especially where mobile equipment is used, if glued down to eliminate friction and air pockets.*

**FIGURE 11-28** *Most frequently voiced objection to glue-down method was difficulty and expense of removing carpet when it had to be replaced. Now problem has been solved by new adhesive informally called "stick and peel." This release adhesive has cohesive characteristics so that on nonporous floors it will pull up—no scraping is necessary to remove worn areas.*

sides to hold the fibers together and provide additional resilience. Sometimes this padding has an animal hair waffle-top and a jute back reinforced with a patterned rubberized application.

**Cellular Rubber.** In addition to cushioning bonded directly to the carpet, foam or sponge rubber is produced in sheet form with waffle, ripple, grid, or V-shaped rib designs. A scrim of burlap fabric usually is bonded to the rubber sheet to facilitate installation of the carpeting; when laid with the fabric side up, this permits a taut and even stretch of the carpet.

Rubber padding is more expensive, but it retains its resilience longer than hair padding. It is highly resistant to decay and mildew and is nonallergenic. It can be used at all grade levels, but the denser cushionings are not recommended for radiant-heated floors.

## Reputable Sources

Because the carpeting and padding of large areas requires special know-how and involves thousands of dollars, it is essential to deal with reputable, experienced manufacturers, dealers, and installers. Beware of too low bids. There are no bargains.

Each commercial carpet installation is a custom job with its own specific set of conditions, from the openings to be provided to the proper placement of seams. The schedule, the methods of installation, and the responsibilities of all parties concerned should be reviewed and clearly understood before bids are written or purchase orders are made out.

The manufacturer of the carpet and the contractor who lays it should be consulted on the method of installation, and every precaution should be taken to obtain satisfactory results. For example, the condition of the existing floor should influence selection of padding and method of installation, and the amount of activity that must be carried on in the area where installation occurs should also be taken into consideration.

Hauling heavy equipment, fixtures, or partitions over newly installed carpet may cause puckering. To prevent this, during the moving-in make sure that the carpet will be protected with plywood or similar material placed over the heavily trafficked area. A guarantee on the carpet, and perhaps on the fibers, should be secured; one producer advertises a five-year wear guarantee on its nylon contract carpet.

## Installation

Generally the dealer takes responsibility for installation. If for some reason he cannot, then he should recommend a reliable installer. If one is not readily available (good ones are busy), housekeepers should not be tempted to take on the job for it is fraught with pitfalls. There should be a written agreement as to when the installation is to be made and what the cost will be. The contract covering installation of corridor carpet may include a clause providing the installer to restretch carpet after six months. This

TABLE 11-4.

| Product | Construction | Face Fiber | Yarn Ply | Machine Gauge (inches) | Stitches per Inch | Finished Pile Height (inches) | Face Weight (ounces) | Primary Backing |
|---|---|---|---|---|---|---|---|---|
| P/1140 Magistrate Coll. | Tufted one-level cut pile | 80% wool 20% nylon | 1 | 1/10 | 15.63 | 0.328 | 34.02 | Typar® |
| P/1543 Intermission | Tufted two-level loop | Antron® III | 1 Co-mingled | 1/8 | 10.42 | 0.182 | 26.03 | PolyBac® |
| P/1670 Artique Coll. | Tufted one-level cut pile | Nylon | 2 | 5/32 | 5.70 | 0.388 | 24.05 | Typar® |
| P/1729 Bankers II | Tufted one-level loop | MILACRIL® | 3 | 1/10 | 6.70 | 0.250 | 41.92 | PolyBac® |
| P/2050 Corp. Square II | Fusion-bonded cut pile | Antron® III | 2 | 1/9 | 7.90 | 0.250 | 28.91 | Vinyl |
| P/4080 Times Square | Tufted one-level loop | Antron® III | 2 | 5/64 | 10.20 | 0.140 | 20.02 | PolyBac® w/ vinyl precoat |

will remove wrinkles and bubbles caused by the traffic of heavy wheeled vehicles—maid carts, Room Service tables, and bellman or houseman carts.

One should try to coordinate the carpet shipment with the installation dates unless the dealer is willing to store the carpet and deliver at the right moment. Usually huge rolls of 15-ft. broadloom are a problem to store, so as soon as the carpet is delivered, the installation should be started. In public areas of a busy institution, late at night or on weekends is the best time unless that involves double pay. One must weigh that against the cost of taking a profitable public area out of service.

Carpeting may be installed over finished flooring, subflooring of concrete, plywood, particleboard, or any other suitable smooth surface (including other carpet). Carpet that is not rubber-backed can be installed over nailable surfaces by tacking the edges or by using a tackless strip around the perimeter.

Installations on concrete can be made by tacking, if wooden plugs are provided in the concrete slab to receive the tacks. Also, tackless strips can be installed on concrete with power-activated fasteners or epoxy adhesive. Regardless of the installation method, the carpet should be stretched sufficiently to eliminate wrinkles.

The question currently debated by installation experts is whether to glue down carpets backed with foam rubber or double jute, or whether to use the traditional method of separate padding and carpet stretched over tackless strips. Even experienced commercial installers admit it is strictly a matter of opinion as to which technique should be used, what type of backing, and what type of adhesive. Each location in

| Secondary Backing | Total Weight (ounces) | Width (feet) | Number of Colors | Traffic Class | Special Features | Flammability: Tunnel Test | Flammability: Radiant Panel | Flammability: NBS Smoke |
|---|---|---|---|---|---|---|---|---|
| Leno-Woven Synthetic | 76.38 | 12 | 11 designs 48 colors | Heavy | Scotchgard® 3M static control | Class B | 0.996 | 266 |
| Leno-Woven Synthetic | 63.98 | 12 | 16 | Heavy | Wear/anti static guarantee 17-lb. tuft bind | Class A | 1.07 | 371 |
| Leno-Woven Synthetic | 65.66 | 12 | 7 designs 25 colors | Moderate | Scotchgard®/ 3M static control FHA | Class B | 1.00 | 310 |
| Leno-Woven Synthetic | 83.30 | 12 | 18 | Heavy | Natural appearance value | Class A | 0.500 | 229 |
| Reinforced Vinyl | 167.70 | 18″ × 18″ | 18 | Heavy | Modular wear/antistatic guarantee | Class A | 0.860 | NA |
| Reinforced Vinyl | 120.21 | 18″ × 18″ | 12 | Heavy | Modular wear/antistatic guarantee | Class B | NA | NA |

each institution may require a different approach, depending on the activity to take place there.

Where the floor is in particularly bad condition and budget does not permit proper filling and patching or replacement, a separate thick pad will help camouflage the unevenness.

## *Maintenance and Training*

If the carpet to be purchased is unlike any that are already in use in the institution, then the floor cleaners, housemen, or maintenance men should be given instructions about how to vacuum, shampoo, and steam-clean it.

If he wants to keep a customer's business, a dealer should expect to have to demonstrate proper care for a new carpet.

The most important point to be discussed are the chemicals to be used on it—what types of detergents, whether to use dry foam, what will happen if a very wet shampoo is used, will backings be affected, and so on.

A case history may help to point up some reasons for selecting rubber-back carpet with the glue-down plan in mind.

One of the most enthusiastic supporters of this method is Ruth Coopersmith, housing manager of the residence halls at Fresno State College, Fresno, Calif. For four years, Mrs. Coopersmith carefully evaluated the performance of the nylon tweed carpet in use. It was tightly woven (10 stitches to the in.), with a rubber backing, and was glued directly to the concrete floors throughout the building.

In one residence hall, 103 suites with living rooms (including refrigerators) and bedrooms house men students who are responsible for their own housekeeping. The carpet there wore well, did not buckle or split, and did not come loose even where water was spilled.

In another residence hall where the conventional installation technique (with a separate padding) was used, there were many problems, reports Mrs. Coopersmith. Buckling, seams splitting, and seams that loosened because of water standing on them were the unhappy results.

Where glue-down carpet has been installed in public areas—halls, lounge, and entry—it is vacuumed daily and spots are cleaned with dry compound. If students spill anything on the carpet in their suites, they are asked not to clean it up but to report it immediately so that the housekeeping staff can do it promptly and properly.

The goal of Fresno State's housing manager was to keep the carpet in the residence halls in acceptable condition for eight years, and she succeeded.

As carpet technology becomes more complex, and produces carpets to fill more needs, the specifications become more complicated. In Table 11-4 one manufacturer, Millikin carpets, included 17 characteristics or qualities that describe specific carpets. Although we include only six different carpets, many many more stock items are available and countless combinations of the above specifications. Note: the term modular here means carpet squares.

# Chapter 12

# Carpet Care

**OVERVIEW** Housekeepers should try to prevent the problem of tracking in outside soil by providing runners and mats at entrances. Carpet care differs from public areas to upstairs guest rooms. Today there are new theories on carpet cleaning but it will be difficult to change old established habits in carpet care or, alas, the lack of it. Equipment and chemicals are reviewed in this chapter that deals with Housekeeping's most urgent concern.

## *Protective Carpet Coverings*

Any cleaning and maintenance program devoted to carpet care must begin with that familiar ounce of prevention—the use of floor mats and runners to clean soil or moisture from the shoes of incoming guests. Before any discussion of cleaning machines or shampoo techniques, let us first consider what floor mats can do.

The ingenuity of a Minneapolis grade-school custodian is instructive as a reminder that common sense and a little imagination can often solve housekeeping problems for which there are no scientific solutions.

During a severe Minnesota winter, the custodian found the task of keeping the corridors clean virtually impossible as hundreds of children entered and reentered with shoes, boots, and galoshes covered with snow and mud.

Wide floor mats placed outside in front of the doorways having proved almost completely ineffective, the custodian cleaned and dried the matting, brought it indoors, and gathered together all the clean rags he could find. Then he stood by the doorway and good-naturedly asked the incoming children to wipe their feet on the mats and then dry their boots and shoes with the rags he had provided. Because of the custodian's friendly approach, but also because most of the youngsters were accustomed to following exactly the same procedure when they entered their homes, their cooperation was both gratifying and effective. The result was a container full of soiled wet rags and very wet mats, but such a minimum amount of moisture that the custodian was able to complete his work in less than half the usual time.

The chances of getting college students to stop and wipe their feet—with or without rags—before entering a building are probably slight, and obviously a hotel cannot insist that its guests wipe their shoes on mats and then dry them with rags before entering the lobby. However, the mere presence of an attractive mat (and there are dozens of styles today from which to choose) will effectively remind guests to pause long enough to wipe off most of the water, snow, or soil. A pair of mats—one inside and one out—is doubly effective.

Matting should be used in connection with grills and should be changed as often as is necessary. The soiled mats should be mopped, wet-vacuumed, or shampooed before reuse.

For optimum soil removal at entrances, however, also use a large piece of carpet immediately inside the entrance, preferably of the same pattern as the lobby or foyer carpet, to pick up soil or moisture not caught by the mats. It is also important that all employees—especially bellmen, carhops, and doormen—be reminded to use the mats themselves, preferably in such a demonstrative manner as to bring into play the power of suggestion to wet-footed guests entering the hotel. If the architect has planned well, there will also be a metal grating or grill to intercept soil, either outside or immediately inside of each entrance.

Automatic doors in institutions often lead directly from the lobby to the outdoors. They are activated by treadle plates covered in matching carpet without padding. These areas receive the damp and soil of entering guests and need daily attention.

### *Floor Mats and Runners*

These are made for five important purposes: (1) to remove dirt, mud, or water at entrances, so that floors will not become heavily soiled (these are often called walk-off mats); (2) to provide comfort for people standing and walking in specific, limited areas; (3) to eliminate danger of slipping or skidding; (4) to prevent wear on floor or carpets; and (5) to control noise. Some mats and runners will fill all of these functions. Some need adhesive to hold them in place, but most are made so that they will cling properly to the floor and not pose any danger of tripping.

Some entrance mats are integral to the floor, fitting down into a recessed frame. Since they are custom-cut to fit their individual frames, they are not easily dislodged by foot traffic. Most, however, are designed to be laid over carpets or hard surface floors.

Link mats, either rubber or vinyl, are designed to withstand heavy traffic, but are not recommended where mobile equipment will be used often. They can be obtained in a number of colors or patterns, and are also available in distinctly corrugated surfaces, or molded with a grid of upright "fingers"—either with beveled edges for surface installations, or squared edges for recessed areas.

Characteristics of vinyl mats include the following major advantages:

1. Color stability and brilliance
2. Abrasion resistance
3. Flame retardance
4. Resistance to acids, chemicals, and ozone

Characteristics of rubber link and corrugated mats include these desirable features:

1. Resistance to deformation at high temperatures
2. Flexibility at low or high temperatures
3. Excellent reaction wet or dry
4. Stability—nonshrinking, dimensionally stable
5. Excellent traction

Light, medium, or heavyweight vinyl runners with or without sponge backing are available for situations in which floor protection, noise control, or foot comfort are needed. Many embossed surface patterns are offered.

Vinyl-backed cocoa mats provide brush action for removing sand, snow, or moisture.

Heavy-duty link mats made of recycled auto tires give extra protection in slippery areas and offer good drainage. They are available as runners, mats, or inserts.

Flexible steel mesh mats give excellent service in kitchens or back bars but are hazardous for women wearing small-heeled shoes.

Flexible duckboards of wood or synthetic rubber that roll up for easy removal to a washing area are excellent for kitchens, back bars, counters where food is served, and in shower rooms, locker rooms, or walkways around pools or beaches (Figure 12-1). Rubber runners with ribbing to prevent slipping, vinyl links, and composition open-grid patterns are used for the same purposes as duckboards.

Mats of cut nylon pile bonded to vinyl backs are used in elevators and to remove moisture or soil at doorways (Figure 12-2). They are especially suited to pro shops and locker rooms, where golf shoes with spikes cause problems (Figures 12-3 and 12-4). Some of these have backings designed to fit into recessed frames; others are for use over carpet and grip the surface, having been engineered not to creep. These are most effective when used at entry from kitchen into dining area to clean grease off shoes.

Some carpet mats are designed to be used indoors or outdoors. Durable, continuous-filament olefin pile is most likely to survive outdoor conditions. This is often made into simulated grass and used year-round. It can be vacuumed or hosed. The "grass" comes in several colors. The famous AstroTurf made of polyethylene blades has a skid-resistant spongeback.

Ridged carpet mat has replaceable treads of 70 percent acrylic, 30 percent nylon. These are mounted in lightweight aluminum rails. As people walk across the mat, mud and water are scraped from their shoes, falling through spaces between the rails to the floor. The spaces between the treads are safe for all types of shoes and boots, including spike heels. In addition, carts and material-handling equipment roll easily across this type of mat.

The scraping-cleaning action and high absorbency of the deep-pile carpet treads keep lobbies, corridors, foyers, and entranceways clean, dry, and nonslippery. In addition, constant mopping is unnecessary.

As needed, the mat is rolled back and the dirt and water that have collected beneath it are cleaned up.

Carpet tiles woven of strips of buffed truck tire material bonded to glasscloth backing can be installed indoors or out, can be recessed or cemented into place, or can have beveled sides for surface use.

Just as hard and resilient floors surface in

**FIGURE 12-1** *Duckboard matting is recessed in kitchen floors to allow spills to run through so that floor will not be slippery. Matting is flexible and can be removed for cleaning. Floor beneath can be hosed.*

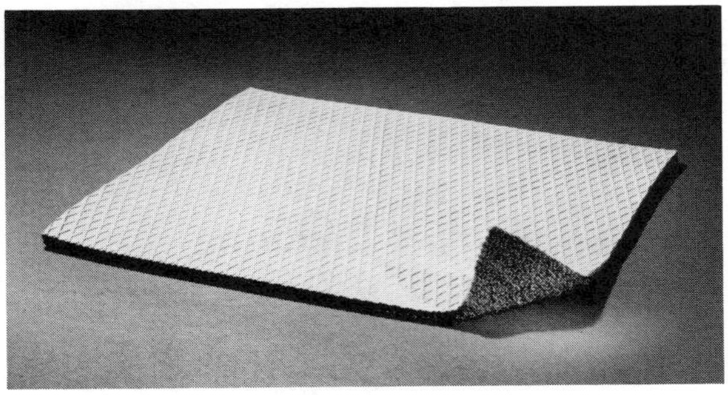

**FIGURE 12-2** *Just a simple piece of carpet with nonslip backing can be positioned at entrances to carpeted rooms, where it will serve as a considerable deterrent to soil build-up, especially if it is changed often.*

**FIGURE 12-3** *Lightweight, durable nonwoven vinyl is used for protecting floors, for comfort, and as a safeguard against slipping. Its porous construction lets dirt filter through keeping the surface clean. Useful in elevators, doorways, behind counters, etc. Credit: 3M Co.*

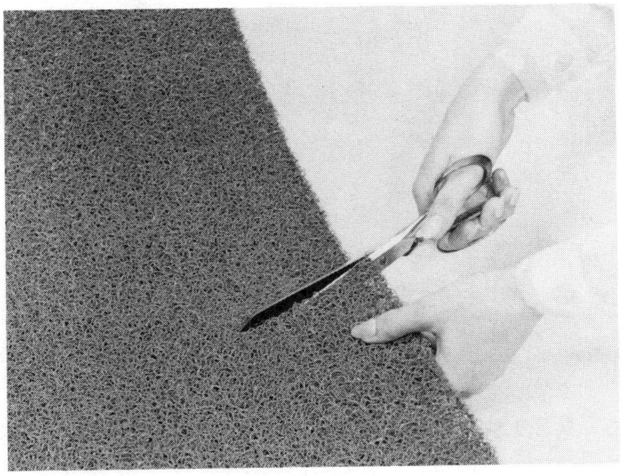

**FIGURE 12-4** *Scissors will cut the curly vinyl surfacing material available with foam backing for dirt pick-up and retention, and without backing for wet areas. Flame-resistant material is cleaned by shaking, vacuuming, or washing.*

new forms every year and carpets have endless new twists, so improved types of runners and mats will be offered.

## *Carpet Cleaning Upstairs and Downstairs*

Unlike the care required for student rooms in residence halls discussed in the previous chapter, carpet care in guest rooms of hotels and clubs and the corridors on the upper floors is a simple matter. These are locations that do not bear the brunt of dirt being tracked directly from outside, or the continual wear and spotting that accompany foodservice activities. Only in hotel hospitality suites are the upstairs carpets a real problem to keep clean, or in areas where Room Service personnel spill food and forget to report it to Housekeeping. For the most part, however, routines are sufficient that include: daily vacuuming or sweeping with electric brooms by maids, thorough weekly vacuuming done by housemen, periodic dry or foam shampooing done by housemen, and annual steam cleaning (usually done by an outside firm).

In the public areas, however, carpet maintenance is an expensive and endless activity. Restaurant carpets sustain so many spills that they must be vacuumed after each meal, or swept up with a toy broom and long-handled dust pan, or tidied up with a tiny, all purpose carpet sweeper (Figure 12-5), and then shampooed with a dry foam nightly. Entrance areas in lobbies usually require some sort of nightly shampooing as well.

**FIGURE 12-5** *Popular all-purpose sweeper is tiny, easily stored, and lightweight. It is especially useful in dining areas where busmen must quickly pick up loose debris before it is ground into carpet. This works on an electrostatic principle. Dust or debris are attracted to the underside. Small metal sweeper works on hard surfaces as well as carpeting.*

The action of walking on gritty soil and embedding it cuts carpet fibers and wears out the pile; if spots and oil-based spills are allowed to remain, it will take powerful detergents and solvents as well as abrasive action with a brush to remove them.

Regular and relentless carpet care is vitally important, not only to keep floors looking clean and attractive but as a means of protecting a major investment.

## *Philosophy of Carpet Care*

As with many matters in housekeeping that are affected by fast-moving technology, a change in the whole approach to carpet cleaning is taking place. For years, conscientious Maintenance or Housekeeping directors used a foam shampoo. Now experts are finding that the foam is as big a villain as the dirt, for much of it remains in the pile yarns leaving a sticky residue that only attracts more soil. While chemists are busy rewriting their formulas to prevent this, housekeepers are having professional steam-cleaners come in to rid the pile yarns of their soiled foam. They are investigating also the purchase of hot-water extraction machines in order to have an ongoing program to purge carpet pile of such built-up residue.

There is a difference between maintenance and cleaning when it comes to carpet. Most people only clean it, and they don't do that until it's really too late to remove embedded soil. Management of most hotels waits until the carpet looks dirty before the shampooing begins and then it can't possibly be cleaned.

About 70% of the dirt in carpets is less than 3 microns in size (the human hair is 50 microns thick, and 10 microns is all that a human eye can see). By the time you see the dirt in the carpet, there is so much of it there that the fibers really can never come as clean as they were when the fabric was new.

If a proper maintenance program were followed to prevent soil from being embedded before it was visibly dirty, the heavy carpet investment could be protected.

Fibers used in carpets once were round, and the soil came off easily. Technology today has developed soil-hiding yarns. Instead of being round, they are amorphous and kidney-shaped so that light does not reflect the tiny particles of dirt. In fact, the dirt actually congregates in the crevices of the nonround yarn fibers. So it's even more difficult to clean the carpet made of soil-hiding yarns, since each fiber has considerably more dirt embedded in it.

A new system has been developed by a carpet manufacturer and maintenance machine maker to ensure proper cleaning and maintenance. When a buyer purchases a carpet, he gets a service contract that offers him regular maintenance on the carpet with a dry-particle cleaning system.

This is more preventive than corrective. In other words, the first soil is quickly picked up so that the carpet never really gets noticeably dirty. Most carpet cleaning involves using an organic detergent or substance which, though it picks up soil, leaves a residue that then attracts more soil. Carpets that have been cleaned with organic cleaning compounds certainly do soil faster. Many a director of housekeeping was reluctant to have a new carpet shampooed for the first time, knowing that it would never quite look the same thereafter.

With the new maintenance system that uses a man-made polymer and water-base system, a bond is created between the cleaner (a dry powder) and the soil so it can easily be picked up in a vacuum. Using this *dry* cleaning system, downtime is negligible and high-traffic restaurant and public areas can be cleaned during regular hours since the public can walk on it immediately. With traditional carpet cleaning, which involves wet solutions, downtime is necessary—the area has to be blocked off so customers won't walk on it, and business can be negatively affected (Figure 12-6).

Using traditional damp-shampoo methods is very expensive because: (1) carpet cleaning usually is done at night on the more expensive labor shift; (2) if done in winter, buildings have to be heated so that the wet carpet will dry quickly; and (3) labor is needed to move and replace furniture.

In the dry system, "Concept 4000," it is not necessary to shampoo the entire carpet, and one need not move furniture except occasionally. The system is particularly useful for frequent cleaning of badly soiled areas or high-traffic lanes.

Since this new system (which uses a soil-release prespray and a powdered soil extractor called Capture) is only really effective on new carpets that have been cared for by this method exclusively, it may be too late for older carpet to benefit from this new approach. If carpet fibers are matted with dirt and soap residue, it's possible that hot water extraction cleaning prior to using the new Capture powder Concept 4000 method will condition your carpet sufficiently so you can start on a new dry-cleaning program.

## *Hot Water Extraction*

Now on the market are uncomplicated extraction machines that are not much larger than a conventional upright vacuum. There are no tanks to fill or empty, no solutions to heat, no separate wet-vac to carry along (Figures 12-7, 12-8, and 12-9).

Incoming water (from a hose connected to a faucet) is automatically mixed with a cleaning agent and shot into the carpet by high-pressure spray nozzles. Dirt is flushed to the surface and is picked up by the vacuum and carried off to a sink drain through the discharge hose (combined with the supply hose). No sticky or oily residue is left behind if there was not too large a build-up of embedded chemicals, and if the operator used the machine properly.

The chief drawback to the extraction machine is that the carpet is far damper than when it is dry-foam shampooed. Unless the area is non-humid, warm, and well ventilated, it may not dry before it must be walked on.

Hot water extractors are not satisfactory for upstairs guest rooms and corridors in a high-occupancy property because the need to connect the supply/discharge hose means guest bathrooms have to be used, and hoses may be trip-

| METHOD | EQUIPMENT/PREP | | | CLEANING PROCESS | | | | | | POST CLEANING AFFECTS | | | | | |
|---|---|---|---|---|---|---|---|---|---|---|---|---|---|---|---|
| | EQUIPMENT | MANPOWER | REQUIRED OPERATOR SKILL | AVERAGE HOURLY COVERAGE | TRAFFIC DOWNTIME | EARLIEST APPRAISAL TIME | ANNUAL CLEANING FREQUENCY | BLENDING ABILITY | CHANCE OF COLOR BLEEDING | AMOUNT OF RESIDUE | RESOIL RATE | SHRINKAGE/ SEAM SPLITTING | MILDEW | SPOT WICK BACK | BROWNING |
| ROTARY WET |  |  | MOST SKILLED | 719 SQ. FT. | 6-24 HRS. | 6 HRS. | 4 TIMES | GOOD | POSSIBLE | HEAVY | RAPID | LIKELY | LIKELY | VERY LIKELY | POSSIBLE |
| ROTARY SPIN PAD |  |  | LESS SKILLED | 1600 SQ. FT. | 1 HR. | 1 HR. | 6-8 TIMES | GOOD | POSSIBLE | MEDIUM | VERY RAPID | UNLIKELY | UNLIKELY | LIKELY | POSSIBLE |
| MIST |  |  | LESS SKILLED | 1200 SQ. FT. | 1 HR. | 1 HR. | 3 TIMES | FAIR | POSSIBLE | LIGHT | SLOW | UNLIKELY | UNLIKELY | LIKELY | POSSIBLE |
| FOAM |  |  | MORE SKILLED | 1000 SQ. FT. | 1 HR. | 1 HR. | 5-7 TIMES | FAIR | POSSIBLE | HEAVIEST | MOST RAPID | UNLIKELY | UNLIKELY | LIKELY | POSSIBLE |
| HOT WATER EXTRACTION |  |  | MORE SKILLED | 638 SQ. FT. | 6-24 HRS. | 6 HRS. | 2-3 TIMES | POOR | POSSIBLE | LEAST | SLOW | VERY LIKELY | VERY LIKELY | UNLIKELY | POSSIBLE |
| DRY SOIL-EXTRACTION |  |  | LEAST SKILLED | 750 SQ. FT. | NONE | IMMEDIATE | 1-2 TIMES | EXCELLENT | NONE | NONE | VERY SLOW | NONE | NONE | NONE | NONE |

**FIGURE 12-6**

ping hazards. Only where a large number of janitors' closets with sinks are located can this kind of effective carpet washing be done.

The steam (hot water extraction) machine that is useful on hotel guest room floors is one that carries a tank for the hot water and a chamber for the soiled water (Figures 12-10 and 12-11).

Before these machines can be effective on carpets that have been repeatedly dry-foam shampooed, a defoamer may have to be sprayed into and removed from the carpet so that hot detergent solution injected under high pressure can penetrate better.

Another means of removing the residual detergent and soil is with an extracting machine. After shampooing, the machine, which looks like a large tank vacuum on wheels and has a hose with a spray head 14 in. wide also on wheels, sprays hot or cold water into the pile, then squeegees the nap and recovers the foam and deep soil. This step is followed with a powerful wet-dry vacuum. This two-step operation is best accomplished with two operators working the machines.

There are many makes and styles of hot water extraction machines on the market today—each with an impressive capability for removing embedded soil and sticky residues. Unless used by a competent operator, they may,

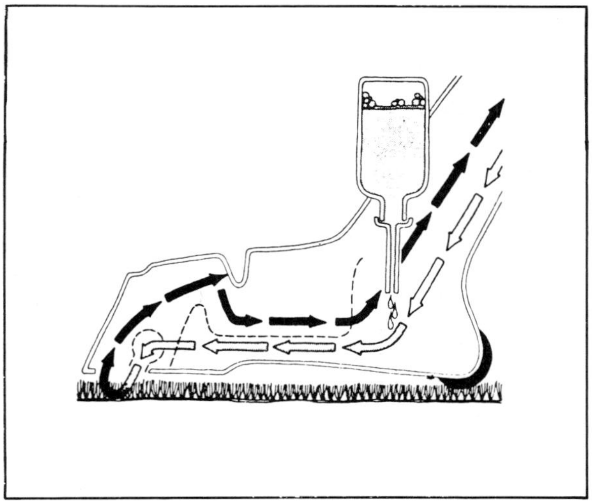

**FIGURE 12-7** *A compact hot water extraction machine. Credit: Clarke Floor Machines*

however, over-wet or set stains. Which one to buy (and whether indeed to purchase instead of contracting the service out) is a matter that the professional Housekeeper will study seriously and then discuss with management.

### Carpet Cleaning Equipment

Although it would be impossible to classify every type of carpet cleaning device, there are certain broad groups of equipment with which the professional Housekeeper must be familiar (Figure 12-12).

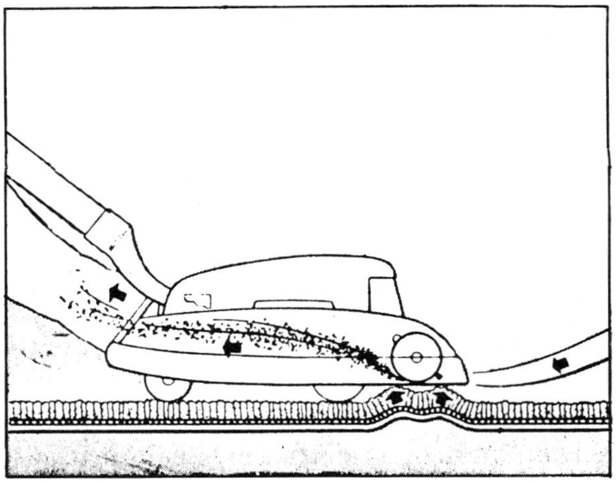

**FIGURE 12-8** *The beater-bar vacuum effectively removes loose soil buried in the pile. Credit: Amer. Hotel & Motel Assn.*

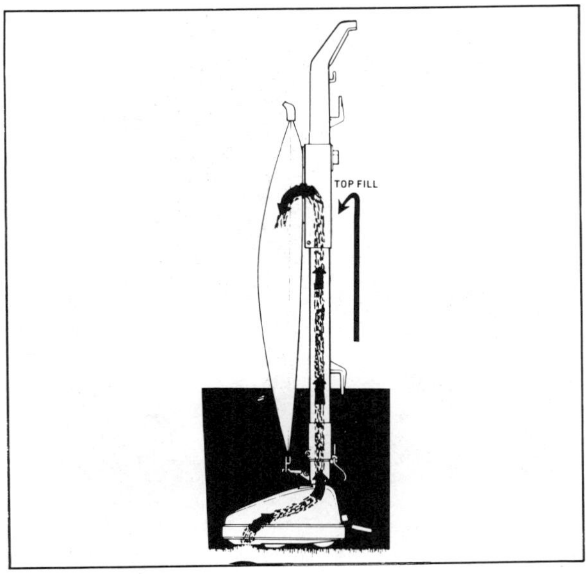


**FIGURE 12-9** *With capacity for cleaning 14 to 25 rooms, this upright vacuum cleaner has top-fill feature that permits soil to be sucked up handle to top of bag, thus there is no interference with speedy soil flow. Credit: Oreck*

### Shampoo Machines and Traditional Methods

*The rotary brush shampoo* method is the oldest and most widely used technique for carpet cleaning (Figures 12-13 and 12-14). The machine itself is versatile as it is used for wet or dry foam

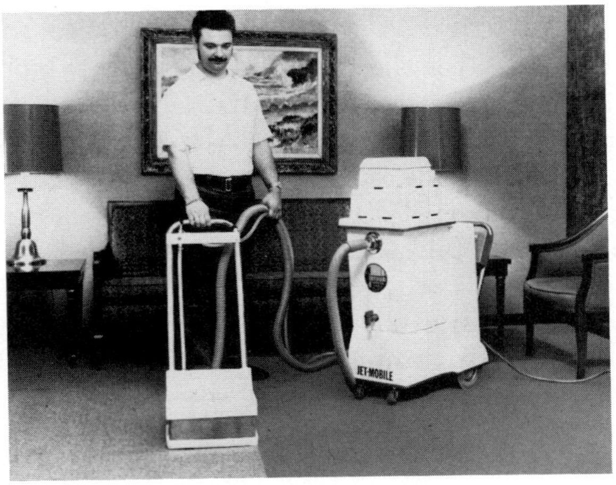


**FIGURE 12-10** *A jet and extraction machine has a self-leveling nozzle that covers more ground than a wand-type nozzle. This steam cleaning principle—low-wetting, high penetration—does not distort carpet fibers if it is performed by a skilled operator.*

**FIGURE 12-11** *"Steam cleaner" does not require hooking up to guest's bathroom, has its own supply of hot water and a tank for holding soiled suds and water. Houseman, trained to be expert with machine, does not over-wet carpet. Most rooms dry in two hours if well ventilated.*

shampooing or for floor scrubbing or buffing depending on which kind of pad or brush is attached to the disc. For shampooing, a solution tank is used for the detergent solution, which is driven into the pile with a flat revolving brush combined with the weight of the machine. This action breaks the oil film on the fibers and releases dirt particles trapped therein. Some procedures call for wet-vacuuming to remove dirt and water. In other procedures the carpet is allowed to dry, after which powdered detergent and dirt are vacuumed away (Figure 12-15). This method can be quite effective when employed by a skilled operator. Drying time can be four to six hours for a badly soiled carpet, two hours on a lightly soiled one. Costs run from $400 to $800 and tanks, brushes, and attachments are extra. Operators do become fatigued as the machines may weigh 100 pounds or more (Figure 12-16).

*The reel brush shampoo machine* can be used for dry foam or with a mist attachment that sprays detergent cleaner on the brushes and carpet. Carpet damage and wetting are kept at a minimum because the mist leaves so little moisture. Vacuuming before and after is necessary but a conventional upright vacuum is effective because so little dampness remains.

*The dry powder* technique has traditionally involved a solvent-saturated, absorbent powder that is worked into the carpet pile by hand or by a revolving brush machine. It is allowed to dry for a period, generally several hours, and then the carpet is vacuumed. This does not distort color or texture and carpet can be walked on immediately after vacuuming. The technique can also delay the need for a wet shampoo. It is *not* effective, however, for cleaning a heavily soiled carpet.

The Concept 4000 previously mentioned is unlike other dry systems, primarily because the dry particles are designed for cleaning carpet fibers, according to the manufacturer. The particle is saturated with specially formulated water-base cleaning solution. Other dry systems may use clay, wood, or flour where less control over particle size, shape, and porosity can be exercised, and, says the Concept 4000 expert, they are generally saturated with a solvent.

## Smaller Machines

Some of the machines designed for daily floor care, if used frequently and properly, will decrease the number of times that the previously described big machines and systems will have to be used (Figures 12-17 through 12-20).

*Carpet sweepers* are almost ineffective on high-pile or multilevel carpets but easier for maids to carry and effective if guest rooms have level carpet.

*All-purpose sweeper* is a tiny, modular tool used most frequently for quick clean-up on carpet restaurant floors. It can pick up sand, dirt, broken glass, and cigarettes.

*Electric brooms* are another lightweight tool (6 lb.), satisfactory if made of metal, but a burden if made of an easily breakable plastic. These can get into small places and under beds.

*Ladylike upright vacuums* range from 8½ to 11 lb. and most models today have enough power to make it worthwhile for a maid to carry one

| | | |
|---|---|---|
| **Steam-Extraction**<br>The steam-extraction system is based on injection of a jet of hot water containing detergents at a prescribed rate and subsequent extraction by a vacuum system. | 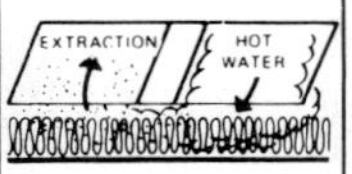 | Does not cause pile distortion or flaring.<br>Built-in vacuum removes soil.<br>There is no residue of shampoo to collect more dirt.<br>Slower, more expensive. |
| **Cylindrical Brush-Dry Foam**<br>This system has a cylindrical brush which scrubs and picks up in one pass the foam generated by the machine. |  | Causes severe pile distortion and flaring of tufts in high plush or shags.<br>Has built-in vacuum which removes shampoo and dirt. |
| **Rotary Brush-Wet Shampoo**<br>A rotary brush cleaner using wet shampoo. A complete line of accessories including vacuum and drying equipment is also employed. | 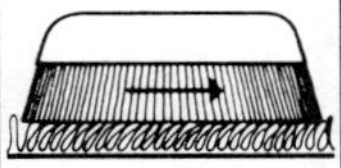 | Causes severe pile distortion and flaring of tufts in high plush or shags.<br>Does uniform cleaning.<br>Does not remove shampoo and soil. |
| **Small Rotary Brushes-Wet Shampoo**<br>This is also a rotary brush cleaner but employs two brushes instead of one and is somewhat smaller than the rotary brush-wet shampoo machine. | 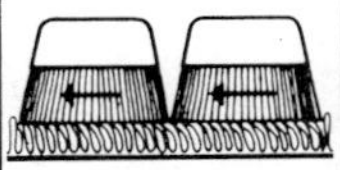 | Causes severe pile distortion and flaring.<br>Does very little cleaning.<br>Does not remove shampoo or soil. |
| **Sonic Cleaning**<br>Sonic cleaner consists of a vibrating plate, which gives a sonic cleaning action, causing activation of the foam.<br>*Credit: Allied Chemical Corp.* | 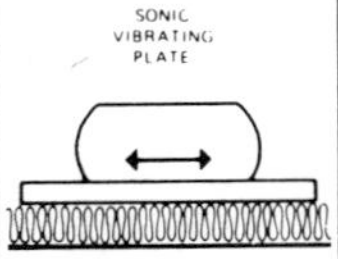 | Does not cause pile distortion or flaring.<br>Incurs less cleaning than above methods.<br>Has no method for removal of shampoo and soil. |

**FIGURE 12-12** *Carpet Shampoo Techniques. Although the above comments (right column) would imply that steam extraction is the only effective method of cleaning a carpet, it certainly is possible for skilled operators to do well with rotary wet shampoo, dry-particle, or dry foam technique IF proper detergents are used and IF vacuum that picks up foam and soil is powerful. To solve pile distortion, certain types of brushes, carpet mops, or yarn pads can be used.*

around on her cart. These come with or without attachments and may have two motors; cost about $170 each.

*Lightweight tank vacuums.* Several of these have sufficient power for maids to use in guest rooms and are small enough to fit on carts, but these are more important as cleaning tools for upholstery, draperies, lampshades, and vents than for heavy duty floor care.

*On-the-back vacs,* called piggy-back, back-packs, papoose, are light and easy to carry into inaccessible areas.

*Heavy upright vacuums* are the famous beats-as-it-sweeps-as-it-cleans machines. These usually have two motors: one to drive the brush, one to furnish suction. There are now so many heavy-duty vacuums made for institutional cleaning that to enumerate their advantages and disadvantages would be a major project. They range from the wet-dry pick-up to the heavy giant that will inhale crumpled papers, cigarette packages, and white mice.

*Heavy commercial tank-type vacuums* are most often the wet-dry pick-up type as the tank can accommodate more of the water that has been extracted from the carpet than the upright type. There is even a top unit that fits over a bucket or a large drum so that in an emergency maintenance personnel can quickly carry the portable unit to an area if the regular large wet-vac is in use elsewhere.

The six types of vacuums described should

**FIGURE 12-13** *Single-disc carpet shampoo machine with fast brush (200 rpm) has multiple-feed solution system that goes into bristles of 16-in. diameter brush. Bumper base saves walls. Rotary machines such as this one can be used with wet shampoo as well as foam, or even on hard-surface floors.* Credit: Certified

be used in daily maintenance programs, and usually before the shampoo machines do their regular work. Just how often to vacuum and shampoo is strictly up to the professional Housekeeper who will base that decision on the amount

**FIGURE 12-14** *Dry foam shampoo is discharged through brush on single-disc rotary floor machine. Operator moves fast, in reverse direction, not lingering lest machine overwet carpet. Wet-dry vacuuming follows.*

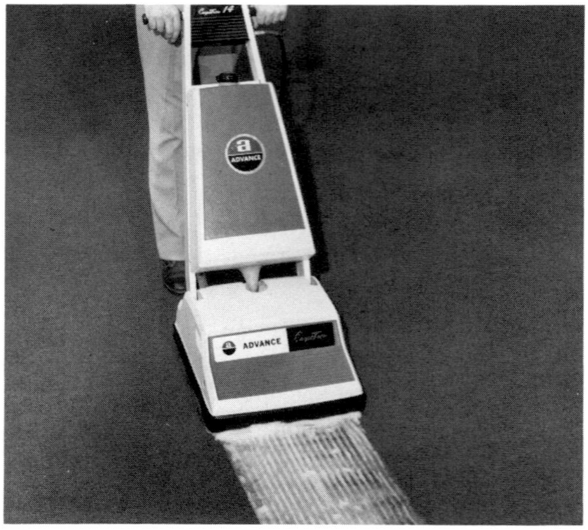


**FIGURE 12-15** *Cylindrical brush combs foam through fibers in a path up to 18 in. wide. Machine is self-propelled, both forward and reverse. Foam with only 6 percent moisture dries in an hour, is vacuumed up.*

and type of soil that appears on the floors of the property in question.

## *Specific Maintenance Problems*

Whatever tools, machines, or methods are used, there are certain maintenance situations about which the professional Housekeeper should be aware with regard to carpet.

**FIGURE 12-16** *To clean carpets at Memphis State University's nine residence halls, maintenance supervisor Norman C. Lee uses rotary brush and pad. Specially formulated shampoo (biodegradable and 1 percent phosphorous) dissolves soil and then frequently changed fabric pads remove it from pile yarns. Note difference in clean and soiled pads.*

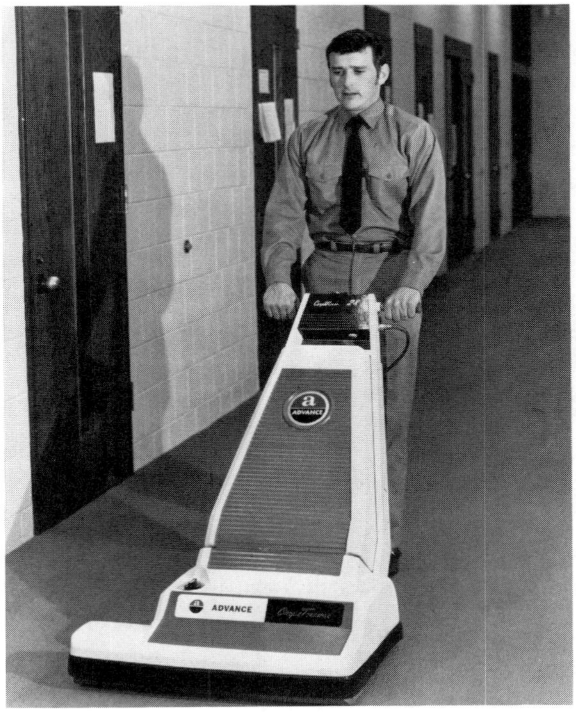

**FIGURE 12-17** *Extra-wide and very quiet, this sweeper-vac picks up litter as well as dirt and soil. Thirty of these are used at Michigan State University in all carpeted corridors of the residence halls. Maneuverable enough for a woman to handle, machine will clean up hall carpet in two passes. Sides are swept with broom before carpet is vacuumed in order to prevent dust build-up along baseboards.*

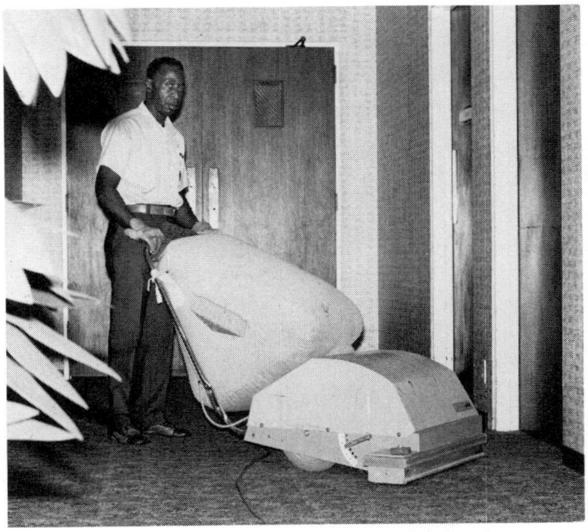

**FIGURE 12-18** *Giant space-vacuum has opening big enough to pick up large bits of debris and deposit in five-bushel bag. Extra-large wheels make it easy to push. Brush action and vacuum suction work together.*

**FIGURE 12-19** *Heavy-duty traditional vacuum that beats-as-it-sweeps-as-it-cleans, is effective in guest corridors where vacuuming is generally done weekly. Only Room Service spills cause major carpet problems in upper hallways.*

**Spot Clean.** As soon as spills occur they should be cleaned up. Using a spotting kit is the handiest way as it contains necessary items for emergencies (Figure 12-21). If spotting cannot be done as soon as a spill is made, it should be attended to before daily vacuuming or dry-foam shampooing. Early blotting of liquid spills with soft absorbent paper towel or cloth reduces chances that a spot will become permanent. All carpet and fiber producers make spotting charts available for housekeeping staff.

**Sprouting** describes long yarn ends that are found in new carpet. These are loose pile yarns and should not be pulled out, but should be cut off even with the regular pile.

**Vacuuming Shags.** As these carpets are so difficult to clean, it is important to vacuum in four directions (despite the fact that it is physically very tiring to vacuum shags) (Figures 12-22 and 12-23). If they have a very long pile, they should be raked to make the pile stand up. In guest rooms, after all other work is complete, the maid can rake as she backs out.

**Amount of Soil** will be greatest at junctions of carpet and hard-surface floors. Mats at those places will help to reduce the amount of dirt tracked in.

**Burns.** A cigarette burn that chars carpet superficially can be disguised by careful clipping of the blackened ends of the tufts, followed with a

**FIGURE 12-20** *In club foyer, quiet, high-powered vacuum cleans 22-in. swath but is easy to handle thanks to 6-in. wheels and 2-in. ball casters in front. One motor drives beater brush, another powers vacuum. Unit has attachments for stairs, walls, draperies. Bag is enclosed in vinyl "clam-shell-type" hood.*

soapless cleaner and water sponging. On very low-pile carpets sandpaper is sometimes used to remove char.

**Cart Marks.** If masking tape is placed on the wheels of maids' carts, it will eliminate black marks made by wheels after hall carpet has been scrubbed.

**Dents.** To prevent dents in carpets from furniture, use furniture coasters, and shift position of furniture now and then. If furniture is rearranged and dents still result, these can be removed by moving a steam-iron back and forth slightly above the dent and brushing up the nap.

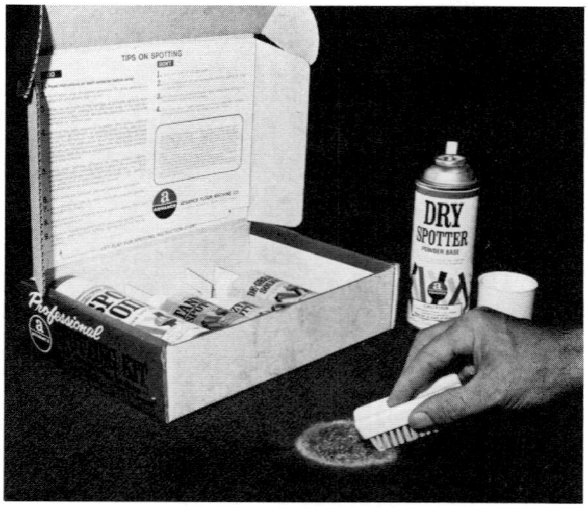


**FIGURE 12-21** *Stains can be removed from carpets and upholstery with a spotting kit. Kits should be kept where spots occur frequently. Kit contains aerosol cans of spotter and dry powder, bottle of tannin spotter, one of dry solvent—degreaser, one bottle enzyme protein digester, and a brush. Simple instructions are also included.*

A major problem in cleaning guest rooms has always been with vacuum cleaners. If they are powerful enough to do the job, they are usually too big and heavy for a maid to carry from room to room. If they are light enough for a maid to carry comfortably, then they normally don't hold up under the wear and tear of hotel use.

Sheraton Hotels people experimented with a cart-mounted vacuum cleaner heavy enough to do cleaning properly and, because it is bolted to the cart, one that did not have to be lifted on and off (Figure 12-24).

The cart, 45 in. high and 67 in. long, made of sturdy 14-gauge steel, turned out to be too large to maneuver in and out of guest rooms and the 30-ft. hose (stored on a fire-hose reel) did not reach far enough from the hall into the rooms. The vacuum hose was powered by twin single-horsepower motors housed right on the carts.

## *Chemicals in Carpet Cleaning*

The nature of the detergent in carpet cleaning, much as it is in laundry machines, is critical. Chemists are endeavoring to come up with just the right formulas to clean various kinds of fibers properly. A few suggestions that follow may help in selecting proper detergents.

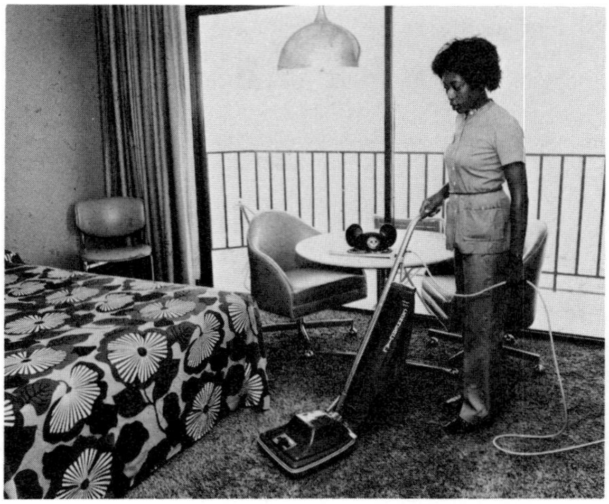

**FIGURES 12-22 and 12-23** *At the Contemporary Hotel in Walt Disney World, Orlando Fla., maids use lightweight vacuums to clean modified shag rugs. This is important as the occupancy is 100% and the average number of people staying in a room is 3.8—usually children included—so debris can be highly noticeable. Finishing up with a rake makes the shag look better and it is obvious that the carpet has been thoroughly attended to.*

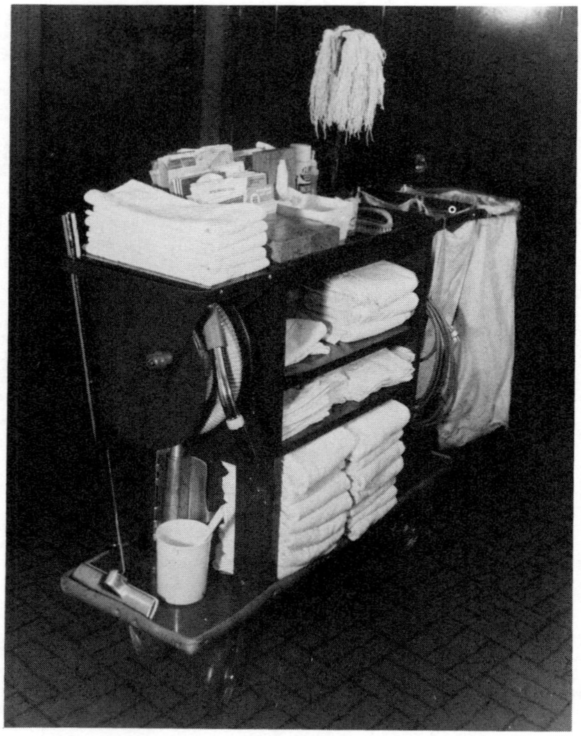

**FIGURE 12-24** *First-of-a-kind maids' cart has a vacuum installed permanently. Developed by Sheraton Hotel manager Willie Armstrong, the cart made it possible for maids to vacuum each guest room daily with a heavy-duty vacuum, yet not have to lift the machine on and off the cart. This experiment was not mass-produced as it was very costly and too bulky to be pushed in and out of guest rooms.*

1. Avoid those with oily residues as they contribute to resoiling.
2. Use detergents with pH of 8 to 10.5 for nylon carpeting.
3. Always follow manufacturer's recommendations.
4. If possible, select detergents that offer soil-retardancy.
5. Determine residue of detergents or shampoo. Simple tests can be run to evaluate resoiling characteristics.

## *Testing Detergents for Sticky Residues*

Use a clean glass pie-plate. Add approximately ¼ in. of the detergent-water mixture just as one would when cleaning a carpet. Allow the liquid to evaporate. It may require several days under normal room conditions, but to shorten the time place the dish in a dry room, in the sun, or in an oven set no higher than 150° F.

After all liquid has evaporated examine the remains to see what is left. If it has dried to a powder that can be vacuumed up there should be no problem with accelerated soiling. On the other hand, to determine whether the residue is sticky, run a finger over the inside of the plate and hold the plate up to a light to check the results; it can then be determined whether or not the detergent is contributing to faster soiling.

## *Typical Carpet Stains and What to Do About Them*

Blot is the main word to heed here. Do *not* rub. Blot with *any* clean very absorbent, natural material—paper towel, tissue, sponge, or cloth, but *not* a synthetic fabric.

**Baby and Pet Urine.** Blot up excess, sponge with detergent solution, blot, sponge with water, final rinse with sour solution,* blot. Repeat if necessary. * Add one teaspoon of a neutral detergent, such as those used for fine fabrics, to a quart of warm water. To this add one teaspoonful of white vinegar. White vinegar is a weak acid and will serve to neutralize any alkaline materials. Do not use soap or allow the carpeting to become overly wet.

**Blood.** Blot up excess, sponge with cold water, blot, sponge with detergent solution, blot, sponge with water, blot. Repeat if necessary.

**Butter.** Scrape up excess (with the grain of the carpet), sponge with dry-cleaning solvent, blot. Repeat if necessary.

**Candle Wax.** Scrape up excess, sponge with dry-cleaning solvent, blot. Repeat if necessary.

**Candy.** Scrape up excess, sponge with detergent solution, blot, sponge with water, blot. If stain persists, sponge with dry-cleaning solvent, blot.

**Catsup.** Scrape up excess (with the grain of the carpet), sponge with detergent solution, blot, sponge with water, blot. Repeat if necessary. If any stain remains, sponge with bleach solution, blot, sponge with water, blot.

**Cola.** Blot up excess, sponge with detergent solution, blot, sponge with water, blot. Repeat if necessary.

**Chocolate Milk or Cocoa.** Blot up excess, sponge with detergent solution, blot, sponge with water, blot. If stain persists sponge with dry-cleaning solvent, blot.

**Cod Liver Oil.** Blot up excess, sponge with dry-cleaning solvent, blot, sponge with water, blot.

**Coffee.** Blot up excess, sponge with detergent solution, blot, sponge with water, blot. If stain is old the last traces may be removed with bleach solution, blot, sponge with water, blot.

**Crayon.** Scrape up excess, sponge with amyl acetate, blot. Repeat if necessary.

**Cream.** Blot up excess, (with the grain of the carpet) sponge with detergent, solution, blot, sponge with water, blot.

**Digestive Fluids (Regurgitation).** Scrape up (with the grain of the carpet) and blot up excess, sponge with detergent solution, blot, sponge with water, blot. Repeat if necessary.

**Egg.** Scrape or blot up excess, sponge with detergent solution, blot, sponge with water, blot. Repeat if necessary.

**Foods (General).** Scrape (with the grain of the carpet) or blot up excess, sponge with detergent solution, blot with tissue, sponge with water, blot with tissue. If trace remains sponge with dry-cleaning solvent, blot with tissue.

**Foods (Greasy).** Scrape (with the grain of the carpet) or blot up excess, sponge with dry-cleaning solvent, blot. Sponge with detergent solution, blot, sponge with water. Repeat if necessary.

**Fruit Juices.** Blot up excess, sponge with detergent solution, blot, sponge with water, blot.

**Any Salad Dressings.** Blot up excess, sponge with dry-cleaning solvent, blot. If spot persists, sponge with detergent solution, blot, sponge with water, blot.

**Furniture Polish (Solvent-base).** Blot up excess, sponge with dry-cleaning solvent, blot. Due to dyes in most furniture polishes stain may persist—sponge with bleach solution, blot, sponge with water, blot.

**Furniture Polish (Water-base).** Blot up excess, sponge with detergent solution, blot, sponge with water, blot. If stain is old and persists sponge with dry-cleaning solution, blot. Due to dyes in most furniture polishes stain may still persist—sponge with bleach solution, blot, sponge with water, blot.

**Glue (Carpenter's).** Scrape (with the grain of the carpet) or blot up excess, sponge with detergent solution, blot, sponge with water, blot. Repeat if necessary. If glue has dried before removal is attempted a damp sponge should be left on it for sufficient time to soften the mass before removal is attempted.

**Glue (Plastic Base).** Scrape or blot up excess, sponge with amyl acetate, blot.

**Grease.** Scrape up (with the grain of the carpet) excess, sponge with dry-cleaning solvent, blot.

**Gum.** Scrape up excess (with the grain of the carpet—may be easier if chilled first with ice cube), sponge with dry-cleaning solvent, blot with tissue. Repeat if necessary.

**Ice Cream.** Scrape (with the grain of the carpet) or blot up excess, sponge with detergent solution, blot, sponge with water, blot. If trace remains sponge with dry-cleaning solvent, blot.

**Ink.** Blot up excess, sponge with detergent solution, blot, sponge with water, blot. If trace remains sponge with bleach solution, blot, sponge with water, blot.

**Iodine (Tincture).** Blot up excess, sponge with alcohol, blot. If stain persists sponge with bleach solution, blot, sponge with water, blot.

**Lipstick.** Scrape up excess, sponge with amyl acetate or detergent solution, blot. If detergent solution is used, sponge with water, blot.

**Liquor and Wine.** Blot up excess, sponge with detergent solution, blot, sponge with water, blot. If stain persists sponge with bleach solution, blot, repeat if necessary.

**Mercurochrome.** Blot up excess, sponge with alcohol, blot, sponge with detergent solution, blot. If stain persists, sponge with bleach solution, blot, sponge with water, blot.

**Milk.** Blot up excess, sponge with detergent solution, blot, sponge with water, blot.

**Mud.** Allow to dry, scrape up excess, vacuum thoroughly. If trace persists, sponge with detergent, blot, sponge with water, blot.

**Mustard.** Scrape up excess (with the grain of the carpet), sponge with detergent solution, blot, sponge with water, blot. Repeat if necessary. If trace persists sponge with bleach solution, blot, sponge with water, blot.

**Nail Polish.** Scrape up excess (with the grain of the carpet), sponge with amyl acetate or polish remover, blot, sponge with dry cleaning solvent, blot. Repeat if necessary. Try lacquer thinner if all else fails.

**Oil.** Blot up excess, sponge with dry-cleaning solvent, blot. Repeat if necessary.

**Paint (Latex).** Scrape (with the grain of the carpet) or blot up excess, sponge with detergent solution, blot, sponge with water, blot. If paint has dried professional attention is recommended, although a paste type paint remover followed by dry-cleaning solvent and then detergent can be tried.

**Paint or Varnish (Oil-base).** Scrape (with the grain of the carpet) or blot up excess, sponge with paint thinner, blot, sponge with dry-cleaning solvent, blot. Repeat if necessary. If paint has dried professional attention is recommended, although a paste type paint remover followed by dry-cleaning solvent and then detergent can be tried.

**Rust.** Scrape up excess (with the grain of the carpet), sponge with detergent solution, blot with tissue, sponge with water. Repeat if necessary. If stain is old, professional attention may be necessary.

**Soft Drinks.** Blot up excess, sponge with detergent solution, blot, sponge with water, blot. Repeat if necessary.

**Shoe Polish (Liquid).** Blot up excess, sponge with detergent solution, blot. If stain persists sponge with dry-cleaning solvent, blot. If trace still persists dye must be removed. Sponge with bleach solution, blot, sponge with water, blot.

**Shoe Polish (Paste).** Scrape up excess (with the grain of the carpet), sponge with dry cleaning solvent, blot, sponge with detergent solution, blot, sponge with water, blot. If trace persists, sponge with bleach solution, blot, sponge with water, blot.

**Tea.** Blot up excess, sponge with detergent solution, blot, sponge with water, blot, sponge with sour solution, blot.

## *Important*

If staining is deep, it may be necessary to repeat cleaning procedure. Avoid excessive wetting at any one time. Call your rug cleaner.

## *Bibliography*


Berkeley, Bernard. The Selection and Maintenance of Commercial Carpet. *The Cornell Hotel and Restaurant Administration Quarterly,* 1970.

*The Executive Housekeeper,* April 1974.

*Floor Covering Weekly,* January 8, 1973.

Hill, Thomas J. Extracting the Human Factor. *Executive Housekeeper,* April 1978.

*Maintenance Manual.* Wilmington, Del. E. I. DuPont de Nemours & Co., Inc. 1973.


## *Floor and Carpet Machine Manufacturers*

Advance Floor Machine Co.
P.O. Box 275
Spring Park, Minn. 55384

Oreck Corp.
100 Plantation Road
New Orleans, La. 70123

Hoover
Commercial Cleaner Division
North Canton, Ohio 44720

Bissell Inc.
Grand Rapids, Mich. 49501

Argo & Company
P.O. Box 2227
Spartanburg, So. Carolina

Clarke
2800 Estes St.
Muskegon, Mich. 49441

Certified Chemical & Equipment Co.
5366 St. Clair Ave.
Cleveland, Ohio 44103

Hild Floor Machine
5339 West Lake St.
Chicago, Ill. 60644

Von Shrader Co.
1300 Junction Ave.
Racine, Wisc. 53403

# Chapter 13

# Linens, Textiles, Bedding, and Beds

**OVERVIEW** Not only is it important that the professional Housekeeper know how to purchase bed, bath, and table linens, but he or she should know how to handle, store, and protect that enormous investment, or else have carefully thought out why linen rental is best. If maids do not have a proper linen supply they simply cannot make up the guest rooms. Bedding, mattresses, and pillows are the basis of the primary product that the hotel or residence hall has to sell, so they should be the very best the institution can offer.

In a large hotel hundreds of thousands of dollars are spent on linens and the organization usually looks to the Housekeeper to purchase, store, issue, care for, and inventory the countless components that represent this investment. With the exception of personnel, only floors and floor care represent a greater responsibility for the professional Housekeeper (Figure 13-1).

This chapter deals with fabrics known as linens. Fabrics for curtains and draperies are covered in Chapter 15 (Windows—Their Care and Treatment), and upholstery fabrics are discussed in Chapter 17 on Interior Design—Guest Rooms. The latter part of this chapter is devoted to a detailed examination of beds and bedding.

All linens required for any type of institution should at least meet the American Standards Association's Minimum Performance Requirements for Institutional Textiles, L-24 established by the federal government. Quality lodging operations should standardize on textile products at least one rung above that minimum. If no laundry is on the premises then linen rental from a commercial laundry should be considered. See Chapter 28, on the Laundry for pros and cons.

Ray Pedderson, author of the book *SPECS: The Comprehensive Foodservice Purchasing and Specification Manual,* published by CBI Publishing Company, has the following good advice on purchasing linens.

> In selecting the proper linens, the three most important features to look for are: *durability, laundry* costs, and lastly, *purchase price.* From these three features the true cost of an item can be determined.
>
> The cost per use is obtained by adding the original cost of the article to the total laundry cost during its life expectancy, divided by the life expectancy, or the total number of launderings that might be expected for the item.
>
> Therefore the following *cost per use* formula is invaluable in determining the difference between operating efficiency and a profit, or incompetency, and a loss.
>
> $$\text{Cost per use} = \frac{\text{Wt} \times \text{Ldr Cost/lb} \times \text{Life Exp.} + \text{Orig. Cost}}{\text{Life Exp.}}$$
>
> For example, a tablecloth or bedsheet weighing 1.4 lbs. which has a life expectancy of 250 launderings and costs $3.50 when new, is processed in a laundry which figures

**FIGURE 13-1** *At Contemporary Hotel, Walt Disney World, maids handle over 25,000 pieces of soiled linen daily, and of course that many pieces of clean linen too. Because 1046 rooms average 3.8 guests per night and hotel has 100% occupancy, linen is used in abnormally large quantities. Laundry at Disney World supplies the hotels and restaurants throughout the park.*

that their laundry cost is \$.10 per pound. Cost per use is derived in the following manner:

$$\text{Cost Per Use} = \frac{(1.4)\,(.10)\,(250) + 3.50}{250}$$

$$= \frac{35 + 3.50}{250} = \frac{38.50}{250} = .1542$$

It is quite evident from the formula that the original purchase price plays a relatively insignificant part in arriving at the true cost of an item. In the above formula you will note that the institution had invested \$35.00 in the laundering of the item during a period of 250 uses. The investment of the original cost of \$3.50 is rather insignificant when compared to the investment of \$35.00 spent on this item to process it through the laundry 250 times. It is the number of launderings an article can sustain through normal processing which is the most important factor in determining the true cost of an item or its cost per use. The finesse in purchasing linen, therefore, is to be able to closely approximate the probable additional uses you might obtain from an item, weighed against its cost to launder and lastly, purchase price.

## *Construction and Types of Weaves*

The durability of a fabric is contingent on construction of the material. Construction of a woven piece of textile is the interlacing of yarns woven at right angles to each other. The lengthwise yarns are called the warp. The crosswise threads are referred to as filling, or weft. The lengthwise sides or edges of a fabric are called the selvages.

There are three types of weaves, but you are confronted with only two basic weaves, namely, plain and twill. The plain weave is the common weave method. It is accomplished by alternately interlacing the warp and filling threads, one warp over and one warp under the filling throughout the construction of the fabric. Both muslin and percale for example, are made from plain weaves. The twill weave forms diagonal lines on the face of the cloth. Each warp yarn does not interlock with each filling yarn. Instead it interlaces with only the second, third, or fourth filling yarn, floating over and under the remainder. This creates the illusion of a diagonal weave. Twill weaves are frequently used in heavier cloths and sheets and work cloths.

The number of threads which are interlaced together in a square inch constitutes "a thread count." For example, the most commonly used sheet in a hotel is approximately 180 threads per square inch. It is made from approximately 94 threads per inch running in the warp, that is, lengthwise direction, and 86 threads per inch in the fill, with a total of 180 threads per square inch. The weight of the average 180 thread count sheet is approximately 3.6 ounces per square yard.

The weight of a fabric is specified in one

of two forms: either ounces per square yard or yards per pound. When the weight is specified in yards per pound, it is important that the width of the fabric is defined. Naturally, the wider fabrics will weigh more per lineal yard than the narrow fabrics of similar construction.

In addition to thread count and weight of a fabric, it is necessary to be aware of the item's tensile strength, which will give a good indication of its potential durability. The tensile strength is determined in laboratory conditions under controlled temperature and humidity. It is the number of pounds required to break a strip of fabric, or any other piece of textile, one inch in width and three inches long.

Another important guide in predetermining the possible durability of an item is an abrasion test, which measures the degree to which a fabric resists surface wear. Institutional linens, including apparel, are very often subjected to more surface wear than strain. The abrasion test itself determines how many rubs the material will withstand under certain circumstances until the yarn weakens.

**Tablecloths and Napkins** Tablecloths and napkins of the best quality are made of cotton damask, two layers thick, weighing 10 ounces or more per square yard. It will have about 200 threads per square inch and shrink less than 10 percent. Combed cotton, broadcloth, and momie cloth are also commonly used. The trend in recent years has been to the use of synthetic fibers due to their resistance to wrinkles and ease of care. Some of these synthetic fibers resist absorption of spills. That may please the restaurant owner but highly displease the customer.

Tablecloths and napkins should be ordered preshrunk. If ordering to the cut size remember that shrinkage will reduce tablecloth dimensions as much as four inches (a 72 × 72 cut becomes a 69 × 69 cloth) and napkins by as much as two inches. Specify a 64-pound or greater seam strength, nonfading color-fast colors, a breaking strength appropriate to the type of material, and a weight of at least 4 oz. per square yard on single-thickness materials.

## *Points on Purchasing*

### *Bath Linens*

One of the less scientific rules of thumb in the hospitality industry is that the room rate rises in reasonably direct proportion to the thickness of the towels. With the recent astronomical increases in the cost of terrycloth toweling alone, that rule of thumb may soon become an exact science!

There appears to be no solution to the cotton shortage, nor is its price likely to move other than upward. Terrycloth will remain readily available only because polyester blends are being used in the ground warp under the pile. This still gives the absorbency of the cotton on the surface, provides the characteristic thickness and weight, and actually prolongs the life of the towel.

Usually two hand towels, two bath towels, two washcloths, and one bath mat are provided in a single room and, surprisingly, in most properties for a double room also. Bath towels now on the market are somewhat smaller than those they are designed to replace because the addition of blend yarns reduces shrinkage. The blend yarns also decrease the weight of toweling, thus cutting laundry costs.

Most of the better hotels provide a bath towel approximately 20 in. by 42 in. The extra large 24-in. by 48-in., or even 32-in. by 64-in., bath sheets are still conversation pieces as they become more of a rarity.

White terry is making a comeback. Colored bath towels, once the height of fashion, are now so commonplace that white represents a refreshing change. Using white towels also reduces washing and bleaching complexities.

Terry towels that hold up longest are those with side stitching that reinforces the vulnerable edges—that part of the towel that usually frays first. If selvages are on the ends as well as the sides, the towels will last longer. Selvage edge ends should also be hemmed, however.

Quality terry has double-thread loops and is

densely woven for more absorbency. It weighs more too, so will cost more to launder.

Huck towels are small hand towels made of carded cotton, rayon, or a combination of both. They are fairly smooth, linen-like, usually have a slight texture in the weave, and sometimes require ironing. They are used only in the more expensive hostelries.

Bath mats (the toweling type for step-out use, not the in-tub shower mat) are both expensive and heavy (a good one will weigh over ⅔ lb.). Maids are sometimes instructed not to change them daily unless they are noticeably soiled, but they should always be changed in c/o rooms.

Towels are less likely to be used promiscuously if they are folded and stacked on towel holders rather than hung on racks. Chrome metal rings hold and separate towels and washcloths in one compact unit. This prevents wet towels being placed over unused ones.

Shower curtain fabrics today are mildew-proof and machine-washable. Those used most often are made of either woven nylon or a nonabsorbent, germ-resistant, heavy-duty vinyl (9- or 10-gauge) that does not flap when the shower is turned on, and has no hems or seams. Both types are available in many colors.

## *Table Linens*

The type of table linens purchased depends on the character of foodservice activities, and the level of service desired. This decision will be made by management and the head of the Food and Beverage Department, with counsel from the professional Housekeeper. The areas for which linens are needed may include: restaurant, room service, and banquet and private dining rooms.

If white napery is used throughout, the size of the inventory can be reduced, and stain removal, storage, and sorting will be immeasurably simplified. But part of selling meals today includes color on the table top, so it may be necessary for the commercial operators to set up a system that includes several colors and patterns. If colors can be keyed to sizes and/or functions, sorting and handling will be speeded up substantially. Where this is done, a specific color is used only for banquet cloths, making them instantly distinguishable from the smaller cloths used in the restaurants. This can be adapted for all areas.

Most napery used today is of two types: damask and momie. Damask is an elegant, reversible, solid-color textile—a closely woven, lustrous satin fabric made of linen, cotton, silk, or synthetics. (Damask is sometimes referred to as jacquard, the name of the loom on which it is produced.)

Momie cloth is a plain weave that wears well and is almost always used in a color. It is far less expensive than damask, being of looser weave and cheaper yarns.

Traditionally, these favorite foodservice linens have been made of cotton, but today they are available almost exclusively in blends of cotton and polyester so that no ironing is needed. Table linens laundered on the premises are removed quickly from dryers, folded, and placed on carts that serve as their storage shelves (wherever they are held) until they are put into service.

Combed cottons have finer threads and wear longer than carded cottons. The strongest combed-cotton table linen will have a thread count of 148 (warp) and 74 (fill), whereas momie cloth made of carded cotton may have a minimum of 62 by 56 thread count.

Damask is also measured by threads per square inch. A single damask should have a minimum of 140 threads per square inch; a double, a minimum of 195 threads per square inch.

Weights of cottons and blends differ, the blends being slightly lighter, but the quality of the fabric also affects the weight (the finer being heavier), and washability. Soil is released from damask much faster and more easily than from momie cloth. The shrinkage of the cottons is about 5 percent in the length; with no-irons it is barely measurable.

## *Bed Linens*

### *Sheets and Pillowcases*

Virtually all sheets and pillowcases manufactured today are a blend of cotton and polyester, whether muslin or percale. Since the late 1950s, there have been three stages in the improvement of no-iron fabrics. Today's third-generation blends have a permanent press finish that requires no ironing and, for many institutions, not even a roller-folder.

It should be noted that if the sources of polyesters, essentially petroleum products, are seriously imperiled there could be a need for a new fiber; but a return to all cotton seems unlikely as all the new textile factories are set up with machinery for blends. Cotton will also continue to become scarcer and more expensive since U.S. cotton fields are fast being replaced by soy or other profitable food crops, or by suburban housing tracts.

So let us consider what seems currently to be the most realistic choices when selecting sheets and pillowcases. The terms T/180 and T/130 refer to thread count, meaning threads per square inch. Obviously a 130 (or 128) muslin sheet will be less densely woven, and thus lighter and less apt to wear as long as a 180-count percale. The percale is made of finer combed cotton fiber than muslin, which has only been carded (separated and straightened).

Blend sheets seem to pick up tensile strength as they are laundered. Like towels with polyester warp threads, they are stronger after 75 washings than when new. (One manufacturer makes an extra-fine combed percale blend sheet of 200-thread count, a luxurious sheet that should give exceptionally long wear.)

The manufacturer or supplier will be able to provide proper instructions as to just how the laundering of blend sheets should be handled, including the important matter of the proper detergents to use.

Most hotels will use some sort of roller-folder or mangle to give the sheets a more finished look before they are folded and put on the shelf to rest—a step that is said to prolong their life.

According to William D. Nolte, laundry and textile consultant, New York, "The wear factor or longevity of no-iron fabrics has completely upset the hitherto accepted life expectancy factor. Polyester sheets are building a history of 500 washings or better.

"An on-premise laundry means the fabrics never leave the house, so inventory levels can be reduced—they average 3 to 1. The money invested for initial stock becomes less, the sheets last longer."

Sheets are available in these sizes: twin, 72 in. by 104 in.; double, 81 in. by 104 in.; queen, 90 in. by 115 in.; king, 108 in. by 115 in. They come packed 1, 5, or 10 dozen to a case. A wide color selection is offered by several manufacturers, but few institutions use patterned sheets.

Where many different sizes of sheets are required, it is helpful to have the colors correspond with specific sizes so that sorting and selection will be easier for maids. If they are all to be the same color, however, the supplier should be asked to have them permanently identified with ink applied at 400° F. This usually includes name of hotel, date purchased, and size.

Because blend sheets can look fresh even after having been slept on, maids are often tempted to remake the beds without changing the sheets. To combat this, some housekeepers have supervisors go into rooms and strip the beds before maids start to clean.

Blend sheets still are measured in the cut (or torn) size, that is, before hemming. Because of their nonshrinkage blend sheets come several inches shorter than did the all-cotton.

Fitted sheets have been frowned on for institutional use because the corners too often rip due to added stress at those points. They also take up about twice as much storage space as regular sheets. They complicate the inventory and sorting, and mean that maids must stack still another item on their carts. It is possible to use double or queen-size sheets on twin beds or double beds, but a fitted sheet has little versatility, slows down bed making—and because it remains fresh looking, it may often go unchanged either deliberately or accidentally.

Top and bottom hems on institutional sheets are usually the same width, unlike those made for home use that have a deeper top hem. Being able to put the sheet on in either direction saves the maid's time, and assures that the sheet will wear evenly at both ends.

If pillowcases are to be made from used sheets, their hems should be two to three inches wide, with the case itself from two to four inches longer than the pillow. Hem ends should be back-stitched for greater strength. Stitching at ends of seams should be checked by using a pin to see if threads can be lifted up.

A third sheet, called a crinkle sheet or blanket cover, has been used over the years in the finer hotels to protect the blanket from soil, and to give guests the feeling of being surrounded by clean linen. With the trends toward washable blankets and the use of washable quilted bedspreads as blankets, this third sheet is now almost extinct.

## *Blankets*

Natural fibers are disappearing from blankets just as they are from other textile products. The new weaves and fibers are producing some extremely attractive and functional coverlets that we today can call blankets. But let's consider first just what are the characteristics of a blanket.

**Warmth.** This is produced by either pockets of dead air formed in the blanket nap and weave that trap body heat, or a tight construction that traps body heat. Both insulate the body from surrounding air.

**Moisture.** Construction and fibers should be porous and/or absorbent enough to transmit any moisture build-up to the surrounding air so the blanket doesn't feel wet and clammy. Undue amounts of moisture trapped under the blanket can cause chills and overheating.

**Weight.** The lighter blanket makes a more comfortable bed cover. A heavy blanket can constrict movement during sleep. Blanket weight is not a true indicator of warmth.

**Durability.** When the nap wears (abrades) and mats with laundering, the blanket's thermal value is greatly diminished. Most blankets shrink 10 percent after only a few launderings and are then subject to undue amounts of tugging and pulling during use.

The manner in which a blanket wears depends on its basic characteristics: fiber content, yarn type, construction, and finish. Proper laundering can increase its useful life. Improper procedures can cause undue matting and abrasion of the nap, excessive shrinkage, and running of the dye or ink used to color the blanket. A blanket should be impervious to natural degradation from mildew (mold), moths, and perspiration, and should not be treated with any finishes, or colored with any dyes or inks that can cause allergic reactions.

## *Types of Blankets*

**Standard.** It is manufactured with tight construction and napped on both sides. The blanket can be manufactured by conventional loom-weaving, needle-punch, or by an electrostatic process. Needle-punch and electrostatic blankets are warmer pound for pound than a woven blanket of the same fiber. Patterns may be woven and dyed into the woven blanket, whereas the needle-punch and electrostatic can only be dyed or printed.

**Thermal.** Thermal blankets are loom-woven with a leno weave, which forms many large air pockets. When used under a tightly woven cover (sheet blanket), it is ideal for cold winter nights. It can be as good an insulator as a standard blanket of the same fiber weighing three times as much. When used alone it is excellent year-round for warmer climates or in the summer for air-conditioned rooms.

**Sheet.** Sheet blankets are used for air-conditioned evenings and as the extra cover for cold winter nights (can be used over thermal or electric blankets). The sheet blanket is a lightweight version of the standard blanket. Its tight construction means it is not as porous (breathable) as the thermal blanket.

**Flame-retardant.** All woven and needle-punch blankets can be made from flame-retardant fibers. These fibers self-extinguish when the flame is removed.

**Electric.** Many people who like great warmth without the weight of heavy blankets prefer the electric blanket. As room temperature changes throughout the night, the thermostatic control automatically compensates for these changes and keeps bed warmth at the desired setting. A sheet blanket can be used over it to help conserve energy, but no other kind of blanket should be used on top of an electric blanket.

## *Construction*

**Size.** The size of a blanket will affect its wear. One that is too small to be tucked in sufficiently at the foot or sides, aside from giving insufficient insulation, may wear out quickly from being pulled by the maid to tuck in or by the guest to protect his shoulders.

Proper blanket size (after shrinkage): *length* of the mattress plus the thickness (7 in.) plus a minimum allowance for tuck-in (6 in.), *width* of the mattress plus twice the thickness (14 in.) plus a minimum allowance for tuck-in (6 in.).

**Fiber and Yarn.** Textile fibers may be spun into yarn or made directly into blankets (needle-punch). Crimped fibers of an irregular cross-section and rough surface make warm blankets because they form many air spaces. Staple fibers, not filament fibers, are therefore used for blanket production.

**Binding.** All blankets regardless of their construction are either whipstitched or nylon bound. The latter is more decorative and feels better when in contact with the skin, however, it will fray after repeated use. The selvage (sides) of the woven blanket need not be bound to prevent unravelling. Nylon binding is being replaced by four whipstitched sides due to the cost of replacement.

**Napping.** Heavy napping tends to weaken woven more than needle-punch constructions. If the nap wears (abrades), the thermal properties of the blanket are greatly reduced. To lessen pilling and shedding, and to give a better washed appearance blankets are treated after they are napped. They are sprayed with a solvent (acrylic blanket) or adhesive (polyester blanket) and heat-treated to bond the ends of the nap fiber. Although this process is standard, each manufacturer uses its own trade name.

## *Woven Blankets*

A woven blanket is made from two series of yarns: warp and filling. The warp runs lengthwise and is generally stronger than the filling, which runs continuously from side to side forming the selvage (self-edge). The fill yarn is slackly twisted to increase bulk (and consequently the number of air spaces). Ply yarn, an assemblage of many yarns twisted together, is used as warp yarn to assure sufficient strength to bear the weight of water during washing. The selvage is usually made with heavier and more closely spaced warps. If the weave is not uniform it may ripple and pucker after laundering. Four weaves are commonly used.

**Plain.** The simplest weave is comparatively inexpensive. A tight weave gives strength. A loose weave is warmer because more air spaces are formed.

**Twill.** This is the most durable of all weaves. Filling yarn is interlaced with warp yarn to form diagonal ridges across the fabric. This weave throws more fill yarns to surface for napping, making it a warm weave.

**Leno.** This weave gives firmness and strength to an open weave (thermal blanket). Adjacent warps are twisted around each other with the filling passing through.

**Double-cloth (Channel-cloth).** Weave used to produce shells for electric blankets. Five sets of yarn are used: two individual sets of warp and filling are woven into two layers of fabric. They are connected by the fifth set that produces channels in which the wiring for an electric blanket is placed.

**Needle-punch blanket.** Nonwoven constructions are a recent innovation. The needle-punch process converts fibers directly into blankets, eliminating the need for yarn-making and weaving. The process involves a device with barbed needles that punches through a thick web of fibers, forcing the fibers from layer to layer until they become entangled.

**Electrostatic (flocked) blanket.** The blanket consists of a layer of foam laminated to each side of a nylon knit scrim. The foam is then coated with a precolored adhesive and passed through a flocking chamber in which short lengths of electrically charged nylon fibers are deposited on the wet adhesive. The electric charge causes the fibers to stand on end on the adhesive. The insulation derives mainly from the many air cells in the foam laminate.

## *Decorative Finish*

Blankets are colored and patterned so they can be coordinated with other items in guest rooms.

Most blankets in hotels must be very simple, however, and although attractive, not so appealing that the guest will be tempted to put it in his suitcase. Since the bedspread is usually the piece of bedding that is immediately seen there is little need to be concerned with purchasing patterned blankets.

**Fiber Dyeing.** The fibers can be dyed before being spun into yarn or needled into blankets, thus assuring deep, even, color penetration.

**Yarn Dyeing.** The fibers are dyed after being spun into yarn. The dyed yarn can then be woven to form solid colors or intricate patterns. This is expensive because the manufacturer must inventory each color and design.

**Jig Dyeing.** Rolls of finished woven or needle-punched fabric are dyed in open vats, and then cut and finished. This is relatively inexpensive.

**Beck Dyeing.** Finished blankets are placed in vats and dyed. This method affords the manufacturer flexibility to produce many colors and patterns without the expense of inventorying them. Styles can be dyed when needed.

## *Natural Fibers*

**Wool.** Wool, a soft protein fiber from sheep, makes a warm absorbent blanket, but needs much care. Wool is the weakest fiber and can lose an additional 25 percent of its strength when wet. It can be strengthened by blending with other fibers or selected grades of remanufactured wool. Because of its weakness, absorbency, and tendency to shrink and mat, wool must be laundered with extreme care.

The quality and price of wool varies greatly. It is difficult to know what quality is used, and consequently what a blanket may be worth. The Merino sheep yield the best quality wool. All others are compared to this standard. Fleece types are as follows:

1. *Lamb's wool.* First fleece sheared from a lamb about six to eight months old; very soft.
2. *Hogget wool.* First fleece sheared from a sheep 12 to 14 months old; very strong.
3. *Wether wool.* Any fleece clipped after the first shearing.
4. *Pulled wool.* When sheep are slaughtered their wool is pulled from the pelt. This is inferior because sheep are not bred for fleece and wool roots are damaged during the pulling.
5. *Dead wool.* Wool that has been recovered from sheep that have died. This fiber is decidedly inferior in grade.
6. *Cotton wool.* Sheep that are exposed to severe weather conditions or lack of nourishment yield a wool that is matted and brittle. It is a poor grade.
7. *Taglocks.* The torn, ragged, or discolored parts of fleece. This is an inferior grade of wool.

Different parts of the fleece yield wool with different characteristics. There has never been a sufficient supply of new wool stocks to meet growing demand. To alleviate this situation, wool fibers are recovered from used wool goods and from the waste of wool manufacturing.

The Wool Products Labeling Act was instituted to inform consumers as to the type of wool that should be used. This is the only description that legally must be given:

1. *New wool.* This comes directly from fleece, never before processed.

2. *Virgin wool.* It is the same as new wool but includes pulled and dead wool.
3. *Reprocessed wool.* Fiber that has been reclaimed and remanufactured from *unused* wool materials.
4. *Reused wool.* Fiber that has been salvaged from all kinds of *used* materials.

**Cotton.** Cotton is a very strong fiber with the unique property that it becomes stronger when wet, therefore no special care need be taken in laundering. It is very absorbent but not a good insulator. Cotton is not dimensionally stable and will shrink about 5 to 15 percent after only a few launderings if it is not preshrunk.

**Rayon.** Rayon is no longer used for blankets. It is a cellulose fiber and will flame and burn after source of ignition is removed.

## *Synthetic Fibers*

Acrylic, modacylic, and polyester fibers are used for blankets. Three general features of all three fibers are:

1. *Dimensional stability.* The fibers will not normally shrink when wet if the yarn and blanket have been properly constructed.
2. *Mildew- and moth-resistant.* Synthetics are naturally resistant. If the fiber appears to be moldy, it is actually the finish that is mildewing. This surface mildew will wash off readily leaving no stain or damage.
3. *Brand identity.* Synthetic blankets are sold as either generic or branded fibers. Branded synthetics assure the manufacturer of uniformity in fiber length and denier (width). Finished blankets carrying a brand identity must meet the fiber manufacturer's specifications.

**Acrylic.** Acrylic is a lightweight fiber, and when crimped, provides considerable insulation. The dog-bone shape of some acrylics (Orlon) contribute additional bulk, making blankets that are warmer than wool of the same weight. Acrylics are not absorbent, but hold moisture on their surface and consequently dry slowly. The fiber is stronger than wool, but not as strong as polyester. Acrylics lose strength when wet and should be laundered with care. They have good abrasion resistance and won't lose nap readily through wear, matting, or pilling.

**Polyester.** When bulked and crimped, polyester is a good insulator but not as warm as acrylic. Polyester is the strongest fiber used for blankets. It does not lose strength when wet and therefore will withstand institutional laundering. Polyester has very little absorbency and dries quickly. Cigarettes will melt a hole in the blanket.

**Modacrylic.** A chemically modified acrylic fiber is used in blankets for its flame-retardant characteristics. Modacrylics can't smolder and reignite after the source of flame has been removed (e.g., cigarettes). Its other characteristics are similar to acrylics.

**Fiber Blend.** Textile fibers are combined to enhance the overall characteristics of the blanket or to reduce the blanket cost.

## *Blanket Selection*

It is perhaps wise to settle on two sizes of blankets—72 in. for twin or double and 108 in. for king- or queen-size beds.

Electric blankets are used by a few of the more deluxe hotels but are a sizable investment, tricky to launder, and even frightening to some guests. Nonetheless, many travelers are so accustomed to sleeping under them that they are uncomfortable under conventional blankets.

If electric blankets are selected, there must be an effective means of anchoring them to prevent pilferage. Checking them regularly to be sure they are always in proper working order can be enormously time-consuming. The most efficient way to do this is to have the maids remove the blankets and replace them with new ones. Granted this causes a bit of confusion in the halls as housemen bring trucks full of new blankets to each floor and remove the used ones to the laundry or housekeeping areas to be checked in assembly-line fashion. If a quality electric blanket is purchased initially, repair and adjustment costs will be quite low.

The Greenbrier, one of America's last bastions of old-fashioned elegance, still uses the luxurious comforter (also known as a puff), which is placed on the foot of the bed by the maid when providing turn-down service (Figure 13-2). The puff is used in addition to a fine wool blanket. Management and guests consider this combination to be far more appropriate to the grandeur of the Greenbrier than introduction of conveniences that tend to have space-age connotations.

### *Bedspreads and Mattress Covers*

There is no piece of furniture in a bedroom that is so obvious as the bed; therefore its cover is perhaps the most important single item in the room from a visual standpoint (Figures 13-3 and 13-4). It can make a correspondingly strong impression (for better or worse) on the guest or resident.

Fortunately today, the synthetic fibers have made it possible to make bedspreads that do not show wrinkles or soil. The most popular of these spreads are the washable quilted prints.

Because beds are now constructed so that the once unattractive box spring and frame can be covered with any of a wide selection of colorful vinyls, spreads no longer have to reach the floor on all sides to conceal what is underneath (Figure 13-5). Being smaller means they can be handled more easily, washed more easily, are less expensive, and take less space to store.

Figures 13-6 through 13-8 show how a well-coordinated fabric plan can be used to create, at low cost, an attractive, washable room. No blankets are used because the washable, colorful,

**FIGURE 13-2** *Guest rooms at Greenbrier, decorated by Dorothy Draper with cabbage-rose decor. Silky comforters are provided as extra covers together with wool blankets on beds. Beds are higher than standard to discourage guests from sitting on the mattress instead of using the comfortable lounge chairs. Because of the success of this psychological maneuver mattresses have lasted since late 1940s.*

quilted spread doubles as a blanket. The attractive pillowcases need no cover-up, and the bed spring is concealed with a dust ruffle that is actually another color-coordinated sheet. Guests have the secure feeling that everything is clean.

Unorthodox institutional spreads are becoming increasingly practical and acceptable. The vinyl spread that has the wet look can be wiped clean or be easily laundered. Another vinyl bedspread is textured to have a tweedy look, and even feels warm, but is waterproof. "Fur" spreads have an ultra-luxurious texture, yet they are no more expensive than the reliable old chenille or corduroy that was once excused for wrinkling because it was so serviceable.

William Wilterding of Innkeepers Supply, Memphis, has advice for housekeepers who have switched to nonquilted, 100 percent preshrunk cotton bedspreads, none of which have yet been washed. Before risking the whole inventory in the washer, he says, "Wash one bedspread in mild detergent, using warm water. Add no bleach. Dry in low-heat, slow tumbling cycle. After it is dry, it should be examined carefully and compared with the original spreads that have not yet been washed. Acceptable shrinkage would be about 2 percent, and color should show no change. In the event of color loss or excessive shrinkage, contact your supplier at once."

Quilted bedspreads with tags inviting users to wash them should be handled the same cautious way (Figure 13-9). Quilting threads are easily broken and filler can break up, too. Many users, however, including the Tucson Marriott, report that after hundreds of washings the quilts are still holding up well.

Figure 13-10 shows how to measure for bedspreads.

To protect the mattress from spilled liquids, sick or incontinent guests, bloodstains, and so forth, it is important that a moisture-proof mattress cover be used on every bed.

Vinyl plastics have been so improved (particularly those designed for hospital use) that they can be machine-washed and dried at any temperature; they are even autoclaved (sterilized) in hospitals. Many of the new mattress covers are stainproof, waterproof, nonallergenic, noiseless, and flameproof. Made like a contour sheet, they stay in place.

**FIGURE 13-3** *Most reliable of all the types of bedspreads is the classic white chenille since it can take repeated washings, is reversible, requires no ironing, and can also act as extra cover if needed. Highly functional room at a Ramada Inn might be more comfortable for guest if headboard were softly padded vinyl.*

**FIGURE 13-4** *A classic hobnail chenille bedspread is used on the charming four-poster canopied beds in the Williamsburg Inn at Colonial Williamsburg, Va. This famous historical restoration attracts overnight visitors who enjoy staying in the authentic old-fashioned settings.*

**FIGURE 13-5** *In elegant suites at Harrah's Lake Tahoe Hotel, bedspreads are designed to fit beds in only one way, so they are easy to put on and do not wrinkle. The beds have no legs and are on stationary box springs covered with vinyl so they will present no problem during vacuuming. Design also eliminates need for dust ruffle.*

Quilted polyester pads are shrinkage-controlled, which means that they must be dried on the low setting, but they add a soft layer between mattress and sheet, and have anchor bands on corners to hold them flat. Mattress covers should be changed with every check-out, and at least once a week.

## *Pillow Purchasing and Care*

Selecting pillows to please every guest is an almost impossible job because too few people can ever agree on what the "right" firmness or content should be.

Even for the most opulently furnished estab-

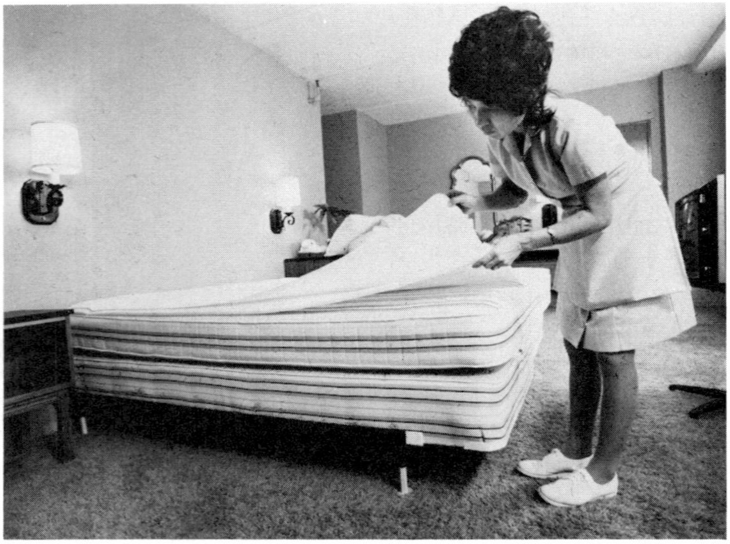

**FIGURE 13-6** *Maid at Marriott Hotel, Tucson, Arizona, adjusts the two pads used on guest room beds; one is a foam pad to protect mattress from moisture, the other a quilted polyester with corner anchor bands.*

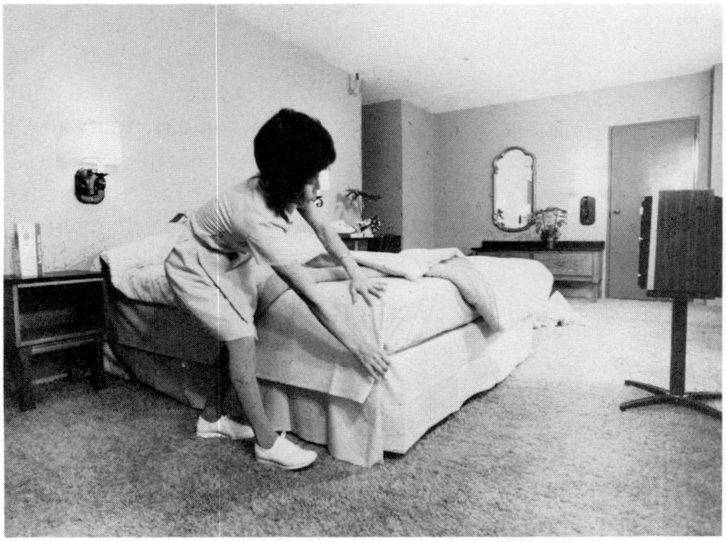

**FIGURE 13-7** *Maid adjusts colored top sheet forming squared corners. Dust ruffle is made of flat king-sized sheet, and two-tone comforter is used for blanket as well as for a bedspread. Comforter is warm enough for 32° F nights; has withstood 150 cycles of washing, remaining colorfast and with polyester padding undisturbed.*

lishments, down is just about out. Pure goose down is prohibitively expensive, as are combinations of feathers and down. Neither type holds up well, and although they can be cleaned and recovered, that may be more expensive than discarding a synthetic and buying a new one.

The synthetic fiber-filled pillow is far and away the most popular in the United States. Feather and down pillows follow, with down (extravagantly priced) being used only in very luxurious properties.

To evaluate the feather and down market properly, it is important to understand that the consumer market is principally located in Europe, specifically in Germany. Traditional German bedding includes featherbeds and comforters as well as pillows. Synthetics in Europe represent only 10 to 20 percent of their sales.

**FIGURE 13-8** *Bright patterned twill is used on headboards and draperies at the Marriott in Tucson. Sheets and comforters are in coordinated colors. Reversible comforters, which serve as bedspreads, are avocado and tangerine. Pillow slips are colored; they are allowed to remain uncovered, thus adding another shade to the color scheme.*

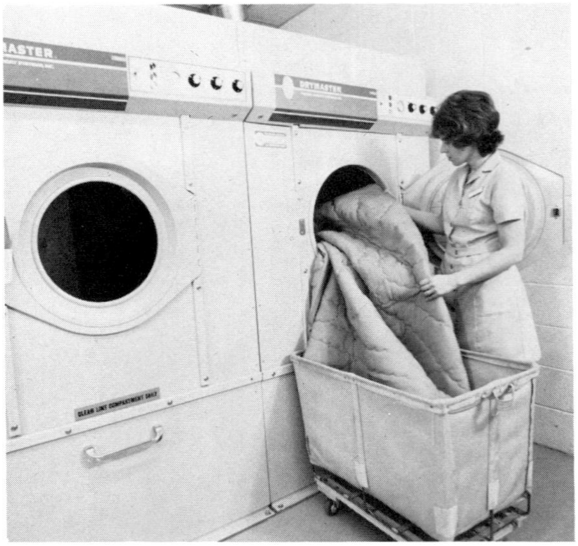

**FIGURE 13-9** *Laundry room supervisor removes one of the unusual puffs used in all 312 guest rooms at Marriott Tucson. They are velour on one side, which helps hold the comforter in place while in use. The European idea of having the spread do double duty is possible here, thanks to polyester-cotton blends.*

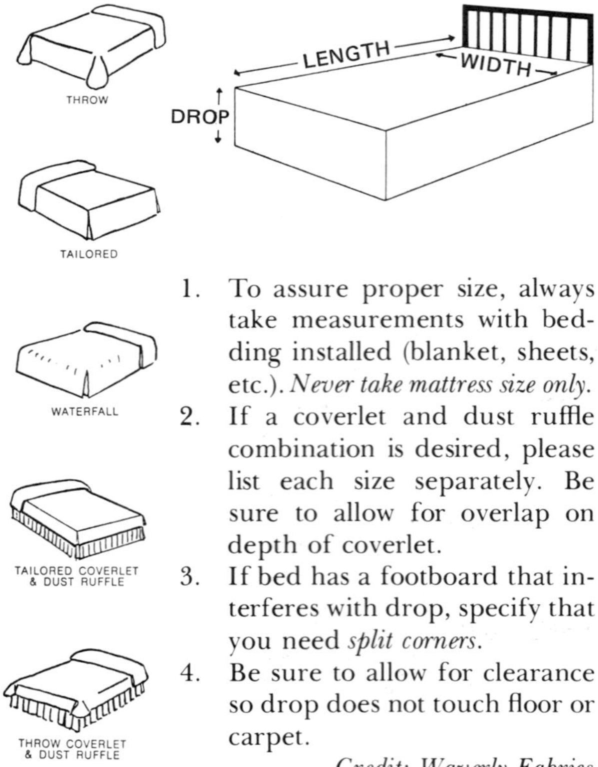


1. To assure proper size, always take measurements with bedding installed (blanket, sheets, etc.). *Never take mattress size only.*
2. If a coverlet and dust ruffle combination is desired, please list each size separately. Be sure to allow for overlap on depth of coverlet.
3. If bed has a footboard that interferes with drop, specify that you need *split corners.*
4. Be sure to allow for clearance so drop does not touch floor or carpet.

*Credit: Waverly Fabrics*

**FIGURE 13-10** *How to Measure for Bedspreads*

Most German featherbeds and pillows are made of half-down. This represents the ratio of feathers to down as the product comes off the bird which is normally 25 percent down and 75 percent feathers.

In the last few years American consumption has risen sharply, with the outerwear and sleeping bag industries placing a premium on down-filled products.

### *Down versus Feathers*

Down is a feather with pin-head sized nucleus to which are attached the down fibers that make up a cluster of down. The basic measurement of quality is the size of the clusters: the larger they are, the better the quality, the higher the price, and the less material it takes to fill a pillow.

The largest clusters of goose down come from mature European birds in winter coat from cold climates. These birds weigh about 15 pounds. The smallest clusters of goose down come from China goslings from a warm climate in summer coats. A mature China goose only weighs about nine pounds.

Duck down is somewhat inferior to goose down. The quality of both runs the full gamut from best to worst, and blending is common.

Feathers are more durable than down. (We are discussing duck and goose feathers, as they are the only ones that make a quality pillow.) Body feathers have a natural curl. Each one acts as a spring, which provides a pillow's resilience. If the feather is left in its original state, it is important that the smallest feathers are used so the pillow does not have a "rough hand."

Although small whole feathers are the best quality and the most durable, they are often unsaleable because they cause the pillow to seem rough, even though this roughness is not felt when sleeping on it. Often a light crushing that breaks the feathers produces a softer pillow.

Duck feathers are pointed and readily distinguishable from goose feathers with their square flat tips. Goose feathers have very little odor before washing, whereas duck feathers have a strong odor, making the washing process more critical with them.

The filling quality of goose feathers is superior to that of duck feathers. Like down, the best feathers fill a pillow with the least weight. It takes two and one-half large geese to make one pound of raw stock (the term for feathers and down as collected or picked). It takes four large ducks to make one pound of raw stock.

### *Synthetics*

Because they are nonallergenic and approximate down, synthetic fiber pillows constitute the largest pillow classification in the United States. In this category there are such synthetics as acrylics, acetates and polyesters. The last has emerged as the leading volume item because of durability, price, and its purported hypoallergenic properties. The U.S. military purchases only feather sleeping pillows, however, and has found very few allergy problems connected with properly processed feather products.

Experts suggest that the term "nonallergenic" should be replaced with the term hypoallergenic (*hypo* defined in *Webster's Dictionary* as "less than normal").

**Red Label Polyester Fiber.** Made by one of the major synthetic manufacturers, certain manufacturing standards are attested to when the red label is put on the pillow.

A second-generation Dacron, though relatively new in the market, has already assumed major status. This pillow is guaranteed machine-washable and machine-dryable. The fiber possesses a slick finish that is provided by the application of silicone.

The ultimate goal is for all polyester fiber-fills to have the hand and softness of a down pillow.

**Unbranded Polyester Fiber.** In the past few years there has been marketed a substantial amount of unbranded polyester fiber-filled bed pillows. The growth of this product has been exceptional, mainly attributable to its low cost. Now, however, it seems the emphasis is switching from unbranded back to branded fiber pillows. This change has been necessitated not only by the desire to improve unit selling price, but also by increased costs for the unbranded items.

**Molded Foam Latex.** Presently there exists a tremendous void of a viable supplier for molded foam latex buns, although there are a few being imported from Canada. The type of foam latex that has been generally considered the best is the Talaley process. This is one-piece construction with thousands of pin cores that allow the pillow to breathe and provide softness.

Another type of foam latex is the Dunlop process, which consists of two buns glued together. It does not have a desirable longevity.

Several companies are exploring polyurethane buns (heretofore unacceptable to the consumer). It is possible that someone will produce a good product with the quality of foam latex at a reasonable cost.

Perhaps the one way to please every guest is to furnish one feather and one Dacron pillow on each bed, thus cutting costs and offering a choice.

Tickings should be of heavy enough fabric to withstand many washings, and it may be important for some operations to have them be fire-retardant, waterproof, and stainproof.

If the label "American Institute of Laundering Certified Washable" is on the pillow (or any linens or fabrics) one can be sure the materials conform to official government regulations for washability.

Check to see if the pillow can retain its shape. When placed over the center of one's arm, it should balance evenly. If it folds, it indicates that the filling is tired (thin and bumpy). A good pillow is not heavy. Heaviness often indicates a poor quality filling.

A good pillow should be resilient. When both hands are pressed into the center of the pillow, it should spring right back into shape as soon as the hands are lifted. It should be buoyant to provide comfortable support for the head. The filling should not shift or settle.

By law, all pillows must be labeled as to content and bear the manufacturer's name and address. Data concerning flammability should explain whether the entire pillow is fire-retardant or just the covering. Some pillows can be autoclaved (sterilized).

Zippered pillow covers that make pillows waterproof, stainproof, and resistant to mildew, are available in washable, heat-resistant synthetics. These are worth the investment if ticking lacks those characteristics.

### *Mattresses*

Most mattresses being used today in the hospitality industry are of the innerspring variety, but constructions differ widely, as do their costs, comfort, and longevity. The coils within the mattress determine its ability to support sleepers comfortably over a period of many years. In lodging establishments, the ability of the mattress to retain its shape under the weight of persons sitting on the edge of the bed will have everything to do with its life expectancy. Unfortunately, it is impossible to construct a cube that will have the properties of both the perfect mattress and the perfect sofa (Figures 13-11 and 13-12).

There are three types of coils inside the mattress that support the sleeper. Each type has a different number of convolutions, a different type of spring steel as well as a different contour.

Called the (1) Marshall, (2) Karr, and (3) Bonnell types, they are rated in that order, with Bonnell being least expensive and most commonly used. The Bonnell coil has only three convolutions, the Karr and the Marshall, five. The Marshall spring, with its convolutions encased in fabric pockets, offers the highest degree of independent coil action. Its heavy-gauge wire provides better spring action and increased durability. When the mattress is pressed down the innerspring construction will maintain a certain buoyancy, and the areas surrounding the pressure point will remain undisturbed.

**FIGURE 13-11** *Built-in king-size beds save over 4 ½ sq. yd. of carpeting, eliminate need for cleaning under beds, and prevent guests from moving furniture around. Although base was custom-built to hide box springs, beds of this type are available now as stock items.*

**FIGURE 13-12** *When bed is in place, room at Las Vegas Hilton (formerly the International) has spacious, luxurious look. Elimination of need to carpet under bed and labor saved in moving beds and finding lost articles justify extra cost for the built-in base.*

Rarely will guests or management of lodging establishments agree on just what is a comfortable bed (any more than they agree on what a comfortable pillow is), but firm support with a certain amount of resilience from the coil springs and a soft surface are considered by experts to be desirable. Just as important to the proprietor and the professional Housekeeper is that the mattress have a long and useful life.

Laura Pollishuke, in her article "A Survey of Institutional Bedding . . . The Standard and the Unusual," *The Executive Housekeeper,* November 1973, explains: "The number of coils in a mattress serves two purposes; increased conformability and tighter construction. There is less space with no visible means of support."

About padding, Ms. Pollishuke says, "It should be inert and non-allergenic, thick enough to eliminate any indication of the springs beneath. It must be securely attached, either to the springs or to the outer ticking.

"Padding, which may act as an insulation between the coils and upholstery, may be an integral part of the outer two-thirds of the mattress. If one thinks of the mattress in terms of three sections: a layer of upholstery and ticking, a layer of coil springs, and a second layer of upholstery and ticking, the picture is clear.

"There should be some means of protecting the padding material from abrasion and preventing its being worked down into the coils. A separate cover can be used or the protection can be part of the upholstery itself.

"The padding and upholstery material in a mattress may be made of cotton, latex, felt or foam.

"The importance of the mattress cover is evidenced by the care and attention that manufacturers give to their selections of materials for mattress ticking.

"Six- to 8-oz. conventional cotton ticking is commonly accepted as adequate for use on conventional mattresses. Most hospitals make use of fluid-resistant or waterproof ticking."

Ms. Pollishuke warns that mattresses should be bought from reliable manufacturers and have a strong guarantee. That alone, she feels, will help to assure that the product is well built.

A mattress for institutional use may cost from $50 to $100 and may last from three to ten years. On the assumption that the cheaper mattress would last the minimum time, it would seem more sensible to buy the more expensive one, as it would ultimately cost less.

Government flammability standards rule that mattresses must be properly labeled. Labels must state that they meet the flammability standard, which means they resist ignition by a cigarette (these have a large T on the labels). Instructions for cleaning the mattress must be included.

When purchasing new mattresses or innersprings, it is important to check the costs and practicability of having them rebuilt when the occasion eventually requires it. If such a long-range plan seems indicated it might be well to have pick-up and delivery service written in the contract. Frequently, the box spring will outlast the mattress three to one, so some provision should be made for storing the springs while the mattresses are being rebuilt.

Most mattresses are of innerspring construction. Mattresses properly constructed of urethane foam that is sufficiently thick and dense, covered by a quality fabric, and supported by a box spring or flat spring wire foundation are, however, not only durable and comfortable, but also less expensive and lighter than innerspring mattresses. Being more flexible, too, they can be more easily turned and rotated by the maids.

Turning and rotating is extremely important as it means mattresses will last longer because they wear evenly, just as rotating tires on a car will increase their mileage.

When and how to turn and rotate mattresses is always a problem in hotels because maids usually do not have the strength to flip and turn them or union regulations do not permit them to do so. The job is usually done by a houseman when he does the heavy vacuuming, but that is only convenient if the bed happens to be stripped when he is there. More often the bed is made and it is difficult to locate a maid with time to remake it. Housemen are not encouraged to strip beds when they are busy with their many vacuuming chores. Nonetheless, attempts must be made to coordinate mattress-turning with other heavy work to be done in the room.

The professional Housekeeper may assign one houseman to work in an area with two maids, rotating and turning mattresses as the beds are stripped. He may also switch twin beds at that time so that wear on mattresses will be equalized, since in a twin-bedded room, the one next to the wall is used much more frequently than its mate. A record of this switching must be kept, together with a record of mattress-turning. Usually it is written on a form displayed in a nearby back corridor.

TraveLodge has worked out the following system as outlined in the company's housekeeping manual:

"The executive housekeeper of each motor hotel will designate four permanent groups of rooms to assure the orderly rotating of mattresses. The mattresses in one-fourth of the rooms will be turned every two weeks. Starting every other Monday, the maids making the beds in the specified rooms will turn the mattresses and record it on the daily worksheet. If only one bed in a twin room requires a change of linen, then only that mattress will be turned and will be so marked on the daily worksheet. The maids will continue to turn mattresses during the week until all the designated mattresses are in the new position."

How will an inspectress know that a mattress has been turned? Labels on each corner will be a sure means if they have "January" and "April" on tags in opposite corners of one side, and "July" and "October" on opposite corners on the reverse side. Then no matter when the mattresses were bought, one can ascertain whether they were changed within the past three months. Mattress makers will put those tags on if requested to; tags should be specified when purchasing mattresses.

## Bibliography

Nolte, William D. It Is Time to Update No-Iron. *The Executive Housekeeper,* March 1974.

Pollishuke, Laura. A Survey of Institutional Bedding . . . The Standard and the Unusual. *The Executive Housekeeper,* November 1973.

Smith, Fran. Flammable Fabrics, *Hotel & Motel Management,* October 1973.

## Sources of Information

West Point Pepperell
111 West 40th St.
New York, N.Y. 10018

Waverly Fabrics, Div. F. Schumacher & Co.
939 Third Avenue
New York, N.Y. 10022

Baker Linen Co.
1101 Macy
Los Angeles, Calif. 90033

# Chapter 14

# Walls—Their Covering and Care

**OVERVIEW** All types of wall coverings that can take the rigors of public areas are considered here, along with the many places for which the professional Housekeepers will have to specify wall treatments. Cleaning techniques are reviewed, and many aspects of vinyls are explained including three grades and where to specify them.

Although decorative appeal is essential, wall coverings should be selected for their functional quality, durability, and ease of maintenance.

Walls must necessarily complement or enhance a designer's concept; however, they must first and foremost be able to endure daily wear and tear for a reasonable period of time. All wall coverings should be cared for regularly in order to maintain their appearance. To choose one of silk, moire, velvet, or other delicate fabric purely for its aesthetic value, knowing that it will be exposed to the long-armed negligence of John Q. Public, is naive and can lead to costly dents in the maintenance budget. All wall coverings should, of course, be regularly inspected and maintained by the Housekeeping Department.

## *Types of Wall Coverings*

**Tile,** long relegated to outdoor areas or limited to kitchen and bathroom walls, has come into its own with ceramic and mosaic patterns that brighten and create a cool airy feeling in rooms. Generally easy to keep clean, tile's main drawback is that grouting may become discolored or loosened.

**Wallpaper** is the old standby, and with the help of modern chemistry that has made it water- and stain-resistant, today's papers need only the traditional mild soap and sponge (with a little elbow grease) to clean (Figures 14-1 and 14-2). A special release paste now on the market facilitates removing wallpaper without the usual scraping, wetting, or steaming.

**Wood's** rough texture generates warmth. Paneling comes in many variations of color, tone, and texture. It can be both rustic and elegant and rarely needs heavy-duty cleaning. Now through the use of carved textures and patterns, a three-dimensional effect has been achieved on hardwood panels. The same effect can be achieved with cedar shingles or hand-split shakes indoors.

**Plastic Laminate Sheeting** or paneling is usually found with a wood grain (sometimes with grooves to make it resemble planks) on lobby, meeting, or guest room walls. In a print or solid, it is particularly appropriate in bathrooms, rest rooms, or foodservice areas where walls are subject to splashes or spatters. Developed to meet most fire codes, the vertical grade 0.035 smooth laminate can be bonded to particle-board for interior panels. Plastic panel producers also make coving and baseboards in matching or harmonizing colors to be used with the walls.

**Foil** coverings tend to illuminate a room, but must be used discreetly to avoid creating a gaudy effect (Figure 14-3). Certain manufacturers have perfected interesting and tasteful foils with

**FIGURE 14-1** *The very best of modern wallpaper achieves the level of at least a minor art form, as exemplified at National Cash Register's San Diego facility, where a papered wall provides main design motif in a lobby created to communicate and reinforce the company's progressive image. Unlike most wall coverings, this one will not show soil or marks. Other easy-care features include vinyl upholstery, resin surface on table, and vinyl tile floors. Credit: Armstrong Cork Co.*

**FIGURE 14-2** *Especially suited to long wall expanses is this pattern that introduced the concept of super-graphics, (bold diagonals to be joined back to back, upside down, etc.) on either side for infinitely varied effects. Credit: Winfield Design Associates*

woven backings that are patterned as well as being washable and durable.

**Cork** offers a dramatic and luxurious effect and is easily installed. Its chief virtue is sound control; its disadvantage is its perishability.

**Inorganic Wall Coating** is a continuous wall covering (sprayed on with a high-pressure gun) resembling a ceramic glaze. It cannot burn or deteriorate through oxidation and exposure to moisture. It is almost permanent and comes in a full range of colors. Some are specifically designed for use in wet areas—showers, steamrooms, and the like. It has a clear glaze that prevents scratching of the surface.

**Fiberglass** walls can be almost indistinguishable from old woodwork and weathered plaster when properly painted. The material can be poured into molds to duplicate any shape or surface.

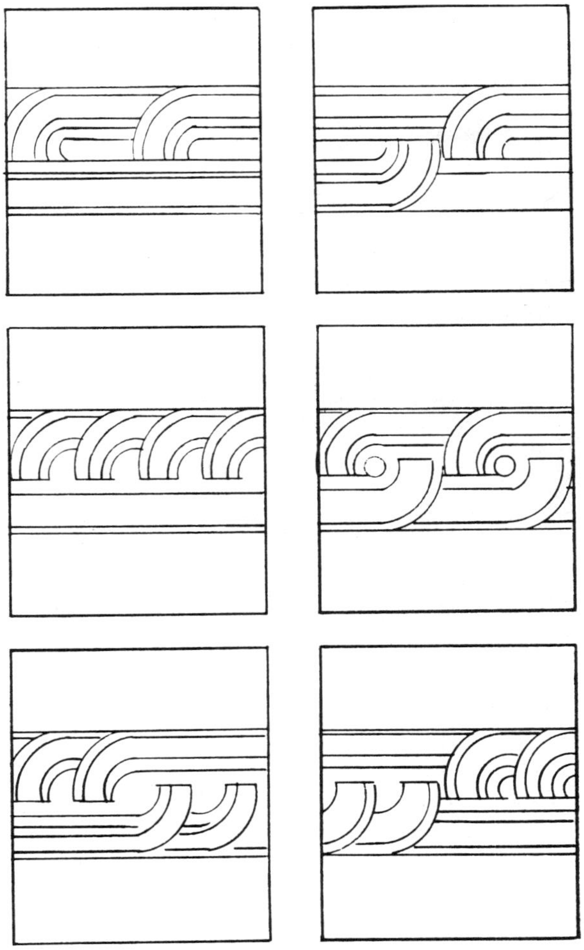

**FIGURE 14-3** *From Howard Johnson motels comes this diagram suggesting coordinated foil super-graphic wall coverings. Sold separately, panels can be used in variety of combinations, applied in any direction. Panels are 10 ft. high, trim to 27 in. wide; printed on metalized mylar.*

Many Old English-style pubs have fiberglass interiors and it is a rare patron who can detect that it is not the authentic material. Just as a fiberglass boat is easy to clean so is a wall of fiberglass, as it requires only dusting or a bit of wiping with a damp cloth.

**Plush-on** is fabric that resembles deep fluffy plush but has an adhesive back so it can be cut and glued to any wall or flat surface to give a soft luxurious look and/or cover a badly bruised surface. It can be purchased in 29-in. or 12-in. widths.

**Vinyl with Fabric Backing** has become a most popular wall covering for institutions (Figures 14-4 and 14-5). Because vinyls are so varied and come in so many different weights, textures, patterns, and colors, many of which have special properties, they have been dealt with separately later in this chapter.

**Painted Surfaces** are of course the most commonplace of all wall coverings. The paints available today are expensive if they are to be counted on to hold up. But to economize on the paint itself is foolish, because the labor to apply it is so costly that the surface should last for years and be easily washable. In hotels, it is important to use a fast-drying paint so that rooms will not be out of order for more than a few hours. Select one that leaves no lasting odor. In residence halls where the rooms may be empty during summer months when repainting is done, the fast-drying feature is not so important.

**Fabrics** are used by impractical designers on some walls, but if the professional Housekeeper has any input into the interior design scheme that includes fabrics, it is worth protesting because when they are affixed to walls they are so difficult to clean. Almost any pattern can be laminated into plastic sheets or can be backed with vinyl to make installation and cleaning easier (Figure 14-6). If one is stuck with fabric walls, check to see if they can be waxed with paper wax or liquid carnauba wax when they are new.

**Clear Plastic.** Acrylic plastic sheeting, either all-purpose or one of the new break-resistant types, is highly useful on walls in elevators or vulnerable spots in corridors where walls are continually being scuffed or marred. It may only be needed as far up as the wainscoting and can be affixed with decorative screws in each corner. It must be cleaned only with soap and water, not with abrasives or solvents. This material *is* combustible, so one should check for fire code approvals.

**Rubber/Vinyl Wall and Corner Guards** are another means of protecting the busy back corridors, elevators, and service areas where carts in constant use may lead to gouged walls (Figure 14-7). Wall guards are simply a continuous trim

**FIGURE 14-4** *Bold striped vinyl wall covering with woven back is the dominant decorating element in the sitting room of a suite at the Marriott Hotel Tucson. Lively accessories such as a handsome antique mirror from South America and framed yarn designs from Mexico are in the same color tones as wall, so that room is not too busy.*

strip that may be run horizontally approximately a foot up from the floor, or just above the baseboard. The vinyl or rubber stripping may also be used vertically to protect protruding corners that normally sustain high-impact damage.

## *Care and Maintenance*

It takes roughly four man-hours thoroughly to clean approximately 1000 sq. ft. of painted wall. Proper methods, however, can prove the time well spent, extend the periods between washings, and possibly reduce the estimated washing time.

Many cleaning experts recommend that walls be dry-dusted before being washed. If the surface dust is obvious, and there are many cobwebs, dusting first will probably speed up the washing process. Dry-dusting will not give satisfactory results, and will cause streaks if done in a damp room, so avoid assigning this task on a

**FIGURE 14-5** *Since vinyl can be applied to curved surfaces, it is used in such older hotels as the Lexington, New York, where the aim is to keep traditional elegance, using modern materials with minimal maintenance. When the attractive columns (above left) with Corinthian capitals crack, they can be covered with an elastic-backed vinyl that blends with concrete so the effect is clean and continuous. Credit: General Tire and Rubber Co.*

rainy or humid day unless the building is warm and dry. The best tool for this operation in unobstructed wall areas is a very soft hair or lamb's wool brush on an extension handle. Pipes, high moldings, and door tops should be dusted with a treated rag.

Walls should be dusted upward as dust tends to cling to walls at downward angles. Action should be gentle as pressure will make smears and force dirt into paint. Change cloths and brushes as they become soiled.

When it comes to washing the wall, most professionals will use one of three cleaning methods—the sponge and pail, and mop and pail, and the wall-washing machine—all outfitted with water-based cleaning solutions. Acoustical material, wallpaper, and other special wall coverings generally require certain variations or combinations of these techniques.

First rule for washing walls: choose an effective cleaning solution, one that is strong enough to do the job without dulling, staining, or otherwise damaging the wall covering. Strong alkaline cleaners such as trisodium phosphate remove the dirt and also part of the paint.

Before using any recommended cleaner for the first time, test it on a small part of the wall. Start with a weak solution and keep adding the cleaning agent until the solution is the right strength. The manufacturer's recommendations are a good guideline, but keep in mind that both the effectiveness of a cleaner and soil resistance vary from wall to wall. Synthetic detergents made of sulfonated hydrocarbons are regarded more highly than soaps, which may leave a residue and hasten future soiling. If soaps are used, rinse the wall very thoroughly after application. Synthetic detergents, on the other hand, act as wetting agents and therefore leave no residue.

Badly soiled wall areas should be saturated and the solution allowed to penetrate the surface for a few minutes. Where dust and smoke have accumulated—over a radiator, for example—apply a little abrasive powder on a wet sponge or pad, but *use it sparingly* because it may dull the finish on the wall.

Rinsing is essential on the heavily soiled areas and so is wiping immediately. Chamois cloths are ideal but expensive. Absorbent rags are an ac-

**FIGURE 14-6** *Not one sheep was shorn to manufacture Wooly Stripe, a new UltraTexture wallcovering. A woolly look and feel was embossed into hardy vinyl that retains its fabric-like qualities even as it stands up to the toughest wear throughout lobbies or offices. Credit: Ultratex/Reed Wallcoverings*

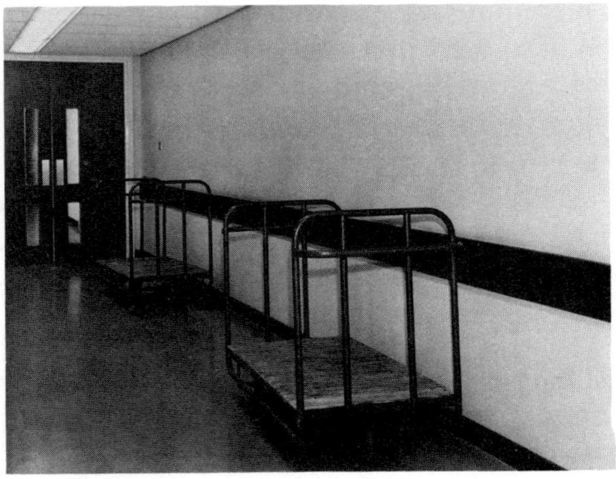

**FIGURE 14-7** *In back corridors where high-railed carts will bump into walls frequently, a wall guard can be installed to save on repairs and to improve the hallway's appearance. Sturdy, the trim will withstand a blow of 30-ft. pressure per square inch at 60° F. It is 8 in. wide, and can be formed in angles and bends.*

ceptable substitute if the chamois supply should disappear.

Where on the wall does one start? Beginning from the top and working your way down may seem logical, but bear in mind that sponges and mops drip, and trickles of solution running down a soiled wall cause stains that are difficult, if not impossible, to remove. Trickles will not stain an already cleaned area, which, in any event, can be wiped off before it dries and becomes streaked.

Wash and rinse the lower half of the wall as far up as possible (Table 14-1). Wipe dry as you go along and work from one corner of the room to the other. Then tackle the upper half. Just be sure to change the solution and rinse water frequently to ensure good results.

If using sponges, use large ones (about the size of a brick) and plan on two three-gallon pails to hold the cleaning solution and rinse water respectively. Cellulose sponges are firmer than natural ones and can be applied with considerably more pressure.

Mops are the answer for rush jobs, but mopping walls has never gained wide acceptance as a regular or preferred method because it is difficult to do a thorough job unless the worker is very energetic and conscientious. Also, mopping cannot be done successfully in crowded quarters such as a guest room. In the corridors, however, this technique has proved highly effective.

Where walls are to be cleaned with mops, as with the sponge method, workers will need two 3-gallon pails (on a dolly to facilitate moving them around), two 10-in. cotton dust-mops (non-swivel), and a long-handled scrub-brush. Telescopic mop handles make it easy to reach high on the wall. Since this job is at best, messy, it is always a good idea to use drop-cloths to protect floors from possible spillage or dripping. A sponge mop with a hand-wringer for squeezing excess moisture is preferred by many when using the mop technique.

First, remove loose dust with a dry mop. Then dip a clean, dust-free mop into the solution and wring thoroughly to prevent dripping. Beginning at the top of the wall, swing mop from side to side in rhythmic strokes as far as possible from that position on the floor. Do not apply too much pressure because this may rub soil into the wall.

When that point of the wall is reached where side-to-side cleaning is impractical, step closer and swing the mop in long vertical arcs from the cleaned upper surface down to the baseboard. Use the scrub brush for more stubborn spots. Before the cleaning solution dries on the cleaned area, follow with the rinsing mop, using the same swinging technique. A third mop may be used to dry the cleaned and rinsed area.

**TABLE 14-1.** Wall Cleaning

| Cleaning Utensils | Cleaner/Sanitizer | Method |
|---|---|---|
| Brush, sponge, bucket, cup | Spray cleaner or institutional all-purpose cleaner | Daily: (1) scrub with solution using brush, (2) dry with sponge |

## *The Great Wall-Washing Machine*

Strictly speaking, the wall-washing machine is not a machine at all. The results depend on trained operators using it properly.

Although there is a machine with a single tank, the conventional equipment contains two cylindrical tanks—one for the cleaner and one for rinse water.

Pour a gallon of cleaning solution into one tank and a gallon of water into the other, allowing enough space for air pressure to build up. Plastic or rubber tubes from each tank connect to metal trowels that hold terrycloth pads. (For rough walls, special brushes or synthetic pads are used.) There generally is a third trowel, not connected to the machine, that is used for drying.

The flow of both fluids is controlled by a trigger on each trowel, with the exact amount regulated by the operator. Trickles are thus avoided.

Begin at the top of the wall with a trowel in each hand. Apply the solution first, followed by the rinse trowel in left to right arcs as far as you can go without stepping ahead. When the solution pad becomes too soiled, change it. You may also switch the rinse trowel to the solution trowel, and the drying trowel to the rinse trowel.

The machine represents a cost-saving investment in that one filling contains enough solution and rinse for one person cleaning for eight hours. Therefore it precludes changing solutions that get dirty, and the necessity of adding more than needed. It is a labor- and timesaver since the operator does not have to change the mops, the rinse, or the solution; nor does he tire himself out bending over to squeeze pads.

**FIGURE 14-8** *A ceiling of colored glass squares in a cocktail lounge gives off a warm amber glow conducive to conviviality. About every six years each of the 5800 pieces is taken down to be placed in a rubber-covered rack and washed in a giant automatic dishwasher. Job is done by a night crew of five people, working three nights. Metal frames were also dismantled, vacuumed, and run through dishwasher.*

## *Special Procedures for Special Walls*

**Marble.** When discolored or stained, apply a poultice of abrasive cleaning powder and hot water. It should have the consistency of mortar and be applied about ½-in. thick with a plasterer's trowel. After 48 hours, remove with wood paddles to avoid scratching the marble surface. Rinse well and dry with a chamois. There are also commercial chemical cleaners that can be used. If marble becomes waterstained, it may have to be ground down to remove the markings. Proper sealing initially is the key to keeping marble beautiful without undue expense.

**Ceramic Tile.** Although generally easy to keep clean, tile does get some stubborn stains. A mild alkali or synthetic detergent is a good cleaning agent on a regular basis, but a mild abrasive cleaner will usually remove stains.

For grouting discoloration, use a whitewash mixture of an abrasive cleaner and hot water. First wash the wall and then while it is still damp, apply the mixture with a thin paint brush. After 24 hours, scour the grouting by dipping a stiff brush into the whitewash. Avoid the edges of the tile as much as possible. An abrasive cleaning powder containing a bleach is preferred. (In the case of coffee or lipstick stains, use a soapy No. 1 steel wool pad.)

Figure 14-8 illustrates one very specialized cleaning need and the procedure used to accomplish it.

**Plastic and Metal Walls.** Because these coverings are often embossed, a soft brush is occasionally required to clean the indented portions (in addition to the three cleaning methods described). A machine with a brush attachment for rough walls is recommended.

**Acoustical Walls.** When acoustical walls or ceilings have been painted one can use the same methods as those on any painted wall. Some acoustical material is enameled, making it easier to wash. A spray paint is advisable if repainting, but take care not to fill the holes in the acoustical material. Also, a flat or water-based paint is preferred over glossy for better acoustical effects.

Acoustical walls and ceilings of all types should be vacuumed (with a soft brush in the nozzle), particularly if they are unpainted. Where unpainted surfaces are soiled, use a putty-type wallpaper cleaner. For small spots, a piece of art gum will do the trick. Very difficult spots are best removed with dry steel wool or a synthetic abrasive pad. Colored chalks are ideal for touching up nicks.

**Wallpaper.** Frequent vacuuming is recommended to delay heavier cleaning. When heavy-duty cleaning is needed, washing will be effective in most cases. Be sure to check the wallpaper for water resistance before you proceed.

Rub a small area with a wet cloth and check whether the color comes off or spreads. If it does, then the wallpaper is not washable. If this is the case, try a dough-type cleaner. Use long sweeping arcs. Keep kneading the dough cleaner as you proceed to provide a clean surface against the wall. When a clean surface is no longer attainable, use a fresh piece.

For grease or oil stains, apply a thick paste of Fuller's Earth and benzine or use a nonflammable dry cleaning fluid. Remove when dry. For small grease marks, press a piece of white blotting paper against the spot with a warm iron and hold for a few minutes.

If wallpaper is washable one can use a cellulose sponge and a mild solution of synthetic detergent without bleach. A moist sponge should be wrung out just enough to be effective. Start at the baseboard and wipe with long sweeping strokes. Keep the sponge clean by squeezing out excess soil. Rinse with a clean sponge and clear, softened water.

**Wood.** Figure 14-9 illustrates method of handling one of the few problems that occur with wood.

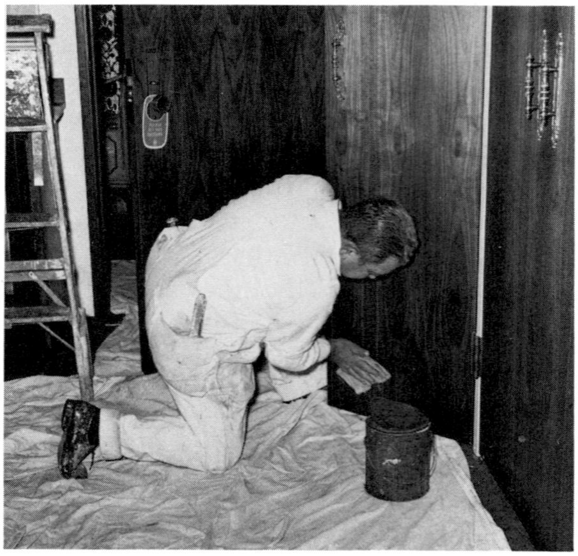

**FIGURE 14-9** *Oiled wood doors require little attention unless damaged or dried out by a heating system. Painters remove marks with steel wool, then use wood sealer followed with Danish oil containing stain.*

## *Vinyls—What You Should Know About Them*

Early in 1974, the federal government issued new specifications on weights of vinyl wall coverings. They may aid designers in specifying light, medium, and heavy weights, however, that in itself does not help in predicting longevity of a wall covering. It is proper installation, amount and type of wear, and quality of maintenance that determine how long a vinyl wall covering will hold up. Constant scrubbing eventually will wear off the pattern and texture, and fumes will ultimately discolor it.

In fact, until now there has been no real way to predict the life expectancy of vinyl wall coverings because the product is still relatively new and untried.

Institutions with low budgets have tended to reject vinyl because it seems extremely expensive. But the truth is, vinyl walls generally pay for themselves within three years because the down-time required for painting is no longer a consideration.

Where hotels and restaurants have invested in costly vinyls, the splendid array of textures, colors, and patterns, plus such practical enhancements as sound-absorption, and fire and water resistance have justified the expenditure in the opinion of almost all operators queried.

Some vinyls now contain germicidal additives to eliminate *Staphylococcus aureus* (staph). Thus the material becomes staph-resistant within 24 hours of contact.

The days of the cheap plastic vinyl are gone. Today, walls in many hotels and colleges, exquisitely decorated with seemingly plush elegant wall coverings, challenge even the trained eye looking for the difference between vinyl and the real thing, for example, moire, grass cloth, shantung, Dutch painted tiles (Figures 14-10 through 14-13).

**Types of Vinyls.** The government's specifications encompass three types of vinyl.

Type I is lightweight (9 oz. per sq. yd.) and should be used in areas where there is virtually no traffic. These vinyls are extremely vulnerable to abrasion from carts, suitcases, dollies, and so on.

**FIGURE 14-10** *Stain-resistant line of vinyl wall coverings offers six versatile patterns in 54-in. widths protected with a Tedlar film coating. Even ballpoint and marking pen ink can be wiped off with mild cleaners. Credit: General Tire & Rubber Co.*

**FIGURE 14-12** *Contract Collection is a lightweight, fabric-backed vinyl designed primarily for commercial and public areas that do not require the durability of heavier grades and, in fact, may require frequent redecorating. Packed in 60-yd. rolls, material has a 12-oz. weight, is 54 in. wide. Credit: General Tire & Rubber Co.*

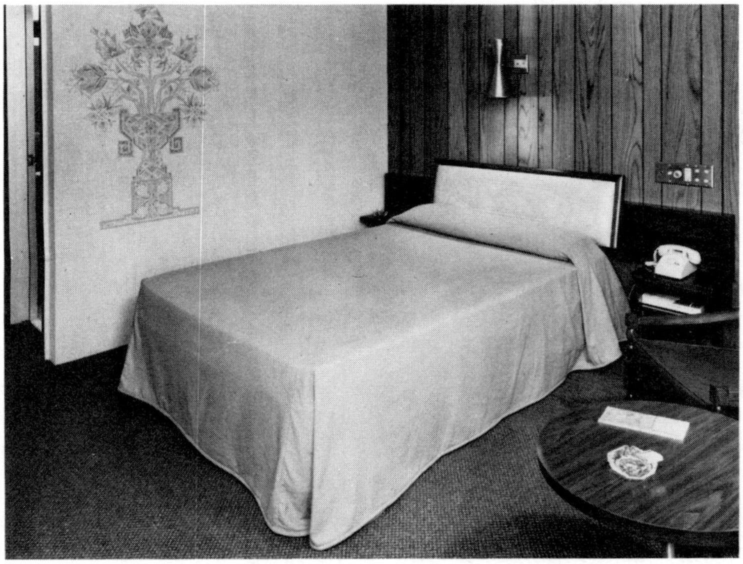

**FIGURE 14-11** *In a Howard Johnson Motor Lodge, random grooved oak plank wall is actually plastic paneling that requires no maintenance except annual wiping. Vinyl wall with flower vase as design motif provides total decor at reasonable cost. Unlike many design elements, it is break-proof and theft-proof.*

**FIGURE 14-13** *Giant photomurals in the dining room of Benson Hall at the University of Santa Clara, Calif., are exciting to students who see them clearly from any place in the dining room. Vinyl coating keeps them clean. Other maintenance features in room include: low-pile, nonstatic nylon carpet (glued down); molded plastic chairs and tables, and recessed spots instead of hanging light fixtures. Credit: Burlington Industries, Inc.*

Type II is the workhorse of vinyl wall coverings and weighs anywhere from 13 to 23.5 oz. per sq. yd. It accounts for the largest volume of sales in the commercial market. Where wall areas receive more than an average amount of staining, a 17-oz. vinyl clad with polyvinyl fluoride film offers the greatest performance for the cost. It is used most in lounges, dining rooms, offices, and in hospital wards.

Type III weighs in at 22 oz. and up, and should be installed in areas where exceptionally heavy usage and impact occurs—cafeterias, elevator foyers, corridors. It is the most expensive of the three.

A rule of thumb regarding cost is to consider installation to be about the same per square foot as the cost of the material. This, however, can jump to as much as three times more in some areas, so it is advisable to obtain competitive bids.

In hotels and residence halls all types and weights are used, but Type I is most popular for guest rooms. Type II with polyfluoride film is recommended for bathrooms and lobbies. Walls in student rooms should be covered with easily removable paneling that can have tacks stuck into it—important for students.

Another version of vinyl is the one designed to give extra sound-absorption. It has a 1-in. fiberglass core and is continuous—there are no seams to show on the panels.

## *Maintenance of Vinyl Wall Coverings*

Vinyls are laminates, which means vinyl film is glued to a cotton backing. If not done properly the material can tear easily. In some cases the film is applied as a liquid and fused to the cotton backing under intense heat. This reduces tearing considerably.

Vinyl has a softening agent called plasticizer that tends to absorb dirt and stains that other wall coverings cannot: ball-point pen ink, shoe polish, lipstick, carbon smudges, scuff marks from rubber heels or soles. Dirt should not be allowed to remain on these surfaces for lengthy periods of time.

Another point to keep in mind is that the backing is cotton, and therefore subject to mildew, especially in humid areas. While most vinyl is mildew-resistant, this is not necessarily guaranteed.

Most manufacturers recommend a mild detergent for vinyl wall covering. For special stains, where you are uncertain of chemical effects, contact the manufacturer directly, describe the stain and what caused it, and rely on his suggestion for removing the stain.

Vinyl coverings are an exceptionally good investment in hotel/motel closets where suitcases and garment bags mar every wall on contact. The same applies to the wall behind luggage racks.

**Cleaning.** Use a hard bristle brush and scrub vinyl with a low-sudsing detergent mixed in warm water. Start at the bottom of the wall and work upward. Rinse thoroughly with a sponge in clean warm water. Then dry with a soft lint-free cloth or towel.

Never use abrasive cleaners, acetone, or strong solvents, since these may discolor most vinyl materials.

The one exception is where polyvinyl film has been applied to the wall-covering. Only in this case may solvents be used repeatedly without any adverse effect on the vinyl material on the walls.

**Patching.** When purchasing vinyl material it is advisable to retain extra rolls, preferably from the same run by the original manufacturer to ensure against any variations in color or pattern.

For small damaged areas (where it is unnecessary to replace an entire panel), patching may be done by placing a new piece of covering over the damaged area and pasting it to the wall. Allow about a half hour for the paste to set, then take a new razor blade and cut through both materials. Peel off the old section underneath and replace with the new. This procedure re-

**TABLE 14-2** Comparative Costs of Vinyl versus Paint

| | | Single Room | Double Room |
|---|---|---|---|
| **TO COVER WITH VINYL WALL COVERING** | | | |
| 50 yd. of 54-in. material | @$5.90 | $295.00 | $295.00 |
| 5 gal. adhesive (mastic) | @$4.50 | 22.50 | 22.50 |
| 8 hours labor to install* | @$6.00 | 48.00 | 48.00 |
| | | $365.50 | $365.50 |
| Loss of revenue for one day | | 20.00 | 22.00 |
| Total cost | | $385.50 | $387.50 |
| **TO COVER WITH 1 COAT LATEX PAINT** | | | |
| 2 gal. paint (white) | @$7.95 | $ 15.90 | $ 15.90 |
| Color added to each gallon | @$1.00 | 2.00 | 2.00 |
| 6 hours labor | @$6.00 | 36.00 | 36.00 |
| | | $ 53.90 | $ 53.90 |
| Loss of revenue for one day | | 20.00 | 22.00 |
| Total cost | | $ 73.90 | $ 75.90 |
| 4 paint jobs in 12 years | | $295.60 | $303.60 |
| **TO COVER WITH 2 COATS LATEX PAINT** | | | |
| 4 gal. paint (white) | @$7.95 | $ 31.80 | $ 31.80 |
| Color added to each gallon | @$1.00 | 4.00 | 4.00 |
| 6 hours labor for each coat | @$6.00 | 72.00 | 72.00 |
| | | $107.80 | $107.80 |
| Loss of revenue for 2 days | | 40.00 | 44.00 |
| Total cost | | $147.80 | $151.80 |
| 4 paint jobs in 12 years | | $591.20 | $607.20 |

* Note: *In many cases where the institution's maintenance staff cannot install it, the wall covering supplier's installation charge will average between $4.50 and $5.00 per yard. $4.50 per yd. labor for installing 50 yd. would amount to $225.00 instead of $48.00. Total cost for material, adhesive, and labor would be $542.50. With even a $16 single room total revenue loss to cover supplier's installation charge becomes $562.50; for double room, $564.50.*

quires a special knack but competent maintenance men can learn to do it very easily.

**Removing Vinyl Wall Coverings.** Fabric-backed vinyl wall coverings, unlike wallpaper that needs to be steamed off, can be peeled back from the corner and pulled off at a 180-degree angle parallel to the wall. Do not pull from the corner at a 90-degree angle because tear resistance at this angle is much greater and loose plaster may be pulled from the wall. Since all adhesives used to apply vinyl are water-based and soluble in water, using a little bit of water for removal of difficult areas will help lighten the chore.

When installing vinyl wall coverings, follow the manufacturer's instructions. Be sure to remove all dirt, grease, and old adhesive paste from the walls. Rough spots should be sanded and all cracks and holes sealed up with spackling compound.

On painted surfaces, sand all enamels and gloss paint. Use a latex primer where vinyl is to be applied to smooth, difficult-to-bond-to surfaces. This helps installation adhesives to hold when applied.

## The Subject of Cost: Vinyl versus Wall Paint

The claim that vinyl is expensive is true in the short run but probably erroneous in the long run. With proper care, a room is repainted every 3 years; with proper patching, when needed, a vinyl wall covering should last from 12 to 15 years.

Table 14-2 represents a room with 240 sq. ft. of floor space and 468 sq. ft. of wall space to be either painted or covered with vinyl material. (The private bathroom is generally finished in ceramic tile and, along with woodwork and window ledges, is not counted in the cost figures.)

The wall covering used is a textured, standard-weight vinyl. The paint is a latex. If it is to be repainted with the same color, one coat might be sufficient. Replacing hardware and eliminating odor usually requires waiting another day to finish and that means additional loss of revenue. If it is possible to get by with only one coat then do, since nothing ages a room so much as a heavy build-up of paint.

## Sources of Information

L. E. Carpenter (Vicrtex)
964 Third Ave.
New York, N.Y. 10022

Deco Gard Products
P.O. Box 451
Cranford, N.J. 07016

General Tire & Rubber Co. (Genon)
Contract Group
979 Third Ave.
New York, N.Y. 10022

## Bibliography

Chase, Mildred. *Basic Interior Design, A Supplement to Supervisory Housekeeping.* Educational Institute of the American Hotel & Motel Assn., 1969.

Paterson, Peter R. Selecting and Installing Vinyl Wallcoverings. *The Executive Housekeeper,* October 1971.

# Chapter 15

# Windows—Their Care and Treatment

**OVERVIEW** Glass-washing is only a part of the major chore of keeping windows clean and attractive. Here are suggestions for ways to treat problem windows, discussion of flammability of fabrics, linings, and backing for draperies, and a detailed look at the various properties of different fabrics.

Individuals entering a hotel room for the first time unconsciously focus on the window and its view before noticing the rest of the furnishings. One guest put it more poetically. He defined a hotel room window as his "lifeline to the outside world."

If we accept this as a general truth, it stands to reason that that "lifeline" should always be attractive and eye-appealing to look *at* as well as *through*. No drapery design or color in the world can compensate for dirty windows.

But keeping windows clean is no easy chore. On the inside it is, at best, difficult. Outside, the task is virtually impossible. Mother Nature has little regard for aesthetic-minded guests and even less for over-burdened hotel housekeepers and maintenance crews. In humid climates especially, the situation is compounded by inside air-conditioning that causes windows to steam up. In areas boasting rugged winters and steamy summers, window ledges (unless they are treated with laminated plastic or similar highly resistant surfaces) become susceptible to these extremes of temperature. Cold weather may result in a good deal of moisture leakage, which when followed by extremely hot weather causes damage to a room's interior. Also, heating systems generate extra soil due to the hot air piped into rooms.

As for dust and dirt, there is little anyone can do about them except to weather-strip the doors and the windows or to use dust filters or screens where possible (particularly in industrial areas).

## Window Cleaning

Laboratory-tested compounds and chemical solutions, guaranteed to clean windows better and more easily and make them brighter as well, have inundated janitorial jobbers' shelves in recent years. All work satisfactorily to a degree, but usually old methods—mild liquid detergent and water—do the trick just as well. Adding a little ammonia softens the water and increases the shine, although it may discolor putty. Bluing is also effective. Alcohol is another agent that helps loosen dirt, particularly in cold weather when grime tends to congeal and adhere to window panes.

Scouring materials are best avoided; they seep into cracks and crevices and their abrasive qualities have been known to mar glass.

Chamois skins, sponges, or lint-free cloths

(including nonwoven cellulose) are ideal window cleaning materials. The squeegee is recommended for large or medium-sized windows as the most efficient and easiest tool to work with. One point about the squeegee: it should be wiped dry of excess water after each stroke. Surprisingly, some maintenance people forget to do so. Wet rubber or wet glass limits the tool's effectiveness.

Other basic equipment includes two buckets, one for the cleaning solution, the other for rinsing; a razor blade for scraping off hard-to-remove dirt or spots; a safety ladder if needed; and coverings to protect the sills.

Prior to actual cleaning, all window dressings, shades, and blinds should be removed. Screens, if they are not removable, should be propped out and away from the frames with rods. They should also be washed or vacuumed free of dust before windows are washed. Sills should be cleaned inside and out and frames dusted or washed to prevent edge streaks on the glass. Metal-framed windows should be kept rust-free and extra-clean by rubbing them lightly with fine steel wool. Wooden frames are subject to rot if too much water or solution collects on them over lengthy periods of time.

The least tiring method of washing windows consists of sponging or washing in continuous strokes from left to right, then back again, right to left, in a downward pattern until the bottom is reached. (Work probably goes from right to left for left-handed people.)

After the side-to-side motion, continue washing—starting at the bottom left corner—but start washing with up-and-down strokes, working toward the right side of the pane and ending up at the lower right corner.

Follow the same pattern when drying, using a chamois skin or lint-free cloth for smaller windows, or a squeegee for medium-sized or large-sized panes.

If working with a squeegee, the operator may go from side to side or from top to bottom, whichever suits him. If side to side, hold squeegee vertically at the upper left of the pane and pull in one stroke toward the right; make a half-turn before reaching the other side, and continue the same stroke back to the left. Wipe excess water off the rubber and follow the same motion until the bottom of the window is reached.

In the end, cleaning windows is a full-time job for one or possibly more employees. However they tackle it or whatever laborsaving tools they use, the job demands a lot of the proverbial elbow grease if it is to be done well.

One last point: when employing commercial glass cleaners or synthetic detergents the directions should be read and studied very carefully before beginning work with them.

## *Window Trimmings*

Window shades, thanks to modern textile science, are available in a variety of washable fabrics. For washing, most types only need to be spread out on the floor or a large table and rubbed with a sponge dipped in warm suds. Use overlapping strokes to wash, then rinse with a damp cloth or sponge.

Venetian blinds respond best to terrycloth. Many enterprising Housekeepers fashion hand mitts from an old towel because, ideally, both sides of the slat can (and should) be dusted at the same time. Also, mitts protect the hand against cuts and scratches from the sharp edges of the slats. A clean mitt or cloth is necessary for rewaxing. Double-thumbed plush mitts that fit right or left hands are on the market today. They are completely washable, as water-absorbent as sponges, and can be used with all-purpose cleaners.

Traverse or curtain rods need a good rubbing with paste wax to prevent excessive soiling. This also keeps drawcords from sticking when pulling draperies back and forth.

The type of rod used depends, of course, on the type of drapery. The handpull is probably more popular than the traverse rod because the latter occasionally slips off the runner.

If one has to contend with beaded curtains, a mild detergent and water solution offers the best cleaning method. A lint-free cloth is best for drying.

Decorative as it is, balled fringe (poms) becomes excessively dusty and is not washable. The only economically practical recourse is frequent thorough vacuuming.

## *Drapery Fabrics*

The man-made fabric probably does not exist that is not resistant in one way or other to something—soil, abrasion, fire, water, sunlight, wrinkling, creasing, and even odors. Yet all fabrics contain different properties in terms of their durability, color-fastness, and washability.

Cost is a decisive factor, but if one is contemplating a really inexpensive fabric, remember, the old maxim holds true: spending less in the beginning often costs more in the end (Figure 15-1). Cheap materials simply will not retain shape, color, or weave over extended periods of time. Nor can they be expected to withstand numerous launderings.

Price should not be the sole criterion where fabric selection is concerned. Yarn quality and thickness (denier), color-fastness, and construction are other salient considerations. Densely woven materials with generous quantities of yarn usually have the superior stamina that is so vital in nonresidential situations.

Another important factor in choosing a fabric is the type and amount of daily abuse it will take, including excessive exposure to sunlight, humidity, soil, and frequent handling or mishandling by guests and employees. How much vacuuming the draperies will have may influence choice of colors and textures. Employees should be instructed to open and close draperies in guest rooms, as this movement will help to dust them (Figures 15-2 through 15-5).

If draperies are going to be subjected to heavy usage and abrasion on a daily basis, tweeds, tapestries, and friezes are rugged enough to take the beating. Pile weaves such as velvet are not only luxurious, they are strong, especially those constructed of chemical fibers. Damask is another durable fabric for all its elegance and prestigious appearance.

Loosely woven materials stretch, lose their

**FIGURE 15-1** *Elegant dropped panels over twin-level curtains, framed by matching tie-backs lend a luxurious feeling to this room at the Broadmoor in Colorado Springs. Note that furnishings are placed away from window treatments. This saves accidental brushing against fabrics when guests pull out chairs to sit down. Handpainted cafe curtains on walls between windows tend to unify wall by connecting windows.*

**FIGURE 15-2** *Not all windows need curtains or draperies. Some in residence hall at "Cal Poly," San Luis Obispo, with curved corners and stainless steel frames complement the building design and leave room for the outdoors to come in.*

**FIGURE 15-3** *The serenity of the chapel at Mercy Hospital, Chicago, is augmented by the ethereal design of floor-to-ceiling sheer draperies that diffuse the light but do not obstruct its passage into the room.*

**FIGURE 15-4** *Austrian shades say elegance and are a decorator's delight, but are subject to a build-up of dust within scalloped folds of fragile fabric. Such curtains should be vacuumed almost every day.*

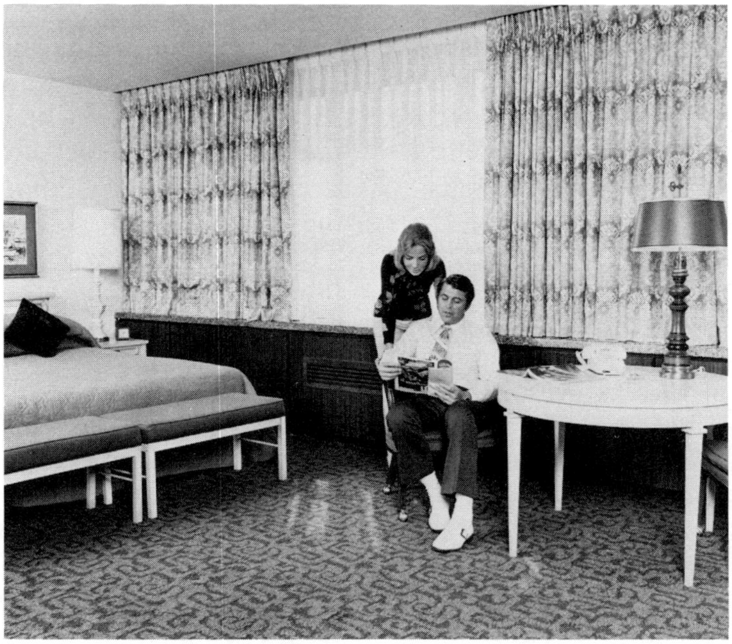

**FIGURE 15-5** *Ceiling-to-sill draperies are effective and warm-looking but curtains should not touch ledges or sills (note left-hand panels) because they ruin line of folds and suffer greater wear at edges. Pleasant guest room at Benson Hotel, Portland, has luxury touches—pillows and alarm clock.*

shape, eventually snag or sag, and are the least practical in a hotel or dormitory.

Good quality vinyl-coated fabrics, on the other hand, are gaining in popularity due to their soil-, spot-, and stain-resistant qualities, and need for only minimal soap-and-water care.

## *Finishing Processes— The "Resistant" Syndrome*

Modern miracles in chemistry make it possible to finish a fabric with properties that make it resistant to almost anything, although no fabric can be made resistant to everything. These coated fabrics are impregnated with invisible chemical finishes that affect neither the fabric's color nor its texture.

Stain repellers are undoubtedly the most popular. Fabrics treated with fluoro-chemicals resist oil, water, and soil stains. When something spills the liquid does not seep into the fabric; it remains on the surface and therefore can be blotted (never rubbed) with a dry cloth or towel. As for soil and dust, they collect on top of the fabric surface and only have to be brushed off. Most stain-resistant fabrics last through five launderings or dry cleanings before having to be retreated.

### *Heat Resistance*

Silicones are responsible for a fabric's heat resistance. Among their outstanding virtues is the ability to withstand temperatures up to 500° F. Other advantages include water repellency, spot- and wrinkle-resistance, longer wear, a softer feel, and greater seam strength. The last improves a fabric's sewability due to the lubricating qualities in silicones that reduce the possibility of needle cuts in the cloth—a frequent problem in high-speed sewing of synthetic fabrics.

Silicone finishes are also a boon to foam laminates (fabrics laminated to polyurethane foam), which may catch or stick to the sewing machine and tend to absorb moisture as well.

Draperies in public areas in many communities *must* be treated to be flame-resistant. Licensed applicators may be contacted to treat and tag draperies so that fire department inspectors can see those fabrics comply with law. Many professional finishers make fabrics flame-retardant at the same time that they apply any of the other finishes described. But in order to have permanently flame-resistant curtains, fiberglass or Saran-type fabrics must be purchased.

### *Fabric Backing*

Fabric backing is a worthwhile investment for several reasons. Thus treated, material, particularly a loose weave, becomes ravel-resistant. It improves seam strength and allows for easier cutting of the fabric. Backing checks shrinkage and increases abrasion resistance.

Fabric stabilizers (chemical treatments) are another kind of invisible backing. These offer more accurate control of shrinkage and stretch caused by atmospheric changes and dry cleaning. Fabrics with stabilizer backings are more resilient and crease-resistant, and hang more attractively than a conventionally sized fabric.

Odor neutralizers are an invisible finish needed on fabrics containing jute or similar fibers that have been treated with an emulsion of aromatic oils causing objectionable odors.

Existing fabrics not treated with any kind of backing or finishing processes may still be protected. Samples of the materials sent to commercial finishing houses will be put through laboratory pretesting at no cost.

When weather is a factor, there are alternatives to protecting drapery fabrics. Blackout linings are helpful in rooms that get prolonged or strong sunlight. They should, however, be hung separately from the draperies. If they are laminated to the fabric it poses great cleaning or laundering problems because these backings pucker and drastically detract from the appearance of the drapery.

Actual blackout fabrics can be laminated to the drapery fabric without detracting from the appearance. Then only one rod is needed and the weight is lessened, cutting cleaning and labor costs.

Aluminum coatings act as insulation against summer heat and winter cold that penetrates through windows. When an aluminum-coated lining is hung several inches from the drapery fabric it creates an extremely effective buffer zone, particularly where heavy insulation is needed to counteract hot sun coming through glass walls in air-conditioned rooms. These coatings make fabrics opaque and prevent fading. They can be applied directly to the back of most fabrics or to a separate panel of heavy cotton lining.

### *Standards for Selecting Fabrics*

The government-adopted American Standard Performance Requirements for Institutional Textiles, L-24, is the best source to consult when purchasing fabrics. There are several things to watch out for:

1. Does the fabric resist spots and stains?
2. Will it wear well, resist snags, pilling, and abrasion?
3. Are the color, weave, texture, and pattern attractive and appropriate?
4. Does the fabric have a tendency to shrink?
5. Does it have a tight weave and/or bonding agent on the back to prevent fraying and yarn slippage?
6. Does it have more than the recommended 80-lb. minimum warp-and-filling yarn-breaking strengths?
7. Does it have self-extinguishing properties so that it will not support combustion?

Other things to remember when purchasing drapery fabrics: the fibers in the material can control the temperature, light, and mood of a room; always check whether linings can be washed or dry cleaned; and be doubly sure of the exact quantity of material actually needed. For extra-long draperies, allow for the fabric to drop as much as two or three inches after installation. Another rule of thumb is to permit a two-inch drop on the valance for every one foot of drapery (Figure 15-6). (Bonus hint: when redoing or installing new draperies, mark the date on the inside panel lining with an indelible pen or pencil

**FIGURE 15-6** *Windows only go up two-thirds of wall, but designer's initiative prevails with draperies extending all the way to ceiling. There are very few rigid rules for window treatments—imagination goes a long way. In Hemet Valley Retirement Residence lounge, the designer felt need to make ceiling seem even higher.*

to have an accurate record of when they were put in.)

## *Fiber Facts and Fictions*

A fabric's life expectancy depends on the amount of exposure it will have. Many believe that the natural fibers are more durable than man-made fibers, yet many of the latter have been tested and shown to endure severe exposure to direct sunlight for over four hours a day. Among these hardy materials are glass fibers, polyesters, Saran/viscose rayon blends, and several types of nylons. All fibers, both man-made and natural, will bear indirect sunlight especially the casement curtains made of heavy-denier yarn. Silk is the one exception. Sheers deteriorate faster except under minimum exposure in heavy shade.

Fabrics must be color-fast to resist ultraviolet rays, the effects of dry cleaning, water spotting, and crocking (rubbing off). Solution-dyed fibers, that is, material that is dyed before being spun into fibers, have a better chance of retaining colors. The best way, if time permits, is to test color-fastness beforehand on a sample on location.

Another thing to watch is yellowing, especially with white, beige, and champagne colors. Such discoloration is more evident in cellulose fabrics such as rayon and acetate. It can be minimized by dry cleaning before the fabric is visibly soiled.

Draperies that are going to rub against the frame of an open window or be handled or brushed against often, should be protected by some means such as tie-backs. When hung they should not touch the floor, but neither should they touch the window frame or sill area.

As for cleanability, most draperies, with the exception of fibrous glass casement curtains, should be dry cleaned at least once a year; however, they should be vacuumed regularly (Figure 15-7). The time and cost of dry cleaning versus laundering are relatively the same. The general belief is that dry cleaning helps retain the life of the drapery for longer periods of time. (See Chapter 28 regarding cleaning processes.) Glass casement curtains, however, are designed for frequent washing.

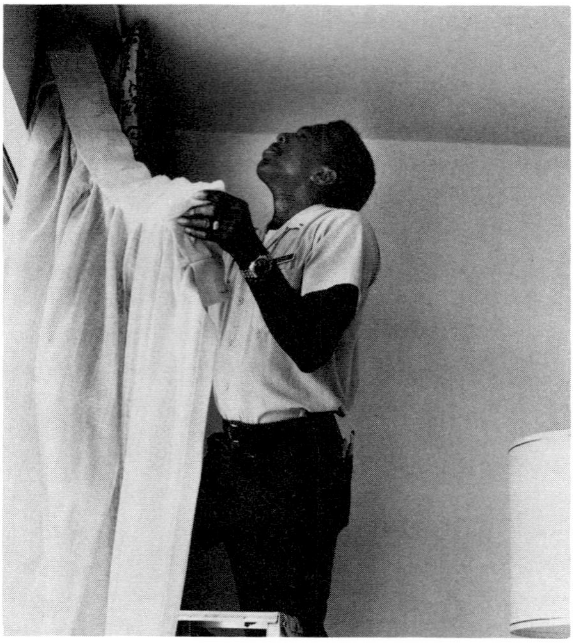

**FIGURE 15-7** *Regular vacuuming reduces need to remove and rehang, although with premise facilities, draperies can be quickly cleaned and replaced so room is not out of order. Casement curtains protect interior from direct sunlight. They should be hung, as shown, on separate rod and not touch patterned draperies.*

## *Types of Fibers*

Fibers come in two groups—natural and man-made. The former include cotton, linen, silk, and wool.

**Cotton** still wears well, and with the special finishes developed over the years, can be made wrinkle-resistant and can withstand mildew, rot, weather, fire, and water. Cotton is easily cared for. It is a natural fiber that can be washed and ironed at high temperatures.

**Linen** is strong, absorbs moisture, can endure hard laundering, and does not collect lint. But linen is susceptible to rotting from too much sun, and to mildew unless treated with a mildew-resistant finish.

**Silk's** natural resiliency renders it wrinkle-resistant. It is a strong drapery fabric that dyes well and can be backed with soil- and stain-resistant finishes.

**Wool** also has great resilience, and resists soil better than any of the others.

**Acetate** is a popular curtain and drapery fabric known for its wrinkle-resistance and excellent draping qualities. It retains pleats and creases when wet or in areas of high humidity. It must be ironed at a low-heat setting. It is both moth- and mildew-resistant. More important, it is inexpensive.

**Saran,** a not-too-well-known fiber, resists flame, chemicals, moths, and stains. Blessed with tough and flexible qualities, Saran may be solution-dyed in various colors. If blended with modacrylic and rayon it offers superior performance over long time spans.

**Acrylic** materials retain shape and resist shrinkage, creasing, moths, mildew, and sunlight.

**Modacrylic,** like acrylic, is known for its softer feel and high bulk. It is durable, dries quickly, holds its shape when wet, and resists alkalis, flame, acids, moths, and mildew. It should be ironed at low heat to avoid shrivelling.

**Polyester** offers strength with high wrinkle-resistance and good crease-retention, and launders easily. Moths, mildew, sunlight, and soil are no problems. Static electricity has been minimized through technical fiber modifications.

**Rayon** is probably the most widely used man-made fiber but is not recommended for sheer drapery fabrics because it yellows and requires ironing when washed.

**Nylon,** on the other hand, has great tensile strength (wet or dry) and high abrasion-resistance, is easy to wash and quick to dry (with little or no ironing), and wrinkle-resistant. It is, however, sensitive to strong sunlight and hot iron temperatures. Its susceptibility to static causes it to pick up and hold soil, so it must often be treated with an antistatic agent.

**Metallic** draperies, usually composed of aluminum filaments covered with polyester or acetate butyrate film, are generally tarnish-resistant. Some types are color-fast to sun, laundry, and cleaning.

**Glass** fibers require no ironing, are easy to wash, and resist sunlight, moths, and mildew. Most important, they are also nonflammable.

Despite all our discussions here about being practical in selecting drapery fabrics, there are some properties that maintain their elegance regardless of cost or complicated upkeep. For example, the windows at the Greenbrier resort are 15 to 20 feet high in some areas. These have special hand-painted draperies that are custom made and virtually nailed in place.

In some cases they have never been dry-cleaned because the fabric is so fragile. They are, however, kept clean by frequent vacuuming and if there is some fading along the edges, those are turned under or replaced in order to continue using the rare and handsome handpainted fabrics. Such an installation falls under the heading of a museum piece and cost becomes secondary.

## *Curtains and Draperies*

The following information is found in "Seng Furniture Facts," now a classic reference on basic interior design, published by The Seng Co.

**Glass curtains** cover the glass of the window assuring privacy, but are thin enough to admit light and air. They may be used alone or in combination with over-draperies of heavier opaque material. They may hang straight or be tied back (if used without draperies); may be severely plain or with ruffled edges. If hung inside the window casing, they should reach to the window sill. If hung on the outside casing, with or without over-draperies, they should hang to the bottom of the apron. Glass curtains may hang even to the floor in very formal rooms.

**Draperies,** also called over-curtains, over-draperies, and (colloquially) drapes. Draperies are generally of heavier material, tailored to fit the particular window. They are usually lined (and, in the case of lighter materials like taffeta, they should also be interlined) and weighted at the bottom.

**Draw draperies.** These are hung on a sliding track so that they may be drawn together at night. These should be made with a fullness ratio of 2 to 1; for installation, special hardware is required.

**Heading** is the top of the drapery, especially that part that extends above the supporting rod. When hung with hooks, the heading should be long enough to cover the rod. Shirred curtains or draperies should extend between one and two inches above curtain pole.

## *Window Treatment*

The following suggestions are of a general nature but may assist in helping one arrive at appropriate window treatments (Figures 15-8 through 15-10).

Valances (Figure 15-11) should not be used in rooms with extremely low ceilings. A valance with curved lines makes the window seem wider; one with square lines makes it seem narrower.

For the window composed of two framed panes of glass one above the other, straight hanging curtains are better than looped-back ones.

Looping back the over-draperies tends to soften the severity of a room.

A pair of curtains looped back make a window seem narrower than a single curtain tied back to give a diagonal line.

Tie-backs should be located either above or below the centers of the window to avoid the appearance of cutting the window space in half. The best spot for the individual window can be found by experimentation. Usually the higher the tie-back, the taller the window appears.

Ruffled glass curtains are best when tied back. Straight, plain ones, as a rule, are never looped back.

Criss-cross curtains usually make a window seem wider.

Straight draperies without valance or cornice tend to make windows seem taller (Figure 15-12).

French doors leading outside should be treated as a window. If between rooms, they should not be draped but fitted with shirred glass curtains if necessary.

All measuring should be done with a steel tape or yardstick. Measure each window separately even when they appear to be the same size. If length varies, use dimension of shortest length. Double-check all measurements to be sure they are accurate.

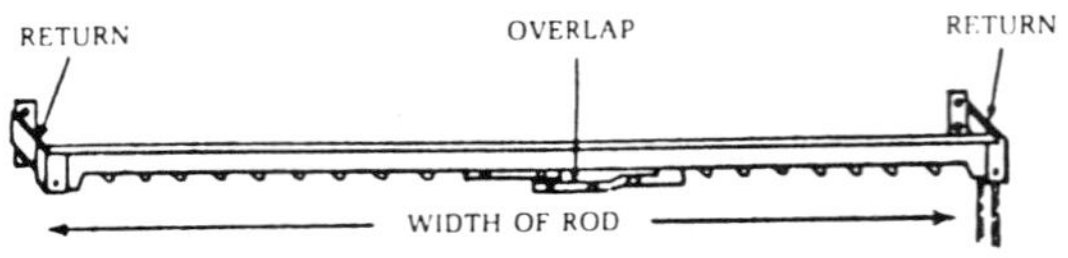


Caution: When both under-curtain and over-drapery are used, be sure to allow for clearance of face drapery. For example, an under-curtain with a 3½-inch return requires at least a 6-inch return on the over-drapery.

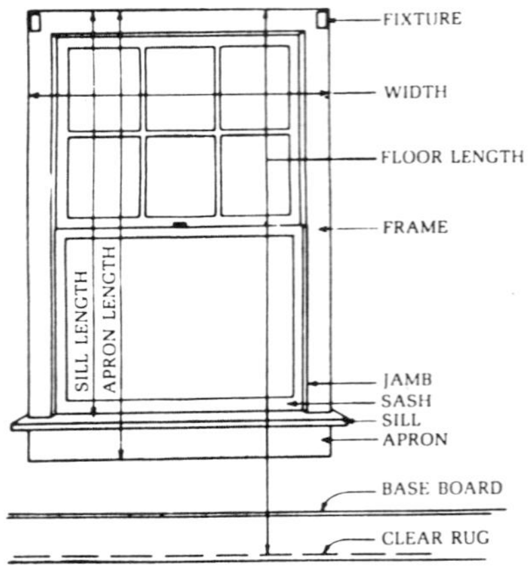


**Drapery Width**

- Install hardware first so that width will be accurate. Then refer to this set of instructions for drapery width figure:
- Measure width of drapery rod from end to end.
- Add to this figure an extra 12 inches to include allowance for standard traverse rod returns and overlap. If ceiling track is used, add only 8 inches overall.
- Standard returns are 3 inches in depth. For over-draperies allow for clearance of under-curtain. A 6-in return should be sufficient.
- Drapery fullness will be 2 to 1 unless otherwise specified.
- When ordering panels that stack (draw) in one direction only, specify if the drapery is to stack (draw) left or right.

**Drapery Length**

- Measure from top of rod to floor or to carpet. (By inserting pins 1 inch from top, drapery will automatically clear the floor or carpet.)
- Under-curtain should be at least 1 inch shorter than over-drapery.
- When floor-length draperies are used, it is best to measure length at each side and in the center. Use shortest figure for your measurement.
- Rod should be placed a minimum of 4 inches above the window so hooks and pleats will not be observed from outside.
- If sill-length, allow 4 inches below sill so bottom hem will not be observed from outside.
- When using pole rings, measure length from bottom of rings.

*Credit: Waverly Fabrics*

**FIGURE 15-8** *How to Measure for Draperies*

Cafe curtains.

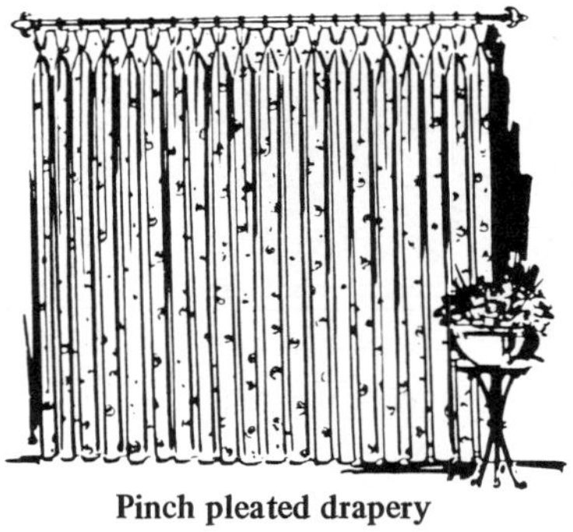

Pinch pleated drapery
suspended on pole.

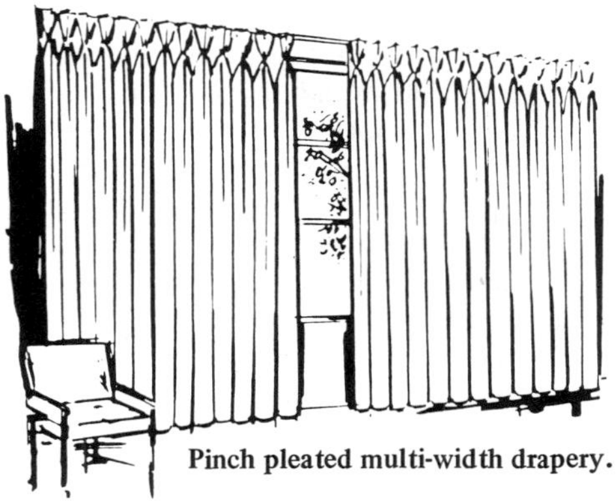

Pinch pleated multi-width drapery.

Shirred glass curtains with rod pocket
(hem and headed) and full.

Pinch pleated one way stack (draw).

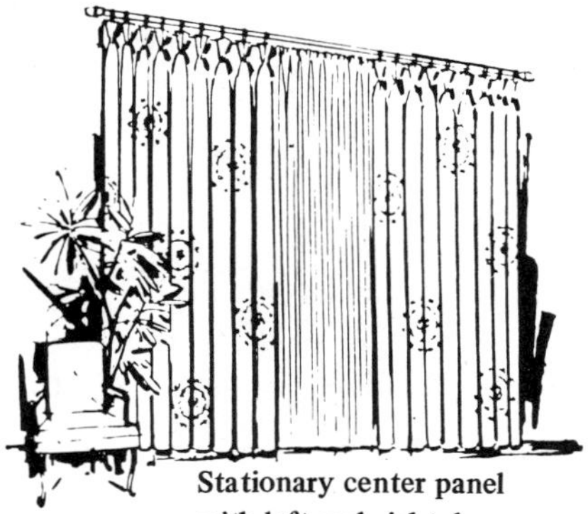

Stationary center panel
with left and right draw.

Pinch pleated valance and
pinch pleated draperies.
Single width each side.

Pinch pleated draperies with
tie-back and shirred under-curtains.

*Credit: Waverly Fabrics*

**FIGURE 15-9** *Basic Window Treatments*

Choose square or mitred corners—pinch pleats, Windsor (French) pleats, or hem and headed panels.
All tops are 4 inches and doubled for strength and appearance. Buckram (or Pellon) headings are available.
Side hems are doubled and all seams are blind-stitched, over-locked, and serged.
All corners are weighted so draperies hang evenly.
Multiwidth draperies are sewn with pucker-free joiner; patterns are perfectly matched.
Each finished drapery is carefully inspected; then fan-folded into a specifically designed carton to assure perfect arrival and wrinkle-free hanging.

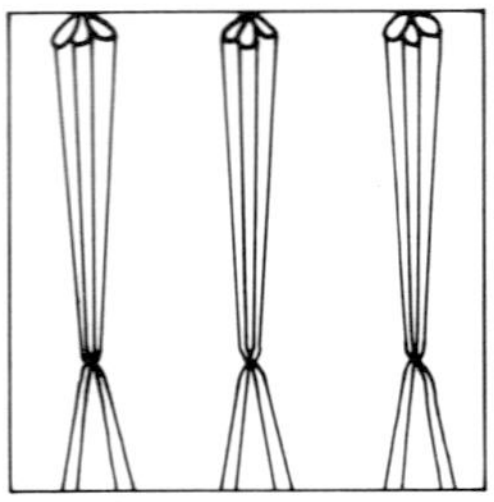

**Pinch Pleats**

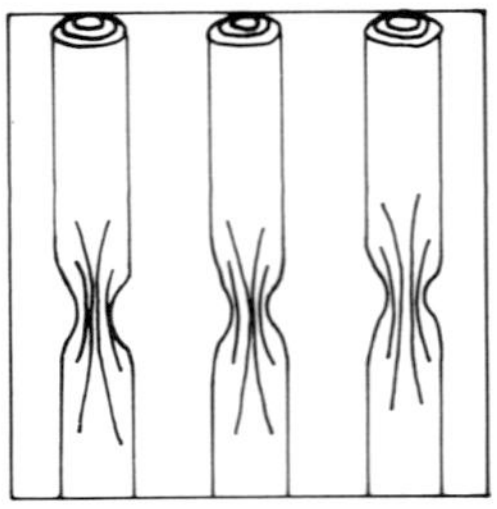

**Windsor Pleats (French)**

*Credit: Waverly Fabrics*

**FIGURE 15-10** *Quality Features of Custom-Made Draperies*

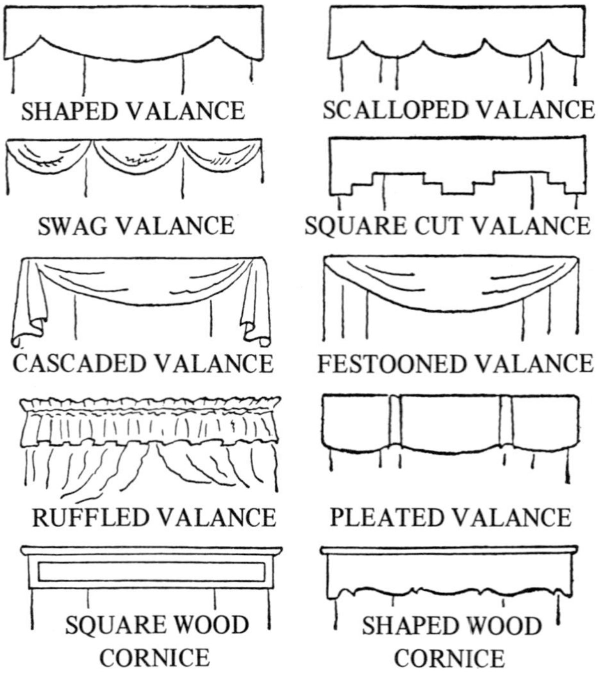


**Valances:** Valances are the top or horizontal portion of the drapery treatment. They were originally used to hide the drapery poles and fixtures but have survived because of their decorative effect.

On heavier materials, tailored or draped valances of various shapes and styles are generally used. Those shown on the accompanying illustration are the most popular. In general, the simpler types of valances are most effective although special treatments may be indicated for rooms of certain periods. Flat, shaped valances are usually lined with heavy material to give them stiffness. In general, they should be from about 8 to 12 in. in depth; they should never exceed one-sixth of the window length.

**Cornices:** Cornices are box-like shapes sometimes used instead of valances to hide the poles and hardware. They are generally from 4 to 7 in. deep and can be made of wood or metal. Some are painted to match the woodwork or to match the draperies. Others are covered with the drapery material, mirrors, carved or molding designs, or applied plaster or painted decoration. On short windows they add to the apparent height of the windows.

**FIGURE 15-11** *Drapery Valances and Cornices*[1]

1. *Seng Furniture Facts.* Chicago: The Seng Co., n.d.

**FIGURE 15-12** *Favorite tactic for making a room seem larger is to make wall-to-wall draperies of same pattern as wall covering. This will not be effective, however, if the pattern is large and the room is small. Draw-draperies with blackout linings laminated to them may develop the problem of puckering.*

**FIGURE 15-13** *Colorful, striped roll-up blinds are made of woven wood, plastic, and velour, with ball fringe. These enable employees to control sunlight quickly, can be wiped off more easily than fabric if food is splashed or spilled on them. Restaurant is in TraveLodge in Lahaina, Maui, Hawaii.*

Bay windows or two adjacent windows are best treated as one with a single drapery at the outer edge of each. This gives the effect of one large window and makes them appear much wider than if each window has its own pair of draperies. If a valence is used it should be continuous across the windows.

Picture windows, used primarily to give the effect of spaciousness or to take advantage of a view, should be equipped with draw draperies.

Casement windows often require special curtain treatment. Those that swing inward are generally fitted with shirred glass curtains fastened to the top and bottom of the window; those that swing out are usually fitted with straight hanging curtains or draperies.

To present a uniform appearance from the street, all windows on the same floor should have the same kind of glass curtains.

Glass curtains and over-draperies should be hung at least one inch apart.

Venetian blinds can be used without curtains or draperies. They are customarily used with tie-back curtains, over-curtains or both. They are seldom used with draw draperies.

Sliding wood shutter panels may be used in place of draperies. This full-window treatment is used very effectively in tropical hotels such as Mauna Kea Beach Resort on Hawaii and at Kapalua Bay Hotel on Maui. These shutters last for years but do require much dusting; only an occasional oiling of the wood.

Small shutters are often effective on bathroom windows, providing some privacy as well as daylight.

## *Sources of Information*

The Seng Co.
P.O. Box 465
Lietchfield, Ky. 42754

Waverly Fabrics
Div. F. Schumacher & Co.
939 Third Ave.
New York, N.Y. 10022

# Chapter 16

# Interior Design—Public Spaces

**OVERVIEW** Ideally, the professional Housekeeper should be consulted on the furnishings used in public spaces, since he or she will be responsible for cleaning and maintaining them. There should be some involvement with the designers and architects from the beginning of a project if at all possible. Understanding how to read floor plans and elevations drawn to scale is very important if one is to communicate with the designers. To furnish large public rooms one must have a feel for scale and be aware of built-in safety factors necessary in areas where very large groups congregate. Rules for selecting seating—especially function and dining chairs—are described here in detail.

Probably there is no individual or department in a hotel or residence hall more concerned with interiors and their design and furnishings than the professional Housekeeper and the Housekeeping Department. Yet few Housekeepers today are actually involved in creating the initial plan and visual impact of a room. That in itself is a full-time occupation requiring talent as well as special training. The professional Housekeeper, however, is likely to be very much involved in selection of items for replacement when the components of the interior become worn or too soiled to save.

The Housekeeper is the continual critic of the interior designer, alternately friend and foe. When the professional Housekeeper is applauding the designer, a hotel's general manager knows a proper balance has been achieved between function and aesthetics. Fabrics selected are practical; the furniture forms and surfaces present no cleaning problems; and the carpeting and wall covering are not supersensitive. He will, of course, be even more gratified if the interiors please the guests and keep them returning.

This chapter covers the design of public spaces, while in the succeeding chapter the emphasis will be on guest rooms, because that is the area in which the professional Housekeeper will most often be called on to exercise his or her own skills at refurbishing interiors. (Chapter 17 on Guest Rooms also contains information on color, fabrics, and wall treatment that may be applied to public areas.) If the Housekeeper understands and agrees with the designer's basic room scheme initially, then there should be few problems when the replacement period eventually begins for either public areas or guest rooms.

Today many interior designers are architects, and the opposite is also true. If the professional Housekeeper can speak their language when the building or remodeling is being planned, and can read blueprints, scale drawings, and mechanical drawings, it will facilitate communications immeasurably.

### *Floor Plans Drawn to Scale*

Reading blueprints and drawing rooms to scale can best be accomplished by learning to use scale rulers and triangles, which enable the user to read or create designs drawn in exact proportion to actual size (Figure 16-1). (The perfect example of drawing to scale is a road map, which is an exact copy of its subject in miniature; every inch may, for example, represent 5 miles or even 50 miles.)

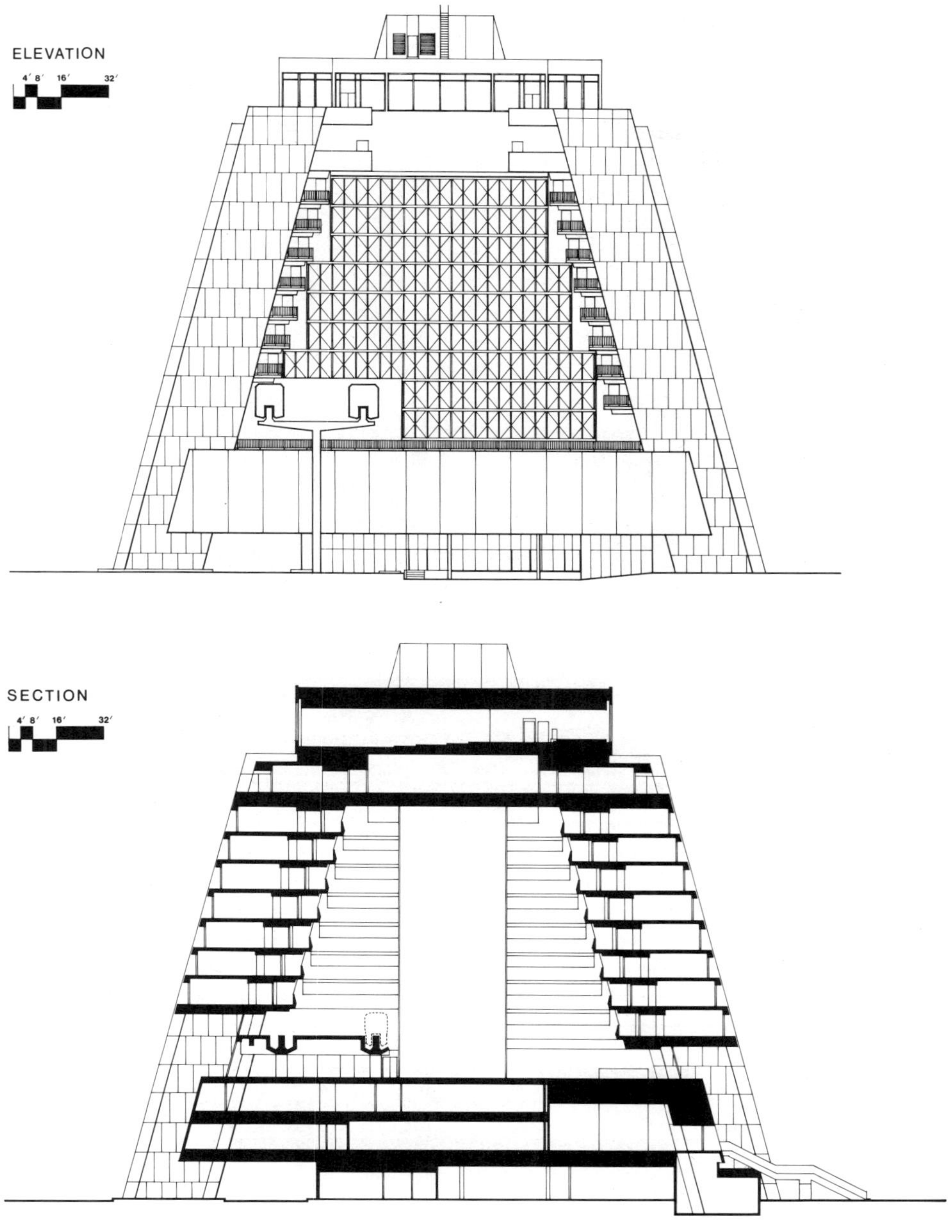


**FIGURE 16-1** *Elevation of one end of the Contemporary Hotel (one of lodging industry's most exciting structures) at Disney World, shows where the monorail goes through the center of hotel just above concourse (lobby) level. Building and interiors were designed by Welton Becket and Assoc. of Los Angeles. Elevation and cross-section were originally drawn to a scale of 1 in. equals 32 ft.; here it has been reduced to ½ in. equals 32 ft. Interior views appear in Fig. 16-2.*

A scale rule enables one to draw a floor plan is similarly exact proportions. Such plans take up very little space on the graph paper on which they are usually drawn. For example, if a 20-ft. by 30-ft. room were drawn to a scale in which 1 in. equals 1 ft., its scaled-down size would be 20 in. by 30 in. Drawn ½-in. equals 1 ft., its scaled-down size will be 10 in. by 15 in., which is still larger than can be accommodated on ordinary 8½-in.-by-11-in. graph paper. Therefore a practical scale in this instance is one in which ¼ in. equals 1 ft., which reduces the size of the room

**FIGURE 16-2** *Immense and cavernous space in hotel lobby is humanized somewhat by wildly imaginative plastic trees that help relate people to the building. Street-lamps also prevent mere 6-ft.-tall humans from feeling lost in 125-ft. height of the great atrium. Towering mosaic on rear wall is one of world's largest.*

being drawn to 5 in. by 7½ in. Everything to be included in this room will be drawn to that same scale: ¼ in. equals 1 ft.

If there is a series of rooms to be drawn, and it is desirable to have them all on one manageable sheet, then an even smaller scale, 3/16 or 1/8 in. equals 1 ft., will be necessary. In any case, to read blueprints intelligently it is essential to acquire a feeling for scale. Only then is it possible to visualize the room size in miniaturized spatial relationships. (See Figures 17-4 and 17-5 for scale drawings.)

## *Vocabulary of Architecture*

In addition to reading plans drawn to scale, the housekeeper can communicate more easily with the designer or architect if the vocabulary is familiar. Examples of terms will be helpful.

**Elevation.** View of an interior or exterior wall looking straight on, with no perspective; when drawn to scale with furnishings or landscaping it will show exact wall coverage by mounted or adjacent objects, as well as shape of completely exposed wall.

**Soffit.** The underside of an architectural element such as an arch, beam, or cornice.

**Atrium.** Open court surrounded on all four sides by the building.

**Arcade.** A series of arches supported on piers or columns. Also a covered corridor, often with shops on either side.

**Clerestory.** That part of a building that soars above room tops and is glass-enclosed or contains opening to illuminate area below this architectural feature.

**Cornice.** A projection, usually horizontal, at the top of a wall or other construction feature.

**Dado.** Decorative horizontal band at the base of a wall or pedestal.

**Engaged Column.** Element forming part of a wall and projecting from it.

**Facade.** Front of a building.

**Fenestration.** Arrangement of windows in a building.

**Flush.** Of the same height as adjoining surface(s); even with.

**Lintel.** Horizontal beam across top of an opening (door or window) that provides structural support.

**Loggia.** Space within the building that is open to the air on one side.

**Modular.** Architectural construction that employs interchangeable components, for example, modular bookcases. Each component is a module.

**Molding.** Continuous narrow surface either projecting or recessed to decorate a surface.

**Monolith.** A single large stone block, stone obelisk, or statue; monolithic means a large continuous unbroken surface.

**Mullion.** Vertical bar that separates a window into more than one light (pane).

**Pier.** Vertical support.

**Pilaster.** Flat rectangular element that projects from a wall, such as an engaged column.

**Pillar.** General term for pier, column, or upright.

**Plinth.** Lowest member of a base, or a block serving as a base.

**Stringcourse.** Horizontal molding to indicate a division in architectural design.

**Terra Cotta.** Hard baked clay used as building material (or sculpture).

**Vault.** Roof formed by arches; an arched or dome-shaped structure.

**Wainscoting.** Continuous horizontal wall paneling that extends from the floor to about halfway up the wall.

## *Public Areas*

One of the important points to bear in mind when designing for public spaces is that large-scale furniture, fabric patterns, and accessories must be used; otherwise furnishings can look lost or insignificant (Figure 16-3).

Breaking up large spaces with interesting arrangements of furniture is necessary (Figure 16-4). Although a certain amount of repetition is needed in the layout of a large room to give it unity, monotony must be avoided to eliminate the institutional look that can cause guests or residents to feel the space is lifeless. Sometimes room dividers—planters, decorative screens, or see-through walls of fabric or beads—will help to make a large lobby less cavernous.

Furniture should be out of the way of traffic and arranged in conversational groupings so that those using the lobby will not be sitting cheek-by-jowl as in an airport waiting room.

Where furnishings are exposed to the public they should be as vandal-proof and tamper-proof as possible. Nothing spindly or delicate can survive in most lobbies or foyers. Accessories must be out of reach, well-anchored, or of such scale that making off with them would be difficult (Figure 16-5).

Upholstery should be easily cleanable (vinyl, if possible) so that a damp cloth with cleaner will be all that is required for maintenance. If an upholstery fabric is to be used, however, it should

**FIGURE 16-3** *An enchanting lobby appears in the Dusit Thani Hotel in Bangkok. Lights hang inside the plastic chains that flow between lotus blossoms. The gigantic scale of these accessories is worth noting. Even table lamps are enormous compared to size of the obviously over-sized lounge chairs.*

**FIGURE 16-4** *Floor treatment is highlight of this lobby where a circle is covered with round ceramic tiles and surrounded by a custom-made carpet with raised textured circles. Tables in center carry flowers and refreshments, and one is a glamorous waste receptacle.*

be one that can withstand repeated washings with a foam shampoo, rinsing, and vacuuming. A fabric spray treatment, professionally applied, is usually worth the extra cost. This protective coating over any fabric can delay its initial dry cleaning or shampooing for six months to a year of continual use in areas where no major spills have occurred.

If the supplier does not have facilities to apply soil-retardant or flame-retardant finishes, fabric may be shipped to companies that specialize in this service. Kiesling Hess in New York, Philadelphia, and Culver City, Calif., is one of several reputable firms.

Safety is another aspect of interior design that should be considered when planning furnishings for a public area. Avoid:

1. A floor near an entry that will become slippery on a rainy or snowy day
2. Rug edges that can trip people
3. Fabrics that are flammable

FIGURE 16-5 *Lobby of the Stanford Court, San Francisco, contains perishable but interesting pieces, such as immense Imari bowl that is filled with fresh flowers daily, brocade upholstery, and sparkling Baccarat chandeliers.*

4. Rustic outdoor furniture that might have splinters
5. Glass window walls or doorways without some mark to prevent persons from mistaking them for open doors
6. Steps in dark areas, particularly dining rooms, unless they are lighted at ends of risers
7. Dining chairs that will tip over when patrons lean back (college students are famous for that position)
8. Floor mats with openings that can catch ladies' heels
9. Hanging planters or light fixtures that are low enough to hit tall people
10. Swinging doors anywhere
11. Ash stands that can be easily tipped over, or non-weighted ash urns
12. Floor lamps that may tip over easily

Large accessories—sculptures, bas reliefs, paintings, clocks, mobiles, tapestries—once played a more important role in interiors than they do today because the popular decorative element now is live greenery. A small tree, hanging ferns or vines, and masses of potted plants attractively soften the austere lines of today's contemporary spaces and bring the outdoors inside (Figure 16-6).

FIGURE 16-6 *Live natural greenery is used lavishly in public interiors as a decorative element; in consequence, gardeners' expertise is needed on the Housekeeping staff. Floor pathway consists of redwood cross-slices embedded in epoxy. Window wall contributes to garden effect.*

Maintaining the live decor is a specialized procedure, however, and is best left to an outside firm that will rent the plants, keep them in good order for a regular monthly fee, and take responsibility for their staying in good condition (Figure 16-7).

FIGURE 16-7 *Planter boxes are often used to break up large spaces. Lounge is broken up into intimate areas by floor-to-ceiling openwork paneling as well as plywood planter that forms room divider.*

**FIGURE 16-8** *Old sandblasted brick is exposed as decorative instead of being hidden as in past. Where patrons sit on floor to eat, carpets must be especially well maintained. Clear plastic sheets with water pouring down them into anodized metal fish tank cause specialized, tricky housekeeping problem.*

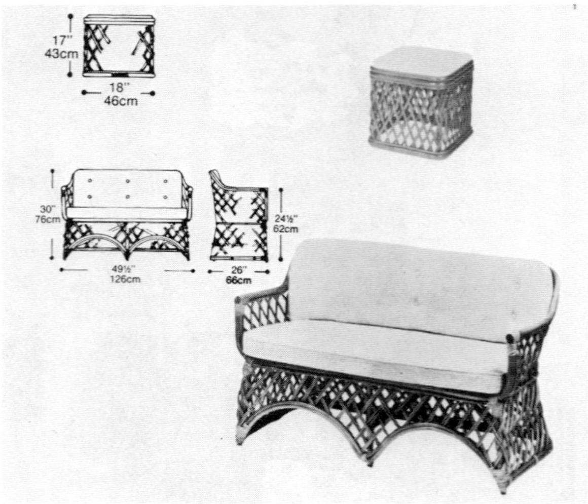


**FIGURE 16-10** *Rattan loveseat and ottoman have the look of lighthearted wicker but are remarkably sturdy, almost indestructible. They are lightweight enough to be easily moved but are not fragile like wicker. Foam-padded cushions snap-fasten to back. Stick rattan is cross-woven and hand-wrapped at joints with rattan peel-binding. Measurements are given in inches and centimeters. Credit: Shelby Williams*

At the Colony Square Hotel in Atlanta, palms form walls for privacy in the verandah, the terraced lounge that connects directly with the lobby (Figure 16-9). Pools and waterfalls add visual interest and background sound that keeps conversations private.

Wicker furniture is appropriate for the setting, as are basket-type planters. Wicker is likely to catch on clothes if it is not properly upholstered along the edges or cushioned to prevent rough places from snagging. Similar to wicker but sturdier and smoother is rattan—an almost indestructible material made of reeds treated under pressure while wet. Note table and love seat (Figure 16-10), which shows dimensions in inches as well as centimeters. (See Appendix with section on metrics.)

For large wall hangings in public rooms a

**FIGURE 16-9** *At Atlanta's Colony Square Hotel, seating areas are surrounded by planters and palms on many levels. Lobbies are being designed today with great flexibility—spaces can be delineated with seating and plants forming cocktail lounges wherever needed.*

**FIGURE 16-11** *Soft sculptured art-work is fabric-quilted and tufted so it has a subtle dimensional quality. Three separate ones are grouped horizontally but can also be hung vertically or in various sequences.*

**FIGURE 16-12** *At the Pima Country Club in Scottsdale, Arizona, a wall is composed of four different suede-like surfaces with art-work tufted in a design with proper scale and sufficient impact to make the entry visually exciting.*

current design approach is the soft sculptured wall. These are available in stock items as well as custom for specific installations (Figures 16-11 and 16-12).

**FIGURE 16-13** *Comfortable upholstered seating is especially appropriate for residence hall lounges or areas that sustain heavy wear because of its easy maintenance features. Sofa and chairs have removable side and back panels that are easily detachable with a special tool so that fabric can be cleaned or replaced. Seat and back cushions can be released and reversed. Seat edge is removable for recovering. Solid oak frames can easily be refinished when fabric is removed. Floor here is slate; table, butcher-block. Credit: Metropolitan Furniture Corp.*

## *Seating*

For public areas the items of furniture that will be most obvious to the visitor are the lounge seating—sofas and upholstered comfortable chairs. These *can* be practical as well as beautiful if the upholstery fabric can be cleaned easily or if they are engineered to come apart for easy recovering and repair (Figures 16-13 through 16-16). The furniture that will be bought in greatest number will be the dining and function chairs. Because there will be so many and their maintenance will ultimately be the realm of the professional Housekeeper, it is advisable to become acquainted with many different types of chairs.

The chair to be cherished is one that can be repeatedly refinished or recovered; one that can keep its strength while its outer "garments" change with the current fashions.

The all-around perfect dining chair has yet to be built. Classics survive the test of time because they are basically sturdy and have universal style. Should one stick with them or choose a new design for its novelty value and impact on patrons?

In chair selection there are nine pertinent points, listed alphabetically here, not necessarily

**FIGURE 16-14** *This new lobby seating system offers limitless solutions for straight-line and right-angle layouts. The series consists of one table, two seating modules (full seat and seat/back units), and one arm unit that fastens to a peripheral tubular steel frame (or polished chrome). Each base component merges with the next frame creating a support leg that is totally integrated. The seating units as well as the arm unit are on-site recoverable, and the resin table modules, available in satin or high gloss, offer a durable surface that is on-site refinishable. Credit: Metropolitan Furniture Corp.*

**FIGURE 16-15** *The basic design, available in chair, loveseat, and sofa versions, consists of an upholstered seating unit suspended between two arm panels. One version has polyester resin arms (available in a wide color spectrum) and another has fully upholstered arms. Both groups are on-site recoverable for easy cleaning and maintenance. Fabric covers on the seat, back, and upholstered arms are removable for repair or replacement thanks to Velcro. Reversible back and seat pillows are attached. Credit: Metropolitan Furniture Corp.*

**FIGURE 16-16** *Where a small lobby needs a smashing center of interest, soft, white, glove-leather vinyl on a stark black plastic sofa rivets attention, especially when combined with black area rug with woven bars of lavender, raspberry, green, and blue. Asymmetrical bands of black, green, and white in the art seem to move across the wall. Although the covering used here may take careful handling, other surfaces and shapes covered in white vinyl can be easily sprayed with all-purpose cleaner and wiped dry.*
***Credit: Wool Carpets of America***

in order of importance: (1) appearance, (2) appropriateness, (3) availability, (4) comfort, (5) cost, (6) function, (7) guarantees and/or integrity of the source, (8) repairability, (9) sturdiness.

**1. Appearance.** Institutions' buyers usually compromise on this because they are afraid of making a mistake on practical aspects and thus repeating their error 100 or 200 times. Many designers, on the other hand, will err on the side of aesthetics and be willing to take a chance on an interesting new chair because it will give some character to a room.

The most important trend in dining area seating today is to mix and match chairs to relieve the monotony and change spatial effects. Tall backs can actually serve as a room divider or enclose the space to form a booth. Having a variety of chairs suggests a living room environment—a popular look in commercial restaurants today.

Although appearance will probably be the most subjective evaluation—based on a strictly emotional reaction—it may also be the most important. A chair that excites and pleases the visitor is, in itself, a silent invitation for him to be seated. A dining chair should make the user anxious to have a closer look and to sit down. Handsome chairs will also give employees pride in an operation.

But the old saw should be remembered: "Beauty is in the eye of the beholder," so what one person considers very attractive may be an atrocity to others. (This is one reason an experienced contract designer is worth his fee; he will save department managers from having to rely on their own untrained taste.)

For function rooms or theater-style seating, the pattern and rhythm that are set up by row after row of chairs is as important visually as the look of the individual chair (Figure 16-17). So one should see those chairs in a row arrangement before purchasing.

**2. Appropriateness.** This is really an inseparable part of appearance. If the chair is suitable for the setting—a captain's chair in a nautical scene, a ladderback in a Colonial room, or a Louis XV in an elegant French restaurant—then it will probably look attractive (Figure 16-18). What is usually unacceptable is an adaptation of

**FIGURE 16-17** *Meeting room chairs should look attractive en masse as well as individually, should stack as well as gang (fasten together at sides) to meet fire regulations for public assemblage. Manufacturers usually provide special dollies for stacking and transport. Credit: Samsonite Furniture Division*

a classic chair in another material. Why make a director's chair in gold "bamboo" metal? Or a formal, tufted upholstery French chair out of aluminum? If one sticks with wood, metal, or plastic in original, successful designs, the chairs will be useful for from 10 to 20 years (with repairs and recovering, of course) (Figure 16-18).

**3. Availability.** No matter how beautiful the chairs or how intelligent the selection, if they are not there on opening night it is all for naught. In these days of shortages, make sure that the chair, upholstery fabric, padding, and the glides or casters are all available to be shipped when needed. When replacing chairs this consideration is not so critical. In case of a catastrophe, if the chairs do not arrive, there is always the chance of filling in with rented seating.

**4. Comfort.** This is one of the prime considerations in seating, especially for executive dining rooms, expensive commercial restaurants, and meeting rooms. Conversely, comfort is often deemed a drawback in fast-service restaurants (or even in some employee cafeterias), where management seeks to prevent people from lingering.

Finding a chair that is equally comfortable for a 6-ft. 4-in., 250-lb. man and a 5-ft., 100-lb. woman is not easy, but there are some on the market; one just has to keep looking.

For the executive dining room or the country club lounge off the men's locker room, a purchaser would naturally choose an outsize, padded model and leave the Victorian ice-cream-parlor chairs for teenagers. If one has selected what is appropriate, it is likely to be comfortable.

Not only shape and size, but climate, can influence how comfortable a chair will be. In warm places where there is no air conditioning, a molded plastic chair or a vinyl seat that encases one is a poor choice. An open weave, canvas, or fabric that breathes is more comfortable.

For outdoor use, a material that does not hold the heat is essential—nylon or vinyl straps or cords serve best for keeping cool (Figure 16-20).

Comfort in a dining chair is based primarily on its relationship to the table. Check to be sure the heights are compatible. Usually, tables are

1. 

2. 

3. 

4. 

5. 

6. 

7. 

8. 

9. 

10. 

11. 

12. 

1. *Sturdy "Stanford University Chair," of wood, leather, and brass nailheads; for dining in rugged splendor. Credit: Shelby Williams*

2. *Classic Charles Eames plywood and chrome for contemporary dining or lounge; a best-seller. Credit: Herman Miller*

3. *Stacking, folding lounge furniture of steel and vinyl zippered cushions; lightweight and portable. Credit: Fixtures Mfg. Co.*

4. *Based on the famous 1870 Thonet bentwood chair, excellent for dining and withstanding abuse. Credit: Shelby Williams*

5. *Cycolac wet-finish plastic shells come in bright colors on ganging, stacking armchair. Credit: Shelby Williams*

6. *Chrome frame with suspended velvet sling seat has rocking action. Conversation piece for dining or lounge areas. Credit: Vecta*

7. *Mate's chair is armless version of early American, indestructible, classic captain's chair. Credit: B. Brody Seating*

8. *Smooth version of captain's chair has no wood turnings on legs, works with modern or traditional decor. Credit: American Chair Co.*

9. *Stacking chair has foam-padded seats and backs; opening at top is a hand hole for easy handling. Credit: Shelby Williams*

10. *Comfortable dining chair has easy-to-clean wood arms; u-shaped stretcher on legs for strength. Credit: Shelby Williams*

11. *Vinyl upholstery material is flame-retardant, made to go on vacuum-formed chairs. Credit: General Tire and Rubber Co.*

12. *Dining function or stacked chair of steel and vinyl is constructed to hold up, despite rough treatment.*

**FIGURE 16-18**

**FIGURE 16-19** *Credit: Krueger*

from 28 to 30 in. high and chair seats from 17 to 18 in. high; one should avoid pairing up the highest table with the lowest chair.

Do the chair arms hit the table and prevent the chair from sliding under the table? If so, there will be uncomfortable people with food in

**FIGURE 16-20** *Practical outdoor seating for dining areas eliminates theft and unauthorized table movement; seats are suspended from table so sweeping and scrubbing can be done without moving chairs. Steel is treated to resist rust, and finish is permanent, weather-resistant, polyester resin mixed with waxes. Credit: Suspended Seating Inc.*

**FIGURE 16-21** *Meeting rooms to accommodate executives in seminars must be well furnished with up-to-date equipment and comfortable luxurious seating. Chairs that are padded and roomy with fabric covering are more comfortable during long sessions than a non-breathing vinyl. Shelves under curved, tiered tables require careful cleaning by housekeeping crew.*

their laps, and scarred chair arms where they hit the table edge.

Chairs with armrests (about 24 in. high is comfortable for most) require slightly more aisle space because patrons have to pull chairs farther out to get in and out of them.

Another important point to check is whether the chair legs hit the table base. This is irritating to guests and hard on both chairs and tables.

As for the width of a chair seat, there is a startlingly wide variation of widths that will comfortably accommodate the multitude of posterior dimensions of the population. A small, slim man can sit in a seat only 11½ in. wide, but his petite female counterpart will take up 13½ in.; average-sized men are perfectly comfortable in 15-in. widths, although the average seated lady's derriere measurements go to over 15¾ in. For the most portly, a 21½-inch chair width should suffice.

The comfortable depth (front-to-back measurement) of a chair is more universal than the width. Sixteen inches is sufficient for conventional needs, although luxury dining and plush meeting room chairs are often 18 or more inches deep (Figure 16-21). The public frequently equates depth with luxury.

**5. Cost.** The quantity of chairs purchased will almost always affect price. The range is very wide, however, and the cost of a utility dining or folding chair may be as low as just under $10; the expensive chairs may go as high as several hundred dollars each if ordered with leather, custom finishes, swivels, and giant brass casters.

The many different qualities of fabrics and vinyls from which one can choose will in themselves account for a wide range from budget to luxury.

If, for example, the original price is balanced with the cost of upkeep over a period of ten years, it is possible to determine whether the chair is going to prove expensive. But one must count on a maximum of wear and tear, with early repairs required.

A dark nylon velvet or heavy plain transportation cloth on an upholstered seat will require continual vigilance to keep it brushed and clean but it may wear for many years. It will be costly initially, but if one has service people on hand who can keep the chairs brushed as part of their regular duties in a dining room, then the investment in the very finest fabric is wise.

**FIGURE 16-22** *Sculptured plastic shell chairs and tables are dramatic in many sorts of dining areas. Basic style, designed by Saarinen in 1958, has been adapted by several companies, comes with bugle or star base, swivel or fixed.*

In contrast, a thin textured vinyl may not require any more attention than a bimonthly scrubbing, but at the end of a year may be losing its color and texture, be badly burned, or ripped. What was saved on the original purchase would be used up quickly by replacement costs.

One should pay enough to assure that the finish on the frame will adhere to the surface permanently (e.g., epoxy paints). Steel must be thoroughly cleaned by such a method as the steel shot blast machine that cleans the metal and prepares it to receive the paint. If an epoxy is applied with an electrostatic spray gun, the bond will be superior and there will be a minimum of chipping, even if it is on a stack chair that is tossed about carelessly most of its life.

Often one hears the question, "How much did the restaurant cost per seat?" Answers such as $356 or $775 do not necessarily mean that the chair itself cost that, rather that the furnishings in that restaurant, if it has 100 seats, cost $35,600 or $77,500. That would include chair costs, lighting, carpet, wallcoverings; furnishings would have been totaled and divided by the number of seats.

**6. Function.** The way that chairs actually perform will affect an operation very seriously, especially if the problem each causes is multiplied by tens or hundreds. There are many points to watch on a working chair.

Do the gliders glide, or do they mark or scratch the floor to get caught in the carpet? Do they fall out?

Should a chair have casters? If the carpet in a dining room is a shag or a deep pile with resilient padding, large casters will greatly assist all patrons getting in and out, and save the carpet.

Wall-saver legs (those that splay out in back so that the chair will hit the baseboard rather than scrape the paint off the wall where its back or arms would normally hit) are often valuable. They do take up space, however, so one might consider a bumper molding around the wall to protect the area the chairs might hit.

Is there a proper weight for a chair? A luxurious dining chair on casters can be very heavy because maintenance men will not need to invert it and pitch it up onto a table. For most table service operations, however, light weight is an advantage for the floor cleaners—whether they simply move the chairs to vacuum or put the chairs up on the tables while mopping or buffing.

Is an armchair really necessary? It costs more initially; it presents two more parts that have to be cleaned and repaired; it often hits the table, requiring refinishing; and it takes up more space around the table. For meeting rooms where participants will be spending long hours in a chair, the arms do offer additional comfort. The fact that chair manufacturers today have been able successfully to engineer a stacking armchair is a definite break for those institutions that offer particularly commodious meeting facilities but are short on storage space.

If hundreds of the same style of chair must be ordered for function rooms, it is advisable to check with a manager or owner who has had experience with that chair. One should not depend entirely on his supplier's word, for there is hardly a chair that can withstand the almost willfully destructive treatment it gets from the crew that dismantles or sets up meeting rooms.

Frank discussions with somebody who has used a particular chair over a period of time may help you ascertain whether the one being considered will work. Better still, see the chair in action and check the stress points on it to see how it is holding up. Check the welds, the corner blocks, structural joints, and corners of the upholstered seats.

Function also includes ease of maintenance. An important design feature that helps eliminate cleaning problems connected with dining chairs is having the seat separate from the back so that dust or crumbs cannot get caught in the back and sides where they are so difficult to vacuum or brush out.

Having an opening at the base of the back often contributes to comfort too, as it relieves pressure at the bottom of the spine for those who must sit for long periods.

**7. Guarantees and the Integrity of the Source.** There is an overwhelming variety of dining and function chairs from which to choose. About 100 manufacturers have developed acceptable models—many almost exactly alike.

**FIGURE 16-23** *Four handsome chrome-based lounge chairs, popular for contemporary settings, can be considered classics. (1) Designed by Robert Hussmann, chair has unusual oval steel tubing. (2) Designed by Charles Eames, chair has tilt-back; is engineered for comfort. (3) Lundquist chair with molded fiberglass shell and aluminum base, designed by Maurice Burke. (4) Designed by Charles Gibilterra, this patented cantilever chair is offered with caned or upholstered seat and back. Credit:* 1. *Stendig,* 2. *Herman Miller,* 3. *Vecta,* 4. *Vecta.*

The chairs shown throughout this chapter are available from many different factories at a variety of prices. Therefore one should purchase from a reliable source that will stand behind the warranty, and will deliver on time and according to instructions. The purchase of dining or meeting room chairs is a very sizable one, and it is just as important to find a good dealer as a good deal.

Usually there is a factory warranty on well-made furniture, and if the chairs are bought from a reputable dealer one can be sure that he will pick up any damaged or defective chairs and adjust the matter with the manufacturer. No matter how carefully a factory may check chairs made on a production line, there are going to be some with defects that get through inspections. Buyers should feel confident, however, that every chair in their shipment will perform properly or be replaced.

Fabrics are usually not covered by the chairmaker's warranty, only by the material manufacturer's claims. A customer's own material is never covered by the chairmaker's warranty.

One should check with his supplier as to whether chairs will be shipped KD (knocked down) and agree on who will assemble them and where. Unpacking alone will require labor, and even minor assembly tasks may cause an unexpected increase in costs.

**8. Repairability.** Can the frame be easily refinished? A hardwood frame is perhaps the easiest to redo and the complete job of sanding and removal of finish can often be delayed by the use of a stain stick to touch up dents and scratches.

A tubular steel chair with baked-on enamel or mirror chrome that has finally become chipped or marred will need to go back to the factory—or to some expensive expert—for refinishing. Check to see what the cost will be and about when that might have to be done. What arrangements can be made for replacements while chairs are out?

Vinyl-padded seats and backs must be recovered after a few years, although for emergencies there are vinyl mending kits.

Real cane is beautiful, but hardly a craftsman is left who can repair that material. Nylon cane is today's substitute for practical buyers.

Brass nailheads and tufting add to costs of repair—perhaps they don't merit repeating when reupholstering.

**9. Sturdiness.** Materials and structure make the chair delicate or durable. Aluminum tubing in contract chairs has been almost entirely replaced by cast aluminum or heavy-gauge square steel tubing for extra strength. A molded fiberglass or polypropylene shell will outlast vinyl-padded seats—given equal treatment. Vinyl seats can be replaced, but there is no cure for cracked or abraded fiberglass.

Hardwood chairs with joints that are both glued and screwed will prove highly durable. Connecting rods (i.e., stretchers) should join all legs together except at the front of the chair. U-shaped stretchers that are curved and span the side and back legs are especially useful in warding off chair wobble. If a stretcher is placed on the front, it should be high enough so it will not be used as a foot-rest.

For strength, rear legs especially should be steam-bent with the grain rather than cut out of cross-grain pieces that can easily split.

Bentwood is formed under extreme heat and pressure to retain its curves, and makes chairs of superior strength. Bentwood chairs have been used in institutions since 1842 when they were first introduced in Europe. Although the first ones, cafe chairs with plywood seats and bare backs, were hardly comfortable, many are still around serving as attractive dining chairs. Modifications today—padded seats, backs, and armrests—make them beautiful both to sit in and to observe.

*(See list of Suggested Readings on Interior Design at end of Chapter 17.)*

# Chapter 17

# Interior Design—Guest Rooms

**OVERVIEW** The professional Housekeeper is likely to be doing the interiors of guest rooms, and therefore must understand color. Here is a review of hue, value, and intensity plus psychological effects of color. Style trends are discussed, and fabrics needed for bedspreads, draperies, and upholsteries are examined. Repairing furniture and cleaning it is integral part of Housekeeping. A glossary of art and design terms provide the vocabulary needed to discuss design. Because guest rooms are the larger share of the professional Housekeeper's responsibility, an understanding of the material in this chapter will be of special value as a basis for making decisions on refurbishing, replacement, and upkeep. Points made in Chapter 16 that also apply to guest rooms should be kept in mind.

## *Color*

For many years the subject of color selection seemed to carry a mystique that inferred it was an esoteric art that few could possibly understand. Furthermore, self-styled experts have laid down endless rules to discourage neophytes from using color with any confidence.

To feel comfortable when discussing the subject with interior designers, all that is really necessary is knowledge of a few basics along with the terminology.

There are three primary colors—red, blue, and yellow. From combinations of these three hues, *all* other colors are formed. To understand this it is helpful to see their positions on a color wheel (Figure 17-1). Secondary colors are orange (resulting from mixing red and yellow), violet (red and blue), and green (blue and yellow).

Tertiary, or intermediate, colors are those that result from mixing a primary with the secondary next to it on the wheel—these are called red-orange, yellow-orange, yellow-green, blue-green, and so forth.

Because fashion has come up with salable, appealing, or bizarre names, people tend to believe that such colors as lime, shocking pink, cinnamon, mauve, or turquoise are definite, identifiable colors. Actually, everyone has a slightly different idea as to just what hue fits these vague, romantic words. Such terms should be used sparingly. Pigment colors—ochre, alizarin crimson, cobalt blue—are truly descriptive, and thus mean much more to those familiar with color.

Color has three important dimensions and it is important to keep them separate and distinct.

1. *Hue* is the name of the color; the three primaries together with the secondaries make up the chief hues of the spectrum.
2. *Value* is the amount of light and dark in a color.
3. *Intensity* (or chroma) is the relative density of the pigment; if unadulterated, it may be said to be at its maximum chroma or full saturation.

Complementary is not a general term when used in relation to colors; it is very specific, meaning those colors that are directly across from each other on the color wheel. Thus blue is a complement of orange, green of red, and yellow of violet. To change the intensity of a color and

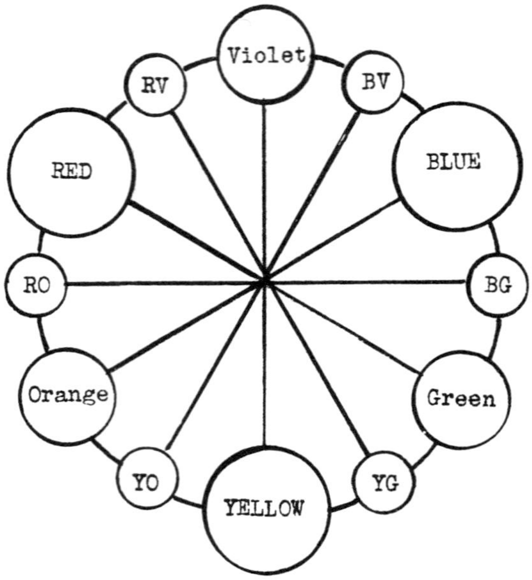


**FIGURE 17-1** *Color Wheel*

Primary colors—RED, BLUE, and YELLOW—are represented by largest circles on the color wheel. Secondary colors—violet, green, and orange—made by mixing the primaries are the medium-sized circles. The smallest circles are tertiary colors—blue-violet, blue-green, yellow-green, yellow-orange, red-orange, and red-violet—made by mixing primary and secondary colors.

make it less vivid, add the complementary color to it.

Black and white are not colors, they are neutrals. A shade is a darkened color; a tint (or pastel), a lightened color. Adding black to a color will change its value, making it a darker shade. To make a color lighter, white is added.

Classic color schemes are of three types:

1. *Complementary*. Using two complementary colors, for example, red and green (but not in full intensities as that would make a Merry Christmas scheme) perhaps a light, light tint of green used next to a deep red, or other tints and shades of red and green.
2. *Analogous*. Using colors next to each other on the color wheel, for example, orange, red, and orange-red, although here again many tints and shades would be used with only accents of the colors in full intensities.
3. *Monochromatic*. One color with many tints and shades, usually also employing black and white in large areas (some contend monochromatic schemes make a small room seem larger).

There is no reason to have every color scheme fit into one of these classic types. One may use a color scheme based on the hues, tints, shades, and intensities within one flower, or perhaps in one painting.

Many designers start with a pleasing motif in a large-patterned fabric and select colors that will work with that.

Most readers are familiar with the fact that there are cool colors—violets, blues, and greens; and warm colors—reds, yellows, and oranges. The former group is conducive to quiet and relaxation, while the warm colors are stimulating and encourage activity. It is well known that rooms on the north side of a building, which get little sunlight, should have light warm colors, while those on the south side should have cooler, darker colors.

Light colors make a room seem larger, while dark colors tend to compact a room. Hot, full-intensity colors come forward; cooler, less intense colors recede. Dark walls should be opposite windows as the light that hits them will brighten and enhance their dusky colors. Walls in which the windows are inset receive little natural light so should be painted or covered in gayer tints.

Dark colors create the illusion of lower ceilings, as do patterned wall coverings. It is usually best to avoid any patterns on the ceiling, except perhaps in a small vestibule if the pattern is not too busy.

If two walls are to be painted and two to be patterned, the two adjoining walls should be painted the same color, and the other two adjoining walls should be patterned. Alternating them would break up the space and give a choppy effect. Where there is a single, intensely bright wall, the other three walls should be as neutral as possible and all in the same shade or tint.

## Color and Pattern on Walls

Mural wall coverings are tricky to use and are most effective on adjoining walls. Often it is interesting to use a mural on a window wall, for if the design is of a suitable type, interruption of the windows may enhance the treatment. Mural designs are usually best used in large banquet rooms.

Small prints belong in small rooms. Large prints should be used sparingly, even in large rooms, and should be combined with expanses of non-patterned surfaces. The trend today is toward mixing many patterns and textures in fabrics, wall coverings, and carpets as long as they have some colors in common. This, however, is a passing fad and the jazzy, jarring effects from these combinations are only suitable in areas where they can be changed to keep up with current fashions.

To use such bright tangles of confusion in a prototype guest room and then repeat it 100 times is hardly economical, because when the fad passes (as it soon will) the rooms will have to be refurbished. Using classic designs in bedrooms is the wisest approach, both as an investment and as a means of pleasing most guests.

## *Guest Room Design*

Howard Johnson's Design Department began work on the motor lodge's guest rooms over 20 years ago, using contemporary styles, quality furnishings, and good taste. The result is that today many of those first guest rooms still hold up—both physically and aesthetically.

In these lodges (not necessarily the franchises) there is constant replacement and design updating (Figures 17-2 and 17-3). Under normal circumstances, carpeting, bedspreads, and draperies are replaced about every five years. Infrequently, walls are redone and beds replaced. Case pieces are replaced in older lodges, according to Elizabeth Cooper, the designer. She explains, "Our operational supervisors frequently visit the lodges to encourage licensees to upgrade and replace before wear becomes too obvious.

"Our design theory in the '80s is toward softer earth-tone colors and lighter wools. Not only does it set the mood for a restful room, but also allows for ease in refurbishing with bedspreads and the drapery treatment. As we now refurbish, efforts are made to replace in-

**FIGURE 17-2** *Characteristic of a new lighter look, this Howard Johnson Motor Lodge guest room uses sturdy rattan (not wicker, a perishable material) in its chairs and table. Notable, even whimsical, touches are the ceiling fan, plants, and ginger-jar lamp. Window treatment employs only a one-sided drapery—a good idea if light or view would be blocked by another panel.*

candescent light with less energy-intensive fluorescent fixtures." An effort is made of course to use warm fluorescent bulbs to avoid a harsh, cold, blue look.

Says Ms. Cooper, "There are certain housekeeping chores which must be performed regularly in order to avert future serious problems. These are: (1) cleaning out the sliding door

**FIGURE 17-3** *Simplicity of furnishings and neutral colors are appropriate in this room where the attractive but busy bedspread pattern is allowed to dominate. Note headboards are padded—a custom of designers who believe in providing guest comfort for reading in bed.*

tracks; (2) cleaning air conditioning filters; (3) dusting tops of pictures, shelves, light fixtures, and furniture surfaces; (4) vacuuming carpet along the wall where dust accumulates.

"These are only a few of the jobs that may be neglected—tasks that housekeeping employees apparently find especially time-consuming or unpleasant, since there is a noticeable tendency to avoid them."

The Housekeeping Department is in the best position to report when specific repairs are needed to preserve furnishings—now more than ever an enormous investment that should be protected.

Experience with furnishings and room design over a long period of time has resulted in standardization of many important features:

1. Vinyl wall coverings (essential for easy maintenance in public areas and rooms)
2. Bumper extensions on bed frames to keep beds 4 inches from walls
3. All bedding 80 inches long
4. Electric control panel at beds with switches for room lighting and television
5. Telephones placed on nightstand between lounge chair and bed
6. Padded headboards at wall for comfort of guests who read in bed
7. Lighting in fixed position, either wall-mounted or ceiling-mounted
8. Wooden arms on chairs
9. Desk chairs with padded backs for comfort and to prevent wood from hitting desk tops
10. Clothes hangers on tracks, not rods
11. Built-in luggage racks with replaceable runners to prevent luggage scratching tops
12. Plastic laminate on surfaces whenever possible
13. Desk furniture wall-mounted to effect clean, contemporary look and to facilitate vacuuming
14. For two-story buildings, use of two different color schemes; for three-story buildings, not more than three, and so on. Gold scheme recommended for first floor because it is light.

Figures 17-4 through 17-7 show some basics in floor plans.

Trends in guest room furnishings today include the following (see also Figures 17-8 through 17-14).

1. Larger, bolder patterns in fabrics
2. More table lamps than floor lamps or ceiling and wall fixtures. (How many lights to be used in a room will depend on its size, but a minimum of three lamps plus overhead or wall fixture in hall is to be expected.)
3. The old Murphy bed has made its comeback in studio rooms used for entertaining or showing samples. It folds into the wall, leaving as its only trace an inconspicuous door with a horizontal bar at the top. Some even have textured vinyl wall coverings to disguise them.
4. There is more painted furniture as wood becomes less plentiful and more plastic laminate parts and panels are used.
5. Twin-bedded rooms may be more salable than king- or queen-bedded rooms, but there is no obvious consensus on what distribution of room styles should be; it depends on market and type of business. Double doubles usually are most in demand in resorts where families are likely to stay. In Boise, Idaho, the Rodeway Inn sells out its executive and bridal suites almost every weekend because they have glamorous, big sunken pools and fireplaces priced to attract local people.
6. Placement of the telephone has been changed from bedside to corner chairside. This saves wear and tear on the mattress; it also provides work space for a businessman making numerous calls. Rooms for single women should have telephones with long cords, as a bedside phone is a security factor.

## *Hilton Hotel's Variety of Rooms*

Henri J. Lewin, senior vice-president of Hilton Hotels Corp., western division and Nevada, explains that guest room decor and furnishings

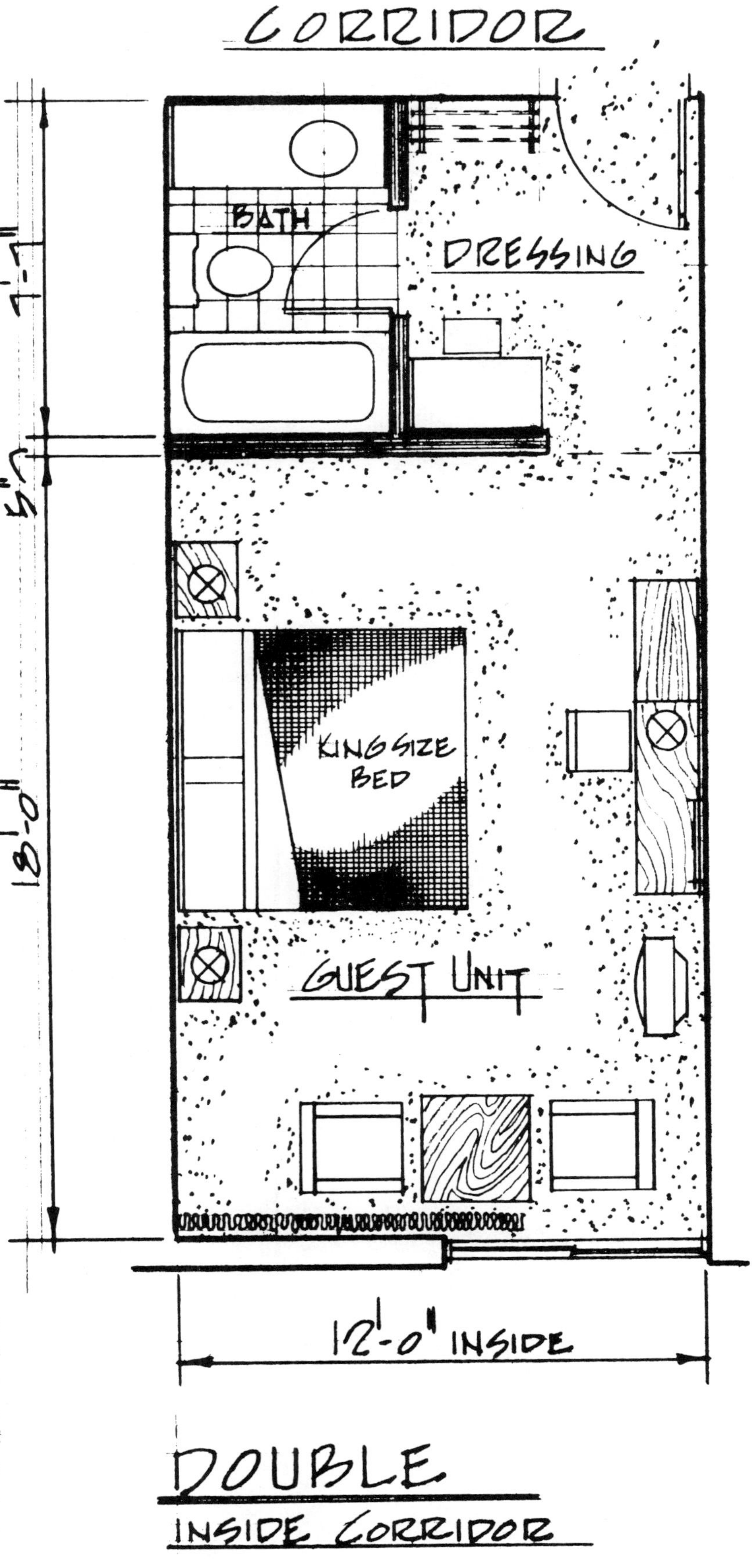


**FIGURE 17-4** *Layouts of Ramada Inn's rooms are drawn here in scale; ¼ inch equals 1 foot. In room at right, window wall is at bottom where measurements indicate width is 12 ft., inside dimension. Door into guest room is from inside corridor, not outside balcony. Note doors are drawn with curve to indicate how they open—in this case, into room. Bed is king-size, measures 6½ ft. across and 6½ ft. long.*

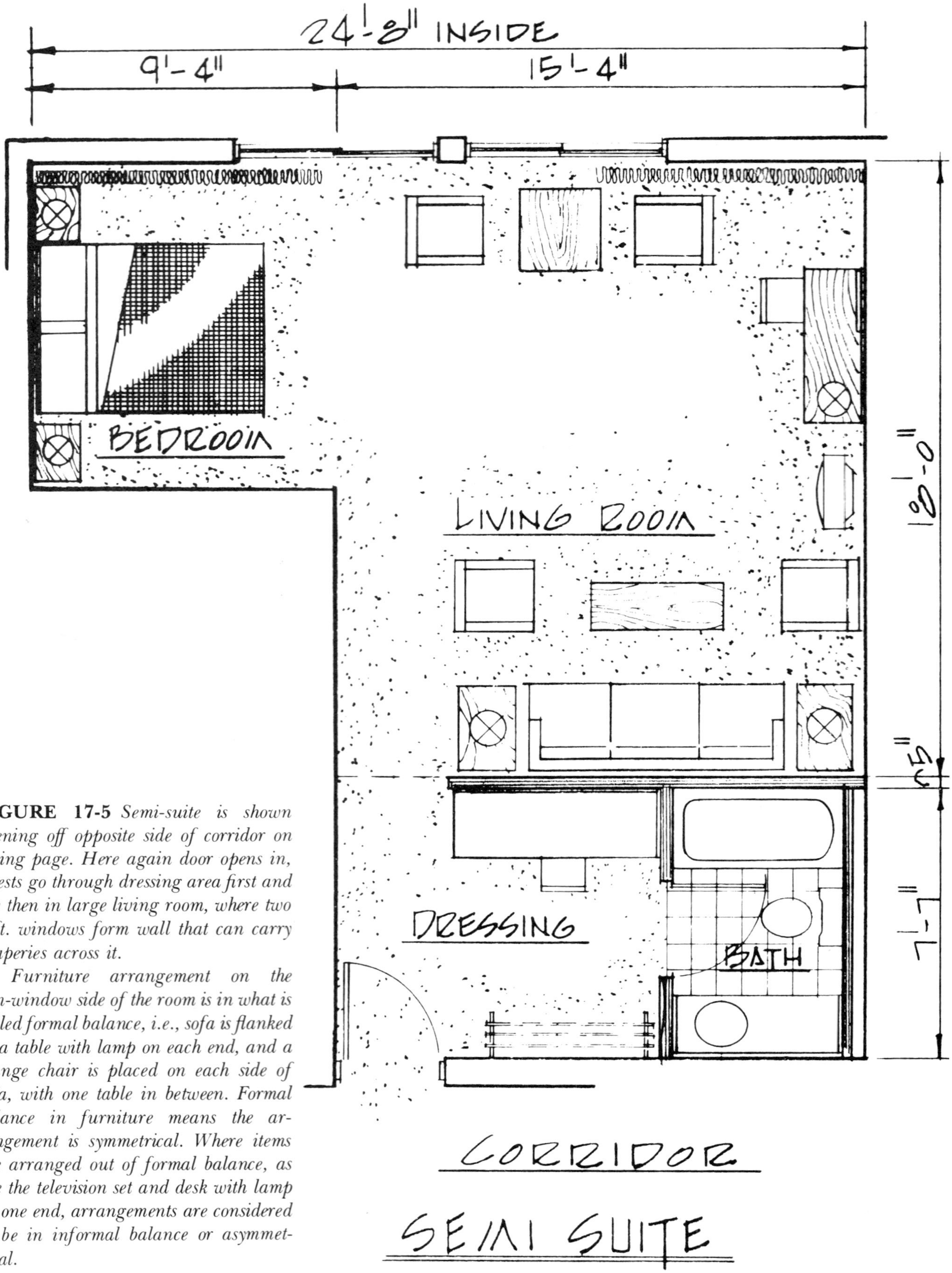


**FIGURE 17-5** *Semi-suite is shown opening off opposite side of corridor on facing page. Here again door opens in, guests go through dressing area first and are then in large living room, where two 6-ft. windows form wall that can carry draperies across it.*

*Furniture arrangement on the non-window side of the room is in what is called formal balance, i.e., sofa is flanked by a table with lamp on each end, and a lounge chair is placed on each side of sofa, with one table in between. Formal balance in furniture means the arrangement is symmetrical. Where items are arranged out of formal balance, as are the television set and desk with lamp on one end, arrangements are considered to be in informal balance or asymmetrical.*

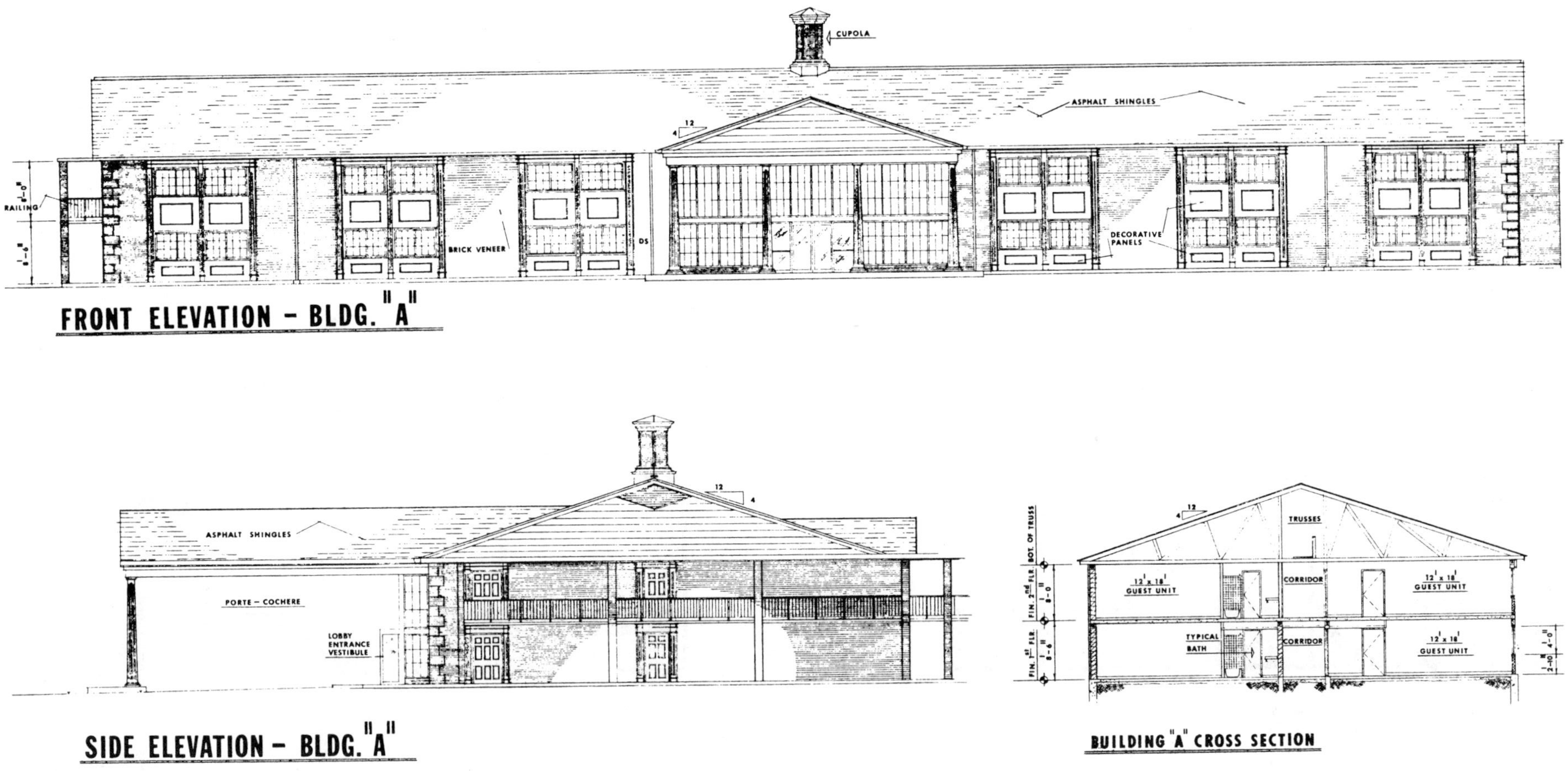


**FIGURE 17-6** *Exterior shown above for standard Ramada Inn includes an elevation of the front of the building and the side, as well as a cross-section. These are drawn to scale and include no perspective, but are simply straight-on views. In Chapter 2 there are photos of the exterior, lobby, and interior of a standard Ramada Inn. Viewing the two sets of figures alternately will give the reader a better idea of how the drawings represent the actual structure and interiors.*

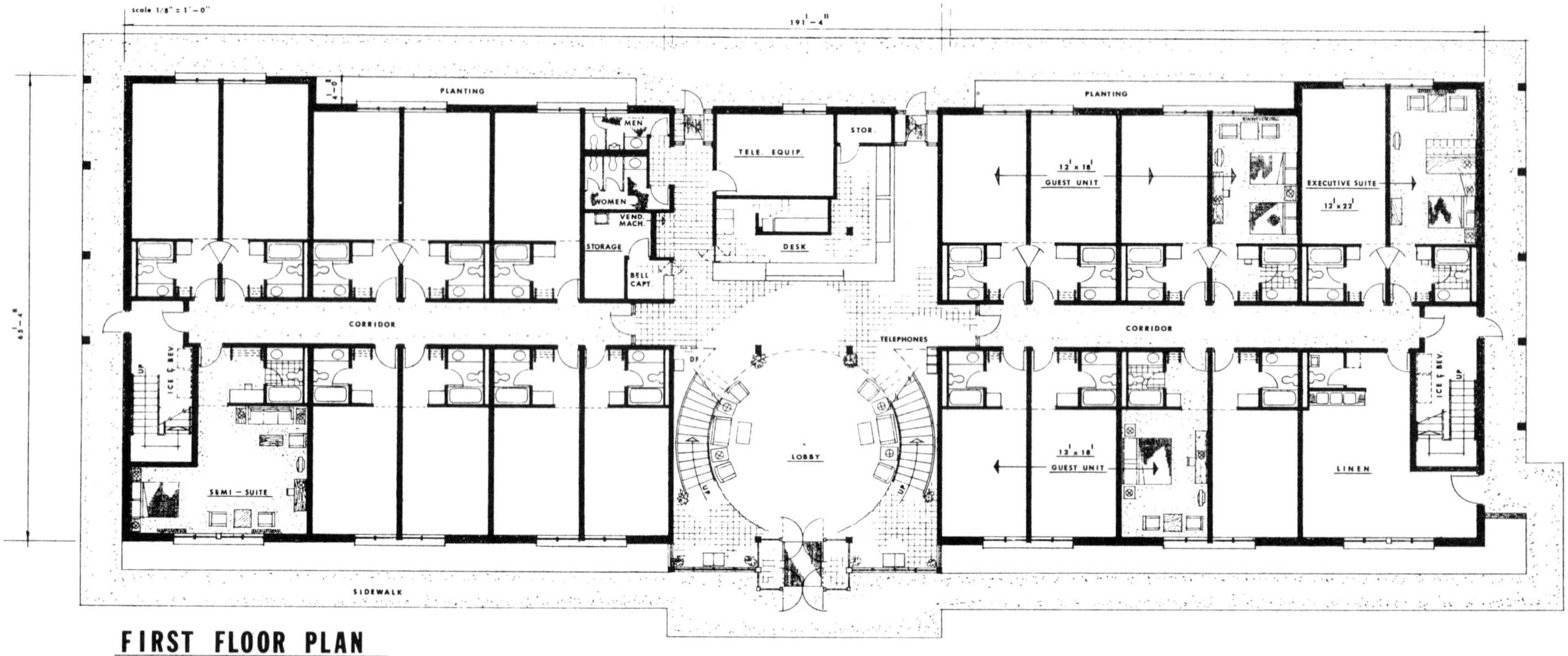


**FIRST FLOOR PLAN**

scale 1/8" = 1'-0"

**FIGURE 17-7** *Scale for the building elevations and this first floor plan was originally ⅛ in. equals 1 ft. but due to space limitations here it has been further reduced. Plans of entire floors such as this do not indicate furnishings in every room because they simply repeat exact arrangements. The floor plan shown here illustrates one semi-suite at lower left, with bed placed in an alcove; and two executive suites with connecting doors, one at each end on the outside of the building. Standard guest room units with king-sized beds and with two double beds are the other room types that have been set up on this floor.*

*Lobby has circular staircase leading up to second floor, which has almost identical layout except that over the front desk and telephone room there is a suite with complete living room and bedroom.*

**FIGURE 17-8** *Typical of new trend toward bold prints, washable quilted bedspreads, and dramatic headboard treatment, this lavish, luxury-priced room at the Los Angeles Marriott Hotel even has throw pillows, rarely used as they are easily packed away by guests. Large room holds two double beds and credenza between.*

**FIGURE 17-9** *Bookcase on wall of student room is custom-designed by dormitory builders. It doubles as a bench when inverted, and students use it on floors or walls. Most student furniture is made of plastic laminate, has drawers that open without hardware. Desk chair is classic captain's chair built to hold up indefinitely.*

in Hilton Hotels often cover a wide range, differing as to style and furnishings.

Guests are offered a choice of rates and types of rooms. Here is their system for designating the different categories.

**Gold Room.** Luxury furnishings in excellent condition; maximum rate can be charged. Rooms are larger and are assigned to VIPs, those who wish to entertain, and favored repeat customers.

**Red Room.** Same quality but smaller rooms.

**Blue Room.** Called multipurpose; used for a sample showroom—usually includes a sofa-bed.

**Black Room.** Less luxurious rooms on lower floors in demand by price-conscious travelers who are content with smaller bathrooms and the like; also used for large groups requiring discounts.

If a prestigious customer who normally would be put in a gold or red room is forced to take a black room because the more elegant rooms are sold out, management usually sends a

**FIGURE 17-10** *Guest room at the Broadmoor was photographed in 1918, shortly after the lavish resort opened. It shows how much softer today's interiors are, despite Broadmoor's plush upholstery, patterned carpet, and antique vases. Placement of furniture apparently stymied early hoteliers who put chest of drawers across from dressing area, and "floated" desk in front of fireplace.*

**FIGURES 17-11 and 17-12** *Comparison of these two rooms will help readers see how a large-scale painting and a headboard treatment can give better proportion to room. Above, in Hyatt Regency, San Francisco, beds seem solidly anchored to walls because headboard relates to entire room, while on left, in 12-year-old room, tiny pictures over bed seem lost. Lampshade at left rests too high above base; though it may help those who read in bed, it looks spindly, out of proportion. Shades on both lamps above are proper proportions. Having telephone table in center of beds, and light tables flanking sides of beds relieves crowding that frequently occurs on center bedside table.*

gift—fruit, liquor, flowers—to make up for the guest's disappointment.

## *Fabrics*

Fabrics cannot be arbitrarily divided into upholstery and drapery classifications because too many materials are used for both purposes. During recent years there has been a growing trend among decorators to cover one or two pieces in a room with the same material as that used in the draperies. Furniture manufacturers, too, have

**FIGURE 17-13** *It is obvious this Sheraton room is up to date with its bold patterns, quilted spreads, white painted furniture, and sculptured pile carpet, but an accent is missing above the bed. Perhaps pattern would be more pleasing if not used on chairs, because when draperies are closed, pattern will be overwhelming.*

**FIGURE 17-14** *"Highly perishable" could characterize this delicate and formal room at the Sands Hotel, Las Vegas. Plain, light fabrics that require dry cleaning, plain carpet that shows every step and spot, marble table tops, silky dust ruffles, brocade bedspreads, crystal chandelier, and ceiling-high wall mirror behind bed all spell upkeep expense. Draperies on room divider dramatize bed area as a stage, can also be drawn to make bedroom private. Room rate must absorb high fabric cleaning charges and other maintenance costs.*

been using more of those fabrics that formerly were restricted in use almost entirely to draperies.

Fabrics can, however, be loosely divided into a number of general classifications according to weave. Almost all are made in either natural or synthetic fibers today.

**Jacquard Weave.** Formal fabrics, used extensively in traditional upholstered pieces and for draperies, including brocade, damask, tapestry, matelasse.

**Pile Fabrics.** Used in both traditional and modern upholstered pieces, especially divans and lounge chairs; include plush, velvet, frieze.

**Rib Fabrics.** Made with firm ribs, both floral designs and plain; used on colonial, provincial, and some traditional furniture. Examples: rep, bengaline, corduroy.

**Plain Weaves.** The coarser examples are used on modern and colonial furniture. Examples: monk's cloth, muslin, homespun. Finer weaves—satin, taffeta—are used primarily for formal draperies.

**Textured Weaves.** Copies of hand-woven fabrics extensively used in upholstery and draperies. Example: raw silk.

**Plastic-coated Fabrics.** Consist of fabric base covered by a plastic coating, which may be resinous or vinyl plastic.

**Plastic Fabrics.** Can be cut and used as any fabric; made in either highly polished smooth finish or with embossed grains or patterns.

**Supported Vinyl Fabrics.** Tough, vinyl plastic face-sheets laminated to a fabric backing; elastic types have a stretchy backing easy to tailor. Come in various weights, many patterns.

**Quilted Vinyl Fabrics.** Vinyl plastic face-sheet, a layer of cushioning, and a fabric backing electronically welded together to form quilt, trapunto, or embroidery designs.

## *Fabrics Must Be Easily Cleanable*

Even the deluxe resorts such as Greenbrier are finding it necessary to use synthetic fabrics rather than traditional ones. For example, on the card tables they are now using plastic covers rather than the felt ones used for many years. The expensive delicate hand-painted draperies in the dining room have clear plastic sheeting part way up to protect them from splashed food spots. The fabric—one that could hardly be duplicated today—may last several more years by being covered, and not having to be dry cleaned so frequently.

Another means of saving fragile or costly slipcover fabrics is to fasten a sponge rubber strip with an adhesive backing on the chair or sofa frame under the skirting where shoes normally hit the fabric and mark it or cause it to wear. The sponge offers a soft surface, preventing abrasive damage.

## *Special Finishes*

Many fabrics for draperies and upholstery come already treated with chemicals that make

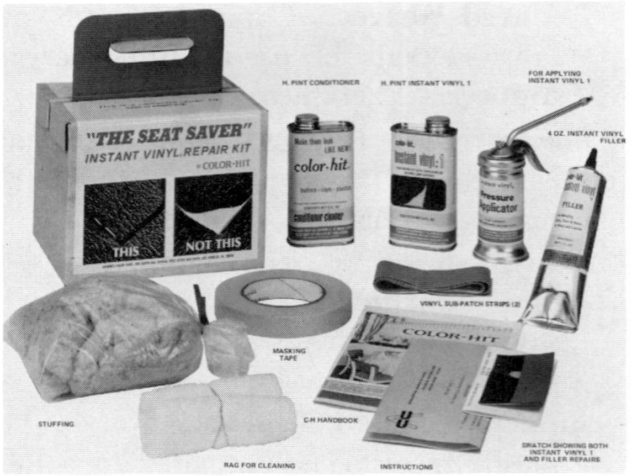


**FIGURE 17-15** *Vinyl repair kit makes it possible to redeem many chairs that would otherwise have to be set aside as out of order early in their careers. If employee is patient, he can use kit to good advantage, so most people will not notice.*

them stain- and water-repellent. Repeated launderings or dry cleanings will destroy this repellency. New chemicals can be applied on the premises, however, or the restoration can be done commercially.

Fabric finishers can process almost any material to make it fire-retardant. One of those processes puts on a most attractive high-gloss vinyl coat, which gives the wet look. The process involves polyurethane film. It costs under $1.00 a yard and considerably less in quantities of 500 yards.

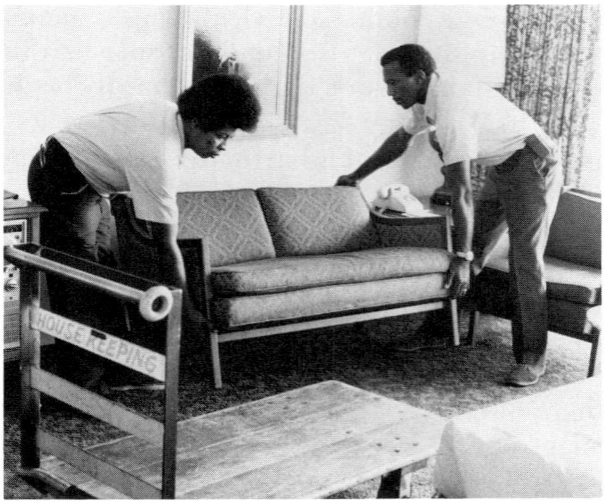


**FIGURE 17-16** *When furniture needs to be cleaned, repaired, or recovered, handymen (those housekeeping workers who are highest paid and able to handle wide variety of tasks) transport pieces on dollies to areas where shampooing, carpentry, or reupholstering take place.*

Sometimes materials that were flame-retardant when purchased can lose their protective qualities after repeated cleanings or launderings. Manufacturer's cleaning instructions should be followed or the manufacturer consulted as whether to retreat fabrics.

## *Upholstery Cleaning*

Shampooing vinyls and plastics is relatively fast and easy, but cleaning woven fabrics is a hazardous matter that should be approached with great care lest upholstery be ruined and require replacement. The personnel given responsibility for shampooing or cleaning upholstered furniture should be conscientious and patient, willing to follow instructions. If this work is to be properly done, there should be a clean, warm, well-ventilated place in which to do it (Figures 17-15 through 17-20).

Before deciding how a certain upholstery fabric should be cleaned, it should be tested. First, stuffing or lining materials should be checked to determine what type they are. This can be done by extracting a small piece with a crochet hook. If latex foam padding is found, the fabric should NOT be dry cleaned. If any dark or red stuffing or thread is found, one must beware of its bleeding during shampooing or cleaning. (This will indicate that fabric must be cleaned with a minimum of moisture.) Jute filling materials may cause browning if they become wet.

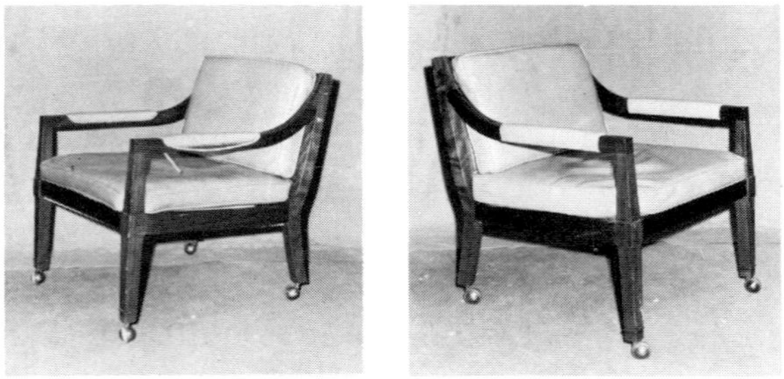

**FIGURE 17-17** *Lounge chair is brought to housekeeping repair department to have arm upholstery replaced. Completed chair has vinyl wrapped completely around arm so it is less likely to have same problem with welting coming loose. Note giant ball casters that must be kept polished. Care must be taken that metal polish is not so strong as to remove finish.*

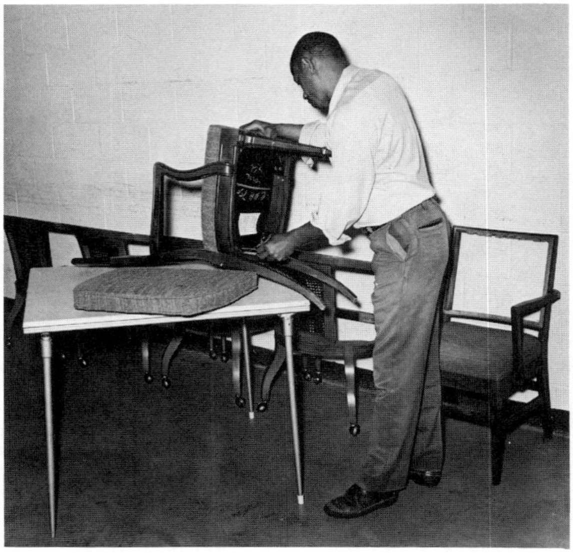

**FIGURE 17-18** *Chair backs are being recaned, seats are removed and replaced, and small casters are being added to these dining chairs. One houseman is in charge of chair repairs.*

A second test is to determine whether the colors are fast. This is done by pressing a rag or blotter dampened with cleaning solution on the fabric's deeper colors on some unnoticeable place *on the front of the chair.* It is not practical to make this test on the back of the chair or sofa, as some fabrics will crock (rub off) more readily after being subjected to abrasion and light. If color is not fast, it should be tested again with a slightly acid cleaning solution. If color bleeds, then the only safe cleaner is a dry solvent.

Even though an upholstered piece—or a natural fiber drapery, for that matter—may have been cleaned several times, it can still run or fade after long exposure to wear or sun.

A third test is for shrinkage. Clean a section on the back or pillow side; allow to dry, and inspect.

There are several cleaning techniques for fabrics:

1. Dry solvent cleaning usually applied with terrycloth and wiped dry with another terry
2. Dry foam applied gently with sponge on fragile fabrics (silky surfaces often are spoiled by using a brush)
3. Shampoo with neutral synthetic detergent and machine that will apply solution evenly and prevent over-wetting

Before any of these methods is used upholstery should be thoroughly vacuumed. Some recommend beating overstuffed pieces and cush-

**FIGURE 17-19** *Studio room in glass tower addition to Regency Hyatt House in Atlanta has wide view. Window walls are features of wedge-shaped rooms in this circular tower designed by famed architect John Portman. With so much light, furnishings must resist fading. Note wood edging along top of chair and on sofa arms to protect upholstery from soiling. Sofa converts into king-sized bed. Credit: Coronet Industries*

**FIGURE 17-20** *In damp tropical climates such as Hawaii, ceramic tiles with a light, expendable, native woven mat are an appropriate answer to flooring problems. Here at Mauna Kea everything is washable—draperies replaced with sliding louvered wood doors.*

ions to bring dust to surface so that it can be removed with vacuum. For pile fabrics, brushing with a stiff nylon brush will bring up soil from bottom of nap.

Spray cleaners and a hand-brush can be used successfully on some fabrics instead of foam shampoo and a rotary-brush machine.

Shampooing furniture, like wall washing, should be done from bottom to top to prevent stains and water marks.

Heavily stained areas, such as headrests and armrests, should be pre-spotted with volatile solvent to remove oily soil before shampooing. Other spots should be identified if possible, and removed with recommended solutions listed on spotting charts furnished by most producers of chemical cleaners. If spot will not come loose easily, employee should not try to remove it with continued mechanical action as the upholstery may become overly worn or its surface permanently marred.

After cleaning, hand-brushing or vacuuming is usually helpful to restore nap. Furniture should dry in a warm ventilated area, preferably with a fan blowing on it, and some fabrics will require occasional brushing during drying.

## *Art or Design Terms*

Familiarity with the meaning of these terms makes it easier for the professional Housekeeper to communicate with designers and architects.

**Batik.** Cloth patterned by the application of wax in certain designs to preserve the material's background color while the fabric is dyed. Usually it is characterized by fine cracks and veins that give it a distinctive character.

**Ceramics.** General term for the art of making pottery, but decorative art objects are also made of clay and fired as pottery is.

**Collage.** Composition made by pasting together various materials; montage usually implies that all materials are photographs.

**Emboss.** To ornament a surface with raised work.

**Engraving.** Process of incising a design on a substance with a sharp instrument; also the impression print made from such a plate.

**Etching.** Process of engraving a design on a copper plate with an acid. Also the impression made from such a plate.

**Filigree.** Delicate and intricate metallic ornament made of fine wire.

**Finial.** Knob-like ornament terminating in a point.

**Fresco.** A method of painting on freshly applied, moist plaster in which pigments become chemically incorporated into the plaster.

**Glaze.** Vitreous coating applied to pottery and fired (baked) to seal the surface.

**Illumination.** Decoration with gold or silver, especially the initial letters of a manuscript.

**Intaglio.** Design sunk below the surface so that an impression made from it is in relief (embossed).

**Lithograph.** Impression (print) of a design made on stone with a grease pencil or crayon.

**Medium.** Material with which an artist expresses himself. Watercolor, stone, wood, clay, batik, and the like are *media*

**Mosaic.** Surface decoration made of small pieces of glass, stone, or ceramic set in cement.

**Patina.** Incrustation that forms on bronze through chemical action, also incrustation or attractive surface on silver, leather, wood, and so on brought about by aging.

**Relief.** Figures projecting from a background; if cut deeply, called high relief; if shallowly, called low or bas relief.

**Repousse.** Process of decorating metal by hammering it from inside.

**Tempera.** Paint mixed with egg white.

## *Bibliography*


Doan, Franklyn E., ed. *Seng Furniture Facts.* Chicago: The Seng Co., 1962.

Dreyfuss, Henry. *The Measure of Man.* New York: Watson-Guptill Pub. Inc., 1967.

Helsel, Marjorie. *The Interior Designers Drapery Sketchfile.* New York: Watson-Guptill Pub. Inc., 1969.

*Interiors Book of Hotels and Motor Hotels.* New York: Watson-Guptill Pub. Inc., n.d.

Klapper, Marvin. *Fabric Almanac,* 2nd ed. New York: Fairchild Pub., 1971.

*Motels, Hotels, Restaurants and Bars.* Boston: Cahners Books, 1960.

Panero, Julius. *Anatomy for Interior Designers,* 3rd ed. New York: Watson-Guptill Pub. Inc., 1962.

*Pentagram: The Work of Five Designers.* New York: Watson-Guptill Pub. Inc., 1973.

Waugh, Alice. *Interior Designs.* Minneapolis: Burgess Publishing Co., 1967.

Whiton, Sherrill. *Elements of Interior Design.* New York: Lippincott Co., 1964.

# Chapter 18

# Bathrooms and Rest Rooms

**OVERVIEW** Housekeeping employees will clean bathrooms and rest rooms only if the proper tools and supplies are provided. They must be trained adequately if these tasks are to be done frequently and thoroughly. Here we present tested procedures. Equipment and suggested ways to remodel plumbing facilities are included, with schedules and supplies for public rest rooms.

Although one procedure for cleaning a guest bathroom in a hotel was reviewed in Chapter 8, there are several other techniques that many motels, hotels, and residence halls encourage workers to use. Figure 18-1 shows how an all-purpose cleaner and disinfectant can clean everything in the bathroom without wasting time or product.

Among the professional Housekeeper's concerns are (1) how to make it as pleasant as possible so workers will actually do the job, and (2) how to keep employees from wasting cleaning products.

## *Proper Supplies and Tools*

To make the work tolerable, proper tools and equipment are vital (Table 18-1)—for example, small mops or brushes that eliminate the need for the worker to put his hands in the toilet bowl, or more complex devices such as the "caddy" make work more pleasant.

In one caddy a quart-sized bottle of sterile disinfectant cleaner has a hose attachment so that liquid can be dispensed into the mop well in the plastic caddy without requiring employees to open a bottle and pour from it each time it is needed. The mop has a retractable plastic cover so the attendant need not touch it. The urinal is cleansed with the solution that remains in the fixture for a few minutes. Flushing then rinses off the chemical.

The caddy also carries a lighted mirror with a long handle for easy inspection of the underside of the toilet rim. Where a toilet brush is used, a squirt-top for the bottle of disinfectant is available so that liquid can be applied directly. For cleaning tubs, a long-handled brush helps prevent backaches.

A regular and frequent cleaning and sanitation program will prevent bathroom fixtures from becoming too unsightly and difficult to clean will also make the job easier.

Wherever new fixtures are to be cleaned, it is very important *not* to use an abrasive cleanser as that will soon destroy the smooth surfaces and cause them to become soiled more quickly. A cream or paste or all-purpose liquid is sufficient to clean new surfaces.

To keep employees from wasting expensive cleaning products several manufacturers have devised packages specifically designed to eliminate that problem. Aside from the dispenser cup on the bottle top (#9 in Figure 18-1), there is an increasing number of "portion control" detergents, wax-strippers, bowl cleaners, carpet shampoos, and germicidal detergents now available.

The containers include polyethylene envelopes, small cans, tablets, paper pouches, and water-soluble plastic pouches. Each premeasured package contains just enough liquid or

**TABLE 18-1** Cleaning Rest Rooms

| Area | Utensils | Cleaner/Sanitizer Dilution | Method |
|---|---|---|---|
| Toilets, urinals, Sinks | Long-handled brush, cup measure, bucket with mop press | Institutional all-purpose cleaner, institutional bleach Dilute: 1 C all-purpose cleaner + 1 C institutional bleach in 1½ gal. water | Daily: (1) clean sink inside and out with solution, (2) clean outside of urinal and toilet with solution, (3) pour half remaining solution in urinal, half in toilet, (4) let stand 10 minutes, (5) use long-handled brush in bowl, (6) flush |
| Floors, walls | Brush, mop, cup measure, bucket with mop press, sponge | Institutional all-purpose cleaner, institutional bleach Dilute: 1 C all-purpose cleaner + 1 C institutional bleach in 1½ gal. water | Daily: (1) wash down walls with solution, (2) dry with sponge, (3) wash floors with solution, (4) dry-mop |

**FIGURE 18-1** *1. Special pouring/measuring spout limits user to 2 oz. of all purpose cleaner per bath.*

*2. Cleaner is combined with warm water in bucket where bottom of shower curtain is soaked in solution.*

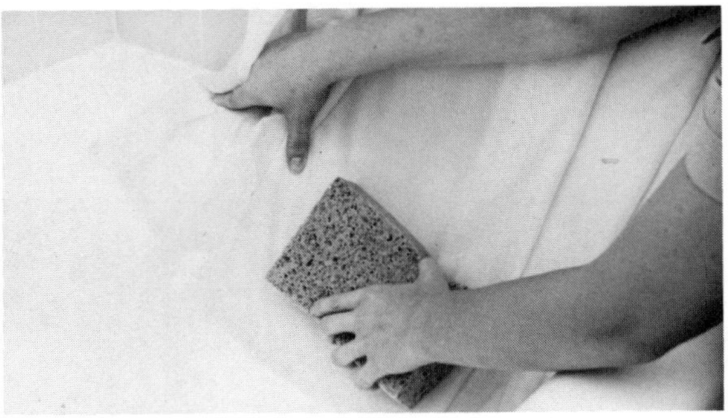

*3. Bottom of shower curtain is sponged completely to remove soap scum and prevent mildew.*

*4. Tub tile and fixtures are sponged with solution, then dried so they shine.*

*5. Small amount of solution is poured into basin so that counter, fixtures, and mirror, can be sponged, and dried.*

*6. Plug is removed from sink and placed in bucket to soak; it is then rinsed and replaced.*

*7. Outside of commode and seat are washed with solution remaining in bucket.*

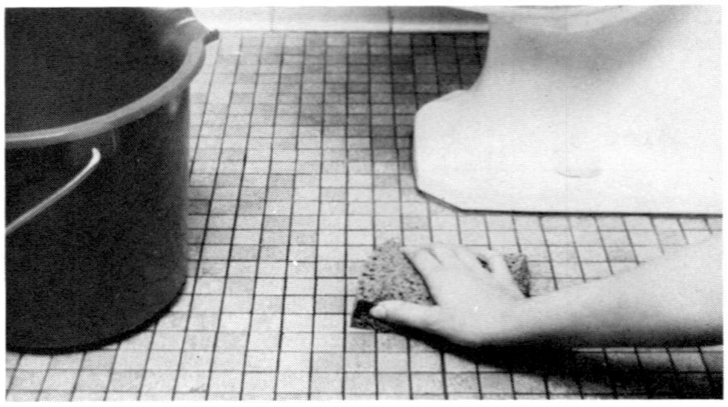

*8. Tile floor is sponged last with the worker using the remaining solution in the bucket.*

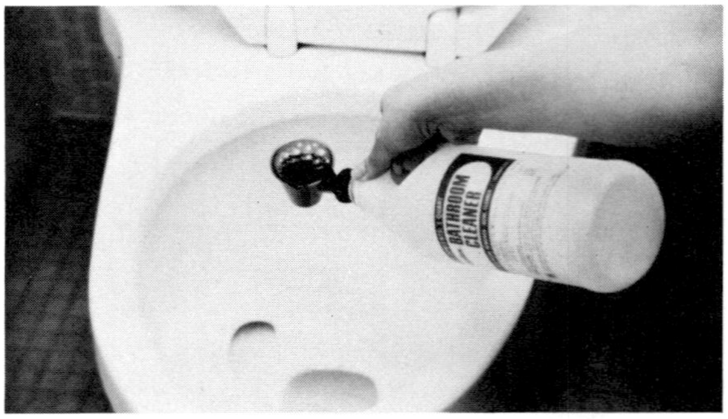


*9. To clean commode, 2 oz. of cleaner, a deodorizer-sanitizer, are poured from measuring spout.*

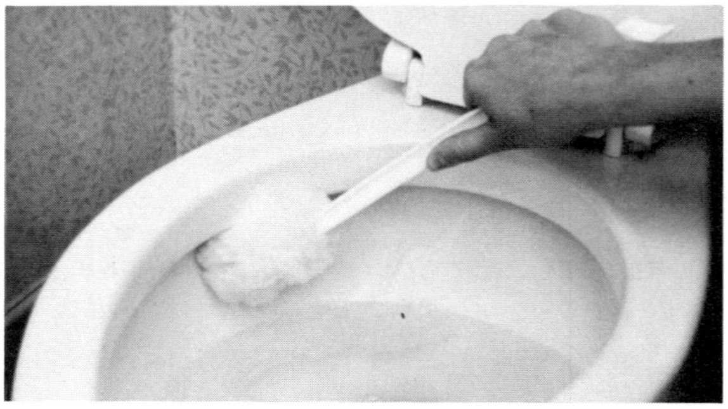

*10. Commode bowl and lip are cleaned with swab. If stained, they should be soaked and the toilet flushed.*

powder detergent for a single bucketful of solution. In some operations the portion-sized packs not only reduce waste, but save on labor required for decanting. (That subject will be discussed later in Chapter 19 on supplies and equipment.)

Where these products packaged for portion control seem too expensive, some supervisors personally put the proper amount of bowl cleaner into each commode shortly before the maid arrives at each bathroom on her duty list.

Maids and housemen should be reminded frequently *not* to throw pieces of guest soap into toilets, as this often causes toilets to clog and overflow. Used guest soaps should be collected for use in employee washrooms if the pieces are large enough. Smaller pieces should be melted down to a useful liquid cleaner, sold to laundries, or given to some charity (Figure 18-2).

**FIGURE 18-2** *Leftover guest soaps are collected in large quantities in hotels where standard soap is 1 ¾ oz.—a luxury-sized bar normally left more than half unused. These can be sold to a laundry to be combined with chemicals, made into laundry soaps. Charities welcome leftover soaps too. Cardinal rule is never throw used soaps down toilets.*

## *Neglected Areas*

Grout between tiles is often neglected. If mildew or a heavy lime deposit has formed it may take time and patience to clean with a brush and a de-liming agent or, in extreme cases, a pumice stick. If the grout is too far gone for these, Novel grout whitener is a remarkable new bathroom cosmetic that will help. It restores like-new whiteness to grout that has been discolored by mildew, hard water, rust, graffiti, and so forth. Used on ceramic tile walls and showers, it provides a coating that stays white for months.

Safe to use, it contains no caustics, bleaches, or harsh abrasives. One two-bottle carton treats approximately 2000 square feet of discolored grout (25 average bathrooms).

Another problem is the shower curtain—whether there is simply one, or a pair (a silk or synthetic curtain protected on the inside by a plain plastic curtain). These must be cleaned regularly with a solution that will prevent mildew from forming. To dry properly, they should then be draped outside the tub, a position in which, unfortunately, the showering guest may leave them. When this happens, it can cause a minor flood as shower water rains straight down the draped curtain onto the floor.

Although told as a joke, the following anecdote shows management's serious concern about the consequences of an improperly positioned shower curtain.

When Conrad Hilton appeared to speak at the opening ceremonies of the Cairo Hilton, his audience included hundreds of national and international dignitaries. They were assembled because this marvelous structure meant so much to Egypt in terms of employment as well as to the international traveler as a place of splendid accommodations. After local officials pontificated through long speeches, Hilton was finally called upon to speak; he strode to the microphone and said to his distinguished listeners, "Ladies and gentlemen, I have but one thing to say on this important occasion—please, please, please, put the shower curtain inside the tub before you shower!"

## *Shower Doors*

Because of this particular problem, hotels able to afford it have often installed sliding shower doors on bathtubs (Figure 18-3). But cleaning the tracks for these doors is a time-consuming procedure that must be frequently repeated. There is no question that conventional shower curtains are less costly to buy, install, and maintain.

The doors installed on a bathtub or the single door on a separate shower stall should be cleaned daily to prevent spotting or soap build-up. If appliance wax is used on the inside of glass

**FIGURE 18-3** *Low-silhouette, one-piece toilet is especially suitable for guest bathrooms as it may be installed with its tank partially beneath a counter. Shower doors pose special cleaning problems on the insides and in door tracks along tub. Dark counters show soap splashes, so should be avoided.*

shower doors, the glass and rails are less susceptible to these. Under no circumstances, regardless of build-up, should a scouring pad, wire brush, or abrasive be used on the glass, aluminum rails, or plastic panels. Where the maid must climb into a bathtub to clean overlapping double doors, the procedure is much more troublesome than opening and cleaning the hinged door enclosing the ordinary stall.

A new type of shower door installation is a single sheet of plastic that swings on a hinge for easy access and cleaning. Double doors are not necessary, but if used, their open tracks will be much easier to maintain than the old fashioned double ones.

In most new constructions the tubs and stall showers are of preformed fiberglass. Their maintenance is less than for tile, and there is no grout to mould, discolor, or break.

## General Bathroom Cleaning

Cleaning faucets and fixtures is routine, and most common liquid cleaners are satisfactory. Again, avoid abrasive cleansers as they may scratch the plated finish.

The faucet aerators should be removed and cleaned when necessary. How often this is done depends on the kind of water and the condition of the pipes. To clean, merely rinse out thoroughly to free the screen of accumulated solids. When replacing the aerator, tighten only until finger-tight.

Dust and periodically wipe the metallic portions of faucet fixtures with a soft damp cloth to keep the metal bright. Do not use abrasive cleansers on the metal sections of the fixture, as this may remove the protective lacquer finish and cause the fixture to discolor.

Detailed, periodic bathroom cleaning can be done during slow periods or when maids do a general clean. The following activities will be involved on various occasions:

1. Wax shelf surfaces to prevent stains on medicine cabinet shelves from spilled medicines.
2. Use a small funnel when putting drain cleaner in sink or bathtub to keep cleaner from coming in contact with porcelain.
3. Remove soap scum on glass shower door with solution of 8 oz. nonprecipitating water conditioner to one-half bucket of hot water; apply with a scrub brush.
4. Coat metal bathroom scale with paste wax to ward off rust that may occur from water splashes.
5. Wash shower curtains, even vinyl ones, in accordance with manufacturer's instructions; rubber bath mats must also be washed periodically.
6. Use brush to clean soap-caked soap dishes.

## Some Special Bathrooms

Most institutions are being required to install washroom facilities for the handicapped. This entails wider booths for wheelchairs, grab bars, pressure water faucets, lower soap dispensers, and open areas under basins.

Some hotels invest in outlandishly lavish plumbing associated in the public mind with Roman orgies, Oriental potentates, and Hollywood superstars so this will be part of the public's image of the hotel. The Rodeway Inns in Boise, Idaho, and Salt Lake City offer an astonishing variety of exotic sunken tubs in their "super" suites (Figure 18-4). The tubs vary in shape, being heart-shaped, circular, club-shaped, but they do pose a problem for maids.

For example, the sunken bath in one bridal suite involves a specific cleaning procedure: an antique wall mirror is the first item cleaned; then gold fixtures, tile wall, and top tile surface around the tub. The maid then sits on the edge of the tub and places a soiled towel or bathmat in the bottom of the tub. She swings her legs down into the tub and steps on the mat, which keeps her from slipping. She begins cleaning the interior of the tub starting with far sides and working her way around. The maid cleans the floor of the tub while backing up and out of the bath. She then lies flat on her stomach, reaches down into the tub and wipes off her last foot prints.

No special cleaner is required. It does, however, take a maid from 13 to 15 minutes completely to clean a sunken bath, where it only takes 3 to 5 minutes to clean a standard bath. The management considers the printed and word-of-mouth-publicity to be worth many times the relatively high cost of maintaining these unique bathtubs in perfect condition.

**FIGURE 18-4** *Luxurious, deep, dramatic bathtubs distinguish the special suites at the Rodeway Inns in Salt Lake City and Boise, Idaho. These require agile maids and take about five times as long to clean as a standard bath, but additional room revenue more than compensates. See text for details on how sunken tub is cleaned.*

**FIGURE 18-5** *The furo, or Japanese sunken tile bath, is used throughout the 172 guest rooms and suites of San Francisco's Miyako Hotel. In special tiled area beside sunken tub, guest sits on stool, soaps, and washes. Low faucets and bucket are used for rinsing. Tub bath is for soaking, often involving entire family.*

## *Bathroom Equipment and Remodeling*

The chances are fairly slim of executive Housekeepers being consulted during the planning stages as to what types of bathroom fixtures should be installed in a new property. When an older institution is about to be remodeled, however, the professional Housekeeper will certainly be consulted; therefore, it is imperative that she be fully informed on the newest developments in bathroom equipment, systems, accessories, and so on.

No matter how elegantly refurbished the guest room is, if the bathroom is old-fashioned or the fixtures are worn or damaged, the guest will downgrade the value of the room to a far greater degree than the reduction made in the remodeling budget by the decision to keep the old fixtures (Figures 18-6 and 18-7).

Upgrading the bathroom is almost always worth the investment. Tearing out all of the tubs and showers and replacing them with completely new fixtures, however, may prove extremely expensive, not just because of high labor and equipment charges, but also because of lost revenue during the nine or ten days the rooms may be out of order. Therefore the professional Housekeeper should be familiar with porcelain refinishing processes and other techniques for reconditioning and remodeling bathrooms within budget limitations that must be imposed if the institution is to be operated on a sound economic basis.

**FIGURE 18-7** *Each bath and powder room at Little America Westgate Hotel, San Diego, features wrap-around mirrors, hand-crafted shower curtains, and solid brass faucets and shower fixtures complementing the decor of each room. Note marble shower walls and wood molding—touches of elegance rarely found in hotels today.*

**FIGURE 18-6** *Bathrooms in Stanford Court, San Francisco, feature tiny TV, hand-milled soap, marble counters and walls, telephone, and electric towel warmer. Somewhat unusual in these days of colored linens are white bath towels, oversized and extra-thick. Soap dishes and ash-trays are fine white china with gold edging.*

Blemished bathtubs and ceramic tile walls can be resurfaced so successfully that all signs of damage are completely eradicated. Resurfacing compounds are applied by professionals in four separate coats that are fused to the original enamel or ceramic to form a permanent molecular bond. The new surface is said to be even more resistant to chemicals, acids, and stains than most original surfaces (Figure 18-8).

Even custom-mixed colors are available so that the original tub can be coordinated with new plumbing fixtures, wall tiles, and so on. The process generally saves about 80 percent of the cost of replacement. It also provides an answer to the problem posed when replacing a tub requires tearing up part of the floor and tile wall, rearranging the piping, and perhaps even having to make compensating adjustments in bathrooms immediately above and below the one in question. Replacing a wash basin, on the other hand, is almost always cheaper than repairing an old one.

Tile walls can also be refinished quite economically with this Thermal-Fusion process. With this method, walls become one continuous surface (grout lines are glazed over, eliminating the need to clean them). There is a year's guarantee on resurfacing—the same as is given with new plumbing fixtures.

## *Re-Covering a Bathtub and Shower*

Just as an upholsterer will re-cover a sofa, so the tub in a bathroom can be covered with a porcelain enamel steel insert, or tub liner, to cover an unsightly old cracked and stained fixture. This made-to-measure liner can be fit over the existing unit in any color specified. Next, a custom-made apron is installed to cover the sides of the worn tub. Finally, seamless sections of custom-cut porcelain enamel wall paneling are mounted directly over existing walls where they are held in place by fasteners that attach to special steel brackets. Joints formed where walls connect are sealed with silicone. Costs for this sturdy cosmetic cover-up amount to about one-third the cost of traditional bath remodeling. Time required for re-covering the tub is about three hours for a two-man crew.

## *Prefabricated Fiberglass Bathrooms*

There is even an inexpensive way to add a complete bathroom in any space that measures at least 6 ft. 2 in. by 6 ft. Prefabricated fiberglass-reinforced plastic bathrooms come in four sections complete with all fixtures, accessories, piping, fittings, even flooring. The entire room has rounded corners that make cleaning easy, and the plastic surfaces are another advantage since they never require repainting.

The original high-gloss finish that makes this bathroom so attractive can be protected by routine cleaning with liquid detergent, soap, or household cleaner. Abrasive cleansers should not be used on these fixtures. Stains from paint, tar, and whatever can be removed with solvents such as lacquer thinner, turpentine, or acetone. Stubborn stains can be removed with bleach or chlorine. Occasional applications of any automotive-type wax will help bring back the original luster and sheen, and will make routine cleaning far easier.

Among gimmicks that glamorize bathrooms and increase the salability of guest accommodations are hydromassage units that make whirlpools out of conventional bathtubs, and portable saunas and steambath units that can be built right into the bathroom with an activating mini-electric steam generator.

Products and equipment to consider for guest bathrooms when planning buildings or remodeling include:

1. A flush valve that reduces noise to a minimum and cuts maintenance and replacement costs
2. Self-cleaning showerhead that saves water, maid's cleaning time, and replacement cost
3. Pressure-equalizing shower controls, based on an average daily use of 25 gal. per minute per 5-minute shower, that provide a significant safety factor in that they prevent steamy blasts or frigid bursts that lead to scalding and falling. Single-handle design prevents the water temperature from exceeding a preset limit (usually 110° F). Guests simply adjust handle to the desired temperature. No need to waste energy heating water to 160° F for such purposes as removing wrinkles from clothing.

## *Cleaning the Public Rest Room*

Cleaning public rest rooms and employee washrooms is quite different from cleaning guest bathrooms (Figures 18-9 through 18-13). For one thing, it must be done more frequently—in some cases every hour—because the public may, in a brief time, make the room very unpleasant for the subsequent user, or actually damage the plumbing equipment. Chores involved should be carefully thought out and delineated for maintenance workers.

The following detailed procedures, schedules, and equipment lists illustrate how a program might be set up.

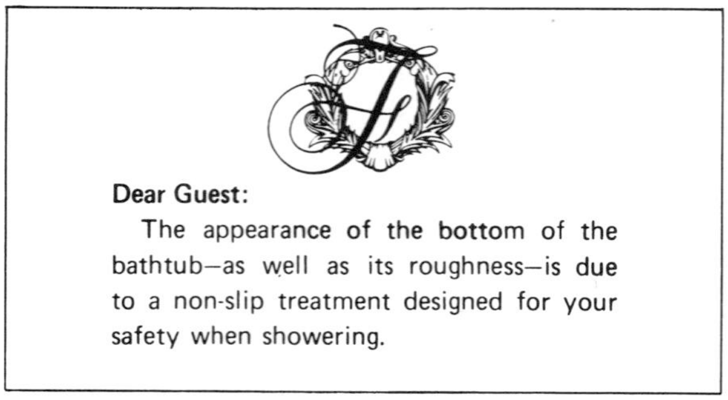

Dear Guest:

The appearance of the bottom of the bathtub—as well as its roughness—is due to a non-slip treatment designed for your safety when showering.

**FIGURE 18-8** *The Fairmont Hotel in Dallas has put a special finish on its bathtubs to prevent slipping. Because it makes the white porcelain appear gray, management decided to have this little card printed. It is placed in the bathroom near the tub.*

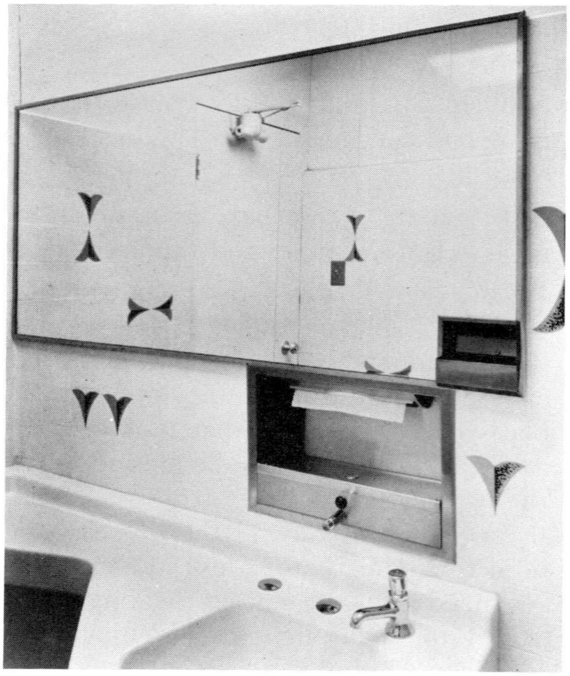

**FIGURE 18-9** *Simple functional rest room in Union Oil Station provides recessed towel dispenser and, opposite it, recessed soiled-towel receptacle. In crowded area, angled porcelain counter provides place for patrons to move in closer to mirror while applying makeup without blocking basin. Credit: Bobrick*

**FIGURE 18-10** *In public washroom an extra roll of toilet tissue is protected in locked rack above roll in use. Maid can open reserve compartment whenever necessary. Napkin disposer is two-sided—reveals another face to adjoining stall, thus saving units and labor required to empty refuse and replace toilet covers. Credit: Bobrick*

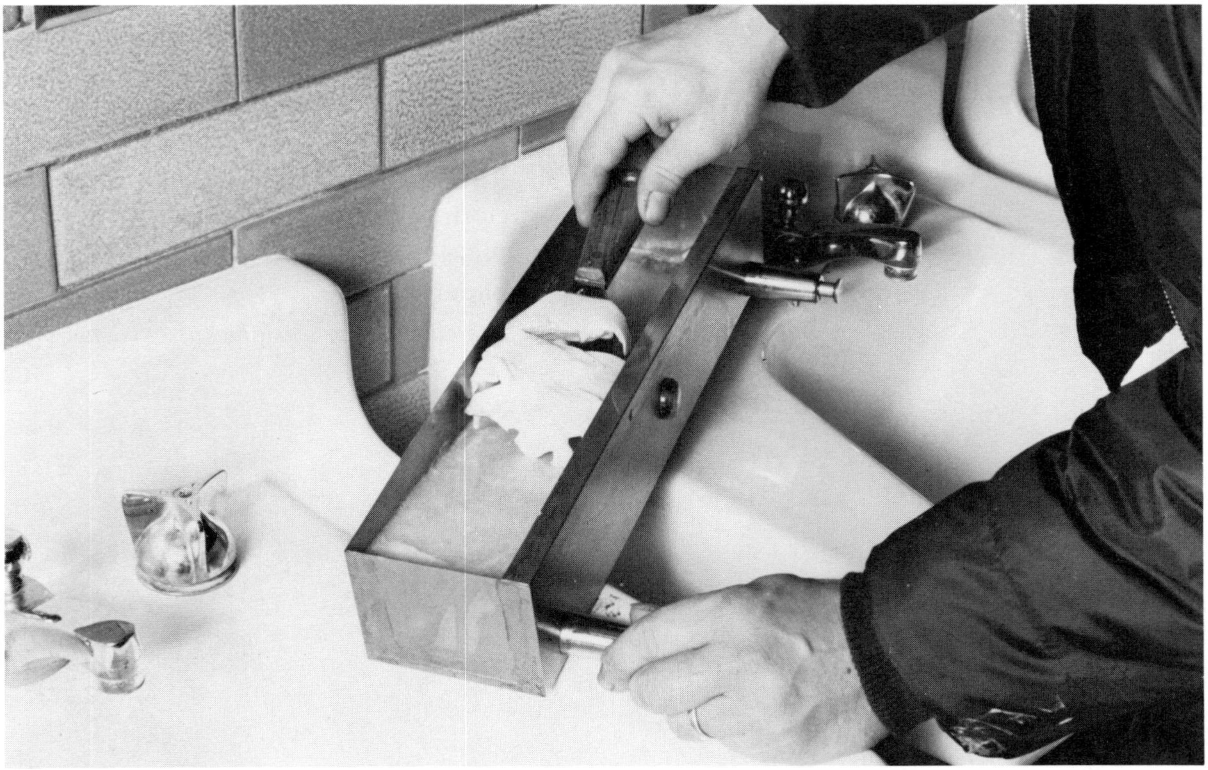

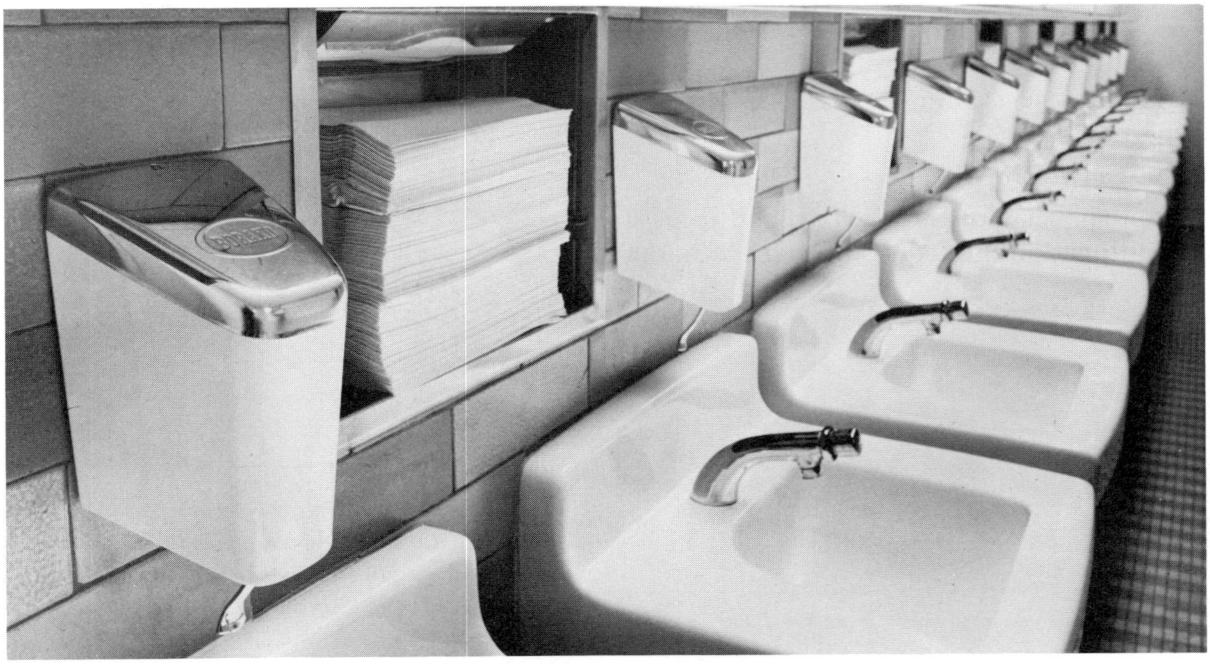

**FIGURES 18-11 and 18-12** *In washrooms at Los Angeles Convention Center, liquid soap congealed in the tanks of the original dispenser openings clogged up and prevented liquid from dispensing. Unclogging, scraping, and cleaning added to high maintenance costs. (Top) Users complained for several years about lack of soap and sticky residue on wash basin until new powdered soap dispensers were installed above each of the 177 basins. (Bottom) Replacement costs were soon paid for in savings effected by lower washroom maintenance. Credit: U.S. Borax*

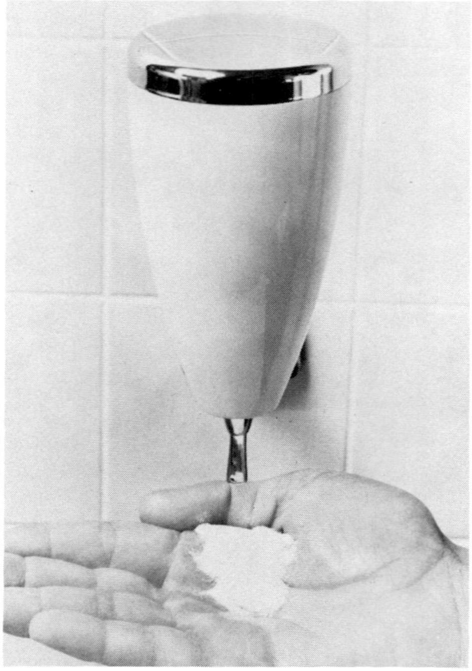

**FIGURE 18-13** *Powdered soap dispensers are preferred by most housekeepers as they do not cause messy areas as do liquid soap dispensers. Powders are available now in a wide range from highly abrasive to gentle sweet-smelling.*

## *Procedure*

1. Refill all soap, seat cover, tissue, and towel dispensers.
2. Empty all trash containers.
3. Put on rubber gloves if desired, and using a cloth dipped in diluted disinfectant solution, damp-wipe all fixtures including mirror, pipes, faucets, and dispensers. Spot-clean cubicle partitions and entrance door handles.
4. Rinse cleaned surfaces, wringing out the cloth under the tap as needed.
5. Polish the mirrors with a lint-free cloth or paper towel.
6. *Use cleanser on wash basins only, not on toilets.*

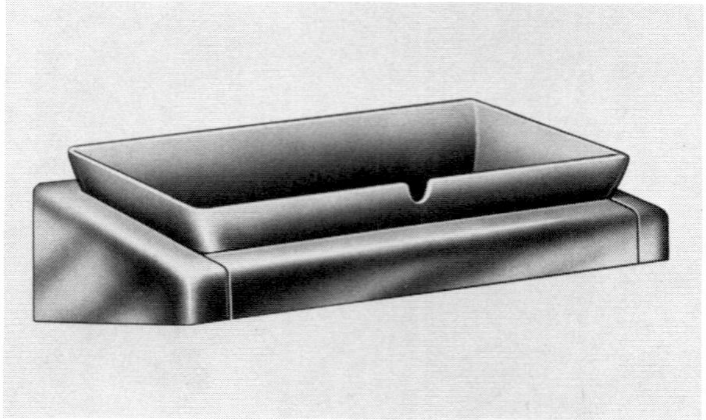

**FIGURE 18-14** *A flippable, stainless steel ashtray was designed to go on the divider in a public washroom because it is so easy to empty. If no provision is made for cigarettes in stalls, they will stain surfaces or set fire to paper or accessories.*

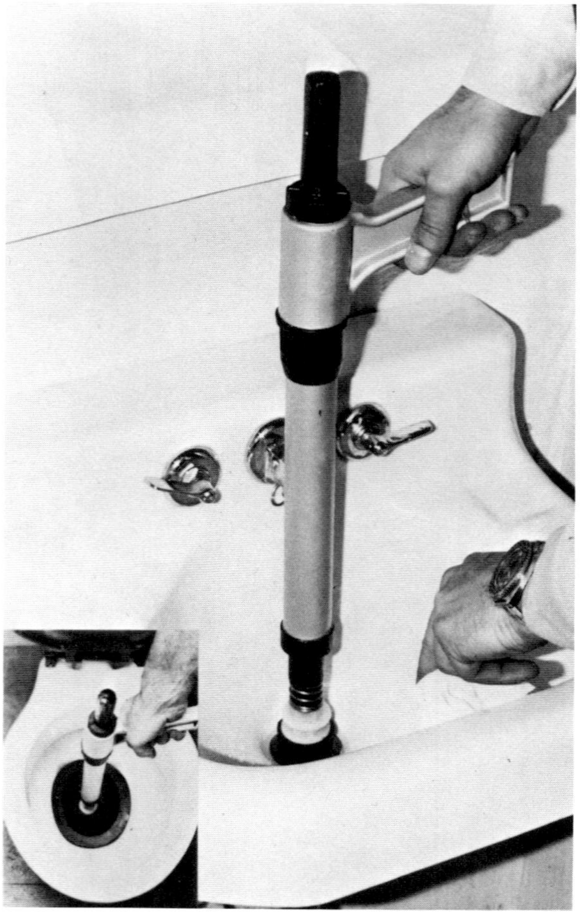

**FIGURE 18-15** *To clear blockage from plumbing, a safe and simple device based on air pressure is now available. In one second an air-gun—the Kyna Colt—clears blocked pipes, clogged toilets, showers, tubs, sinks, and fountains. It works on the principle that water cannot be compressed. It transmits shock waves at the speed of sound, 4700 feet per second. The device triggers the shock impact by a $CO_2$ cartridge that is ecologically safe. This impact follows a column of water acting like a flexible rod. Even though the impact is tremendous, there is little risk to any weak point in pipes or fittings. Government has granted approval.*

7. Using the same diluted disinfectant, immerse an old disposable cloth and wring it out. Wipe down the tops and bottoms of the toilet seats and all exterior surfaces, including pipes. Wipe all exterior surfaces dry with a soft cloth. Wipe the exterior surfaces including pipes of the urinals, and dry.
8. Use bowl cleaner with applicator in toilets and urinals but do not let applicator drip on floor, chrome, and so on. Throw away diluted disinfectant solution in toilet and flush all toilets and urinals (Figure 18-15).
9. Clean and put away equipment. *When you clean your second rest room, use clean diluted disinfectant solution, water, and a clean disposable rag.*

## *Equipment*

The following material must be provided in a tray located in the janitor's closet in each assigned area:

1-qt. container of disinfectant marked with proper formula for diluting

Rags for cleaning everything except toilets and urinals

Disposable rags for washing toilets and urinals

Good rags for wiping dry—if lint-free also use for mirror, otherwise use paper towel

1 can cleanser

1 qt. liquid hand soap

1 qt. creme cleanser

1 qt. bowl cleaner with carrying caddy and swab

1 spray-bottle with glass cleaner

1 pair rubber gloves

1 2-gal. plastic bucket

Towel supply and key for waste container

Liquid hand-soap dispenser key

*If any pieces of equipment or supplies are missing from your rest rooms report them to your supervisor—you will be held responsible.*

Janitor supply closet *must* be kept clean at all times (floor mopped—well stocked—sink clean—walls and doors spotted). Remove any material or equipment not listed for rest room cleaning.

## *Schedule*

Each night of the week special cleaning must be given to a particular portion of the rest room (Figures 18-16 and 18-17). Activities are added

**FIGURE 18-16** *Girls' bathroom in Residence Hall, serving Cal Poly students at San Luis Obispo, Calif., has easy-to-clean counter of plastic laminate. Drawer bases are recessed to eliminate scuff marks. Continuous vinyl sheeting on floor is water-tight. Showers, tubs, and toilets are in individual stalls. Students even put artwork in washrooms.*

**FIGURE 18-17** *Parlor maid at Century Plaza Hotel restores order to powder rooms every few hours. She replaces tissue, cleans ashtrays, basins, and toilets. Supplies are kept in closet or, in some cases, immediately adjacent to ladies' room so that she does not have to carry equipment through lobby. If no closet is available, locked "box seat" will hold cleaning items. Attractive basket (with disposable foil lining) artfully conceals refuse that maid will dispose of.*

to the schedule as follows (item numbers refer to duties described under Procedure):

| | |
|---|---|
| *Monday*-after item 5. | Polish all chrome with creme cleanser and soft cloth. |
| *Tuesday*-after item 3. | Instead of spot-cleaning cubicle doors and partitions, wash completely and dry with soft cloth so as not to leave streaks. |
| *Wednesday*-after item 6. | Specially scour and polish sinks, including undersides. Clean toilets and urinals thoroughly. |
| *Thursday*-after item 3. | Instead of spot-cleaning entrance doors, if walnut, polish with prescribed polish. |
| *Friday*-after item 3. | Wash all tile walls. |

## *Floor Cleaning*

**Equipment**

- 1 bucket with detergent-disinfectant solution and 2 gal. water
- 1 wet-mop for applying solution
- 1 bucket with mop wringer attached for rinsing
- 1 wet-mop for clear-water rinsing
- 1 putty knife
- 1 dust-pan and counter-broom
- 1 squeegee with handle
- 1 12-in. push-broom

**Operation**

1. De-gum floor with putty knife whenever necessary.
2. Sweep floor with 12-in. push-broom. If floor has much debris, gather it up with dust-pan

and counter broom, deposit in trash container.

3. Clean floor beginning at far left hand corner. Do not wring out mop, let excess solution drip off into bucket and apply very wet. Return mop to cleaning solution. Squeegee floor, pushing excess solution down drain.

4. Using clear water (from bucket with mop wringer), damp-mop as dry as possible, picking up any solution that is left.

5. If not too soiled, detergent-disinfectant solution may be used in second rest room.

6. Once a week dump one bucket of solution water down floor drain to prevent odor.

# Chapter 19

# Equipment and Supplies

**OVERVIEW** Establishing sources of supply is important for the professional Housekeeper. Mechanical equipment poses dangers unless personnel are well trained. Here the reader will find a kind of shopping list of all the machines, materials-handling and waste-handling equipment, tools, containers, guest room supplies, and cleaning agents, and a detailed description of the chemicals used in today's Housekeeping Departments.

The assortment of products used by Housekeeping is by far the largest used by any department in the institution. To become expert on everything from the best steam-cleaning machine to the finest quality down pillow is almost impossible; however, prospective professionals should become accustomed to the idea that there are many different types of products about which they will have to make decisions—whether for a club, a hotel or a residence hall. Finding knowledgeable and reliable suppliers is the first step toward making wise selections.

## Sources of Supplies

Janitorial supply houses in large cities have goods in every category that Housekeeping might want—heavy floor machines, bedding, guest room supplies, as well as many cleaning chemicals. In smaller communities, however, to keep abreast of what is new it may be necessary to work with a greater number of suppliers in more remote locations, attend trade shows frequently, and study trade journals carefully.

The professional Housekeeper should make sure her supplier understands the level of her employees' intelligence. In a great many properties the housekeeping personnel who will use machines, equipment, and cleaning chemicals will be less educated, less likely to understand English, and less likely to follow instructions in their effort to get the job over with, than blue collar workers in other industries.

## Training Users

Because new machinery and chemicals can be dangerous (even though more satisfying because they make the job easier), it is of utmost importance that employees be trained to use them. Most companies that sell products train users but one should see that this training is more extensive than a demonstration. It should involve actual use by the operators of the machine or the users of the chemicals. The workers' actions should be observed by the trainer to ensure that the products are being handled properly.

Training sessions should be followed up by a review *on location* in places where the work will be done. On-the-job situations often present different problems than what is shown in the ideal or contrived demonstration area. (As is mentioned in the chapter on purchasing, this kind of training session should be written into the purchase contract.)

The products used by a Housekeeping Department can be divided into four categories: (1) equipment, (2) tools and containers, (3) supplies, and (4) chemicals.

## *Equipment*

There are many challenging aspects to Housekeeping's ownership of large expensive machines (Table 19-1). First, there is the problem of where to put them. They must be stored where they are protected—not just standing in a hallway waiting for a thief in an official looking uniform to make off with them. Locked storage is indispensable for floor machines, wall washers, and so forth.

Second, the equipment must be operated by responsible people who not only use it according to instructions, but are also careful to maintain it properly—greasing, replacing parts, or whatever is necessary. Finally, a heavy, costly machine should not be considered a toy for other persons who have not been trained in its use to try just for fun.

An equipment inventory must be kept that will include purchase dates, sources, prices, and other pertinent information. This record may be a copy of the purchase order or purchase requisition that the Housekeeping office retains, or it may be a carbon of the bill of sale from the supplier. The record should also include when repairs were made and what those cost. This will help determine whether that brand of machine performed properly and should be repurchased.

To be sure about ages and numbers of machines, it is advisable not only to date the equipment itself, but to include the number of machines bought during that particular transaction. For example, if "76-2-6" were stenciled on or marked in permanent ink in an inconspicuous place this would mean that in February 1976, six of these vacuums were purchased.

There are three types of equipment that we will consider here: machines, materials-handling equipment, and waste-handling equipment.

### *Machines*

In the housekeeping world, floor machines by far outnumber all the other mechanical workhorses. As you discovered in Chapters 10 and 11 on flooring and carpeting respectively, there are many different versions of each floor machine—vacuums, buffers, wet-dry pick-ups, steam-cleaning machines, high-pressure sprayers, dry-foam shampoo machines, carpet-sweepers, sweeping machines. But there are other machines—wall washers, fogging machines, insecticide sprayers, all-purpose vacuums for furniture, draperies, and vents (Figures 19-1 and 19-2).

**TABLE 19-1** Suggested Number of Machines Needed Based on Number of Rooms

| Machines | Number |
|---|---|
| Guest room vacuum | 1/room attendant (plus 6–8 spare) |
| Shampoo machine | 1/400 rooms |
| Extraction machine | 1/400 rooms |
| Wet and dry pick-up | 1/200 rooms |
| Back-pack vacuum | 1/200 rooms |
| Carpet-sweeper | depends on use |
| Floor-scrubber | 1/kitchen |
| Pressure hose | 1–2 depending on building |
| No-motor vacuum to pick up trash in public area quietly | 1 |
| Space vacuum | |
| For banquet rooms, lobby | 1 |
| For halls | 1/400 rooms |

The machines in Sewing Rooms and Laundries are usually easiest to keep track of and maintain in good condition as they stay in one place and are usually operated by the same persons daily. All the other machinery is used everywhere in the institution—and alas, sometimes "nowhere" if the operator is not alert.

### *Materials Handling*

This is anything from modular shelving on wheels to a mechanized vehicle such as a golf cart. The latter, one can see from Figure 19-3, is actually the maid's cart used in such sprawling tropical paradises as Kona Village on the island of Hawaii.

Most frequently encountered items in this category include: maids' carts, linen carts or bins, and housemen's carts. There are several types of dollies designed specifically for moving such things as trash cans, pails, and furniture. There

**FIGURE 19-1** *Using centrifugal force tank-type vacuum with many attachments, specific jobs can be quickly done, such as cleaning a lampshade with a dust-brush. Vacuum takes air in, drops dirt on bottom of container, and clean filtered air exits to room. Credit: Filter Queen*

**FIGURE 19-2** *Using same tank vacuum attachments, upholstery can be quickly cleaned with suds concentrate that changes into dry suds to remove stains from upholstery. Suds are made with special attachment that feeds liquid into nozzle and produces foam that is applied to fabric; then separate brush loosens soil. Credit: Filter Queen*

are also steel and wood hand-trucks useful for moving stacks of boxes, and platform trucks for transporting sizable items.

Maids' carts, as discussed in Chapter 8, must be sturdy enough to carry a great many things; not only all the heavy bed and bath linen, but also heavy-duty laundry bags and trash bags that are quickly filled with additional weight. In addition, there are baskets of cleaning supplies, containers with guest giveaways, and often a vacuum cleaner. By the time the cart is loaded it is extremely heavy. Thus when they are purchased consideration should focus on their being lightweight, with large well-engineered wheels, factors that will affect carts' maneuverability. Next, one should consider whether shelf areas can accommodate all the maids' supplies, and if there is enough room on the ends of carts for suspended bags and the vacuum (Figures 19-4, 19-5, and 19-6).

Carts are usually designed with closed backs so that they look neat standing in the corridor and do not expose goods on the aisle side where they can be seen by guests or employees. If carts do not have closed backs, it is wise to use vinyl covers, both to keep supplies and linens clean and to discourage people from helping themselves.

Room attendants' carts today seem to be very heavy and the many items they hold are often tempting to the guests as well as employees. Storing these giant carts takes time and space, and they are a hefty investment for any property. Because of all these drawbacks housekeepers still consider the idea of letting room attendants work with a caddy or basket for cleaning supplies, and carry linen for a few rooms.

Having a cart saves time; in some facilities one more room per day can be done with the time previously taken going to and from Linen Rooms.

A guest is never surprised to find a maid in his room when there is a cart outside. If that big mobile "billboard" is not there to warn guests then maids will have to carry a sign to put on the door alerting guests that "Room is Being Cleaned."

For many years laundry and trash bags were made of heavy washable white duck or canvas. They showed soil and often ripped out around the grommets because they supported a great deal of weight. Also, rips and stains were usually present even if the bags came straight from the laundry. Today, a heavy-duty nylon vinyl material, double-laminated to withstand hard use, is

**FIGURE 19-3** *Instead of maids' carts, Housekeeping at Kona Village on the big island of Hawaii uses electric vehicles on which two maids ride. They clean 20 rooms (actually separate houses) daily. Each house has individual locked supply closet so maids need carry only linens.*

used for laundry and trash bags. They are easily washed or wiped clean and will not mildew or retain unpleasant odors. Since they are available in colors, they can be keyed to the color scheme in the corridors or simply selected in a color that helps make carts inconspicuous. Nylon mesh laundry bags are the latest innovation, as laundry can be washed in them and sorted out afterward, saving double handling.

Some maids' carts are designed to fold in half for storage. The end or ends that carry the bags are hinged so that when bags are removed the structure will fold up, thus saving considerable space in floor linen rooms—crowded places at best.

As was pointed out in Chapter 12 on Carpet Care, one maid's cart has been designed with a high-powered vacuum built right into it. This relieves the maid of the physical task of lifting the upright vacuum off the cart and pushing it around the guest room and then lifting it back onto the cart again. This built-in unit is probably one of the most useful advances in equipment for housekeeping. With the exception of chemi-

**FIGURE 19-4** *Wire maid's cart, although perhaps less attractive when loaded than those with solid metal back, has the advantage of being lightweight; open design leaves no place for dust and lint to collect. Shelves are adjustable; finish is baked-on epoxy to prevent rusting. Credit: Metropolitan Wire Goods*

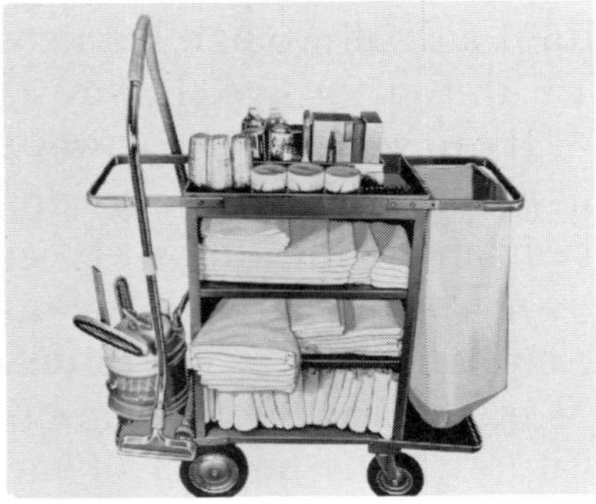

**FIGURE 19-5** *Compact maid's cart carries tank-type vacuum and attachments so that maid is prepared to do general clean any time. Lightweight unit can easily be lifted from cart. Trash bag that normally hangs at end is not needed if maids bag trash in plastic wastebasket liners. Credit: Filter Queen*

**FIGURE 19-6** *What looks like a wire animal cage is actually a convertible linen truck that can be a bin for soiled laundry (as shown above) or a three-shelf cart when hinged shelves (in vertical position here) fold back. Aluminum tubing on sides and back lighten cart; rubber bumpers prevent damage to walls. When used as laundry bin, unit is easily emptied at bottom. Credit: Metropolitan Wire Goods*

cal cleaners and a few floor machines, very little else has been invented recently to advance the profession.

Most of the carts used in an institution should have bumpers or some other resilient guard on the front to protect corridor walls. Wrap-around bumpers are often optional extras, and well worth the added cost.

Cleaners' carts used for public areas usually have space for a broom and a 22-gal. waste receptacle or mop bucket.

Large casters with brakes are highly useful on a heavy cart. These locks are operated by foot and the positions vary from rigid to swivel, and back again by touch of the toe. The wheels can be locked in any of four positions. This is especially useful if a small woman happens to have a heavy cart with four or five shelves stacked high with linens. It can very easily get away from her if there is any slope at all in the corridor—as there well may be in an old institution's basement floor. Also, she is not burdened with the task of holding the heavy cart while trying to unload it.

Today's metal shelving units are flexible and can be set up with any number of shelves at the desired height. It is possible to keep shelving on hand for whatever storage situation arises or whatever size and configuration is needed. Mobile shelves are especially useful when housekeeping linens must be transported from the in-house laundry to the linen storage or the floor linen rooms. No extra handling is necessary; the maids remove linens from the mobile units (previously loaded in the laundry) and place them directly on their carts.

For collecting soiled linen or trash the fiberglass pick-up carts have proved satisfactory as they do not chip or fade and can easily be washed out. These carts also have large 6-in. casters that enable one to move them easily. These are not only very useful for housemen when picking up soiled linen on floors, they are also highly functional in banquet areas where they can be used to pick up soiled banquet cloths. If carts are lined with discarded mattress pads they can also be used to carry dirty dishes to the dishwashing section. Those with 26-bu. capacity are big enough to hold many dishes so that banquet stewards do not have to make numerous trips back to the dishwashing area.

## *Waste Handling*

The subject of modern waste handling is so complex that it would be impractical to take more than a superficial look at the various alternatives here. This is a serious and important topic and it is imperative that it be studied in every institution, unless administrators are so farsighted they have already arrived at a logical waste-removal program that will work for every department and relate to the commmunity's systems.

The professional Housekeeper will study this subject and try to make logical suggestions as to how it should be handled from Housekeeping's point of view. Such problems as the polyethylene cleaning bags that guests throw away (which will not go down the laundry chute) should be outlined so that management can begin to arrive at a system that will work throughout the establishment. But there are many other departments that will have even more significant input, for example Foodservice, or Buildings and Grounds.

There may be many things being thrown out that should be saved (e.g., fat in the kitchen is valuable, as are aluminum, glass, and tin). Perhaps the Housekeeping Department has something to dispose of that could bring in a bit of revenue—consider the left-over guest soaps, the lost and found articles that are never claimed, discarded furnishings, or similar items.

Because trash receptacles are bulky and take up valuable space, the folding metal refuse receptacle frame that supports a reinforced nylon bag is practical. It is on casters and folds flat when not in use, yet its framework opens up to hold a 6-bu. capacity bag. This can also be used for soiled linen or laundry.

The system or device used to dispose of the total refuse from an institution will, of course, determine just what course the Housekeeping Department should take to dispose of its refuse. Perhaps it would all go in plastic bags, mini-compactors, or directly into pneumatic chutes. If Housekeeping needs its own machine for reducing the bulk of its waste before delivering it to a central disposal area, it might be well to consider one of the compactors that takes a bag full of trash and crushes it down to one-third its previous size. There are small portable units that might be used throughout the institution that will accept many throw-away items, including large boxes.

If no system has been decided on by management, it is well for the professional Housekeeper to consider the matter seriously and be responsible for making suggestions on devising a suitable one for the entire institution. New types of incinerators not only can burn all refuse completely without any air pollution but can, as a bonus, provide enough heat to produce hot water or heating for the facility. These new heat transfer systems are the least damaging ecologically and they are a bright new hope amid the gloomy energy shortages.

## Tools and Containers

Housekeeping has come a long way from the broom-mop-and-pail because of chemicals and high-powered machines; yet it is surprising how many of the same tools used in great-grandma's day are still being produced, although with significant improvements. Some of the old improved standbys, still available because no suitable substitutes have been developed, are:

1. *The whisk-broom.* For use with a dust-pan to get up bits of floor debris too large for a vacuum; it is also still best for sweeping out armchairs in dining areas where it is not practical to go in with an upholstery vacuum after each meal.
2. *The treated, washable dust-mop.* For areas too small for a wide push-broom.
3. *Push-brooms.* For areas where a floor machine is too noisy and the soil does not warrant mopping.
4. *Chemically treated cleaning cloth.*
5. *Buckets.* Today these are of plastic, so they are lightweight; they may have two compartments—one for detergent, the other for rinse water. Many have wheels attached or fit on detachable platforms with wheels. Many have wringers built in that can be operated by foot to save bending over and squeezing out mop heads.
6. *Mops.* Still not glamorous, they are now designed with looped ends to prevent yarn from untwisting and linting in laundry, and headbands made of nylon mesh to eliminate mildew and rot; disposable mop heads are also available.

It is easy to see that one cannot simply order "a mop" today; there are choices and decisions to be made just as when selecting expensive machines. For example, should the mop head be "slipthrough," "breakdown," or "tie-on"? How many ounces should it be and is a double tailband needed? Is the mop top cellulose and absorbent enough for the jobs to be done? Is a cotton top needed?

It is not the task of this book to explore the benefits and disadvantages of every aspect of each tool that will be needed by the many workers in a Housekeeping Department. It is only important to explain here that selection and purchase of all these tools will take time. Before purchasing or reordering, one should confer with the user of that tool to see how it performs—what he likes or does not like about it; what improvements could be suggested.

Beware of individual employee bias, however. A maid may think a gong brush is an inefficient tool, far too stiff for washing grout in bathroom tile or for cleaning out shower door tracks, but in reality she does not like those jobs for which she must use the gong brush because they require elbow grease. For her a feather duster or an aerosol can is much more fun.

When interviewing employees about their tools, it is wise to ask both the least enterprising, nonenergetic ones, as well as the efficient productive ones because the way a tool performs depends on the person. If it is difficult for most employees to use, then perhaps a change should be made.

The squeegees, the putty knives, even the miniature tools that maids' supervisors carry with them, all require research and evaluation. The professional Housekeeper will be able to make wise purchasing decisions if she listens to employees and supervisors and has a knowledgeable, honest supplier.

## Supplies

"Supplies" is an umbrella term that encompasses a heterogeneous mix of items that can perhaps be classified as guest room giveaways, guest room supplies, cleaning supplies, electrical and mechanical supplies, and extra furnishing supplies.

### Guest Room Giveaways

These include:

| | | |
|---|---|---|
| soap | laundry bags | plastic utility bags |
| shoe mitts | stationery | disposable slippers |
| shower caps | pens | emery boards |
| bath beads | note pads | post cards |
| sewing kits | magazines | table tents |
| matches | books | portion packs of coffee |

Some luxury establishments include candy and/or fresh flowers; Oriental-style hotels may include a kimono or ukata and tabis or zoris or paper slippers.

Soap is number one giveaway item from the Housekeeping Department. Most wise housekeepers will not provide anything smaller than 1-oz. bars as it often irritates guests to have to bathe with a "sliver;" ¾ oz. is the minimum size to be considered. Usually, better hotels provide both regular facial soap and a deodorant soap.

Shoe mitts, either flannel or disposable, are an important giveaway and should be placed in a conspicuous spot so that guests will not seize on washcloths as shoe polishing rags. Almost every other item on the giveaway list is dispensable, though it is almost inevitable that a guest will call Housekeeping if one or more of them are not in the room.

### Guest Room Supplies

This list includes:

| | | |
|---|---|---|
| hangers | bath mat | Do Not Disturb signs |
| toilet tissue | glasses or cup dispenser | telephone book |
| plastic tray | ice bucket or water pitcher | ashtrays |
| facial tissue | coffee maker and cups | wastebasket liners |
| Bibles | | toilet seat bands |
| shower mats | | fly swatter |
| towels | | |

These items are supposed to stay in the room or be used in the room, but they are often taken as souvenirs, so maids must carry replacements on their carts.

How many hangers and what type are unsettled questions in the hotel business. In residence halls there will never be enough for the women's quarters, so students are often left to their own devices on this matter. In better hotels, where wooden hangers have a smooth waxed finish and are imprinted with the hotel's name, they will cost about 25¢ each, and it is very expensive to keep a dozen or two in each closet—an open invitation for guests to help themselves. The knob head type fits into a slot on a ring that is permanently attached over the rod in the closet. Although these cost slightly more than conventional hangers, they will not disappear as they cannot be hung elsewhere. To satisfy the requests for extras, wire hangers can be kept in stock or can be obtained from an on-premise guest laundry or cleaner.

Nonskid shower mats versus safety treads in the tub or shower are also debatable. The more economy-minded housekeepers who do not have to maintain a 5-star image will settle for the practical safety tread that is affixed more or less permanently to the bottom of the tub or shower. This requires no extra work on the maid's part, whereas the shower mat that collects soap residue and hair must be cleaned after each use to prevent mildew and to assure guests that all is sanitary.

Glasses to be provided in bath or guest rooms are another matter that will be argued pro and con by management. The options are to offer a clean (real) glass, sparkling and uncovered; have it wrapped in cling film or packaged in a paper holder; to use a plastic glass packaged in one of the ways described.

The problem, of course, is to please the guest. Those paying a high price for the room want a real glass and some feel that having it wrapped is no assurance that it is clean; seeing it shining and spotless is more convincing. Others feel secure if it is "sanitized" and in a proper cover; others care not a bit one way or the other. Thus the question becomes what is the most economical way to handle it?

For maids to have to place real glasses in a rack and take them to the Foodservice Department to be washed in the dishwasher is complicated and involves scheduling and transport. Having a small, just-for-this-purpose glasswasher in the Housekeeping Department (or in some centrally located linen room) also poses logistical problems. Having the maid wash the glasses in the bathroom in a germicide, rinse, and polish them dry seems easiest.

Housekeepers can only hope, however, that maids use the proper disinfectant solution and a clean cloth or facial tissue to polish the glasses. With all these complex alternatives (to a seemingly trivial matter), it is easy to see why so many institutions simply provide a disposable plastic glass bought prewrapped. Elegance and gracious living cannot often be espoused in this age of high labor costs.

The toilet seat band also gets a guffaw sometimes from knowledgeable hotel or maintenance people. Its presence may help the patron psychologically to feel the bathroom is sterile, but in reality it is no guarantee that the commode is immaculate and germ-free.

## *Items Loaned to Guests*

In smaller operations where there is no all-night switchboard operator, alarm clocks will be the most frequently requested item. Nonelectric, unattractive ones that go off with a disagreeable sound are recommended as least tempting to guests with taking ways.

In any large urban hotel there will be many guests, particularly women, who will want to borrow a wide variety of items such as hair dryers, irons and ironing boards (being built into some guest dressing rooms now), heating pads, hot water bottles, razors, electric shavers, radios, ice packs, first aid kits, foam pillows, and so on. As discussed in Chapter 26 in the section on security, there should be some system for seeing that these loaned items are returned. Most can be easily fitted into suitcases or inconspicuously placed in guests' cars, but a bold mark with the hotel's or motel's name will discourage theft. Storage for these items should be clean and locked, with places for each category marked. The log in the Linen Room will have a record of dates requested and returned.

The most cumbersome items that guests request are bed boards (for people with back problems), baby cribs, and extra beds (called rollaways or folding beds instead of cots, a term that suggests an uncomfortable night's sleep). These all require sizable storage space and all together represent a considerable investment.

Actually, the fold-up beds available today have extremely comfortable mattresses, many even have box springs, some of which have a counterbalanced frame that permits them to stand upright, taking up as little as 16 in. in depth. These are on rollers so that maids can easily take them to floor storage areas. Plastic covers for rollaways in storage are indicated if the extra beds are not often used.

## *Electrical and Mechanical Supplies*

Depending on whether maintenance in the institution is handled by Housekeeping or another department, these supplies may be stored and inventoried by the professional House-

**FIGURES 19-7 and 19-8** *Almost every institution needs special ladders. At Century Plaza, a ladder attached to scaffolding at base with platform atop is used for relamping chandeliers in lobbies and ballroom; specially designed extension ladder is safe, easily movable on dolly (with giant casters that lock). Items like these require extensive storage space planned prior to construction.*

keeper, engineer, director of maintenance, and so forth.

Light bulbs may come in as many as 40 different shapes in a 500-room hotel, especially if the interior designer has ordered a number of interesting fixtures or chandeliers and has used a variety of styles throughout the property. It is usually a houseman in the Housekeeping or Maintenance Department that does the relamping job, so he will need to keep the inventory of different bulbs and collect necessary ones before his night's work. He does this job on the "graveyard" shift in the public areas so as not to disturb guests, and usually in the back of the house too, so as not to interfere with other employees' activities. (See Chapter 9 for details on relamping.)

The extra-long-life bulbs, despite their higher prices, are best for institutional use because they require such infrequent replacement that they will cut down on labor; the costs saved by eliminating one night's relamping will buy a goodly number of the long-life bulbs.

To reach the high places in public areas, ladders are simply not the proper solution. To make it safe and easy for housemen to work at heights ladders cannot reach, portable, versatile scaffolds are available (Figure 19-7 and 19-8). The Housekeeper and Maintenance Director must decide which or how many of these should be owned by the property (as opposed to being occasionally leased). For safety purposes the scaffolds rise automatically (no need to plug in cords). They are of aluminum construction and powered by self-contained tanks of carbon dioxide or nitrogen to rise to heights of 36 feet, but telescope to 30 inches in width by 41 inches in depth so can be handled and stored easily. These are designed to hold one or two people.

Other types of scaffolds are made for instant set-up manually, using safe locking hinges. They can even be adjusted to stand sturdily on uneven surfaces such as stairs.

Casters are another item that Housekeeping is concerned with, both for equipment back-of-the-house and for furniture throughout the institution, including dining chairs and beds.

There will be a large inventory of casters, in many sizes and types—heavy-duty rubber wheel, swivel or stationary, ball-bearing or roller-bearing axles, or the large round decorative ones in various metal finishes. These are used for guest room occasional chairs, and increasingly, in the dining areas to save wear on the carpet as guests pull up or away from tables. Chair glides for function chairs and outdoor furniture have to be kept on hand, as do the plastic tabs that go on the insides of stacking chair legs.

Bottle openers are vulnerable, both because they come loose and because guests sometimes take a fancy to them, so a supply should be kept handy.

### *Cleaning Supplies*

The chemical products used for cleaning are considered later in this chapter and in the Cleaning Chemicals chart, but the products used to carry, apply, or remove these solutions must be selected with care, and stored, inventoried, and issued properly.

Carriers for use on maids' carts are either plastic or metal caddies (or baskets) and contain small guest giveaways and supplies as well as the cleaning items needed, such as toilet swab, all-purpose cleaner, sanitizer, paste cleaner, gong brush.

Gloves are not often worn by maids or housemen but must be available on request or if they are needed for jobs where solvents or highly alkaline or acid cleaners are involved. The thin, disposable, clear plastic gloves will probably not withstand many harsh chemicals so a sturdier disposable or durable washable type should be purchased.

Steel wool may be needed in small pieces to remove stains from floors. Maids may want to carry these and supervisors most certainly need some type of mildly abrasive pad to keep in their mini-tool kits to do minor spotting jobs as they inspect.

Floor-machine pads will have to be kept in good supply and there are many types needed for the disc or rotary machine: polyester, steel wool, lamb's wool, new yarn pads, to name a few. The machine supplier will advise on which types are needed and in what quantities.

Rags are almost always available in established properties since there are usually discarded sheets or towels to supply them. The blend fabrics do not make satisfactory rags so perhaps more effort should be made to retain absorbent fabrics on hand. Having maids and housemen deposit them in specifically marked containers to be laundered daily will help to keep the inventory up. In a new property, one might arrange a contract with an industrial supplier who picks up and delivers cleaning cloths as part of a regular service.

Sponges are preferred by some workers, but surfaces cleaned with damp synthetic sponges usually require polishing with a dry cloth too.

### *Extra Room Furnishings*

In any property that expects to have 100 percent occupancy, it is advisable to have extra room furnishings on hand for use as replacements in emergencies. A few extra beds, dressers, mattresses, chairs, desks, bed tables, are difficult to store, but may be needed when the house is sold out or excessive damage has been done to a suite.

Extra lampshades, lamps, wastebaskets, draperies, spreads, pillows, toilet seats, luggage racks, and shower curtains should all be within easy reach in case something happens to these necessities. Of all these items, the lampshades should probably be in greatest supply because these can be so easily soiled if, for example, a guest opens an explosive pop bottle. Any mark on a lampshade may warrant its being removed to the Linen Room or Laundry for special cleaning in a liquid solvent or a shampoo with detergent, so replacements should be easily available.

As for items placed in the public areas, perhaps there should be a few substitute coffee tables or sofas, usable in living room suites, too. Of course, extra carpet for all areas should be available in case of emergencies; just how much is difficult to predict, but enough to replace carpet at entrances and to be used as "walk-off" matting to clean shoe soles would be advisable.

### *Chemicals*

As we have said before, Housekeeping's biggest boon in the last decades has been the

**TABLE 19-2**

| | |
|---|---|
| **NUMBER 1 DEVELOPMENT IN CLEANING CHEMICALS:** | |
| *THE ALL-PURPOSE LIQUID CLEANER* | |
| *For use on:* | Glass . ceramic tile . wood . leather . vinyl . metals . plastics . painted surfaces . rubber . porcelain . resilient tile floors . linoleum . quarry tile . terrazzo . cement. |
| *Content:* | Concentrated liquid detergent, usually neutral on pH scale of acid to alkaline; some contain small amounts of ammonia but formulas are mild enough not to stain or bleach if accidentally splashed on fabrics. |
| *Applications:* | In varying dilutions—with spray or squirt-bottles, sponge, cloth, brush, or mop. |
| *Packaging:* | Ranges from 1-gal. plastic bottles to 55-gal. drums with pumps for decanting. |
| *Benefits:* | Using one multipurpose product decreases inventories, storage space, cart, or caddy space; consolidates purchasing and bookkeeping transactions; cuts training time. These fast-working liquids remove organic and inorganic soil, deodorize, require no rinsing (except in foodservice areas on counters or in refrigerators). Convenient for routine cleaning, especially for maids who can mix solutions in plastic wastebaskets in guest bathrooms, or carry in buckets on carts. |
| *Disadvantages:* | Not chemically formulated to remove heavy soil build-up (e.g., high acid content bowl cleaner needed for neglected toilets; degreasers needed in kitchens). |

development of chemicals for cleaning, because of the increased effectiveness of the formulas and also because of the reduced number of products it takes to clean guest and bathrooms (Tables 19-2, 19-3, and 19-4). The summary that follows discusses new formulas for more effective and economical housekeeping, maintenance, and sanitation.

One executive Housekeeper was very much in favor of all-purpose cleaners. She tried many of them and found that the nationally advertised brands were highly effective but also very expensive. She advises housekeepers to check with their local janitorial supply houses to see whether they have their own private-label all-purpose cleaner and disinfectant cleaner because she

**TABLE 19-3**

| | |
|---|---|
| **NUMBER 2 BESTSELLER FOR CLEANLINESS** | |
| *THE DISINFECTANT CLEANER* | |
| *For use on:* | Foodservice equipment . kitchen surfaces, walls, floors . shower rooms . rest rooms . bathroom fixtures, walls, floors, counters . rear entrances and trash areas . throughout hospitals and nursing homes. |
| *Content:* | Concentrated liquid detergent usually combined with quaternary ammonium compounds (quats) that control all bacteria except TB. A few manufacturers produce phenol-based detergents that control TB. |
| *Applications:* | Same as for all-purpose cleaners. |
| *Packaging:* | Same as for all-purpose cleaners. |
| *Benefits:* | Products clean, deodorize, sanitize (reduce bacteria by 50%), or disinfect (kill almost all infectious organisms). Most have lasting residual bacteriostatic effect. Quats are nontoxic, nonirritating to skin, noncorrosive. |
| *Disadvantages:* | Effectiveness varies depending on hardness of water. Phenol-based products, although excellent cleaners, irritate skin, leave unpleasant residue in foodservice areas, and thus require rinsing. |
| *Availability:* | All major janitorial supply houses distribute private-label all-purpose and disinfectant cleaners. Almost every national manufacturer of cleaning products has developed one of each of these miracle formulas. As competition is keen to perfect these solutions, users can be sure of finding one of each type satisfactory for their needs. |

**TABLE 19-4** Chemical Cleaners for Bathrooms and Rest Rooms

| Product | Benefits | Disadvantages |
|---|---|---|
| Cleansers | Contain chlorine bleach that kills germs; scours heavily soiled areas | Unless finely ground abrasive is used, surfaces can be quickly worn, especially fine porcelain; employees tend to use too much |
| Deodorizers | Effective where no ventilation or fresh air is available; many kill germs as well as odors | May have an aroma of its own that will be offensive |
| Disinfectants | Discourage some germs, correct some unpleasant odors | Have unpleasant odors or may damage grout in tile or metal surfaces |
| Drain cleaners | Speed up clogged drains in emergency situations | Engineers prefer that this not be done as it may do damage to plumbing or complicate repair |
| Marble cleaners | Make marble shine and discourage spots | Time-consuming to apply |
| Metal cleaners and polishers | Clean and polish in one operation | Must be applied according to directions or may streak and damage metals; delimers or grease cutters often necessary first before cleaner-polisher effective |
| Paste cleaners | Will not scratch surfaces | Require more effort to use than abrasives |
| Tile cleaners | Clean grouting also; usually cream, easy to apply | Expensive; may attack and loosen grout |
| Toilet bowl cleaners | Clean and deodorize, dissolve heavy scale | Irritates hands; may damage surfaces |
| All-purpose cleaners | Only one product needed; most require no rinsing | Different dilution needed for various surfaces so instructions must be followed |

finds they are usually able to provide an almost equally effective product at an appreciably lower price.

Maids can mix their solutions in plastic wastebaskets in bathrooms, thus eliminating need to carry buckets through halls. They use 1½ pints of all-purpose cleaner daily for 14 rooms, cleaning everything in bath and guest room with it. Kitchens are cleaned with the same all-purpose detergent, with degreasers used on ovens, vents, and grills.

## *Dilemma on Dispensing*

Whether the miracle cleaners should be bought in 55-gal. drums and decanted into quart-bottles for use with 2-oz. measuring cups or squirt-tops, or whether they should simply be purchased in aerosol spray bottles is a dilemma (Figures 19-9, 19-10, and 19-11). Many housekeepers who have extra inexpensive labor such as teenagers working after school or in the summers, can economically decant the liquid cleaners from the large drum with a pump into the quart-bottles. Sometimes employees scheduled to work only part of a day can fill small bits of time with the decanting activity.

There are those professional Housekeepers who believe unequivocably that giving an employee an aerosol spray-can will inevitably waste the product because it is so pleasant to apply it—just shaking and spraying. The argument in favor of the aerosol can, of course, is that it saves labor because no handling is necessary. The very fact that it *is* pleasant to use is a plus in terms of keeping employees happy on the job.

Portion-control cleaners (packaged in foil or

**FIGURE 19-9** *If Housekeeping finds it economical to decant its own detergents and liquid chemical cleaners, an upright pump as pictured will be needed. Here houseman is diluting concentrate for use in spray-bottles. There are also metal structures available to hold heavy 55-gal. drums of liquid in prone position but up off the floor so that faucet only is needed, not pump as shown.*

**FIGURE 19-10** *For cleaning mirrors, spray-top bottle that requires pumping is less conducive to wasteful use of cleaning solution than an aerosol spray-can. Plastic bottles of this type are refillable from large drums; aerosol spray-cans must be thrown away—a problem in some waste removal systems. Credit: Economics Laboratories*

**FIGURE 19-11** *Although some housekeepers consider aerosol spray-cans expensive because employees tend to overuse the product (spraying can be fun!), this system for cleaning upholstery in times of emergency is most expeditious and effective—especially if a VIP suite must be made ready immediately and the chair to be used there has an obvious spot. Credit: Argo & Co.*

plastic envelopes or containers that hold just enough to clean one bathroom, or 1000 sq. ft. of floor) are another way to keep products from being wasted. This is more expensive than buying in bulk until the figure of actual usage versus amount needed is eventually revealed. A story has been told about one executive Housekeeper who dramatically demonstrated the advantages of portion control by adding one package of premeasured wax-stripper to a bucket and showing how the floor finish was removed with minimum effort.

Then he added two more envelopes to the same solution. The result: no better cleaning, but more suds, more mopping, and more residue. The staff saw that a little more is not better . . . at least when it comes to detergents.

As to getting a package program underway, an initial control system is recommended until use becomes routine. Issuing the premeasured envelopes on a daily or weekly basis provides an opportunity to acquire more control over job performance in floor or wall washing. For example, if four water changes are desired per day, issue four packages of cleaner and spot-check periodically.

Some firms using portion packs report dollar savings of 40 percent on product costs. This may seem high but it does not even include the time and space savings made possible by premeasured detergents.

When estimating cost of material, remember the difference in concentration. Most purchasing agents understand this and now buy detergents based on cost per usable gallon of solution. For example, if one is mixing scrubbing solution 30 parts water to 1 part concentrate, and paying $3.00 a gallon for the concentrate, one can determine that each gallon of solution costs ten cents ($3.00 divided by 30 = 10¢).

Then check the premeasured product (if a package costs 32¢ and makes 4 gal. of solution, the comparable cost per gallon of solution is 8¢). Thus one can determine which system is more economical.

A test program can be initiated in a limited area or application. Use the portion control cleaner in one building, on one floor, or with one crew. Then after a few months compare the results with the present system.

In addition to the chemical products described in the accompanying charts, another product worth noting is an automated aerosol dispensing system that controls insects, odors, and bacteria. It dispenses mist into the air at set intervals—as often as Housekeeping deems necessary—by a battery-operated unit that protects up to 6000 sq. ft. for 30 days. The fragrance can be one of a number, peppermint, orange, and the like, or a neutral. (It is not so noticeable as the smell from deodorizers installed in commodes that permeate washrooms with a revolting, sweet, disinfectant smell that simply masks other odors.) This is the same unit that is used to stimulate appetites by sending out such aromas as cinnamon apple, or popcorn.

## *Maintenance and Sanitation*

(As outlined in Institutions Magazine Feb. 15, 1980 p. 208.) Actions used to removal soil are physical, chemical, or a combination of both. Physical action may mean:

**Force.** Pressure used in a dishwashing machine or friction from application of a brush are mechanical force.

**Colloids.** Substances put into cleaning agents that attract soil and react with one another to penetrate and remove soil.

**Dissolution.** Water will break the surface tension of some matter like sugars and dissolve the matter from the surface to which it is clinging. Heat makes many materials soften or even liquify, and thus they are more easily removable. Organic solvents such as alcohol, naphthaline, carbon tetrachloride, and acetone dissolve other materials.

**Emulsification.** When oils and fats mix in solution with other liquids they form an emulsified solution. Cleaning agents with emulsifiers in them help remove fats and oils.

Chemical action may mean:

**Chelation.** Chelating agents pick up soil and salts such as lime and hold onto them for rinsing without returning them to the object that was soiled.

**Ionization.** The electrical charge of a molecule is half positive and half negative; such is called an ion. An anionic (negative charge) cleaning agent will attract the cationic (positive charge) in soil to remove the soil. A cationic agent attracts bacteria, making a good disinfectant. A nonionic agent does not separate its charges easily.

**Peptization.** Proteins changed in chemical structure are said to be peptized. Thus changed they become suspended in the liquid solution as colloids and wash away.

**Saponification.** When an alkaline material combines with an oil or fat to form a soap the materials have saponified.

**Sequestration.** A sequestering agent such as a phosphate collects and holds soils and salts away from the surface of the soiled object.

## *Types of Cleaning Agents and Cleaning Problems*

**Alkalines.** High-pH (7 to 14) compounds that saponify fats and oils work by reducing water's surface tension; corrode aluminum.

**Brown Soap.** A high-pH soap used for tough cleaning jobs.

**Buffers.** Agents added to alkalines to reduce corrosion of aluminum, bronze, or brass.

**Chlorinated TSP.** Trisodium phosphate with chlorine—a good cleaner but corrosive.

**Glycerine.** A substance added to soap to give it smoothness.

**Oxalic Acid.** A bleach.

**Phosphoric Acid.** A descaling agent.

**Soap.** A cleaning agent made of a fat and an alkali.

**Soda Ash.** A corrosive detergent that may leave deposits.

**Sodium Bicarbonate.** A mild low-alkaline cleaning compound often used as a buffer.

**Sodium Carbonate.** See soda ash.

**Sodium Hexameta-phosphate.** An efficient chelating agent used to condition water.

**Sodium Hydroxide (Lye).** Very high pH, highly caustic, and corrosive. Hydrolyzes proteins and saponifies grease.

**Sodium Metasilicate.** High-pH silicate that peptizes, saponifies, and emulsifies; inhibits corrosion.

**Sodium Orthosilicate.** High-pH alkaline silicate sucks up moisture; corrosive.

**Sodium Sesquisilicate.** High-pH alkaline that absorbs moisture on solid metals; corrosive.

**Sodium Tetraphosphate.** A chelating agent used to condition water. (Also sodium tripolyphosphate and tetrasodium pyrophosphate.)

**Trisodium Phosphate (TSP).** Corrosive; likely to leave deposits but a good cleaner that rinses well; pH of 10.5 to 12.

# Chapter 20

# Uniforms

**OVERVIEW** To the people who wear them, uniforms mean many different things. For some, they register authority; for others, they symbolize prestige and status. In the lodging industry, uniforms identify employees as well as enhance them, and suggest that an establishment is efficient, prosperous, and conscious of style.

Recent years have seen an evolution in uniform fashions. Hotels, restaurants, and even hospitals are veering away from what is characterized as the institutional look. Today's range of styles, designs, patterns, and colors is endless. Indeed, even a casual glance at any manufacturer's catalog reveals an infinite variety of truly chic, attractive outfits from which to choose (Figures 20-1 and 20-2).

But appearance and style, while important, are not everything. If they are badly designed, uniforms can spoil a hotel's image; poorly tailored ones can impair an employee's job performance, affect his morale, and ultimately hurt a hotel's high standard of service. Comfort and practicality—not to mention durability and easy maintenance—should therefore be prime considerations in uniform selection, because a hotel's reputation depends on its service.

Since the employees who must wear the uniforms will spend a minimum of eight hours in those working clothes, we owe it to them to make their uniforms as comfortable and attractive as is humanly possible. Good-looking, well-made outfits inspire confidence and also a sense of responsibility, as an employee should take as good care of his uniforms as he does of his personal wardrobe. Furthermore, wearing uniforms does save wear on his own clothes.

Uniforms should be comfortable, practical, and functional. They should also be able to withstand frequent launderings. One other point concerning style: the outfit is still a uniform, and the best design is one the employee cannot and will not wear at home. Doing so can result in exposing the garment to greater damage risks. In the end, the uniform represents a greater economy for the employee than it does for the hotel, so it should not be subjected to undue wear or to activities other than those for which it was purchased.

While uniforms may cut down employees' wardrobe costs, the investment accounts for a sizable portion of the institution's budget. Where expense is a major factor, many institutions rent certain uniforms. This is worthwhile primarily in areas or departments where turnover is common.

Uniform rental saves storage space required when stocking different sizes and quantities. Because most rental firms take care of the laundering, maintenance, and repair of the garments, renting can save money. Purchase is a capital investment, rental is an expense. Many hotels prefer not to tie up costly capital in items such as uniforms that must be amortized in a year or two.

Custom-designed uniforms generally reflect a hotel's taste and desire to preserve its own individuality and thematic concept. However, unless the order exceeds at least ten dozen outfits of the same design, the cost can be devastating. Also, custom designs take a great deal of lead time before orders can be processed. This applies to reorders as well. Custom-designed uniforms

**FIGURES 20-1 and 20-2** *Uniforms such as tunics, smocks, jumpers, and pantsuits (above) are popular for housekeeping personnel. Outfits that allow easy movement at the neck, under the arms, and across the back and front are important for waitresses constantly on the go.*
*Credit: Shane, Angelica*

often may require special care and maintenance and that, too, can be costly.

More and more luxury hotels, restaurants, and fast-food chains are going in for distinctive images. For a while, manufacturers' answer to this was custom-designing almost exclusively. While an enormous amount of custom work is still done, now in-stock selections of wide assortments make this no longer necessary. That's good for the customer too, as he can order in smaller quantities and get faster deliveries.

Since turnover and promotion within hotels are frequent, there is the problem of determining what sizes to order. Larger sizes are sometimes more expensive—which is another cost consideration to file away for the budget. The standard guideline used by many goes as follows:

- 28–42 for maids
- 6–14 for waitresses
- Medium/large for engineers, janitors, and parking attendants
- Small/medium for housemen
- Small/medium/large for stewards and runners

Three uniforms per employee is considered a minimum order, but the wise buyer requests extra lots for replacement.

## *Style—The Aesthetic versus the Practical*

If they place excessive emphasis on concept and image, designers tend to get carried away in creating uniforms, putting their priority on appearance rather than function. The fact remains that most employees who will wear the creation want comfort first. They must still perform their jobs efficiently, and to do so they need freedom of movement in garments that should not be confining.

Those whose jobs entail carrying heavy suitcases, balancing food trays, or any amount of reaching, lifting, and stretching—waiters, waitresses, bellboys, maids, maintenance men—need

**FIGURE 20-3** *Simplicity is an eye-catcher particularly if uniform falls loosely and evenly, has a smooth crisp appearance, and added bonus of deep, easy-to-reach pockets. Concealed front zipper offers easy slip-on and slip-off; cuts down employee's changing time. Credit: Angelica*

ample stretch room under the arms, across the back, around the chest and waist. Clothes that do not have that "give" impair worker's movements and can reduce job performance and efficiency.

Pocket, collar, belt, or sash designs also need careful consideration. Pockets are to restaurant and hotel employees what jackets and ties are to executives—a necessity. They should be roomy enough for a waitress to slip checks into easily and sturdy enough to hold tip change. Straight pockets are best in this instance, preferably on the side, since checks can be crumpled in front pockets when waitresses lean over to clear tables or serve food (Figure 20-3). Slanted pockets sewn in the side seam are suitable for hostesses who rarely have occasion to use them in the course of their work, but it is difficult for waitresses to maneuver checks into them.

Snaps under a collar and at the waist help hold smoothness and prevent puckering or sagging. Tight fitting or stiff collars are not as appropriate for busboys as they might be for waiters, but it is difficult for a waiter to function comfortably if his collar is too stiff.

Housekeeping maids need loose-fitting outfits because they do a great deal of stooping, bending, and reaching while cleaning (Figure 20-4). Their physical labor will be more tiring and frustrating if they must constantly smooth or straighten out their uniforms. Beltless zippered

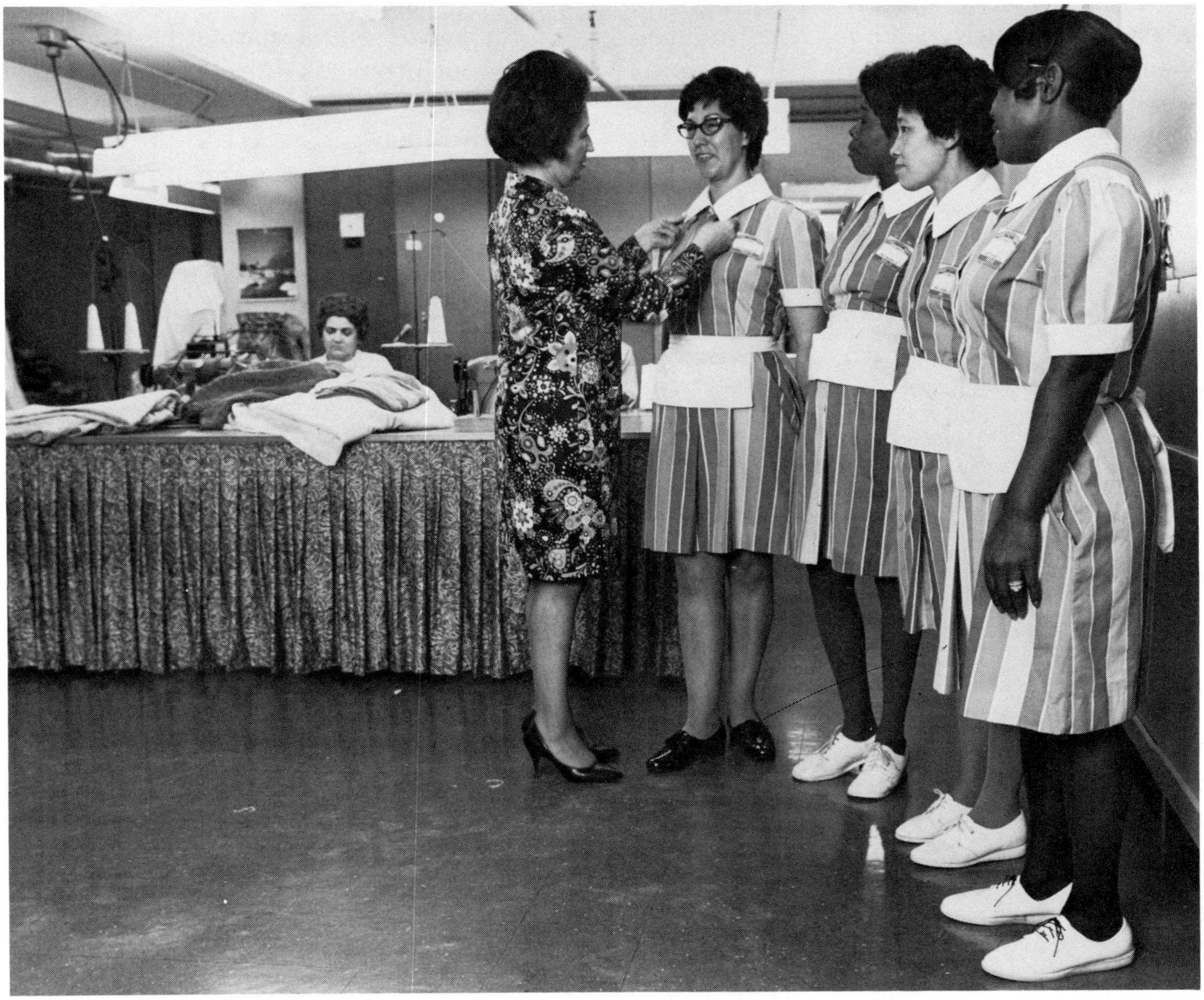

**FIGURE 20-4** *While checking to see how the maids look in their new uniforms, San Francisco Hilton's director of Housekeeping, Henrietta LaBorde, has a personal comment for each of her 200 newly attired employees. Sewing Room personnel hem uniforms; hotel launders them; each maid has two, with time clock numbers on size tags for identification. Rainbow-striped uniforms are part of rainbow theme in all Hilton properties.*

dresses with short raglan sleeves are probably the most comfortable and practical for them, and for restaurant waitresses as well.

Many hotels have switched to pantsuits for both restaurant and housekeeping staffs because they offer comfort and ease plus the bonus of color and style, which elevates morale. The main problem here is that not all women's figures look attractive in pantsuits. An alternate choice might be tunics, smocks, and jumpers, which in some cases can be purchased in only a few sizes because they *are* loose fitting and can be worn comfortably and attractively by large and small figures.

Other refinements for uniforms include concealed step-in zippers that permit women to slip in and out of garments quickly and easily, and without smearing their make-up or disturbing their hair. One brief stroke with a zipper beats fumbling with snaps or—Heaven forbid—buttons! These are time-consuming, and added labor cost may be involved if an employee's changing time is included in her hourly rate. Sewing buttons on is an unnecessary activity for the Sewing Room or the employee.

**FIGURE 20-5** *Big pockets for checks, pencils, and tips are musts. Bellboys, busboys, and Room Service employees wear uniforms that distinguish them from bell captains, waiters, and others. Durable, easy-to-launder jackets and pants come in a variety of colors and designs. Ribbing and insignias add flair, are optional extras. High round collar eliminates need for wearing a shirt under the jacket. Credit: Shane, Angelica*

Belts enhance a uniform's appearance but they are a tricky accessory, easily lost unless securely fastened to the uniform. Tunneled draw string belts should be knotted at the ends to prevent them from getting lost inside the folds of the garment. Belts on maids' uniforms really impede their movements.

Easily alterable skirt lengths are another requisite when buying uniforms. It is highly unlikely that stock uniforms will ever be a perfect fit for all; and employees or in-house seamstresses should not have to use needle and thread more than necessary, or struggle with ruffles, double stitching, fringe, and such.

Shoes are seldom part of basic uniforms because foot problems—like shoe sizes—come in a wide variety. Most employees function better in their own footwear, especially if they are on their feet all day. Custom shoes run into money, particularly if there is a turnover problem. Housekeeping employees should not wear sandals, thongs, houseslippers, or scuffs. High platform soles or heels are also forbidden.

Men's uniforms should have reinforced side slots that expand the openings for easy access to pockets (Figure 20-5). Waiters' jackets need at least one, and preferably two, roomy inside pockets for pencils and checks. Convertible collars fixed with permanent stays are preferred neckwear. Waistbands should be reinforced. Eton jackets worn by waiters and busboys should have interlined collars and fronts for extra body and wrinkle-resistance. Cuffs, decorative embroidery, and epaulets are attractive, but there is the possibility of their catching on things and interfering with performance. Such accessories are also susceptible to abrasion, fraying, or simply coming loose. These extras not only cost more to purchase, but to clean as well.

## Fabrics

Comfortable, functional, attractive uniforms are a big investment and the cost does not end with the purchase. There is also a financial burden of maintenance and replacement. Fabric selection is critical and deserves careful consideration in any buying decision. Modern, permanent-press fabrics have proved that they stand up for a long time if cared for properly.

The use of 100 percent polyester uniforms has increased greatly. Both knits and texturized woven fabrics offer fine appearance, wrinkle-resistance, and good draping qualities. A soil release that adds "wicking" ability to the fabric not only eases laundering but makes the garment cooler and more comfortable by promoting evaporation of perspiration.

Cotton, or more accurately, 100 percent cotton items have all but disappeared in the world of uniforms. Chefs and cooks of the old school prefer all-cotton outfits because they are porous and more absorbent than polyester and cotton blends (Figure 20-6). Working under extreme heat in a limited area causes chefs to perspire even more than very physically active maintenance men or floor maids. Thus they need material that will help keep them cool. However, blends have been increasing in popularity, because cotton has become so very expensive, and also because "breathing" blends have better soil-release qualities.

Wool is important as a uniform fabric where the climate is raw and cold; perhaps, too, in areas where status is the order of the day. Wool, however, must be dry cleaned and this of course is expensive.

Rayon is another durable fabric when combined with other synthetic materials. However, rayon blends should be ironed, which is an added expense. Triacetate blended with cotton and nylon is popular for uniforms because it is a combination that resists wrinkling, moths, and mildew.

**FIGURE 20-6** *Although for years chefs preferred cotton to synthetic fabrics because it was cool and absorbent, modern textiles have made the blends more breathable and excellent for soil-release. Chef uniform designs differ very little; checked trousers are standard. Only properly qualified chefs would wear the high white hats called the* grande tocque blanche.

## *Laundering Techniques with Synthetic Fabrics*

The most widely acceptable uniform fabrics are those having cotton with polyester threads. They offer all the requisites: crisp, smooth outfits that are easy to fashion, comfortable to wear, durable, and easy to maintain.

Proper laundering is the key to a uniform's longevity (aside from the treatment the wearer gives it). No longer is intense heat needed to wash permanent-press blends. Cooler temperatures produce equally clean garments and save precious fuel.

For many years it was considered necessary to have the "break-suds" phase of washing over 160°. With improved fabrics most uniforms can be washed and rinsed below this temperature. The well-established manufacturers provide instructions for washing their uniforms, and the enlightened Housekeeper will certainly work with the purveyor so as to launder the uniforms under most satisfactory conditions.

Chlorine bleach affects the cotton content more than the polyester in blended fabrics. Gauge the amount of bleach to be used by the proportion of cotton in the fabric. The more cotton, the less bleach.

Blended fabrics are subject to thermal shock—the setting of wrinkles through sudden variation in washing temperature. Gradual cooling from one wash cycle to another should be limited to a 15-degree drop. More than that may damage the fabric.

Careful handling in laundering should be followed up by similar concern with finishing methods. Polyester reacts differently to each type of finish. Tumble drying is the most acceptable for the majority of synthetics, but there are a few rules to follow:

1. Dryers should be loaded to only half of their capacity. Excessively high temperatures should be avoided (as already mentioned) and care taken not to expose polyester fabrics to the flame of the tumbler's gas heater.
2. When the fabrics are dry, heat should be shut off but the tumbler allowed to run another ten minutes or so to cool the load. This prevents thermal shock.
3. As soon as the garments are finished in the tumbler, they should be removed and hung on hangers.
4. The new steam-tunnel finisher for blends and polyesters is the best to date. Washed garments on hangers go through the steam tunnel on a motor-driven rack. On emerging they are finished—no wrinkles, no pressing.

The one caution needed with steam-air finishing is, again, temperature. Intense heat causes fuzzing and pilling of the fabric and should be watched carefully. If the garments have been improperly washed, the heat can cause stain accumulation on the fabric. A steam-air finisher may be safely operated, at the highest, between 250° and 300° F. It should never exceed 300° F. When in doubt, be guided by the recommendations of the manufacturer.

A heat-seal machine is handy to have in the Sewing Room for torn or ripped materials. These machines enable the operator to make inconspicuous permanent repairs on cottons, polyesters, flame-retardant materials, and special patterns. Some models offer a heat-seal mending surface that prints permanent property marks and other identification on uniforms. The daily wear and tear on uniforms make this a cost-saving and practical investment. (See Chapter 21 on the Linen Room.)

## *Dry Cleaning Blended Fabrics*

The proponents of dry cleaning for blended fabrics claim that it enhances a garment's life because it eliminates bleach effects by reducing exposure to mechanical action. Dry cleaning takes one-half to one-third the time needed for washing. Some experts suggest that a large operation would do well seriously to consider installing dry cleaning equipment on the premises to remove grease and oil spots from synthetic fabrics. Also, equipment can be used to take care of guests' clothes and thus it can become a profit center in an establishment.

The most favored method of keeping uniforms clean from the practical standpoint (providing on-premise laundry facilities are available), is to wash polyesters every other day. This system requires only three uniforms per employee. One is worn; one is laundered, and one is exchanged for the clean one, so that each employee has a spare should the garment being worn be accidentally soiled. Ideally, then, the uniform distribution room should be set up on a one-for-one basis. This, however, entails extra labor because the area must be staffed daily. Additional manhours, complicated by comprehensive record keeping and accounting to maintain control, add to costs.

## *Distribution of Uniforms*

Frequency of uniform change, except in special cases, is governed largely by frequency of laundering. This also affects the number of uniforms needed per employee.

For example, if garments were to be laundered on the unlikely basis of once a week, an employee would have to be issued five changes per week. This means at least 11 uniforms would be needed for a single employee. Theoretically, he would have six in his possession and the laundry would have five. The employee would turn in five at collection time (keeping the one being worn that day) and receive five clean ones in return.

Needless to say, this would require an extensive inventory and a lot of space, and would be a permanent headache for the record keeper.

Weekly washing is more costly and impractical than daily laundering. With polyester fabrics, the problems are increased.

By its very nature, polyester has an affinity for oily substances. Dirt that settles on collar linings, under the armpits, or elsewhere becomes more difficult to remove the longer it is allowed to stay. The laundry would be forced to use chemicals to remove such soil, and these can ultimately affect the life of the fabric.

## *Uniform Control*

In a large hotel, the business of employees' uniforms can become very complex and will require one person in charge of issuing, inventorying, and ordering hundreds of different sizes and styles that are needed. Where a number of costume uniforms are used, the attendant comes close to being a wardrobe mistress (Figures 20-7 and 20-8). Where doormen are Conquistadores or Beefeaters, there is more to be concerned with than a simple greatcoat. And if the cocktail waitresses are Gay Nineties' ladies, complete with garters and beauty marks, there can be seemingly endless detail required to keep track of them as well as numerous unexpected losses of needed items.

**FIGURE 20-7** *Costumes have become almost standard as uniforms in specialty restaurants. The English "serving wench" uniform is one of the more popular styles but its initial cost is high and it is complicated to clean and launder. Usually, these are custom-made and tailored to fit the individual waitresses.*

**FIGURE 20-8** *Custom-designed uniforms are often dramatic (as is doorman's costume) and chosen to project an image or concept. When designing custom uniforms, consideration should be given to employee comfort. If a uniform lacks "stretch" room, it can affect efficiency and job performance. At the Del Monte Hyatt House in Monterey, Calif., the doorman tolerated his uncomfortable hat and vest for the sake of "show biz."*

Uniform Rooms should be operated on the basis of a daily schedule; that is, the employee brings in the soiled uniform and exchanges it for a clean one during a given time period. The difficulty here is that employees work different shifts and the Laundry or Uniform Room with its limited-hours schedule may be accessible to some employees but not to others.

An alternate procedure is to let Laundry Room personnel be in charge of substituting clean garments for dirty ones by giving them direct access to employee lockers. The person responsible would be the only one with a master key to all lockers. If uniformed employees resent this form of intrusion on their privacy, the other solution would be for them to hang soiled uniforms outside their lockers and let Laundry Room personnel pick them up and substitute clean ones as their schedule permits. Generally, however, uniform control is a simple matter of issuing uniforms to new employees who will have them altered—usually only the lengths. Each employee then turns a soiled uniform in each time a clean one is needed.

Employees should be held responsible for their uniforms. They should be asked to sign for the number of garments received on the day they start. When they leave, final paychecks may be withheld until they have returned all outfits.

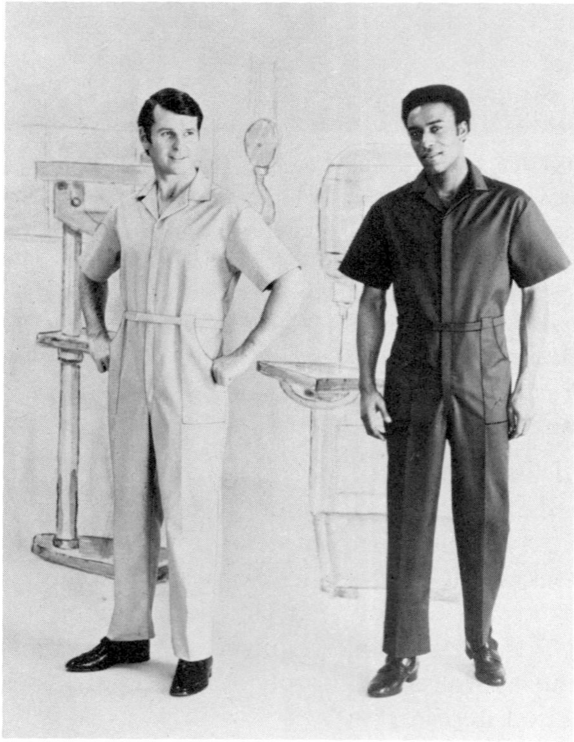

**FIGURE 20-9** *Engineers, janitors, and other maintenance personnel are hard on uniforms. Large deep pockets for tools, concealed zipper front for easy slip-on/slip-off, short open collars, and plenty of sleeve room provide comfort and easy mobility for those who are active and handle heavy-duty chores.*

Where possible, new employees should be issued new uniforms. This is not always practical due to the ever-present turnover problem. An employee's permanency should—as nearly as possible—be determined before investing in new outfits for him.

## *Storage and Stocking*

Some uniforms will be hung up on hangers, so space for racks must be provided. Other outfits will be folded and put on shelves according to departments, and then within that area, according to sizes. Where uniforms must be altered for the individual, that person's name or ID number will be put under that spot on the shelf and on the uniform. The person in charge of uniform control should be pleasant and patient because he or she can have a decided effect on employees as they come into work or leave at the end of a long day.

Determining the number of uniforms that should be kept in stock is probably tantamount to interpreting the Einstein theory. The uncontrollable forces of turnover, anticipating proper sizes, and irreparable accidental damage to garments are all hard to calculate on a day-to-day basis, let alone well in advance, but there are givens to take into consideration.

Employees in public contact areas should appear neat and clean at all times; yet they are probably the most vulnerable to the unexpected accident that may mean a change in uniform. They may need more changes than employees working behind the scenes. Back-of-the-house employees, rarely seen by guests, can more than likely get away without changing their uniforms more than once daily (Figure 20-9). Therefore uniforms for employees in the public eye should be ordered on the basis that emergency changes will be more frequent.

In the case of chefs, sous-chefs, and carvers, two daily changes of coats, pants, and aprons must be available every day for obvious reasons. Anyone dealing directly with food should be hospital-clean; spotless uniforms are a must for them.

It has been estimated that polyester-cotton blends will survive roughly 75 launderings. For maintenance personnel, the life span is shorter since their uniforms are subject to greater exposure to soil and dirt. On the other hand, they are not necessarily bound to daily uniform changes as are other employees. But, in general, the total life expectancy of these fabrics ranges from one year to about 18 months.

## *Disposable Uniforms*

Using disposable uniforms on a regular full-time basis is relatively new. Many managers, however, have come to regard them with an appreciative eye. One reason is that there has been greater effort in the last few years to improve style, fashion, and color. Another is that certain disposable items, aprons in particular, are great garment savers. What is more, they are extremely sanitary—a major plus for kitchen use. Disposable aprons protect fabrics against accidental soiling and from excessive grease and grime build-up. Because of improvements in color and style, many hotels are using them to complement and supplement the uniforms worn by maids, waitresses, attendants, and kitchen employees.

Many new materials have characteristics not formerly attributed to disposables. Spunbonded olefin has proved a nonlinting, water-repellent, abrasion-resistant material. Manufacturers insist there is little chance of tearing, shredding, or fraying edges.

Some disposables have a tough, liquid-resistant plastic finish that allows multiple usage before they must be discarded. Other manufacturers guarantee their products against basic damage from water, oil, grease, and other stains. A simple wiping will remove liquids from the material's surface quickly and easily.

Most disposable uniforms require more storage space because they have to be ordered in such large quantities. In one week an employee might get by with one pair of pants and one change, while in that same week another might require five or six pair of disposable pants. Disposable clothing doesn't flatten down on the shelf. Furthermore, the Uniform Room generally has to stock several weeks of items on its own shelves to be sure of a constant flow.

From the Hyatt Hotels training manual on housekeeping, we quote the following:

**Repair and Condemnation**

1. The Executive Housekeeper will be responsible to ensure that all uniforms are kept in a state of good repair.
2. The Executive Housekeeper should condemn uniforms upon recommendation from the seamstress that a damaged uniform is totally irreparable. Condemnation records should be maintained.

**General** At some hotels, the Security Department will ensure that uniforms are not worn or taken off the hotel property. The quality, appearance, custody, and control of our hotel staff uniforms is very important to the morale of our staff, to good cost control and to the image presented to our guests. It therefore becomes every employee's responsibility to take care of his or her uniforms and present the best possible appearance to our guests.

**Name Tags** Name tags are considered a part of all uniforms. Every employee will be issued a name tag which will be worn over the left breast pocket when the employee is on duty or in uniform. Name tags should be kept in lockers when not being worn.

**Banquet Uniforms** A par stock of banquet uniforms should be established and maintained by the Housekeeping and Banquet departments. A banquet usage sheet can be used as a control, listing the uniforms issued and returned. A banquet sign-in sheet can also be used to note the uniform number issued to an employee. All returned uniforms can be cross-referenced to trace missing uniforms.

# Chapter 21

# The Linen Room

**OVERVIEW** The center of all Housekeeping activities is where linens are stored, where employees check in and out, where the professional Housekeeper normally has her office. Keys and supplies are dispensed from here—sometimes uniforms or clean laundry—and Lost and Found operates from the Linen Room.

Linen room is a misleading description of the area; actually it is far more than a place to store and issue linens. It is in reality the Housekeeping headquarters and many different activities go on there. In some institutions there are, in fact, few if any linens in the Linen Room because they are all stored in the linen rooms (also called linen closets) on individual floors or are en route to and from the laundry. In a college or university housing system there may be several Linen Rooms, or headquarters for Housekeeping and Maintenance because of the many residence halls on the campus.

For a small to medium-sized property the description of the Linen Room outlined in the TraveLodge manual is helpful. It is the heart of the Housekeeping Department, where:

1. All Housekeeping personnel report
2. Inspectresses, maids, and housemen receive their station assignments, work orders, and pass keys
3. All Housekeeping telephone calls are received
4. All check-out and in-order rooms are processed
5. All guest rooms not cleaned by a specified time are reported
6. All household supplies are issued
7. Handling of commercial laundry—counting outgoing linens, counting and checking all incoming linens and billings—is done
8. Table linens are issued (to the Food and Beverage Department)
9. All uniforms and costumes are issued on a one-for-one exchange basis
10. Working records are kept
11. All Housekeeping pass keys are kept and controlled

In almost every institution whatever its size, the Director of Housekeeping will probably have an office adjacent to or within the Linen Room in order to stay informed about current problems and be available to answer questions—really just to be sure of being a part of the mainstream (Figure 21-1). There are many executive Housekeepers who insulate themselves from the everyday happenings and do not realize the kinds of situations that arise hourly with both guests and employees. Simply having the office in the midst of the activity will keep the professional Housekeeper aware.

Lost and Found is normally handled in the Linen Room because of its closed storage area (Figure 21-2). All the found items will be taken care of by the department's secretary, as that

**FIGURE 21-1** *The heart of most Housekeeping Departments is the Linen Room. There the key person answers the phone, records guest's requests, and communicates on "electrowriter" with Engineering Department and Front Desk, and keeps track of room status. At the San Francisco Hilton spacious Linen Room contains Uniform Control, Sewing Room, and storage of extra room furnishings.*

operation generally involves correspondence, wrapping, and mailing.

Storage areas for Housekeeping supplies, guest request items in a hotel, guest room supplies, and equipment are usually in or close to the Linen Room.

The Sewing Room, a subcenter of housekeeping activity is often connected to the Linen Room.

Other activities that will go on in the Linen Room if there is sufficient space are:

1. Filling stationery holders
2. Filling maids' caddies with cleaning supplies
3. Filling maids' second caddies with guest room supplies
4. Removing stains from linens

| | |
|---|---|
| DATE ____________ | PERSON TAKING REPORT ____________ |
| WHO LOST | WHEN LOST ____________ |
| NAME ____________ | WHERE LOST ____________ |
| ADDRESS ____________ | CHECK-OUT ROOM NO. ____________ |
| ____________ | WHAT FUNCTION ATTENDED? ____________ |
| TELEPHONE NO. ____________ | ____________ |
| WHAT LOST ____________ | DATE FOUND ____________ |
| DESCRIPTION ____________ | DATE MAILED ____________ |
| ____________ | DATE PICKED UP ____________ |

**FIGURE 21-2** *Record of Lost and Found*

*This form is used when people attending functions at the institution have lost something, or when a former guest calls in to inquire whether his lost possession was found.*

A deep double sink, table, chemical supplies, and a stain removal chart will all be needed for the last activity. It may be possible to do this in the Laundry or Sewing Room if a sink is available there.

Of course, in some properties it is not possible to find enough space to carry on all those activities plus have all the storage in that same area. If it is at all feasible, however, it is recommended, because these different phases of housekeeping are all so closely interrelated. Psychologically it is better, too, if employees in a department work near one another so that they have a sense of being members of a team.

In the usual course of their work, maids and housemen are scattered about the building, so if it is possible to set up a "home room" for the department, the employees who work throughout the property will feel a stronger sense of attachment to a specific group. When the various segments of Housekeeping are isolated, with the Sewing Room perhaps located on the top floor, the Laundry in an annex, and the Lost and Found on a mezzanine, there is less sense of belonging to the Department.

Let us now examine the Linen Room activities one by one, starting with the obvious—linens.

Linens must be stored in a relatively humidity-free room. In the event the Linen Room is adjacent to the Laundry, it is obviously vital that the door between the two rooms form a tight seal, and that it be opened as seldom as possible. Linen storage must also have adequate ventilation. Shelving must be free of rough edges and protrusions.

In small properties—fewer than 100 rooms—all linens can be stored in the central Linen Room from which they are issued to maids who pick up the exact number of sheets, pillowcases, and towels, required for the rooms in their own sections. Maids will probably have carts to put this linen on, although where distances are not too great some institutions get along without carts, especially if management considers them difficult to maneuver, unsightly obstructions, and a problem to store.

Maids in smaller properties are often expected to return the same number of linens in the afternoon that they checked out in the morning. Some Housekeepers disapprove of this, for several reasons: (1) the maid may need an extra sheet because her supply includes one with a rip or bad stain; (2) there is no point in having a maid push soiled linen around on her cart until lunch time and then again all afternoon, when it could be put down a chute, picked up by a laundry houseman, or placed in a divisional hamper; (3) to keep soiled linen moving is the object of most Housekeeping Departments both because it may mildew or stains may set, but even more important, because there is rarely enough linen in an establishment to allow soiled linen to remain unprocessed for an entire day.

Speaking of *enough* linen—how much is enough? In institutional jargon, the word par is the amount of linen that must be inventoried per bed (including bath linens for one person). Par ranges from three to five (changes) depending on the operation, occupancy, the laundry situation, budget, and the need to service guest rooms over a weekend when there is no laundry delivery.

Minimum par would mean that one set of linens would be on the bed (and in the bath), another in the laundry, and a third in the Linen Room. A more realistic situation would be one on the bed (or in the bath), one in the floor Linen Room or on a cart, one in the main Linen Room, one in the soiled linen en route to laundry, and the fifth set at the laundry.

## *Linen Storage*

Many Housekeepers believe that it will prolong the life of bed linens—particularly the blend sheets—to give them a rest on the shelf before putting them back into use. That is why it is rarely advisable for sheets to be taken out of the dryer and put directly on beds.

There is perhaps nothing that can throw off the work schedules of a Housekeeping Department faster, and serve to irritate guests more, than a shortage of linens (Figure 21-3). Maids cannot make beds so rooms cannot be rented; employees have to be sent to linen rooms on various floors looking for extras; handymen are sent to storage to open boxes of new linen; the Front Desk is calling for ready rooms, and guests are fuming without clean towels or the beds made.

Therefore, the professional Housekeeper will keep the linen supply literally up to par, devise effective linen controls, and keep an accurate linen inventory.

## *Linen Controls*

Linen Room staff will keep a careful count of the linens sent and received from the laundry each day, unless the charges are based on occupancy, as explained in Chapter 30, or are on a per pound basis. Then spot-checking from time to time is all that is needed. Linens from a section or floor can perhaps be counted as they go out in a hamper from that floor or come through the central Linen Room. This may reveal a shortage or an over-supply of the amount needed for the maid's quota of rooms.

Extra sheets discovered in the soiled laundry may be those with holes or bad stains, which the maids simply threw in dirty laundry rather than keeping them separate to be given to the Sewing Room or supervisors. (Of the two alternatives, putting clean torn sheets into soiled laundry is far preferable to putting them on beds, as guests should not have to sleep on worn linen, and it is well known that a restless sleeper's toe caught in one small hole can soon rip a sheet to shreds.) Actually, receptacles should be placed in floor linen rooms for items needing repair or re-washing.

Although some institutions keep injecting a predetermined amount of new linens into circulation at set intervals, in most properties that deal with the public it is a better policy to introduce new linen *when it is needed.* Otherwise it is often difficult to know whether the reserve supply in a large hotel will last for six or nine months.

Checking the linen cost per occupied room each month will keep the professional Housekeeper informed if there is anything drastically wrong.

New linens should, some feel, be stamped to identify the institution and be dated prior to use; on large orders, suppliers will often do this free of charge. However, there are many who feel that it is pointless to mark each piece in a new property, because *if it needs to be replaced for any reason it should be, regardless of its age,* and the bulk of the original order will be around for some time. The year it was bought would be known from purchase orders anyway. If one is using several different brands or types of linens as a test, then, of course, marking is essential.

**FIGURE 21-3** *Cart with bed linens is removed from laundry, sent on its way to linen storage areas. With on-premise laundry, linens for one floor (or section) can be stacked on a cart and remain on it in floor linen closet until maids fill their carts directly from this rolling unit, thus eliminating a second handling.*

## *Floor Linen Closets*

Linen controls should be put into effect in floor linen closets. Pars for each closet should be set and posted. These rooms should be stocked daily according to those pars, and be locked at all times. Locks should be checked frequently to ascertain that they are working properly.

Floor linen rooms should be kept clean and in a state of repair, with adequate shelving. Chutes for soiled linens must be checked often to be sure they are not jammed or that they have not had some foreign substance thrown into them. Chutes should have doors with locks in good working condition.

The floor on which soiled linen lands should be of vinyl asbestos tile, and should be cleaned and sanitized frequently. The area where soiled laundry accumulates should be locked at the end of the work shift.

The same controls and precautions applicable to floor linen rooms also apply to the central Linen Room, and access to the area should be rigidly controlled. Night watchmen or a member of Security should be instructed to check floor linen rooms nightly.

In some operations, TraveLodge for example, maids must mark on daily work sheets the number of soiled linens, by type, that they remove from the guest rooms. At the end of the day they total each column and turn in their work sheets at the central Linen Room.

Distribution equipment should be adequate, in good working condition, and of the right type. For example, one cart (see Chapter 19 on equipment) has wire shelving for transporting stacks of linens upstairs (or out) to guest room areas. When linens have been unloaded, the shelves fold flat and become sides so that the cart can be used as a laundry bin for soiled linens.

Reserve storage areas must be dry and ventilated (to prevent mildew), with linens properly wrapped and labeled. Perpetual records should include all new linens added to storage.

## *Restocking in Central Linen Room*

For any institution using either an off-premise or on-premise laundry with daily deliveries, this idea may help to control linens.

Each room attendant brings her cart to the main Linen Room at the end of her shift. There all carts are cleaned and restocked with an adequate supply for the next day. The top shelf usually holds guest room supplies such as toilet tissue, soap, facial tissue, and shower caps, which can also be restocked by the late shift.

This system accomplishes several goals in a linen control program:

1. All linens are in central area.
2. All carts have same amount of linens.
3. Any tendency to keep foreign items on carts is eliminated.
4. Employees personal carts are eliminated.
5. Linens in floor linen rooms no longer get shop-worn on shelves.
6. Linen inventories can be reduced, as the main Linen Room attendant knows exactly what is in circulation.

Although this system requires the service elevator to be available at the end of the shift, it does eliminate the need for men to deliver linens to floor areas. Each maid brings her cart to the Linen Room 15 minutes before the shift ends. This system also ensures that each employee remains on the premises until completion of shift.

The clean loaded carts are delivered to the floors each morning before the shift begins. This has the advantage of enabling maids to produce clean rooms immediately on each shift.

Another means of keeping careful track of linen is to use the linen pack—that is a complete set of guest linens individually packed for each room (Figure 21-4). This saves the maids' having to count and carry, and perhaps drop, individual pieces. This system is usually employed in hospitals as an extra assurance that linens will stay clean.

## *Preventing Misuse of Linens*

One of the most difficult aspects of linen control is to prevent employees from using linen in carrying out their work routines. To a housekeeper's horror, a maid may use a clean colored pillowcase as a surface on which to drain glasses that she has just disinfected. This not only takes a clean pillow slip out of circulation, but the ammonia in a quaternary disinfectant removes the color and renders the pillowcase useless.

Washcloths make tempting cleaning rags, and terry towels are handy to dry off counters or tile walls, or even, Heaven forbid, a wet floor. Guests will, of course, also be tempted to use washcloths as shoe polishing rags, and if they are marked, to take them as souvenirs. A large property will lose between 1000 and 1500 washcloths per month! All of these abuses will increase the lost and discarded linen totals and boost the already high cost of linen.

**FIGURE 21-4** *The linen pack is a complete set of room linens prepackaged in film or cellophane by the laundry. Often they are color-coded so that linens for twin-bedded rooms can be easily recognized. This system keeps linens cleaner and saves maids' time, simplifies handling and storage in central and floor linen rooms.*

Discarded linen is, of course, very useful to any institution because it provides cleaning cloths for employees throughout the property and cooking cloths for chefs, and can be made over into many other useful items; for example, sheets can be made into crib sheets or pillowcases, quilted bed pads into cooks' potholders or pads for silver burnishing. Poly-cotton rags are not ideal for cleaning, however, as they are not so absorbent as all cotton or sponges.

There are some institutions in which the Housekeeper feels that all discarded items should be dyed an obvious color so that if an employee is seen using a piece of linen that isn't the discard color, the breach of policy will immediately be noted and corrected. It is critically important that the dye be permanent, however; many good linens can be ruined by coming in contact with damp discards that run.

## *Housekeeping Responsibilities*

In a 100-room or smaller operation, the Housekeeper may be personally in charge of linens. In a medium-sized establishment it may be part of an assistant's duties. In a 500-room and larger hotel, one person will be in charge of linens full time, with responsibility for purchasing or requisitioning needed items (with the approval of the executive housekeeper); marking, dating and issuing new linens; salvaging discards; inspecting floor linen rooms; evaluating rewash linen, sorting and counting it before returning it to laundry; checking laundry chutes. This person also keeps all records including the following:

1. Semiannual or annual inventory (including circulating and reserve linens)
2. Linen budget based on current needs (reserve linens)
3. Requisitions for new linens
4. Record of discarded linens (this may be furnished by the laundry)
5. Record of rewash linens sent to laundry.

Techniques for taking inventory differ, and it is only a matter of opinion as to which works best and where. Naturally, the older the property the more complicated the situation may be—with linens of various ages having varying life expec-

FLOOR 3rd DATE July 5 MAID Carrie Adams

| Room | Sheets | | Slips | | Spreads | | Blankets | | Face Towels | | Bath Towels | | Bath Mats | | Wash Cloths | | Soap | Kleen-ex | Toilet Tissue | Bed Pads | | Remarks |
|---|---|---|---|---|---|---|---|---|---|---|---|---|---|---|---|---|---|---|---|---|---|---|
| | In | Out | In | Out | In | Out | In | Out | In | Out | In | Out | In | Out | In | Out | In | In | In | In | Out | |
| 306 | 2 | 2 | 2 | 1 | — | | — | | 2 | 1 | 2 | 2 | — | | 2 | 2 | 3 | 1 | 2 | — | | clean pillowcase & blanket left |
| 308 | 4 | 4 | 4 | 4 | 2 | 2 | 2 | 0 | 2 | 2 | 4 | 3 | 1 | 1 | 4 | 2 | 4 | 1 | 2 | 2 | 2 | stole wash rags + towel |
| 310 | — | — | — | — | — | — | — | — | 1 | 1 | 2 | 2 | 1 | 1 | 2 | 2 | 4 | — | — | | | Sleep out |
| 305 | 4 | 4 | 4 | 4 | 1 | 1 | 1 | 1 | 3 | | 2 | 1 | 1 | — | 3 | — | 4 | 1 | 2 | — | | found wallett |
| 311 | — | | | | | | | | | | | | | | | | | | | | | OK |
| 309 | 2 | 2 | 1 | 1 | — | | 1 | -1 | 1 | 1 | 2 | 2 | 1 | 1 | 2 | — | 3 | — | — | — | | |
| 317 | — | | | | | | | | | | | | | | | | | | | | | Do Not D. |
| 307 | 2 | 2 | 2 | 2 | — | | — | | 1 | 1 | 2 | 2 | — | | 1 | 1 | 1 | — | 2 | — | | |
| 316 | — | | | | | | | | | | | | | | | | | | | | | ok |
| 312 | 4 | 4 | 4 | 4 | 1 | 1 | — | | 4 | 4 | 4 | 4 | 1 | 0 | 2 | 2 | 4 | 1 | 2 | — | | Bath mat stole |
| 314 | — | | | | | | | | | | | | | | | | | | | | | OK |
| 301 | 4 | 4 | 4 | 4 | 2 | 2 | 1 | 1 | 0 | 0 | 2 | 2 | — | | 4 | 4 | 4 | 1 | 2 | 1 | 1 | |
| 410 | 2 | 2 | 2 | 2 | — | | — | | 1 | 1 | 1 | 1 | 1 | 1 | 2 | — | 2 | — | 1 | — | | wash rags gone |
| 419 | — | | | | | | | | | | | | | | | | | | | | | ok |
| 423 | — | | | | | | | | | | | | | | | | | | | | | ok |
| TOTAL | | | | | | | | | | | | | | | | | | | | | | |

REMARKS guests stole towels & washrags + blankets gone 308

Keeping track of linen reveals pilferage.

**FIGURE 21-5** *Linen and Supply Sheet*

tancies. Figures 21-5 through 21-7 will be useful as guides but each institution should develop its own forms.

Actual counting of the linen has a psychological effect on employees for they realize that management is on the lookout for theft or careless loss. But the inventory catches shortages that the laundry or floor supervisors might not notice, and will reveal how rapidly linens do get used up, which items need to be replaced, and how much they will cost.

## *Food and Beverage Linens*

Linens for the Food and Beverage Department are often stored in the Linen Room where they are cared for by the same employee respon-

| ITEMS | 1976 | 1977 | 1978 | 1979 | 1980 | 1981 |
|---|---|---|---|---|---|---|
| Sheets— | | | | | | |
| On hand | 10,500 | 12,159 | 15,176 | 16,006 | 15,497 | 17,193 |
| Purchased | 3,600 | 3,600 | 3,600 | 3,600 | 4,000 | 4,000 |
| | 14,100 | 15,759 | 18,776 | 19,606 | 19,497 | 21,193 |
| Loss | -1,941 | - 583 | -2,770 | -3,109 | -2,304 | -1,890 |
| | 12,159 | 15,176 | 16,006 | 15,497 | 17,193 | 19,303 |

**FIGURE 21-6** *Sample Linen Inventory for 800-Room Hotel*

| Item | Par | Rooms | Floor Closets | Linen Room | Laundry | Last Count | Added from Reserve | Total | Discards | Known Thefts | Count Should Be | Actual Count | Over | Under |
|---|---|---|---|---|---|---|---|---|---|---|---|---|---|---|
| Sheets | 800 | | | | | | | | | | | | | |
| Cases | 1600 | | | | | | | | | | | | | |
| Spreads | | | | | | | | | | | | | | |
| Blankets | | | | | | | | | | | | | | |
| Face Towel | | | | | | | | | | | | | | |
| Bath Towel | | | | | | | | | | | | | | |
| Bath Mat | | | | | | | | | | | | | | |
| Bed Pad | | | | | | | | | | | | | | |
| Huck Towel | | | | | | | | | | | | | | |
| Wash Cloth | | | | | | | | | | | | | | |
| | | | | | | | | | | | | | | |

**FIGURE 21-7** *Linen Inventory Sheet*

sible for Housekeeping linens. Banquet linens are kept separate from restaurant linens as they are usually different sizes, shapes, and colors. Each specific dining area has its own table sizes, and napery is often selected just for that setting and color scheme so must be reserved for that particular room.

One of the important points in selling banquets or dinner meeting rooms today is that the tables can be set up in whatever color or colors might be preferred by the host who is arranging the meeting. This means a large inventory is needed, with special records and marked storage that will designate whether the cloths are for round or oblong tables and whether linens are for banquet, restaurant, or Room Service use. Napkins, like pillowcases and washcloths, are easily borrowed for cleaning tasks, which usually makes their future use in an impeccable table setting impossible.

A bin for soiled table linen should be provided in the kitchen or near the dining area, but it should not be accessible to kitchen employees and waiters.

## *Coordination with Laundry*

The system of management for the linens and the activity in the Linen Room itself will depend largely on whether the institution has its own laundry. If it does and there are fewer than 500 rooms, chances are the Housekeeper will be in charge of it. In any case, Housekeeping and Laundry have to mesh their activities, and because each needs the other's cooperation, it is usually more feasible to have the Laundry supervised by the director of Housekeeping.

If a hotel or an institution controls its own laundry, it is often able to change schedules at the drop of an order—"Stop washing sheets and do all red banquet cloths and white napkins," may be the urgent request during Christmas week when occupancy is low but banquet and party business is up, with everyone wanting festive red and white table tops.

If something has suddenly gone wrong with a formula and the blue towels are fading or spotted, one can quickly stop the machines, whereas in a commercial laundry communications might not be so direct.

## *Employee Check-In and Check-Out Areas*

Although there may be a time clock where employees check in, it is advisable to have them sign into the Housekeeping Department also (Figure 21-8). Situations have been discovered where employees punch time clocks and leave the building until it is time to punch out. In a very large establishment it is difficult to prove they did not work that day unless there are other definite checks on attendance. Signing in is done as employees pick up their assignments and/or keys and their supplies or carts. The check-out areas, where keys and empty supply baskets are brought at the day's end, should also be the place where bonuses and paychecks are distributed.

## *Key Control*

As pointed out in Chapter 26, which deals with security, the control of keys is a matter of utmost importance. Keys that housekeepers, supervisors, maids, housemen, and many others use are master keys and can open at least a section of a floor, and sometimes open all guest rooms. Keys are kept in a storage cabinet—usually in the Linen Room—and supervisors sign for them as they are taken out. Then maids sign for the keys that they receive. If an employee fails to return a key and it is not noted until after she leaves, then a thorough search of her locker is made by the Security Department. If this does not produce the key, the employee is called at home. Discussions are held with employees about the important responsibility of having a key to 15 to 30 rooms, and it is remarkable how few unpleasant incidents have occurred in the hotel business because of the employees taking advantage of having pass-keys.

## *Rooms Control*

This high-sounding expression "rooms control" refers to the person in charge of keeping track of status of the hotel's rooms. That person has a chart with the current house count, or room status, and is continually keeping the Front Desk informed about which rooms have been cleaned, are vacant, and ready for selling. Rooms control is the station notified by maids, floor supervisors, or assistant housekeepers when additional rooms are ready.

One of the cardinal sins in the hotel business is putting a new guest in a dirty room, and this is why many people in the hotel must be aware of which rooms actually are checked and ready. Even with this check there still will be slip-ups; late check-outs, for example, will not get a second tidying up and replacement of linen if the Front Desk does not notify the Housekeeping Department when a person checks out late. (In a later chapter there is comment on new types of

**FIGURE 21-8** *Floor housekeepers at San Francisco Hilton bring blue slips that contain work orders to Linen Room attendant. She stamps them in time clock and sorts out jobs for Engineering and for Housekeeping. Sign in background informs employees that no uniforms are issued on weekends. That cuts labor costs.*

room status systems that facilitate communication on ready rooms to the Front Desk.)

### *Lost and Found*

All articles left behind or apparently lost are brought to the Linen Room where they will be listed in the Lost and Found book, with the date found, description, room number, and package number. The articles should not be forwarded to guests unless written requests are received.

All items should be wrapped and labeled to indicate date found, where found, name of article, name of finder, and number of the package. Packages are numbered in sequence.

The housekeeper will report valuable items to the assistant manager and handle them according to instructions.

### *The Damp Sheet Saga*

The numbers in Figure 21-6, linen inventory, which cover six years, tell an interesting story of what actually happened in an 800-room Texas hotel in a damp, humid climate.

At the opening of the hotel, in 1976, a total of 10,500 sheets was on hand. In the 800 rooms there is a total of 1200 beds, plus 75 rollaways stored in floor linen closets. Colored queen-size sheets are used on both the out-sized twins and queen-size beds. There are no beds in any other sizes in the hotel, so the inventory need not be broken down by size.

Par for this property is five, which means that 6375 (1275 beds × 5) sheets were taken out of storage, unwrapped, and put into service. The balance, 4125, were stored on pallets, in the only available space—the basement.

Linen loss the first year the hotel was open was extremely high (1941) and certainly it was *not* from wear. Because extra help was hired before the opening, and because control systems were not set up until after the hotel was in operation, and because the cool blue sheets were of such high quality, there was considerable pilferage.

Damage was frequent too, and often unexpected, as management and department heads were not familiar with the treachery of a very humid climate, and were obviously uninformed about proper linen handling.

New linen carts had sharp edges that tore sheets and sent them to the repair pile to wait until the Sewing Room (busy making table covers and altering hundreds of uniforms) could get to them. A concrete floor in the soiled laundry area under the chutes stained over 100 sheets and left them prematurely discarded before the problem was discovered and the floor could be covered with vinyl asbestos.

Maids sometimes got trash and laundry chutes confused and put bedding into trash, or vice versa, which stained more of the colored sheets that could not be bleached.

A laundry workers' strike left bins full of soiled linens standing long enough to become mildewed. Extreme dampness in one part of the store room, undiscovered until it was too late, caused several dozens more sheets to be hopelessly damaged from mildew.

Thus expensive lessons were learned about the hazards of opening a new property, the devastating effects of a human climate, the dangers of concrete, the unreliability of some employees, the desperate need for training housekeeping and laundry personnel.

Quite different are the figures in the second year when controls were implemented and rigorously enforced. Most of the original problems were corrected. The Sewing Room even caught up and was able to make repairs. Losses were cut more than half. But the same number of new sheets was purchased.

By 1978, the normal wear factor began to show—as most sheets used daily are only expected to last about two to three years. Occupancy of the hotel was up about 23 percent, running 90 percent and above, and this meant that all the linen was subject to maximum use and wear. Additional maids who had to be hired were not accustomed to handling linens properly, and losses were thus up to 2770.

In 1979, the bulk of the first 6375 sheets wore out, while occupancies continued high. In 1980, occupancies remained high, and the forecast for 1981 was so optimistic that the Housekeeper ordered an additional 4000 sheets, afraid of being short.

In 1981, the head of the Sewing Room died and there was little promise of finding a replacement as capable as she. Thus the possibility of having sheets mended became rather remote. That, coupled with the prospect of more convention business and the prediction that shortages of both cotton and synthetic fabrics (made from petrochemicals) would drastically affect the supply of sheets, prompted the Housekeeper to order another 4000 in 1981.

We can draw some conclusions from this story and avoid the pitfalls that this hotel Housekeeper encountered.

The hotel was extravagantly over-supplied. (This, of course, is far better than the other extreme alternative—being under-supplied.) But now there is an excess that, in that damp environment, may prove to be expensive. Par at five is high, so having a quantity equal to another par—six—in inventory should be enough to make one feel secure. That would be 7650. Even if the worst year were repeated—a loss of 3109—that is still far less than the hotel has in inventory.

Our advice—stop purchasing and keep the linen control procedures strictly in effect. In three years the situation will be normal, bed linen expenditure will have been cut to nonexistent, and the Housekeeping budget will be far more presentable.

## *Mending and Marking*

Marking machines along with heat-seal machines help the Housekeeping Department to identify, code, and date linens, bedding, uniforms, and in fact, any fabric within the property. Mats and rugs can even be mended with heat-seal patches applied on the backs.

Sometimes the marking machine will be needed for labels but more often it is used to print directly on garments or linens. One real labor-saving machine is available that does all these activities—it prints, cuts, and seals identification tapes or mending patches onto fabrics. Depending on the size of an operation and number of people who will be working on these tasks, different machines may be needed:

1. A heat-seal machine with different sizes of platens and bases for putting identification tapes on wide items such as sheets or smaller pieces such as towels or aprons (this also will apply transfers emblems and mending patches)
2. Direct printing (marking) equipment (this is dry ink ribbon on which each print takes one-quarter of a second) activated by a foot pedal

Different models, even portable versions of some, are available.* Some of the machines represent an investment of thousands of dollars, but in a few years will have justified the expenditure in positive benefits of linen control and replacing seamstress time. Tapes, transfers, patches (all soft goods) may be ordered in any color from the same suppliers who sell the machines. The mending and marking items are designed for both cottons and blends, even flame-retardant modacrylics.

* Heat sealing and marking machines: Thermopatch Division of Rospatch Corp., 3505 Pinkney Avenue, Bronx, New York 10475.

# Chapter 22

# Sewing Room and Storage Areas

**OVERVIEW** Sewing Rooms are more important than ever, as repairing fabrics is well worth the investment. Here we outline the many tasks performed by the hotel's seamstresses, explaining what equipment and supplies are needed, and including a job description. Safety precautions are important in this facility. Storage is usually handled in or near the Sewing or Linen Rooms, so ideas and record-keeping related to storing all types of materials are part of this chapter.

## Sewing Room

There may be one part-time seamstress to shorten uniforms or mend an occasional drapery, or there may be a professionally operated, fully equipped Sewing Room with a battery of seamstresses—whichever management and the professional Housekeeper decide is necessary. If there is a competent seamstress who has been with the institution for many years, it is worthwhile to make good use of her. If, on the other hand, the labor market in a particular area just doesn't yield a capable and reliable person to take charge of the Sewing Room, then perhaps the whole function should be minimized.

A well-run Sewing Room can very definitely be an economy for a large institution.

## Sewing Room Tasks

Here are some of the most frequent Sewing Room tasks:

1. Alter uniforms.
2. Patch table linens.
3. Repair curtains, bedspreads, slipcovers—whatever is an expensive item in itself and requires a minimum of repair to put it in order.
4. Make pads from used mattress pads for burnishing silver.
5. Make cooks' cloths. These are needed by the dozens daily. Although they can be rented from linen supply, it is possible to make a highly satisfactory, amply thick one by stitching three discarded napkins together along the edges and with a large X across the center. If made from different-colored napkins these are easily recognizable, and will not fade on other laundry as they have already been washed many times.
6. Examine bed linens on a lighted table (Figure 22-1). This will quickly reveal the holes and rips so that sheets can be set aside for mending. This process may seem time-consuming but it actually saves the maid's time and steps, as she may otherwise have the bed practically made, find the sheet is not usable, and have to stop to locate a replacement.
7. Make crib-size sheets, blankets, and pillowcases.
8. Make some draperies and bedspreads, if fabric is on hand and time permits. Making glass curtains for expansive window walls often represents a substantial saving.
9. Repair hems in sheets.

**FIGURE 22-1** *A light-table placed between two folding tables in the Linen and Sewing Room at the Brown Palace Hotel in Denver, allows Housekeeping workers to open all linens coming in from the laundry and examine them over the lighted glass top. Although this is time-consuming, it saves maids' time when they are making beds because if they find a stain or hole in a sheet, they must stop working and get a replacement. Linen and Sewing Rooms such as this one are kept sparkling clean and well lighted.*

The shortages and slow deliveries of cottons, polys, and all forms of fabrics make the Sewing Room a practical necessity.

Training employees to notice linens that need repair, and having conveniently located closed receptacles to put these items in will help to round up some of the items that, with five minutes of a seamstress' time, may be made useful for years to come (Figure 22-2).

About uniform repairs one authority explains, "Once an item is in the Sewing Room, the seamstress must completely examine it. She knows better than anyone which areas are most subject to damage. In uniforms, under-arm seam weakness must be explored and repaired beyond the actual immediate need. It is far less costly to sew around the sleeve once than to have the same uniform taken out of use several times, or—worse yet—allow the weakness to develop into a tear that damages sleeve or body. The open stitch in the corner of a top pocket must be caught immediately.

"Far too many hems on blend sheets open up. This often happens when the thread used in manufacturing is either cotton or polycore with cotton cover. Neither has sufficient seam strength. All repairs, including the resewing of sheet hems—regardless of the fabric content of the item—must be made with 100 percent polyester thread. The great strength given to seams, the lack of shrinkage, the colorfastness, and the correct elasticity of this type of thread makes it the main one that should be used in the professional Sewing Room.

"Nylon thread (monofilament) should be used for button sewers and occasionally on blindstitch machines, but *never* on a sewing or overlock machine."

## *Equipment*

Equipment for the Sewing Room includes:

1. Heat-patching machines for neat repairs of holes in table and bed linen.

2. Sewing machines able to stitch varying thicknesses of all fabrics (thread release on looper is necessary for tension control on a wide range of fabrics).
3. Blindstitch machine especially useful for uniform hems.
4. Overedging machine essential for remaking discards such as towels into washcloths. It cuts, seams, and overcasts in one operation.
5. Zig-zag machine highly useful for mending, darning, button-sewing.
6. Grommet and gripper tools.

Although any professional seamstress will want to keep the machines in good condition, this is more likely to be done if the oil, lint brush, and machine cover are easily available to her. Machines should be oiled at the end of the day so oil can soak in; dust in bobbin cases must be brushed out; work tables should be treated with silicone spray (to make fabrics slide). Extra parts should be handy and plenty of bobbins provided.

Safety precautions in the Sewing Room include not only having eye guards and finger guards on the machines, but proper grounding of electrical connections. Sewing machines must be equipped with a guard permanently attached to the machine, so that the operator's fingers cannot pass under the needle. It must be formed so that the needle can be conveniently threaded without removing the guard. Because dust and lint from synthetics can be highly distressing to some workers, it is imperative that the Sewing Room be cleaned nightly. That means vacuuming vents and shelves with a long-handled backpack vacuum, dusting with treated rag where possible, and proper floor care—either dampmop or vacuum.

### *Sewing Room Operations*

Some hotels have remarkable Sewing Room operations, almost resembling those of a swank couturier. The Fairmont Hotel in San Francisco is an example. The Sewing Room has a select group of industrious women who customsew all draperies, sheer curtains, and bed skirts. A look at their work reveals perfection in hems,

**FIGURE 22-2** *All sewing is not done in a Sewing Room, as it is sometimes far easier to do repair job on the spot. For example, if a section of guest room drapery lining comes unsewn, seamstress can quickly sew it up in place rather than having entire drapery removed.*

seams, cording, and pleating. One woman is employed exclusively to make skirts for the beds. The hotel normally has 2500 of these speciallymade skirts in use and 2500 more are being made up.

Four seamstresses are assigned to repair house linen and uniforms and three women to drapery creations. Since the draperies are custom-made, the women must cut their own patterns. This involves not only applying their combined skills but adapting them to a particular style: regency, provincial, or whatever. Many of the Fairmont's 700 rooms carry out a period decor, and the harmonizing draperies are essential to the overall scheme. The exquisite work of all the seamstresses assures products vastly superior to anything that could be purchased on the market today.

The women in the Fairmont Sewing Room also hand-wash all the sheer curtains (Figure 22-3) and bed skirts, and in addition, the aprons and collars of waitresses' uniforms.

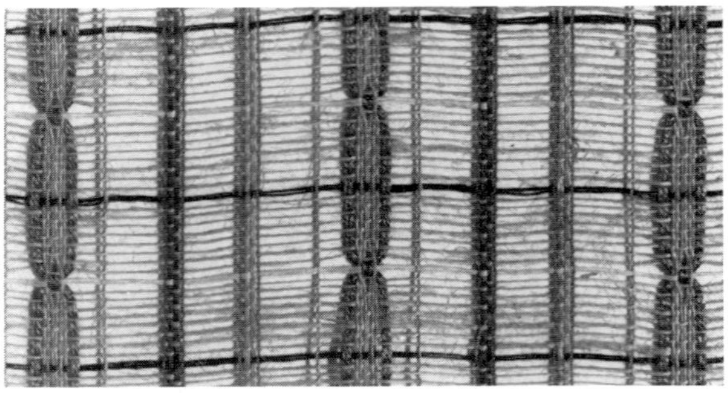

**FIGURE 22-3** *Loosely woven casement fabrics such as this one spend much of their time in institutional sewing rooms, as they often stretch and must be rehung and rehemmed repeatedly. If casement sheer being considered is not guaranteed to have dimensional stability, then Sewing Room personnel should be alerted to overshorten it considerably, lest it be on the floor within weeks.*

Sheets are also repaired when possible. Checked individually every day as they come through the folders, damaged sheets are identified by turning up a corner. Unusable linen is never discarded. White cloths are used for dusters and colored cloths for kitchen cleaning.

Carrying the custom-made tradition even farther, the Fairmont also maintains its own upholstery rooms where two men are permanently assigned to cut and sew chair and sofa coverings (Figure 22-4).

If it should become necessary to make a concerted effort to conserve sheets because of a shortage of cotton and/or polyester, very worn sheets can be mended. They can also be split down the center with the selvage edges on the sides, then sewn together to double the life of the sheet. However, the perfectionists will ask, *would* anyone tolerate seams down the middle of a sheet? Perhaps yes, as a compromise: a secondary standard in hotels will come to be accepted if room rates are compatible.

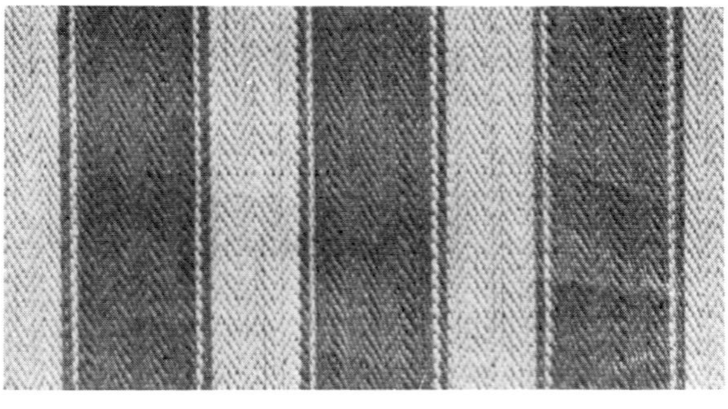

**FIGURE 22-4** *To eliminate the high cost of reupholstering it is wise to consider slipcovering certain traditional kinds of overstuffed or padded furniture. This Belgian linen twill can take heavy wear and is easy to work on. To achieve a coordinated interior in areas where slipcovered furniture is used, this twill is available with acrylic backing so that it can be used for wall coverings.*

## *Job Description—Club Seamstress*

According to the *Club Manager's Association Manual,* the seamstress in a full-facility club would have the following job description:

**Job Summary.** Performs skilled tasks in making and repairing such articles as draperies, curtains, slipcovers, lamp shades, and towels.

**Work Performed.**

1. Makes such articles as draperies, curtains, lamp shades, and slipcovers.
2. Mends torn articles following established methods like the following:
   *Patches:* cuts patch to fit section to be covered; may boil or bleach material to match color of article being repaired; removes basting threads with fingers; if patch is on seam, replaces broken hem stitches using sewing machine.
   *Tears:* places pieces of torn cloth beneath needle of sewing machine, lapping one edge over the other, and mends tear by moving cloth back and forth both lengthwise and crosswise beneath needle of machine.
   *Lace or net curtains:* cuts off worn sections or spots and follows pattern of curtain with loosely arranged hand-sewn stitches; hems curtain by hand or on sewing machine.
3. Installs snap fasteners and grommets.
4. May receive and issue linen to maids.
5. May maintain cost and production records.

**Equipment Used.** Needles, thimble, scissors, tape measure, electric iron, sewing machine, snap-fastener machine, grommeting machine, uniform.

**Working Conditions.** Hazards: possible injury to hands from sewing-machine needle.

**FIGURE 22-5** *John A. Mesmer, Missouri Athletic Club's housekeeping and laundry manager, directs an assistant in rearranging linen storage for guest rooms. Towel requirements for an athletic club are far higher (five times) than would normally be needed for the number of sleeping rooms because many members use locker room facilities after participating in various physical activities, and require extras.*

**Relation to other Jobs.** Promotion to: head seamstress, linen room attendant, inspectress, linen room supervisor, housekeeper.

**Job Breakdown.** The duties of this job may be divided between a seamstress of draperies and curtains (or drapery woman), and a seamstress of fine linens such as napkins and tablecloths.

**Job Combination.** The duties of this job may be included with those of linen room maid, or linen room supervisor.

## *Storage*

It is almost a truism among professional Housekeepers that there is never enough storage space for anything, anywhere, at any time (Figure 22-5).

To some extent this may reflect poor planning and lack of understanding on the part of the architect or of management. Management often waits until too late to hire an executive Housekeeper who would have been able to predict storage requirements at the time the structure was being planned. On occasion it may reflect false economies effected by management to pare down the estimated cost of the contemplated structure.

More likely, however, it reflects one or more breakdowns in housekeeping procedures. The professional Housekeeper will face the reality of the shortage of storage space and do the following:

1. Continually cull out and offer for sale to employees, or otherwise dispose of items no longer needed.

2. Devise techniques for arranging stock compactly rather than wasting space through inefficient packing.
3. Determine what storage areas not assigned to Housekeeping are seldom if ever used, and thus in all probability, available.
4. Remodel storage areas to increase utility (add or remove shelving, increase or decrease space between shelves).
5. Take advantage of space economies inherent in furnishings that fold, telescope, or stack.
6. Negotiate storage of bulk orders at supplier's warehouse, from which weekly or monthly requirements can be delivered automatically.
7. Use trucks, carts, and dollies for storage and delivery to points of use, thus eliminating double handling while at the same time increasing storage space substantially.
8. Use available ceiling space for storage of otherwise hard-to-store carpet. It can be rolled on tubing that slides onto metal rods suspended from the ceiling.
9. Have bare walls converted into storage walls, fitted with permanent or temporary shelving, or at the very least, wall surfaces can be used for hanging brooms, mops, or brushes from sturdy hooks.
10. Consult with management when all of the foregoing measures have been effected and storage space still remains inadequate.

## *Finding Storage Space*

It is particularly timely to discuss serious storage inadequacies with management when renovation or expansion of the facility is contemplated. At that time, it may become feasible to construct additional storage space in conformity with an almost ideal set of specifications. At all other times it may be necessary to make serious compromises between the ideal and the practical.

It is not unheard of for Housekeeping to become so desperately short of storage space that it is necessary to use the crawl space between floors, or even a portion of access space to the air-conditioning equipment. Penthouse storage space has even been erected on rooftops. Where it has been possible to conceal them from the public, prefabricated storage buildings have been erected adjacent to rear entrances of Housekeeping Departments. Even subterranean expansion of lower levels is not an infrequent (or inexpensive) means of acquiring urgently needed additional storage space.

A more desirable and economical source is the dead-end corridor. Every corridor that has solid walls on three sides of its terminus (i.e., uninterrupted by room entrances, escape, or inner stairway) already forms three-fourths of a storage unit, lacking only a front enclosure with access door—plus, perhaps, interior shelving. In a hotel or high-rise motel, storage units of this type are ideal for holding cleaning supplies, all room effects changed or replaced regularly (soap, stationery, ashtrays, electric light bulbs, coat hangers, etc.), plus the supply of floor linens.

One authority estimates the Housekeeping Department's spatial requirements, including an on-premise laundry, at a maximum of 15 sq. ft. per occupant. This completely overlooks the fact that the amount of storage space used by Housekeeping cannot possibly be measured in square feet, but only in cubic feet.

Designing a storage system to make use of virtually the entire height of the cube is the most critical of all factors in determining its efficiency. For example, 20 stacking chairs can be stored in just a little more than two sq. ft., while making total use of available height. Twenty folding chairs may occupy 15 sq. ft., perhaps rendering the roughly 105 cubic feet of air space above the chairs almost totally useless. On the other hand, chairs that neither stack nor fold may use 45 sq. ft. of ground space, wasting over 300 cubic feet of air space.

Trucks, carts, and dollies that are stored unloaded are highly wasteful of space. So, too, are bulky pieces of furniture, which often waste even the space they occupy because they are irreparable or obsolete. As a general rule there is no quicker or sounder answer to this kind of problem than to get rid of useless inventory. Given management's approval, such merchandise should be offered for sale at modest prices to the employees.

With few exceptions, the heaviest items in inventory should be stored at floor level, the lightest items on the highest shelves. Drapery rods, ladders, and other elongated items can be stored most efficiently by supporting them in a horizontal position on walls pegged from floor to ceiling for this purpose.

## *Meeting Room Equipment*

Visual aids intended primarily for use in meeting rooms, and including heavy projectors and cumbersome screens, are best kept in storage walls built into the meeting rooms in which they will be used. Tables, chairs, lecterns, blackboards, and all other pertinent paraphernalia ideally should also be stored within the rooms in which they will be used or nearby. The same rule applies to materials and furnishings for use in banquet rooms. Needless to say, all sound equipment and visual aid supplies should be kept securely locked.

Where check-room facilities are inadequate, nonexistent, or too remote from rooms for which special functions are scheduled, portable facilities must be provided by Housekeeping. Where the function is relatively small, a portable hat-and-coat rack positioned at the back of the room will usually suffice. Where the function is a large one, it may be necessary to line up a series of such racks along the corridor wall opposite the entrance to the room, and conceal them discreetly behind folding screens. This applies, however, only when an attendant will be on duty from first arrival to last departure. In the absence of an attendant, it is an essential security precaution to locate hat-and-coat racks within the function room.

One of the more onerous and specialized chores that may be the responsibility of Housekeeping in institutions that provide foodservice concerns trash removal and garbage handling and disposal. To eliminate offensive odors and potentially troublesome sanitation problems, if there is no automatic electric garbage disposer it is essential that a refrigerated storage room be provided for all garbage, no matter how short a time it may be held. Adequate sanitation requires daily cleaning of the garbage room, preferably with germ-destroying steam. The cans used to transport garbage to the point of pick-up should also be hosed daily, and again, preferably with steam-cleaning equipment.

Trash storage areas should be enclosed (but ventilated) and treated with disinfectants and insecticides, as well as have proper rodent control by outside professionals. Trash from guest room areas should not be immediately fed into a trash compacter. Due to employee carelessness, linens often show up in trash, as well as guests' treasures that inadvertently land in wastebaskets. Searches for paper, airline tickets, and souvenirs among the trash are common.

## *Other Space Needs*

Some additional Housekeeping storage areas frequently neglected in planning a hotel or residence hall include the following:

1. Space for outside furniture or seasonal equipment during off-season.
2. Space for items waiting to be repaired—mattresses, springs, guest furniture, radios, and television sets.
3. Accessible storage for outside and inside maintenance supplies and equipment, such as lawn mowers, rakes; salt or other supplies for walks and driveways; tools of all kinds, paint and other inflammable materials; oils; fireplace wood.
4. Storage on the floors for rollaway beds that guests may request.
5. Areas for baby cribs.
6. Maids' or janitors' closets. These should be conveniently located and readily accessible both to the guest room section and the chutes for soiled linen. Closets may also contain extra cleaning and guest room supplies, storage space for vacuum cleaners, brooms, pails, and other maids' utensils, and a cabinet with a lock for lost and found items, especially liquor.

All storage units must bear clear identification of their contents. Where appropriate, the quantity of items to be maintained in inventory should also be marked. Boxed or crated material

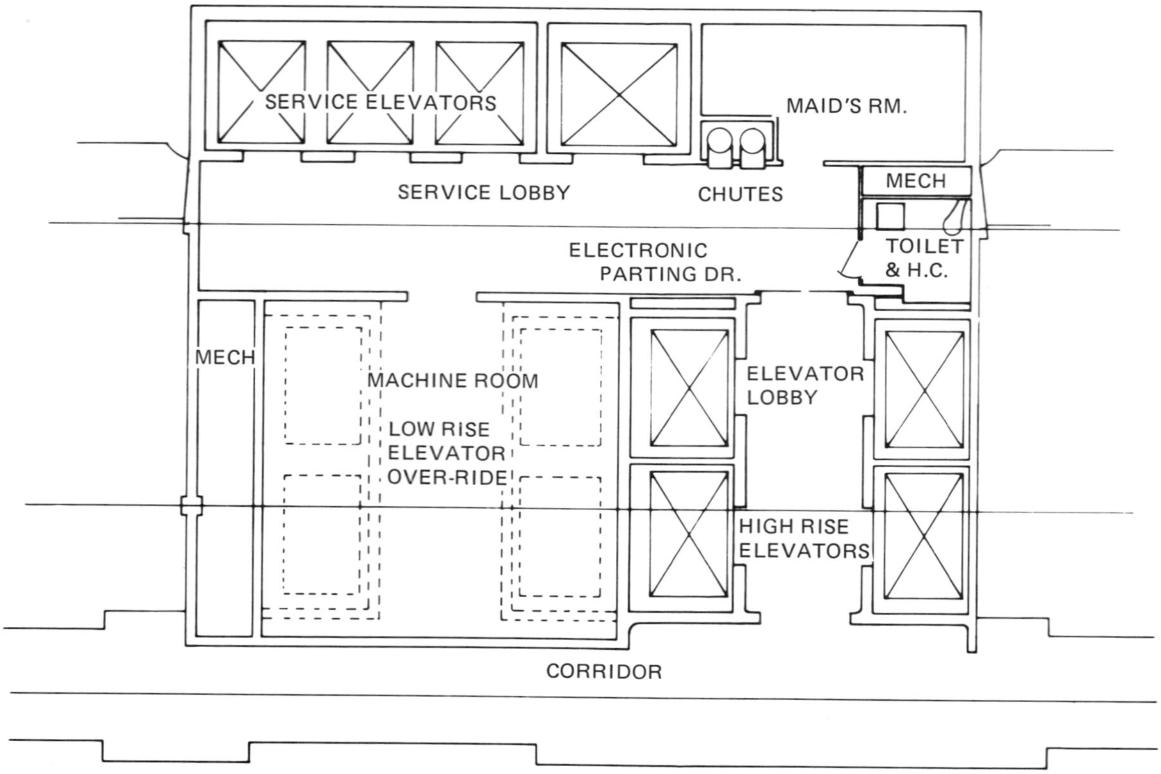


**FIGURE 22-6**

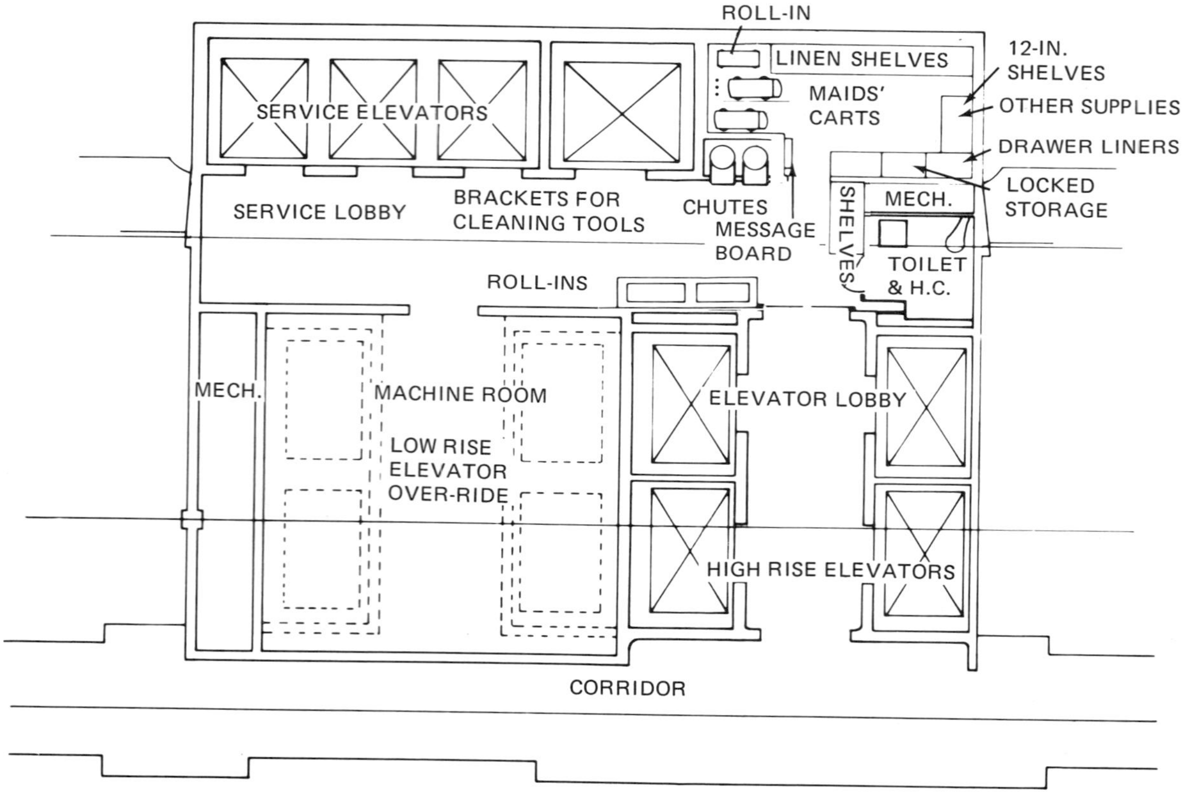


**FIGURE 22-7**

that is not to be opened immediately should be clearly identified and stored in the appropriate area. A complete record of the total inventory of the department should be kept by the executive Housekeeper.

On occasion the job becomes so cumbersome that it is only practical that employees in charge of individual storage areas keep separate sets of records for all of the materials for which they are individually responsible. These individuals will also provide the data required for keeping the master inventory up to date when it is being maintained by the professional Housekeeper.

Like the furniture, furnishings, supplies, and equipment they so carefully itemize, inventory records will also be carefully guarded. Housekeeping employees are, on the whole, conspicuously trustworthy. Otherwise they would long since have been parted from their profession. Inevitably, however, there are rare exceptions. While it behooves the housekeeper to maintain a public posture of complete confidence in the reliability as well as the capability of the entire department, both collectively and as individuals, it is equally important to make security measures so obvious, and so obviously effective, that temptation will be forced to find another department in which to rear its ugly head.

## *Making Best Use of Existing Space*

An architect's plan (Figure 22-6) simply left a room 6 ft. by 8 ft. with an area occupied by laundry and trash chutes. To make best use of space (Figure 22-7), the following were added:

1. 18-in. shelves for linen storage (adjustable heights)
2. 12-in. shelves for small supplies
3. Holder for drawer liners built with angled slots as stationery holder
4. Locked storage cabinet under counter (for lost and found items, especially liquor, until supervisors can deliver to executive Housekeeper)
5. Desk for floor supervisor or inspectress
6. Message board
7. Maids' carts overnight parking area
8. Wall bracket to hold long tools, such as sponge-mops or push-brooms
9. Roll-ins are extra beds for guest rooms
10. Extra roll-in space
11. Cabinet with shelves for buckets, cleaning supplies adjacent to janitor's closet

# Chapter 23

# Procedure Manuals

**OVERVIEW** All the benefits of manuals are explained, with examples of different types written for various tasks and levels of employees, including bed-making, turn-down, general cleaning, and so forth. Job description for a club's executive Housekeeper is also included.

A procedure manual is not to be confused with a job description. The latter—primarily intended for supervisors and management—states the extent of an employee's authority and responsibility. Except in the most general terms it makes no attempt to tell the employee *how* to perform a job. The procedure manual, on the other hand, tells each employee exactly how to perform each step of his job; is written in language that will be clearly understood; and, ideally, is illustrated with close-up photos (or drawings) of those job functions that can be far better explained by one or more pictures than by a complicated description.

Needless to say, the work of the Housekeeping Department cannot be set forth in a single procedure manual. An individualized manual is required for each position in the department. The master manual is comprised of all the procedure manuals for the entire department, plus all the job descriptions, including that for the executive Housekeeper.

## Master Manual

This offers two main advantages. For the Housekeeper assigned to a new property who must recruit, train, and break in a completely new Housekeeping Department, the master manual serves much the same purpose as a recipe file does for a chef. It is also valuable for a new Housekeeper when a predecessor retires or is promoted out of the Department.

Individual procedure manuals are really indispensable in all but the smaller lodging facilities. Without such a document an employee will probably perform his job in a slightly different way each day, and on some days may omit certain job functions altogether. If the matter is called to his attention, he will probably claim, and quite rightly, that he cannot possibly remember the exact order and procedure to be followed in executing literally scores of job functions. Even more serious, in view of the historically high turnover in Housekeeping Departments, is the extra time and expense involved in training replacements in departments without these booklets.

## Continual Revision

In writing a procedure manual it is important to bear in mind that it is never final. New supplies, equipment, and techniques are always being discovered; the employee will often suggest better ways of performing certain aspects of his job; the lodging property itself may be altered in such a way as to necessitate new housekeeping procedures. In any or all of these instances it becomes highly advisable to revise one or more manuals. Each one therefore should be viewed as nothing more than the latest version of a manual that will inevitably be revised.

Because of their impermanence, it is recommended that they be prepared in a reasonably inexpensive form. It is also recommended that they always be given a final revision by a professional skilled in the art of written communication with persons of sometimes limited reading skills. While it is often more economical to reproduce the manual by multilith, mimeograph, or even copying machine on 8½-in. by 11-in. pages, special consideration should be given to printing the manual as a small booklet that the employee can carry in his pocket for reference until such time as he has actually memorized every step of his job.

Where English is a second language for many employees, it is advisable also to publish the manual in the department's principal native tongue. It is also important to have a substantial supply of replacement manuals available. Each, incidentally, should bear the date of its latest revision.

## *Time and Motion Studies*

The most complex and valuable manuals are those that are based on actual time and motion studies of each housekeeping position. These are quite lengthy, since they not only break each position down into every single function, but break every single function down into every individual action that comprises the total function. They can help work minor miracles when used by reasonably well-educated employees, but otherwise serve little purpose.

An excellent example of a highly detailed, but basic, clear, and eminently useful instruction that might be included in a procedure manual is the following, prepared originally by TSgt. Lerow W. Jennings, Jr., housekeeping instructor at Sheppard Air Force Base.

## *Cleaning Stainless Steel in Elevator*

EQUIPMENT NEEDED:

(1) Wooden stick (such as tongue depressor)
(2) Three clean cloths
(3) Sponge
(4) One bucket—one-gallon size

MATERIALS NEEDED:

(1) Water and germicide-detergent (specify product to be used)
(2) Scouring powder
(3) Metal polish or clear white mineral oil (specify which to use)

BEFORE CLEANING:

(1) Mix ½-oz. germicide-detergent with warm water.
(2) Put on rubber gloves.
(3) Dampen one cloth and sprinkle with scouring powder.

DOING THE JOB:

(1) Using the wooden stick, remove any foreign articles stuck on the stainless steel (such as chewing gum). Gently "scrape" only in the direction of the grain scratch marks in the stainless steel.
(2) With the damp cloth and scouring powder clean any "spots" on the stainless steel (also clean over places you have "scraped" with the stick) being sure to rub only in the direction of the stainless steel grain.
(3) Rinse (clean) the entire surface of the elevator with the sponge and germicide-detergent solution.
(4) Using one clean cloth, apply metal polish to all stainless steel surfaces.
(5) Using another clean cloth, polish (by rubbing with the grain) all the surfaces that have had metal polish applied to them.

BEFORE LEAVING:

(1) Check floor—wipe up any water or solutions that may have splashed onto the floor while cleaning . . . using the sponge and germicide-detergent for the wiping.
(2) The elevator should have an overall clean appearance—there should be no "bright" spots in the stainless steel and no fingerprints or powder.

CARE OF EQUIPMENT:

(1) Sponge will be washed clean in germicide-detergent solution, rinsed and placed on rack outside to dry.
(2) Cloths will be rinsed and sent to laundry.
(3) Bucket will be wiped clean with germicide-detergent solution and turned upside down to dry.
(4) Wipe off rubber gloves and hang to dry.

If you do choose to include "Frequency:"

FREQUENCY:

(1) On Monday and Wednesday steps one, two, and three (under "DOING THE JOB") will be accomplished.
(2) On Tuesday and Friday all the remaining steps will be done.

## *Greenbrier's "General Cleaning" Manual*

A more typical example is the following instruction for general cleaning (as opposed to daily cleaning) a guest room as it appears in the Floor Housekeeper's Manual published by the Greenbrier resort:

"All wood furniture is to be washed inside as well as out. Use furniture polish on dark furniture.

"All leather bed heads and white leather chair seats are to be washed with Ivory soap and water and wiped dry.

"Have all brass switch plates polished as well as door plates and metal of any description in the room and bathroom. New chromium plates are not to be polished. Wipe off with Kleenex or damp cloth only.

"Have all mirrors, pictures, and frames washed and rehung. (Housemen do mirrors over chests.)

"Have housemen clean closet ceiling and room ceiling lights and vacuum all upholstery, box springs, mattresses, and carpeting.

"All lamps and shades must be washed; seam of shade must be placed to center back. See that light bulbs are wiped off and glass reflectors removed, washed, and replaced. Lamp cords are to be washed, and when too long, an order given to shorten. Gooseneck lamps are to be washed and new brass lamps also. Do not scour.

"All glass tops must be washed—both sides. Old hotel chests and desk glass tops are to be lifted out and area underneath cleaned.

"Have closet floor, shelves, and hooks washed; also door mirrors.

"Radiators (open) are to be brushed.

"Baseboards are to be either washed or wiped dry.

"Have bathroom wall tile thoroughly scrubbed and polished dry.

"Have bathroom floors washed and special scouring and bleaching done where stained.

"Toilet bowls are to be thoroughly scoured and the mirror-test used to make certain there is no corrosion under the rim of the bowl.

"Make certain all paint spots are scraped off medicine cabinets and lights, switch plates, and bath floors and tile.

"Shades of medicine cabinet lights are to be removed, washed, and replaced.

"Hinges of toilet seats must be cleaned.

"Marble saddles are to be thoroughly scoured and bleached if necessary.

"Have towel hampers washed inside and out.

"White wastebaskets are to be washed inside and out.

"Have window sills, casings, and ledges washed.

"Wastepaper baskets are to be washed inside and out and placed in line with vanity or chest.

"Window shades are to be pulled ¾ of the way down the window.

"Draperies are to be pulled toward center and to hang straight.

"Pencils by phones must be sharpened.

"All beds are to be made with comforters at foot ends of beds.

"Telephones and cords are to be washed and connected. All 'ragged' dials or directions under receiver are to be replaced.

"Connect all lamps.

"Lock all connecting doors except those to parlors after saddles are painted.
*"Your last look will be the guest's first look at this completed room."*

## *Century Plaza Night Turndown*

The following, from the Century Plaza in Los Angeles, is a representative example of well-written instructions for a single function performed by the hotel's maids; this is actually a time and motion study (Figure 23-1).
How to give night or turndown service

**1.** Pull spread gently back, exposing pillows.
**2.** Fold spread in 3-way fold:
   a. Top toward bottom of the bed approx. ¾ down;
   b. Bottom toward the head of the bed approx. ¾ up.
**3.** Fold right side to the center.
**4.** Fold left side to the center.
**5.** Fold one more time, and you have a folded spread that will fit:
   a. In dresser drawer;
   b. On closet shelf;
   c. On luggage bench.
**6.** Pull back second sheet, blanket, and third sheet in one operation into a triangle. *Double-bedded room occupied by two persons:* Follow same turndown procedure on both sides of bed. *Twin-bedded room occupied by one person:* Turn down bed nearest bathroom. Turn down side of the bed facing nightstand. *Twin-bedded room occupied by two persons:* Turn down side of each bed facing nightstand.
**7.** Place "Goodnight" mint on pillow.

General: Night service consists of:

A. Air out guest room.
B. Empty ashtrays.
C. Empty wastebasket.
D. If guests have had bar service, remove glasses.
E. Check bathroom for cleanliness.
F. Replace used towels.
G. Check bathroom supplies.
H. Empty bathroom wastebasket.

**FIGURE 23-1** *Night turndown service is offered only in those hotels, such as the Century Plaza in Los Angeles, that want to maintain a deluxe standard. Housekeeping has a very structured procedure that saves maid's time so each can give night service to about 60 rooms. In some hotels, maids can turn down as many as 120 beds, but at that rate they cannot give any other service.*

I. Empty bathroom ashtray.
J. Close draperies.
K. Turn down beds.
L. Turn on nightstand light.
M. On cold nights, place extra blanket on foot of bed.
N. If extra bed has been put in room, make sure there is enough bathroom linen and soap for an extra person.

Some hotels give guests a polite notice explaining that if they want turndown service they should clear belongings off beds. Hilton Hotels automatically give nightly turndown service to studio rooms but ask guests in regular rooms to call Housekeeping and request night service if they wish it. Management there has ascertained that many guests prefer that their beds not be readied for sleeping so early in the evening.

Theoretically, procedure manuals describe the ideal routine to be followed in the performance of every housekeeping function. In practice, minor deviations occasionally save time; at worst, they slow down the procedure only negligibly. It is important to remember that certain employees find the routine bearable only if they are permitted to make such minor deviations to break up the monotony of their jobs.

It is not enough, of course, to give a new employee a copy of the correct procedure manual, and send him off to his job—unless, of course, he has had previous experience under competent Housekeepers. There are few exceptions to the fact that the new employee will benefit from detailed discussion with his supervisor of all aspects of his new job.

Procedure manuals are most effective for the very complicated jobs. The employee should be taken through the job step by step, with the manual in hand, so that he understands it completely. Then giving him the manual each day when he comes to work will help to remind him of every step and he will not overlook anything. For those institutions in which the staffing is done by square footage and employees are scheduled to work in different areas, sometimes physically far-removed from their supervisors, the "how-to" manual is extremely helpful. It will eliminate the worker having to seek out his supervisor to refresh his memory on the proper procedures.

## *Audiovisual Training*

Audiovisual aids can prove invaluable in accelerating the training of a new employee. These can be as simple as a flip-chart presentation describing and illustrating the new job, or may involve a slide presentation accompanied by narration that is prerecorded as read by the supervisor. An audiovisual aid may be a motion picture that shows every aspect of the job as actually performed on location.

Training aids are available in abundance from audiovisual specialists, management, and consultants, and publishers of manuals and textbooks for specialized industries. There is virtually no phase of or job in the lodging industry for which standardized training materials are not already available. (Chapter 24 includes specific examples.)

Granted that audiovisual aids and procedure manuals designed specifically for the particular establishment in which they will be used are usually superior to prepackaged materials, the latter are often done more professionally and are relatively inexpensive because they are mass-produced.

## *Textbooks*

Most valuable of all, perhaps, are the specialized textbooks published by companies that deal almost exclusively in publications for industry. Almost all of these publishers will provide copies of their catalogs upon receipt of requests either from present or prospective members of the lodging industry. These publications are so highly specialized that they can seldom be found in public libraries.

Bear in mind, however, that they are in no way a substitute for procedure manuals, since they are for the most part, suitable only for employees at the supervisory level or higher. For the hourly-paid housekeeping employee, there is absolutely no substitute for a properly prepared procedure manual.

## *"The Correct Maid"*

One of the classic manuals in the hotel industry is a 64-page booklet called "The Correct Maid."* It has been written with general instructions but includes ample space for specific instructions that apply to the particular institution. The manual is well illustrated and written clearly on a level that will neither confound the uneducated nor bore the more intelligent maids.

The series of drawings in Figure 23-2 that accompany these instructions on making beds are the kinds of material contained in the still timely booklet, "The Correct Maid."

## *Making the Beds*

Turn the mattress as directed, observing the safety instructions to avoid backstrain. Regular

* Pfeiffer, William B. and Voegele, Walter O. Rochelle Park, New Jersey: Hayden Book Co., Inc., 1965.

**FIGURE 23-2** *How to Make a Bed According to* "The Correct Maid"

days are usually scheduled to turn the mattress side over side or end over end. Since there are various methods for making up beds, always follow your housekeeper's instructions. The method illustrated here has been found to save time.

Start at the upper left side and work counterclockwise around the bed as is shown. If you prefer, you can work clockwise, starting at the upper right side of the bed.

1. Place mattress pad properly on the bed. Put the underneath sheet on the bed.
2. Miter the upper left corner of the sheet.
3. Miter the lower left corner.
4. Put the top sheet on the bed.
5. Put the blanket on top of the sheet.
6. Turn the top sheet over the blanket.
7. Lightly tuck both the top sheet and the blanket along the side at the head.
8. Miter the top sheet and the blanket at the left foot of the bed.
9. Miter the underneath sheet at the right foot of the bed.
10. Tuck the sheet in.
11. Miter the top sheet and the blanket at the right foot of the bed.
12. Miter the underneath sheet at the head.
13. Fold the top sheet over the blanket and tuck it in to secure it.

14. Put the pillows at the head.
15. Put the spread in place.

Note that this approach to bedmaking eliminates walking back and forth around bed, thus saving time and energy.

## *Increasing Longevity of Beds*

Hotel beds will last 50 percent longer if wear is distributed evenly. One company has developed a patented method for controlling mattress rotation and improved its product actually to program each mattress for the optimum sequence of rotation. Functional yet inexpensive, this system consists of two iron-on labels that "remember" the last rotation and tell exactly which way to do it next time—in both English and Spanish (Figure 23-3).

There are also instructions for the periodic care of bed frames and foundations. A built-in calendar shows when rotations should be made, thus eliminating the need to keep records. The system also provides a positive means for confirming that rotations have been made correctly. It saves labor by eliminating confusion—no more unnecessary or nonproductive rotations. The system is completely self-contained and is guaranteed effective for the life of the mattress.

**FIGURE 23-3** *To help mattresses wear evenly and thus prolong their life, two different labels clearly define alternate rotations. A built-in calendar system shows when to rotate—quarterly. Completely self-contained, the tagging requires no record keeping. Labels guaranteed effective for the life of the mattress, have directives in both English and Spanish. Consider the savings possible if mattresses are rotated. Credit: RotaGuide Systems*

## *Job Description: Executive Housekeeper*

Just as executive Housekeepers make up manuals with job descriptions for their employees, so management has its own specifications for

**TABLE 23-1** Cost Savings Possible if Mattresses Are Rotated Properly

| Number of Rms. | Number of Beds | Total Cost of Beds @ $200. ea | Annual Cost of Sleep Equipment with avg. life (8 yrs.) | Annual Cost of Sleep Equipment with +50% more life (12 yrs.) | Savings with Rotation Annual Savings First year (less cost of system) | Savings with Rotation Annual Savings Following years − 11 (each yr.) | Total (12 yrs.) |
|---|---|---|---|---|---|---|---|
| 50 | 75 | $ 15,000. | $ 1,875. | $ 1,250. | $ 550. | $ 625. | $ 7,425. |
| 75 | 113 | 22.600. | 2,825. | 1,883. | 829. | 942. | 11,191. |
| 100 | 150 | 30,000. | 3,750. | 2,500. | 1,100. | 1,250. | 14,850. |
| 125 | 188 | 37,600. | 4,700. | 3,133. | 1,379. | 1,567. | 18,616. |
| 150 | 225 | 45,000. | 5,625. | 3,750. | 1,650. | 1,875. | 22,275. |
| 200 | 300 | 60,000. | 7,500. | 5,000. | 2,200. | 2,500. | 29,700. |
| 250 | 375 | 75,000. | 9,375. | 6,250. | 2,750. | 3,125. | 37,125. |
| 300 | 450 | 90,000. | 11,250. | 7,500. | 3,300. | 3,750. | 44,550. |
| 500 | 750 | 150,000. | 18,750. | 12,500. | 5,500. | 6,250. | 74,250. |
| 1000 | 1500 | 300,000. | 37,500. | 25,000. | 11,000. | 12,500. | 148,500. |

the department heads. From the Club Manager's Association Manual the following job description is quoted.
(Note that in clubs, staff will almost always have fewer maids and less guest room cleaning, but there will be considerably more cleaning in public areas.)

## *Executive Housekeeper*

**Job Summary:** Supervises and is responsible for cleanliness, order, and renovation of the Club; orients and plans work of housekeeping personnel; hires, discharges, disciplines, and instructs employees.

**Work Performed:**

1. Supervises Club housekeeping: receives notification regarding prospective social functions and makes necessary arrangements with HOUSEMAN and MAIDS for cleaning; in cooperation with CATERING MANAGER, instructs HOUSEMEN to set up banquet and private dining rooms; receives room numbers of departing guests and makes list for each ROOM MAID indicating the number of vacant rooms on floors assigned to her; examines INSPECTRESS' report to determine rooms wherein cleaning was inadequate, supplies lacking, furniture broken, or painting and papering needed; inspects such rooms, making detailed memoranda of work to be done; notifies CASHIER of breakage or missing articles attributed to guests, suggesting approximate cost of replacement; assigns HOUSEMEN to regular work and any special tasks that may be required, such as setting up chairs and tables, moving furniture, or hanging curtains; inspects cleanliness and order of rooms, banquet halls, halls, lobbies, foyers, and other public spaces to check on work of housekeeping staff, looking particularly for cleanliness and order in places likely to be overlooked in the daily routine; adjusts complaints regarding the housekeeping service or equipment; checks employees in and out determining whether they are on duty in fit condition for work. Orients and plans work of employees: informs new employees of Club regulations, breaks in new employees by assigning them to work with experienced workers, conducts periodic housekeeping discussion groups for employees; may compile and maintain a blackboard work schedule for edification of housekeeping personnel such as MAIDS, HOUSEMEN, and INSPECTRESSES.
2. Disciplines, hires, and discharges employees; checks on work of employees through personal observation and reports of INSPECTRESS and HEAD HOUSEMAN; distributes work equally as a means of maintaining morale and discipline.
3. Supervises linen room: takes periodic inventory of linen and supplies in linen room; supervises mending and stamping of linen by SEAMSTRESS and LINEN ROOM WOMAN, and issuance of supplies and materials to HOUSEMEN and MAIDS; checks laundry bills against entries by LINEN ROOM WOMAN to ascertain correctness of charge assessed; forwards requisitions for supplies or materials to MANAGER or PURCHASING AGENT.
4. Supervises remodeling and renovation: confers with MANAGER on color scheme or arrangement of rooms to be decorated; selects materials to be used in redecoration, such as paints, fabric, light fixtures, furniture, and carpets; supervises activity of PAINTERS, PAPERHANGERS, and other employees engaged in work to insure progress and quality.
5. May buy materials and supplies: purchases linen, cleaning materials and housekeeping supplies in person, by correspondence, or over the telephone; selects and tests wall papers, upholstery, carpets, furnishings, fixtures, fabrics, and other articles to be used in redecoration. In some establishments, all purchasing is performed by a PURCHASING AGENT.
6. Writes out reports for management: records expenditures, makes out payrolls, reports number and condition of rooms cleaned, ar-

ticles missing and articles broken, and amount of laundry used.

(The worker must constantly exercise care to safeguard Club property against theft or loss.)

**Relation to Other Jobs:**
Promotion from: parlormaid, housekeeper assistant, inspectress, linen room maid, linen room supervisor

**Job Combination:** The duties of this job may be combined with those of room maid, inspectress, seamstress, or head houseman

**Special Qualifications:** This job is usually filled by promotion from within the establishment, or with an individual who has performed the same job in a smaller club or hotel. Decoration of rooms and public spaces requires not only a knowledge of materials and their values but also the ability to give an appearance of comfort and dignity to the establishment.

## Maid's Manual

*The Greenbrier's Maid's Manual is a small, 24-page booklet, about 4 in. by 5 in. that can fit easily into maid's pocket. It is both a procedural manual, telling the maids just how to do their jobs in cleaning guest rooms and bathrooms, and also includes certain rules and regulations that apply to the Housekeeping Department personnel. It even includes rules about appearance, cleanliness and grooming.*

## Correct Uniform

Green check dress with head band and apron furnished.

Black comfortable shoes; medium or low heel. Closed toe white shoes may be worn at all times if a maid prefers them. However, these must be kept clean.

Hose may be worn or not, as preferred. However, if they are not, shoes *must* be closed toe oxfords or loafers.

## Neat Personal Appearance

Uniform must be neat, clean and worn as directed.

Jewelry is not permitted while on duty. (Wedding ring and wrist watch only exceptions.)

Perfume and bright or dark nail polish are not to be used.

Heavy makeup is not permitted.

Hair must be well groomed and if longer than collar line, worn in a net.

## Personal Cleanliness

Bathe daily.
Change underclothing regularly.
Keep hair neat and clean.
Use a deodorant.
If necessary use mouthwash to guard against unpleasant breath.
Gum chewing is not permitted.
Keep fingernails trimmed.

## Maid's Rules

Report: *To the Floor Housekeeper or to the Housekeeping Department the following:*

1. Overnight damage to a room, equipment or furnishing.
2. All money, jewelry, clothing or any articles found in room upon departure of guest.
3. Baggage left in a room given to maid as a check-out room to be cleaned.
4. Illness of a guest.
5. If not able to enter a room by 2:00 PM.
6. Expressions by guests in regard to food, service, room rates, and other matters affecting the hotel.
7. Check-outs put in order.
8. Lights and plumbing fixtures not working.
9. Birds or animals in rooms.
10. Vermin in rooms. This includes mice, cockroaches, bedbugs, silverfish, waterbugs, ants, wasps, flies and spiders.

**11.** Room service tables or trays found in guest rooms or corridors.

All Housekeeping department employees must train themselves to pick up matches, cellophane, cigar or cigarette butts, papers or anything strewn on corridor carpets. Maids are to keep the carpet clean in front of their closets.

Maids and housemen are responsible for closing all windows in rooms and corridors in the event of rain or snow storms.

Employees should never say, "I do not know"—state, "I will be glad to find out and let you know"—and call Housekeeping department for information.

Remember, your immediate supervisor represents the management of the hotel and must be shown respect and courtesy.

We are all working for the same purpose—to keep the Greenbrier standards at the very top. Let's work in peace and harmony, cooperate and help each other, and this will make it pleasant for everyone.

## *Your Work*

Go to Floor Housekeeper's office, get floor keys, note bulletin board for special instructions, get lists of arrivals and departures, and be on your section at 8:30 AM in *Full Uniform.* Note: The key issued to you is to be regarded as a special trust. Employees must *never unlock* a door for a guest without first being sure the guest belongs in that room. If uncertain, ask the guest's name and check with the Housekeeping department.

## *Taking Room Report (on your section)*

If you find the indicator protruding on a door, do not disturb the guest for this means the room is locked from the inside.

On other rooms, where the indicator is not protruding, rap gently and if there is no answer, enter room and mark on report whether vacant, occupied, a check-out or out of order (ooo). In an occupied room designate the number of guests in the room.

This report is then taken to Housekeeper's office on your floor.

You are now ready to start your work. Go to the Linen or supply closet and get necessary equipment.

*First*—put check-outs in order, so they can be turned in for your next guest.

*Next*—make up the occupied rooms. If you have a request from a guest, *that* room must be made up before your checkouts. Do not make up while guest is in room except upon request. Leave door open if guests are in room.

Vacant rooms are left until last, unless maids cannot get into occupied rooms. They should be dusted thoroughly, aired and equipment and supplies checked. If necessary rinse out tub, basin and toilet bowl.

# Chapter 24

# Training and Incentives

**OVERVIEW** Here are examples of different types of training being done throughout the lodging industry on all levels—entry-level through management. Audiovisual training aids currently available are reviewed and sources provided. Incentives, both material and psychological, are discussed. Motivation, possible through enlightened management, is proving to be as powerful a force as financial gain.

People are the lodging industry's single greatest investment. To protect that investment management must provide: (1) the training programs that make it possible for employees to perform at peak efficiency, and (2) incentive programs to make it economically worthwhile for employees to exert themselves.

Let us look first at training, an activity that less enlightened executives tend to neglect because of its high cost, especially in terms of employee time spent in training classes rather than in the performance of functions. Aware that employees of any institution with a solid reputation for progressive programs are especially attractive to companies that scrimp on training, unenlightened management often prefers to raid other organizations for trained personnel rather than sponsor programs that might make its own employees the target of the personnel raiders.

However, most of the highly successful lodging chains have become persuaded that they must, indeed, protect their investment in people, and consequently establish training programs. The majority start at the top level.

## TraveLodge Training

Typical of the better-managed motor hotel chains, TraveLodge International, Inc., El Cajon, Calif., conducts a two-week training school every other month throughout the year. It is mandatory that all new partners, managers, franchisees, and housekeepers attend. Participants are given many hours of on-the-job training at one of the company-owned properties located near the suburban San Diego home office.

In addition, TraveLodge International has area managers who work on-premise at their individual properties throughout the United States. These men are well qualified and knowledgeable about the hotel/motel business. They visit TraveLodge properties in their respective areas every 90 days, and normally stay at each property 2½ days during which time they assist operators in all phases of motel management. They make a thorough inspection of the property both inside and out.

Often the area managers hold training classes for the housekeeper and maids at a property, at which time they review TraveLodge Housekeeping standards and screen up-to-date training films.

## Holiday Inn University

Probably the most famous, as well as the most heavily attended of all the training schools operated at the lodging industry, is Holiday Inn University (Figures 24-1 and 24-2). Here new innkeepers and department heads obtain their basic training, while their seasoned counterparts return to the campus for postgraduate courses in the latest management and operational developments. It is essential to good management that innkeepers gain a thorough understanding of Housekeeping's contribution. Although many of the lodging chains are intent on protecting their investment in middle and upper management by providing formal training programs, this policy is not always applied to hourly employees. The irony is that the performance of these employees ultimately determines whether the standards set by corporate policy-makers are truly practiced or merely preached.

Enlightened management, however, knows that appropriate training repays its cost in increased productivity alone. It also pays for itself in reduced turnover because competence breeds confidence, pride in one's work, and job satisfaction. These attitudes contrast dramatically with the demoralized, disgruntled outlook typical of the marginally competent worker who has been given little opportunity to learn his job, and, even more discouraging, has almost no opportunities to earn promotion to more attractive positions.

Competence, in short, is the basis for the high morale that is essential to the development of the esprit de corps that characterizes well-run departments. Competence also has other and even more tangible byproducts. The competent (i.e., well-trained) employee is unlikely to damage expensive housekeeping equipment because nobody bothered to explain its proper operation. Nor is the properly indoctrinated and trained employee likely to offend guests because of obvious incompetence or objectionable behavior.

Joseph Migyanka, manager of the Training Division of Wisconsin Associates in Pleasant Prairie, Wisconsin, offers the following motivation tip:

"The way one starts a skills training program can often determine its success or failure. Frequently, a training department or someone in a capacity to initiate a training program develops and implements one within their own resources. Then they announce who is to be trained and when the training will begin. Now it is management's program, not one for those who are to be trained. On the other hand, early involvement by trainees in the design and implementation of

**FIGURE 24-1** *Holiday Inn University is located on an 88-acre campus in Olive Branch, Miss., just across the border from Memphis, Tenn. where Holiday Inns are headquartered. HIU continually holds classes for innkeepers and department heads such as executive Housekeepers.*

a training program has proven very successful. The more involved a trainee is, the greater the impact of the training system.

"A useful strategy is to ask for volunteers from the work force to act as a Training Committee for the program administration. Key individ-

**FIGURE 24-2** *At Holiday Inn University, housekeeping systems are continually revised and improved. Inspectress is checking to see that everything needed is on cart. Efforts are being made to devise system to ensure that each maid has what she needs to service room.*

uals can provide valuable input into exactly what training is most wanted, what rewards can be offered, how to run the program, etc. Also, by involving a select group of intended trainees in pre-training, an incredible, but effective communication takes place . . . the grapevine. If one's program is intended to be helpful for both the company and its people, then the grapevine will render valuable support for 'their' program.

"Utilization of constructive feedback from the workers, quick reaction to their concerns and involvement from the beginning will mean the difference between success and failure."

## *Four Types of Training*

There are four distinct groups for whom adequate training should be regularly provided by the Housekeeping Department: (1) entry-level employees, (2) potentially promotable employees, (3) supervisory employees, and (4) management trainees.

### *1) Entry-level*

In an ideal situation, the institution employs a full-time training supervisor who conducts all classes in a suitable room; in a hotel this may be a sample furnished room reserved expressly for this purpose. Initial training for new maids should cover all of the following:

- How to make a bed
- How to clean an occupied guest room
- How to clean a bath
- Preparing a check-out room for a new guest
- Care and use of housekeeping equipment

For maximum efficiency, the employee should be taught to perform the same functions in the same way, in the same order in guest rooms and/or public areas. Formation of strong habit patterns minimizes both waste motion and human error (e.g., varying the number of towels placed in rooms, failing to replace the soap or provide sterilized glasses). Nevertheless, certain employees will deliberately deviate from the recommended routine to reduce the boredom caused by endless repetition of precisely the same procedures. A recent survey indicated that the professional Housekeeper might be well advised to tolerate minor deviations if the end result is satisfactory and the individual's work quota is not adversely affected.

Other subjects to be included in the training program are:

- Good work habits
- Good health habits
- Daily work schedules
- Appearance and actions
- House rules
- Safety
- Property orientation

There is considerable evidence of a correlation between property orientation and morale. The employee who at the outset is made familiar with the entire property, and subsequently is invited to be one of the first persons to preview redecorating or improvements and additions, will take pride in gaining recognition and being accepted as a full-fledged member of the total organization, not considered merely an obscure member of the Housekeeping Department. The real importance of the maids, for example, can be most eloquently communicated by inviting them to inspect rooms before, during, and after completion of scheduled refurbishment. Nor would it be amiss to solicit their often highly informed and practical opinions.

Showing maids how the Front Desk operates also gives them a better understanding of where their work fits into the total purpose of a hotel. It will demonstrate clearly that when guests are waiting to check in, room clerks need rooms that are ready for immediate occupancy.

Training for maids and housemen must be keyed to *them,* according to Florence Cunningham, Sheraton Hotel's former corporate director of housekeeping who says, "I've seen a lot of supposed housekeeping experts who try to run training programs without understanding that our employees are at a different level than personnel of other departments, and so the trainers end up talking 'at' the trainees. The trainers simply do not understand the language of the people in the Housekeeping Department, who are primarily from minority groups and are not easily motivated by rewards other than higher wages or days off."

Institutions without a training supervisor adopt the buddy system. This teams an experienced employee with the newcomer as friend, confidante, "tour director," and trainer. This system works particularly well where experienced workers possess not only the requisite patience and genuine desire to assist the newcomer, but have at least some small measure of a teacher's ability to communicate.

Many institutions depend on floor supervisors or floor housekeepers to train maids. This is considered one of their essential functions.

"Maids," according to Henrietta LaBorde, San Francisco Hilton's executive Housekeeper, "behave the way the floor housekeepers train them. The latter must set a good example, have keen powers of observation, and common sense. The floor housekeeper must know each maid's idiosyncracies and be on the lookout for certain neglected items when inspecting the rooms her maids have cleaned.

"The floor housekeepers should not complain about their maids—they should be training them. They have meetings with the maids once daily and have a chance to give them instruction as well as to work with them individually on problem areas."

In addition to basic training for entry-level employees and newcomers in any Housekeeping category, there are several other types of activities that can benefit both the institution and its employees. Regular meetings of the staff can profitably be conducted by the Housekeeper and assistants to:

- Explain changes in policy or recommended procedures
- Allow suppliers to demonstrate and discuss new equipment and other products to obtain opinions of staff members who would use the products if purchased
- Stimulate two-way discussion between the workers on the one hand and supervisors and managers on the other (wise housekeepers never underestimate how much they can learn, year after year, from their own employees)

At the Broadmoor, Valeri Richardson's training consists of a one-week period during which the maid will work swing shifts and be assigned to any floor where she is needed. At the end of that week is a mini-evaluation by Richardson herself. She emphasizes the fact that this should include compliments as well as constructive criticism.

For the next six weeks the maid is assigned on a semipermanent basis to floor, followed by an evaluation reviewing seniority, attendance, appearance, performance of duties, and attitude. "If all points are AOK I say 'congratulations, you've made it', and offer the maid her choice of floor assignments," says Richardson.

By allowing the maid to choose her favorite floor (with or without a lot of stairways, the decor she likes, view, etc.) Richardson rewards her and also makes her feel a sense of responsibility for the rooms she has personally chosen to take care of.

## 2) *The Potentially Promotable*

Advanced training to prepare employees for additional responsibilities should be administered with caution. Candidates must have demonstrated such outstanding potential that their selection is not likely to generate jealousy or suspicions of favoritism. Other qualities being equal, preference should be accorded on the basis of seniority, which is the most reliable evidence of the employee's desire to build a career within the department, or at least within the institution.

In most of the larger properties in big cities it is cause for celebration when a maid is promoted to floor supervisor (if that is the next rung up); or, in a 100-room motel, if she is made assistant to the head maid.

Very often a maid would be qualified to become a floor supervisor except that she is not proficient in her use of English. Perhaps if management would pay for night school to improve her language skills she would be able to assume the higher duties within four to six months.

As to having the actual duties of the floor supervisor communicated to her, that may best be done by another supervisor who works on the buddy system. It is often easier to learn from someone who is new enough in the position to remember what was difficult for her to learn, or just what memory joggers and checklists made her more effective.

Because night crews earn a higher hourly rate than day crews there may occasionally be an ambitious night houseman who should be carefully observed to evaluate his potential as head houseman and eventually night foreman. Day housemen might be tried in the role of handyman on a day when the regular man (with whom the matter will have been discussed) is off, just to see if the trainee may be resourceful enough to fit in there. If so, he can be trained by the experienced handyman.

Another big factor in motivation of the Broadmoor Housekeeping Team is the role model Ms. Richardson provides for all under her supervision. She started as a maid and makes sure everyone knows it. Since most of the supervisors also begin as maids, newcomers get the message that there is upward mobility.

"I encourage them to take courses at the community college, or correspondence classes . . . or if there are no appropriate classes scheduled, I'll take evenings to help girls who are interested." Richardson says all the extra time she spends pays off in the end: "I wouldn't do it if it didn't."

## *3) Supervisory Training*

Selection of persons to be trained for supervisory (not hourly) positions, such as executive assistant housekeeper, must be based on much more than such measurable factors as past performance, general intelligence, and proper attitude. Such intangibles as leadership qualities and the ability to command respect without resentment must also be considered.

Preparation and administration of the actual training program is greatly simplified by the availability of pertinent materials from such groups as the American Hotel and Motel Association, and the National Executive Housekeeper's Association. As a matter of fact, there is virtually no limit to the quantity and quality of aids and assistance available from audiovisual firms, educational consultants, management consultants, motivation experts, and other individuals and companies that specialize in training employees at all levels.

A refresher course for housekeeping supervisors has proved very helpful in some hotels. At the Bonaventure Hotel in Los Angeles, supervisors' workshops are held periodically to maintain awareness of guests' needs and to upgrade standards of proficiency.

## *4) Management Trainees*

The following is a reasonably representative prospectus for a long-range training program designed to qualify employees for junior management positions.

**Housekeeping Program for Westin Hotels.** The candidates for this program are usually hired by the director of housekeeping at one of the individual properties. The length of time spent in each post in the department is adjusted to fit the individual's needs. Trainees who have earned a college degree are usually scheduled to complete the program in one year; those with less than four years of college may take up to two years.

During the first phase the trainee should learn the basic functions and responsibilities of the Housekeeping Department as they occur on a day-to-day basis. During the second phase, the trainee should act as a junior assistant housekeeper. This should get the trainee involved in administrative duties as well as direct supervision of housekeeping staff.

*Work experience* is acquired in this order.

**1. Lost and Found.** To be clerk in this division of Housekeeping gives the trainee a chance to see all the activity within the Linen Room, to see maids come and go, to get a feel for the action. In addition to keeping the log on found items as well as those possessions that guests or visitors report missing, this clerk takes care of department correspondence and most important, the payroll. Having to keep track of different employees and their hours and figure salaries and deductions will certainly test this individual's business acumen. The way the trainee handles these duties, answers the phone, and takes messages for the director and assistants will reveal whether the trainee is management timber.

**2. Night Floor Supervisor.** It is usually easier for an employee to start on the night shift because it is less busy and the trainee can take more time to learn. This is, of course, affected by the assistant housekeeper on duty at night. If she is not capable of explaining procedures properly, then it is senseless to burden her with such tasks.

**3. Night Linen Room.** In this position the trainee will have a chance to answer a few guest calls with special requests; learn how to keep the log book; will not be rushed in learning to make up the Front Office room report; will have numerous opportunities to observe how the assistant housekeeper handles keys, issues supplies, fills in forms, and the like.

**4. Day Floor Supervisor.** This job gives trainees a chance to see how maids actually work, how they are trained, with an opportunity to inspect their clean rooms, and to communicate with Rooms Control in the Linen Room.

**5. Rooms Control.** Working at the telephone in the Linen Room to fill in the house count form, the trainee will be in touch constantly with floor supervisors who call in numbers of ready rooms; then will communicate periodically with the Front Desk to tell them which rooms are ready to rent; and may also give instructions to supervisors based on guests' requests for special items.

**6. Inspectress.** This is a very responsible job that involves checking all rooms where VIPs, MIPs, and pre-reg guests will be staying. These rooms are inspected not only to see whether maids have cleaned them properly, but to determine what repairs may be needed.

**7. Night Assistant Housekeeper.** This position is one of the most responsible because in the absence of the executive housekeeper, all important decisions must be made by the night assistant housekeeper, who also learns how to handle employee grievances, guest complaints, and emergencies.

**8. Scheduling and Training.** This work will be important to trainee now because he or she knows what maids and floor supervisors need to know and why, and understands the techniques for scheduling maids.

**9. Costume Control.** Only the assistant housekeepers relieve the main costume control person, so learning how to issue uniforms and to keep all the proper records is valuable to understand, as the trainee may be needed there at any time.

**10. Sewing Room.** This can be an enjoyable experience if the trainee is an experienced seamstress, but it can be discouraging if she is not, inasmuch as the heavy commercial machines are particularly hard to handle. Scheduling the work in and out and keeping the load moving will be the most important part of the stint here, which is the briefest of the twelve stops along the training route.

**11. Day Assistant Housekeeper.** This job is most difficult because it entails covering the entire building in order to have close contact with maids to be sure things are going smoothly. Having the assistant executive housekeeper take a personal interest in their work is psychologically rewarding to the maids. Trainee also learns how to interview.

**12. Orientation to Other Departments.** As explained in Chapter 4 (Areas of Responsibility) and Chapter 26 (Interdepartmental Relationships), there are many other departments in hotels that Housekeeping must understand to work with them effectively (Figure 24-3). This final step in the training program primarily involves observation, but as always in a hotel, there are unexpected busy times when all hands are needed on deck.

Other involvements planned for trainees include: making out monthly progress reports, reading selected articles from trade magazines and books, touring mattress factories and laundries, visiting housekeeping facilities in other institutions, attending professional association meetings such as NEHA, Home Economists in Business, and attending workshops.

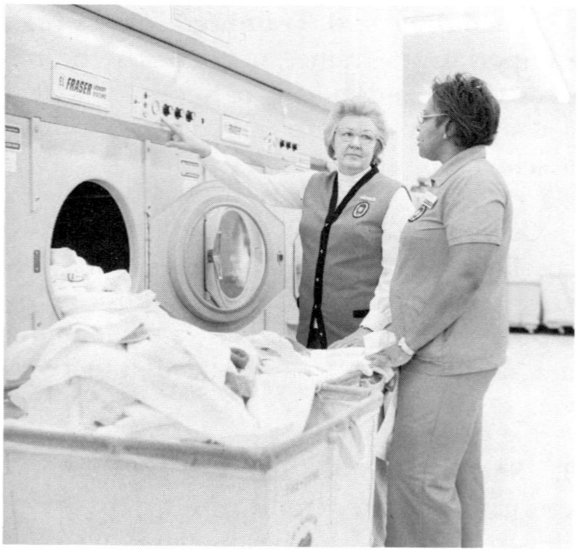

**FIGURE 24-3** *Mrs. Mattie Bowman, (left), executive Housekeeper at Holiday Inn University, trains one of her maids to use the University's laundry equipment. On-the-job training is more effective than any classroom theory, especially where expensive equipment is involved.*

Quite apart from in-house training is advanced training for executive Housekeepers or those assistants who may soon be ready for promotion. For example, once a year three-day Housekeeping seminars are offered in Sheraton's various regions around the country. Emphasis is on management skills, including the psychology of effective interpersonal relationships with subordinates, peers, and superiors.

Westin has a single annual Housekeeping seminar for all of its executive Housekeepers and many of their assistants. It lasts for several days covering, as does Sheraton's, mainly "people problems." Recently it was conducted with the Rooms Department seminar so that there would be an exchange of ideas and information between these two departments that are normally in close contact.

In addition, directors of housekeeping attend workshops and short courses to improve their executive abilities.

## *Training Investments that Paid Off*

To capture the attention of hourly employees and stimulate their interest in training, many different institutions around the country have made investments in new kinds of programs. The case studies that follow exemplify some creative training programs that paid off.

### *1) Instant Replay*

**The Investment.** A training director (or line supervisor) who is responsible for training trainers (who will in turn train hourly employees) and also for the initial training of new employees is part of the staff of each unit in the Host International chain. After the basic orientation, new employees are given preliminary classroom on-job-training sessions with job descriptions. Then the tasks are demonstrated with a verbal review.

Next, new employees are turned over to on-the-job-trainers—experienced employees who have been trained to train. They help trainees to learn tasks, usually during nonpeak hours. New personnel are trained either in groups or individually, but all wear a badge that identifies them as trainees. All trainees are filmed on closed-circuit TV so that they can see how they performed and then discuss it with supervisors. Prior to going on the floor, floor supervisors often do role-playing to check themselves and gain confidence; these sessions are also filmed on video tape.

**The Payoff.** Employees perform well, feel confident, learn to be self-critical, maintain high standards, and (most important to the company's profit picture) turnover is cut impressively, reportedly about 60 percent. Before training in this manner, the company's turnover was over 400 percent!

### *2) Role-Playing*

**The Investment.** Ramada's Management Development School (begun in the early 1970s) not only trains managers, but helps them get hourly employees involved in decision making. According to Ramada's dynamic and outspoken director of training, Robert Petrie, "Hoteliers are usually conservative and autocratic. The employee of today wants involvement."

One of the exercises is a role-playing situation, called The New Truck Dilemma. Employees are split into five groups of four. They are told that they are no longer room attendants, they are all truck drivers. A new truck has been purchased and each group is to decide among its members who will get to drive it. (In each group the decisions vary: one group decides that Harry should get the truck because he has seniority; another decides that George deserves it because he has always had the oldest truck with all the problems—and so on.)

After each group has made a decision, they all get together and vote. Each group ends up voting that it had the right answer. (There is actually no right answer, just proof that a group thinks its decision is best because its members decided together and each individual has had input.) The New Truck Dilemma is then used as a tool in group decision making for actual problems arising in foodservice/lodging situations.

**The Payoff.** Increased employee participation and greater morale result, building up a feeling of responsibility towards the job. Employees resolve their own differences, therefore excluding jealousy and the feeling that the manager made the wrong decision. This kind of decision making can be used in scheduling, production, and layoff areas. Most important, the employee feels that he can voice his own hopes or fears to his supervisor rather than to his union.

## 3) Communications: Fishbowl Technique

**The Investment.** An innovative training method is referred to as The Fishbowl and the objective is to develop a communications link between middle management and hourly employees. Here is how it works. Managers sit in a circle on the floor; hourly employees surround them. Employees are told, "All you can do is look in on this management discussion. There will be no interdiscussion." Managers are then asked, "What do you think workers want out of their jobs?" A discussion among managers ensues while employees listen. The answers from managers are usually: hourly employees want bigger salaries, better working conditions, and the like.

Workers then trade places with managers and sit in the middle. They are asked, "Are these things what you want?" The answers are usually emphatic "NOs." The group leader says to them, "Then, what *do* you want?" Answers vary, but the major consensus is: more recognition, room to grow, flexibility to meet personal objectives. (An aside: the manager usually wants the same things the workers want, but the manager usually thinks that employees only want more pay!)

The group is then split for discussion sessions of one manager to two to three workers. Manager speaks about those points the workers made, and worker does same to the manager.

Then the entire group convenes. One worker from each group reports what they found in common and the differences between managers and workers. Conclusions are drawn from the "audience" as to how to better the needs of workers and the needs of managers, and link the objectives of both.

**The Payoff.** Managers know employees' needs, employees know managers' needs, and each has finally had the chance to speak out. Managers realize that salaries are not the most important thing to workers, and can therefore offer new job techniques and opportunities to those who show promise. Most important, employees eventually find out that management really cares, and with that discovery something important has been done to bridge the communications gap.

## Audiovisual Aids

Many good slide films and motion pictures are available from several different producers. Perhaps the most prolific that deals with the subject of lodging is National Educational Media. Their films can be rented or bought.

For Housekeeping employees:

| | |
|---|---|
| H 201 | *Cleaning the Bathroom* |
| H 202 | *Making Up the Room* |
| H 203 | *The Front Desk* |
| H 204 | *The Bellman* |
| H 205 | *Room Service* |
| H 206 | *Courtesy Is the Answer* |

For middle management and the professional Housekeeper, a few include:

**Eye of the Supervisor M 301.** Employee evaluation; communications; performance appraisal. Principles of employer relations; nondirective communications; self-development. Employees' basic job needs—approval, recognition, consistent performance standards; perceiving total performance; listening and evaluating.

**The Supervisor—Motivating Through Insight M 302.** Employee motivation; communication. Principles of motivation; organization development; team building. Motivation through recognition, participation, creativity, job challenge, dignity; myths about motivation; value of self-motivation.

**Discipline—A Matter of Judgment M 304.** Employee evaluation; motivation; discipline; counseling and guidance; employee relations; EEO; collective bargaining. Leadership; organization development. Dramatizes the imaginary courtroom trial of a supervisor who took disciplinary action against three employees. Emphasizes the need for objectivity and impartial judgment.

**The Training Memorandum M 305.** Orientation to personnel development; introduction to training. Organization development; team building. A mind-opening film to motivate managers to change attitudes of resistance and indifference to training.

**Increasing Productivity M 306.** Principles of productivity; manager as change agent. Maintaining status quo; coping with change; problem solving; creating productive change; managers' responsibilities as change agents. Shows energy and intelligence as the essential resources for productivity.

**Delegate—Don't Abdicate M 308.** Principles of delegation; working with subordinates. Leadership, decision making, organization development, team building, management by objectives. Five principles of delegation: setting realistic goals, communicating goals to subordinates, sharing background thinking and problems, giving clear instructions, checking progress.

**Problem Solving M 315.** Orientation to problem investigation; pitfalls in the problem solving process. Leadership; team building; organization development. Problem recognition and investigation; control; human interaction; psychological pressures of management.

**Managerial Control M316.** Introduction to directing and controlling; orientation to communication with subordinates; leadership; problem solving; cost management. Team building; leadership; organizational goals; planning. Study guide: Includes checklist for establishing a basic control system; list of prerequisites to control.

## *Other Sources*

*TPC Training Systems* include films but also will put on training workshops for you. Their specific housekeeping materials include:

351 *Floor Care*
352 *Rest Room Care*
353 *Powered Equipment*
354 *Cleaning Chemicals*
355 *Carpet and Fabric Care*

Another important film from International Film Bureau is *The Uninvited Guest.* Although it is directed primarily to hotel management and staff, the rules for fire prevention, control, and escape are also applicable to other areas.

The film opens by pointing out little-known facts about fire and explains briefly how fire is fed and increases. Prevention is the basic principle: fire-prevention responsibilities of watchmen, maids, housekeepers, kitchen staff, and management are explained.

The second is control—both of physical surroundings and human frailties—in the event of fire. Examples of structural control such as fire doors and various types of sprinkler and alarm systems are shown and explained. Human control involves sounding the alarm and attacking the fire, if feasible. Various fire-fighting equipment is demonstrated and the use of each is outlined.

Fire drills involving escape routes help to ensure that a control station does not become a panic station. The film points out that knowing escape routes is particularly important in hotels, as guests are often unfamiliar with their surroundings.

### *Incentives*

Management tends to think the opportunity to earn a raise or a promotion is an incentive for all employees. This is true only to an extremely limited degree in Housekeeping, where the pay scale for maids and housemen is low even in unionized properties, and where raises are small and infrequent. (Figure 24-4).

Many housekeeping employees lack the ability, and sometimes even the will power, to improve their job status. These handicaps can affect the employee's morale adversely, which will in turn affect the quality of job performance. Thus incentive programs are vital to the maintenance of morale and productivity in the Housekeeping Department.

Many lodging properties reward their maids for exceptional performance with prizes, money, or even trips. Others find that free meal-tickets for lunch in the employee cafeteria are effective incentives. Among the financially pressed Housekeeping workers, a dollar per day above the union scale and a daily free meal-ticket make the position of floor supervisor even more attractive.

Perhaps the most effective and widely used incentive is the daily cash bonus for maids who exceed their regular quota of 12 to 17 rooms per day, depending on the property. Bonuses range from 50¢ per additional room cleaned to $2.50, which New York's Waldorf-Astoria pays in accordance with its union contract.

Incentive plans may involve competitions among individuals, but more likely among complete work teams. Contests that give everyone a chance to win are even more popular.

"Incentive programs in the Housekeeping Department," says a spokesman for Sheraton's many properties, "are becoming far more important than higher salaries. Sometimes we give employees money as a prize but it still isn't as great a motivator as recognition, so we have many rec-

**FIGURE 24-4** *The Country Store was the In-House Flea Market at the Century Plaza Hotel. There, used furnishings are sold at discount prices that delight employees. To increase inventory of store, employees bring items from home either to contribute or be sold on consignment. Even discontinued uniforms and canned goods no longer used due to menu changes are part of the stock. Money earned goes into the hotel's refurbishing fund.*

ognition awards. They are based on how the workers relate to each other and to the guests—the Courtesy Award, Maid of the Month, Houseman of the Month, etc., and always the winners' photos are posted in the lobby or in the employee cafeteria and the house organ carries publicity."

### *Awards Help Morale*

Some Sheraton Hotels give a U.S. savings bond to the winner of the Housekeeping Award of the Month, whose picture is posted on the employee bulletin board, and who also receives a bountiful supply of free coffee and doughnuts on the morning break.

Hyatt Hotel Corp. has Employee-of-the-Month and Employee-of-the-Year plans that involve all employees in each property. Every month each of eight departments selects one of its members as a candidate for Employee-of-the-Month. Management makes the selection, and at year-end selects Employee-of-the-Year from the 12 monthly winners.

At the Mayflower Hotel in Washington, D.C., the executive Housekeeper checks the house carefully to see which details of guest room cleaning are being neglected and then sets up the Goals-of-the-Week contest. This might, for example, culminate in naming the Champion Bed Maker or Most Expert Grout Cleaner one week, and lead to neater carts or cleaner bathroom floors the next week as all contend to be best in those special categories.

Marriott is famous for its excellent health and insurance programs and all the employee benefits including profit sharing. Stories are legion about the Marriott dishwasher who retired with $481,000! "An employee who started with us at a salary of $2.00 an hour can build up his profit sharing and retire after 33 years with $385,000," according to Foster Kunz, VP, industrial relations, Marriott Corp.

One of the team awards given at the Century Plaza Hotel is the Best Floor Award, a rotating trophy. All maids on the winning floor are awarded prizes and have their names and photos posted on the departmental bulletin board.

The Ilikai Hotel in Honolulu has a white-glove inspection, the winner of which receives a cash prize of $25.00.

Incentive plans must be simple, fair, and understandable. They must be discussed personally with all eligible employees, and perhaps even individually with employees who are not sure how they work. Rules should be posted on the bulletin board. Scores or other appropriate records should be kept accurately, and names of daily winners (or names of leaders when prizes are not awarded daily) should also be posted. It is equally important to avoid embarrassing non-winners or low-scorers.

### *Involve Maximum Number of Employees*

Prizes should be prominently displayed. It is far more important that the program enable many employees to win inexpensive prizes than that it enable only a few to win more valuable items. The more difficult it is to become a winner, the more likely it becomes that the incentive program will be an incentive only to few and demoralize many.

Needless to say, it has been proved repeatedly that employees need rewards other than material ones. Personal recognition is extremely important to the worker's sense of personal worth.

One of the best ways to tell a maid that she is doing a good job is to let her work without supervision. Established properties often select those who have proved consistently reliable, provide them with Honor Maid badges, and assign them to floors that are not checked by supervisors. The Honor Maids feel a sense of increased personal worth because of the trust placed in them; the property gains the advantage of requiring fewer floor supervisors in Housekeeping.

At the Doubletree Inn in Tucson, Valley of the Sun, there are several programs that are proving effective for motivating room attendants.

**Five Dollar Bill.** The executive Housekeeper goes into a check-out room that is not yet cleaned, and places a $5 bill stapled to a business card in one of those typically ignored cleaning areas such as the corner behind the blackout curtains, the top of the closet shelf, on the pipes under the sink vanity, and so on. The Housekeeper returns after the maid has thoroughly cleaned the room and it is ready for inspection. If the reward is gone, the maid is congratulated for her thoroughness. If the reward is still there, the Housekeeper brings the maid back and shows her that if she had cleaned the room with a degree of thoroughness as established by the standards and cleanliness of the hotel, she would have been $5 richer. A total of $100 per month has been allocated toward this program, which, given a maid staff of 19, permits the Doubletree to give this opportunity to every maid at least once a month.

**Silver Dollar Program.** The general manager of executive Housekeeper takes ten silver dollars to a maid station and invites the maid into a freshly cleaned check-out. She is handed all ten silver dollars and then an inspection ensues. For every major discrepancy found in the cleaning process, the manager requires that the maid returns one silver dollar, thereby reinforcing the fact that this money was hers, and had she done a more thorough job of cleaning, she would have had more tangible cash in hand at the end of the inspection. "Never do we ask the maid to return more than $8 to management," explains general manager J. Rudi Heater.

**Quality Control Inspection Program (departmental).** Every maid is inspected five times per month on a surprise basis. A chart is posted in the Housekeeping lunch room with scores and month-to-date averages kept as a running total, thereby creating peer pressure from the other maids for anyone who is lacking. The entire department must maintain an overall average of 90 or above for any incentive payments to be made to individual members. The four highest-scoring maids are rewarded with cash bonuses in their paychecks of $40, $30, $20, and $10 respectively. Housemen have their own rating sheet depending on the condition of the public areas, corridors, hallways, exterior door facings, and the like. Assuming that their aggregate average score is also above 90, the floor housemen as a group are entitled to a $20 bonus per month, the program then totaling $180 potential in any given work month.

These programs, although totally financial in nature, reports J. Rudi Heater, are compounded by intense management involvement in the cleaning habits of each person in the Housekeeping Department, and follow-through is mandatory. A birthday card sent to the employee's home is often shown to other family members as evidence that the individual is recognized by the employer as more than a mere payroll number. Other recognition symbols include: giving birthdays off; having a monthly birthday cake for all employees whose birthdays occur that month; giving flowers left from banquets to the maid with the cleanest cart; free afternoon tea or morning coffee for maids with perfect attendance records.

Says Bill Pistilli, president of the Alameda Plaza Hotel in Kansas City, "James Nassikas at the Stanford Court is the best hotel manager in the country. He believes in the theory that I believe in—if you want to run a hotel right, take the organizational chart, tip it upside down, and manage from the bottom. Our hotels are as good as our maids and our dishwashers."

When asked how you get the maids that you want, keep them, and train them, Pistilli replies, "Training is the biggest job that a manager has. How do you motivate a maid? You do it with money. It's very difficult to say to somebody in today's society 'go clean my john', but that's what we're doing, is it not? But also to motivate the troops, find out when their birthdays are. We send a card to every single employee and I sign them all. I'm a morale doctor. Take some interest in your people. That's hands-on management. I'm sure the Harvard Business School guys would say it's not good, but it works for me—I don't say it works for everybody."

One company, Hospitality Management Corporation, feels so strongly about morale that it requires managers to take a group of maids to lunch in the banquet room at least once a month.

The Idea Contest conducted throughout one of the major chains proved to be highly effective. A $25 reward was given for every idea that might improve efficiency or morale. In addition to money, winners received a letter of thanks from hotel managers, and publicity about the employee and his idea appeared in the chain's monthly house organ. All ideas were acknowledged by management, even if they didn't win.

One winning idea was that heavy disposable trash bags, used for dry refuse or paper, could be reused several times. Implementing this idea actually saved several dollars a week. As employees reused bags, they realized that one of their peers had very profitably suggested that simple procedure, which in turn encouraged them to use their own ingenuity and experience to devise still other money-saving ideas.

Lodging chains sometimes offer middle management attractive stock options, for which the Payroll Department will deduct as little as $5.00 per month from the individual's paycheck. For most supervisory personnel, however, gift certificates from a leading store in the community, free hairdresser appointments, and tickets to sporting events and theaters are typical of the rewards used by management to recognize both longevity and superior performance.

To help morale in some of the Hyatt Hotels, management encourages the formation of small groups within Housekeeping Departments to pursue such interests as golf, bridge, bowling, needlepoint, or theater. Through Hyatt classes, Housekeeping employees have the opportunity to learn how other departments in the hotel function and relate to Housekeeping.

**FIGURE 24-5** *The maid's card is a controversial subject among Housekeepers. Many hotels use them on the theory that a maid will take more pride in her work if she is known to guests, and that guests will feel they are being personally attended to. Others think guests object to being reminded to tip for services that should be part of room rate.*

## *Maids' Tips*

One important incentive for maids, over which the hotel or club has little control, is the tip left by the guest for cleaning his room (Figure 24-5). Very few guests today leave tips unless they asked the maid to do something extra or have left an extremely messy room. What pleases maids even more than the money is a note of praise left by a guest.

If a tip is left in the form of a present, the maid must report it to the supervisor so that she can get a package pass. If the gift is a bottle of liquor, it will have to be turned in to the supervisor. If it has not been opened, the housekeeper may give it to the Beverage Manager, and if it is usable in the hotel the maid will receive the price that the hotel would pay for such a brand.

Many guests hesitate to leave a tip in the room for a maid because they are not sure it will get to the maid who has taken care of them for the better part of their stay. In that case they may occasionally mark the room number and dates of their stay and leave it at the Front Desk when they check out.

Just how much money maids do make on tips is a subject rarely discussed; we can only surmise it is actually of little consequence and should not affect the salary scale as it does in the case of waiters and waitresses.

## *Free Room Bonuses*

Many multi-unit hotel companies give free or discounted rooms or meals. After one year of service, employees of Westin Hotels are entitled to a free room at another property in the chain for three days, plus a 25 percent discount on all meals. After three years, the period is extended to five days. Members of the 25-year club may stay in Westin properties a total of 30 days a year, a maximum of seven days at one hotel, with all meals discounted 50 percent—which surely beats a gold watch! As a by-product of this program, employees begin to visualize their own functions from the guest's perspective and have an opportunity to compare their own performance with that of their counterparts in other properties in the chain.

These free rooms are, of course, on a space available basis so many employees, to take advantage of this desirable bonus, will arrange their vacation dates to coincide with times when a certain hotel can accommodate them.

For an entry-level person on his first vacation it is a tremendous thrill to go with his spouse to a fine hotel and be given the VIP arrival treatment with fruit or gifts and a letter of welcome from the manager of the Housekeeping Department. He suddenly takes great pride in his work and his company and often improves productivity and performance.

Incentive programs of any complexity must always be carefully evaluated in terms of the supervisory time they will require. Occasionally, of course, a member of the staff has a particular talent for arranging and supervising innovative incentive programs, and is even eager to work on such programs during off-duty hours. Where such a person is available and willing to volunteer, even the more complex incentive programs may well justify the time and expense involved.

Although some Housekeeping managers feel that extra pay and incentives are not necessary, contending that employees are paid to do their jobs and should not require an extra "bribe" for simply doing the very thing they were hired for, it is obvious that cleaning halls, bathrooms, back service corridors, and guest rooms left in a shambles by careless guests is actually unpleasant and monotonous work, and there must be some extra thank-you's.

Whatever makes coming to work more fun is an important component of an entry-level job. Many of these employees are lonely or live in unhappy environments—if their working world can be interesting, they will have a different outlook on life. A few contests or surprise rewards can keep them wondering what is coming next and they will look forward to being on the job tomorrow.

## *Sources of Training Materials*

National Educational Media, Inc.
21601 Devonshire
Chatsworth, Calif. 91311

TCP Training Systems
131 South Grove Ave.
Barrington, Ill. 60010

Continental Film Productions Corp.
4220 Amnicola Highway
Chattanooga, Tenn. 37406

International Film Bureau, Inc.
332 S. Michigan Ave.
Chicago, Ill. 60604

Educational Institute of
the American Hotel & Motel Assn.
1407 S. Harrison Rd.
East Lansing, Mich. 48823

# Chapter 25

# Turnover, Evaluating, Promoting

**OVERVIEW** Studying statistics on turnover will convince readers how it must be prevented; here some solutions to the expensive problem are presented. Reasons for discharging employees are given but disciplinary actions are also suggested. Evaluations based on standards of performance help both employees and supervisors to keep track of how well-adjusted and how competent workers really are.

Employee turnover is probably the most serious problem currently plaguing the lodging industry, and almost certainly the problem that has proved least susceptible to solution.

The accompanying chart, Figure 25-1, documents the gravity of the problem under even the best possible conditions—in this case, a relatively new, glamorous, well-managed hotel whose enlightened employee policies and programs are acknowledged in the industry.

## The High Cost of Turnover

Neither the employer nor the employee can afford the high cost of turnover, which is estimated at well over $1 billion annually for the foodservice and lodging industry. This cost can be broken down into four phases: (1) recruitment, (2) selection and placement, (3) on-the-job, and (4) separation.

Recruitment costs include advertising, printed materials, public relations activities, college recruiting, employment agency fees, and other costs associated with the mere act of attracting applicants. Selection and placement costs include interviewing, testing, checking references, arranging for medical examinations, and so on. Major on-the-job costs include wages, training, equipment, bonding, and overhead. Separation costs may include related interviews, severance pay, increases in social security and unemployment-tax payments, possible loss of goodwill, and even more important, the possible adverse effect on employee morale and productivity.

Even under the superior working conditions that prevail in this particular property, turnover for the year analyzed added up to 998 departures among 1182 employees, or 84 percent. This does not mean, of course, that only 184 employees remained at their positions for the entire year. In many instances the same position was vacated repeatedly during the course of the year. As a matter of fact, 11 percent of the employees remained on the job less than one week; 30 percent less than one month; 65 percent less than six months. Turnover in six departments actually exceeded 100 percent:

| | Percentage |
|---|---|
| Restaurant | 145 |
| Property maintenance | 133 |
| Housekeeping | 122 |
| Stewards | 120 |
| Coffee shop | 111 |
| Telephone | 103 |

## Turnover Analysis

With 178 employees, Housekeeping is by far the largest of all departments. Not surprisingly,

| | | Banquets | Coffee Shop | Emp. Cafeteria | Catering | Restaurant | Main Din. Room | Bar | Room Service | Nightclub | Culinary | Stewards | Beverage | Front Office | Housekeeping | Reservations | Service | Accounting | Engineering | Garage | Laundry/Valet | Property Maint. | Purchasing | F & B Cashiers | Telephone | Other | Total |
|---|---|---|---|---|---|---|---|---|---|---|---|---|---|---|---|---|---|---|---|---|---|---|---|---|---|---|---|
| RESIGNED | Death | | | 1 | | | | | 1 | | | 1 | | | | | | | | | | | | | 1 | | 4 |
| | Home | 2 | 1 | 2 | | 3 | 2 | 1 | 2 | 1 | 2 | 6 | | 7 | 24 | 1 | | | | | 4 | 3 | | | 2 | 2 | 65 |
| | Hours | | 2 | 1 | | 1 | | | 1 | 1 | | 4 | | 3 | 2 | | 1 | | | 1 | | 2 | | | 1 | | 20 |
| | Ill Health | | 2 | 1 | | 4 | | | 2 | 3 | 4 | 3 | | 1 | 17 | | 2 | | | | 2 | 9 | | | 3 | 1 | 54 |
| | Marriage | | | | | 3 | | 1 | | | | | | 5 | 4 | | 1 | | | | | | | | | | 14 |
| | Opportunity | 3 | 3 | | 3 | 14 | 5 | 1 | 3 | 6 | 9 | 8 | 4 | 7 | 19 | 6 | 2 | | 1 | 6 | 3 | 10 | 1 | 1 | | 2 | 117 |
| | Residence | 5 | 10 | 1 | | 17 | 7 | | 5 | 1 | 4 | 77 | 4 | 11 | 27 | 2 | 2 | 2 | | 5 | 8 | 3 | | 1 | 7 | 1 | 200 |
| | Return to School | 1 | 3 | | 1 | 2 | | 1 | 4 | 1 | 2 | 1 | 1 | 5 | 6 | 2 | 2 | | | 4 | 1 | 3 | | | 1 | | 41 |
| | Supervisor | | | | 2 | 1 | | | | | 1 | | 1 | | 5 | | | | | | | | | | | | 10 |
| | Transportation | | 2 | | | 2 | | | 4 | | | | | | 6 | | | | | | | 1 | | | | | 15 |
| | Unknown | | 3 | | | 3 | 3 | 2 | 6 | 1 | 11 | 17 | 1 | 4 | 37 | | | | | 5 | 2 | 19 | | | 4 | | 118 |
| | Wages | | 1 | | | 4 | 2 | | 2 | | 4 | 3 | | 1 | 8 | | | 1 | | | 2 | 1 | | | 3 | | 32 |
| | Work | | 4 | | | 5 | 1 | 1 | 4 | | 1 | 4 | | 3 | 32 | | 2 | | | 2 | 3 | 3 | | | 3 | 2 | 70 |
| | No Show | | 1 | | | 5 | 1 | 1 | 1 | | 3 | 2 | | 4 | 1 | | 1 | | | 1 | 1 | 3 | | | 1 | | 26 |
| | Miscellaneous | 4 | 3 | | | 1 | | 1 | 2 | 3 | 8 | | 1 | 3 | 8 | 1 | | | 1 | 1 | 1 | 4 | | | 4 | 3 | 49 |
| | **Total Resigned** | 15 | 35 | 6 | 6 | 65 | 21 | 9 | 37 | 17 | 49 | 126 | 12 | 54 | 196 | 12 | 13 | 3 | 2 | 25 | 27 | 61 | 1 | 2 | 30 | 11 | 836 |
| DISCHARGED | Associates | | | | | | | | | | | | | | | | | | | | | | | | | | |
| | Attendance | | | | | | 1 | | | | 3 | 4 | | | 3 | | | | | 3 | 1 | 3 | | | 2 | | 20 |
| | Conduct | 2 | 3 | | | | | | 4 | 2 | 3 | 3 | 1 | 1 | 8 | | 1 | | 1 | 2 | 2 | | | | | 1 | 34 |
| | Dishonesty | | 1 | | | | | | 1 | | 1 | | | 2 | 4 | | | | 2 | 1 | | 1 | | | | 1 | 14 |
| | Personality | | | | | | | | | | | | | | | | | | | | | | | | | | |
| | Staff Reduction | | | 2 | | | 1 | | | | | 5 | 5 | 1 | 2 | 1 | 2 | 2 | 11 | | 5 | 1 | | | | 2 | 40 |
| | Supervisors | | 1 | | | | | | | | | | | | | | | | | | 1 | | | | | | 2 |
| | Work | 1 | | 1 | | 2 | 1 | | 6 | | 9 | 3 | 2 | 1 | 2 | 4 | 1 | | 1 | 4 | 3 | 1 | | | 1 | 1 | 44 |
| | Miscellaneous | 1 | | | | | | | 1 | | | | | | 2 | | | | 1 | | 2 | 1 | | | | 1 | 9 |
| | **Total Discharged** | 4 | 5 | 3 | | 2 | 3 | | 12 | 2 | 16 | 15 | 8 | 5 | 21 | 5 | 4 | 2 | 16 | 10 | 14 | 7 | | | 3 | 6 | 163 |
| LAID OFF | Temp. Reduction | | | | | | | | | | | | | | | | | | | | | | | | | | |
| | Miscellaneous | | | | | | | | | | | | | | | | | | | | | | | | | | |
| | **Total Laid Off** | | | | | | | | | | | | | | | | | | | | | | | | | | |
| | **TOTAL EXITS** | 19 | 40 | 9 | 6 | 67 | 24 | 9 | 49 | 19 | 65 | 141 | 20 | 59 | 217 | 17 | 17 | 5 | 18 | 35 | 41 | 68 | 1 | 2 | 33 | 17 | 998 |
| LENGTH OF SERV. | Less than 1 Week | | 7 | | | 16 | 4 | 1 | 3 | | 12 | 15 | 1 | 7 | 29 | 1 | | | | 1 | 2 | 10 | | | 8 | | 117 |
| | 1 Week to 1 Mo. | | 8 | 1 | | 15 | 5 | 2 | 5 | 1 | 10 | 34 | 1 | 9 | 48 | 5 | 2 | | 3 | 6 | 6 | 18 | | 1 | 9 | 2 | 191 |
| | 1 to 6 Mo. | 1 | 15 | 1 | 2 | 27 | 6 | 4 | 22 | 5 | 19 | 42 | 6 | 22 | 80 | 3 | 6 | 3 | 11 | 13 | 15 | 26 | | | 7 | 8 | 344 |
| | 6 Mo. to 1 Year | 7 | 3 | 4 | | 7 | 5 | | 4 | 2 | 5 | 25 | 4 | 9 | 27 | 3 | | | 3 | 8 | 9 | 9 | | | 4 | 3 | 141 |
| | 1 to 2 Years | 4 | 4 | 1 | 2 | 1 | 4 | | 8 | 3 | 6 | 17 | 6 | 6 | 22 | 3 | 2 | 2 | | 3 | 4 | 3 | 1 | | 3 | 3 | 108 |
| | 2 to 5 Years | 7 | 3 | 2 | 1 | 1 | | 2 | 7 | 8 | 12 | 8 | 2 | 6 | 11 | 2 | 7 | | 1 | 4 | 5 | 2 | | 1 | 2 | 1 | 95 |
| | Over 5 Years | | | | 1 | | | | | | 1 | | | | | | | | | | | | | | | | 2 |
| | **Total** | 19 | 40 | 9 | 6 | 67 | 24 | 9 | 49 | 19 | 65 | 141 | 20 | 59 | 217 | 17 | 17 | 5 | 18 | 35 | 41 | 68 | 1 | 2 | 33 | 17 | 998 |
| | **Average No. On Payroll** | 69 | 36 | 12 | 13 | 46 | 29 | 12 | 58 | 40 | 99 | 117 | 41 | 57 | 178 | 18 | 42 | 27 | 27 | 67 | 44 | 51 | 9 | 20 | 32 | 38 | 1182 |
| | **Turnover Percentage** | 27 | 111 | 75 | 46 | 145 | 83 | 75 | 84 | 47 | 66 | 120 | 49 | 103 | 122 | 94 | 40 | 18 | 67 | 52 | 93 | 133 | 11 | 10 | 103 | 45 | 84 |

**FIGURE 25-1** *Labor Turnover Analysis For 1 Year in an 800-Room Hotel With an Average of 1182 Employees*

then, the department leads in number of exits by a wide margin, with a total of 217. A startling 72 percent of these occurred within less than six months of the day the employee had been hired. On the plus side, less than 10 percent of Housekeeping exits occurred because the employee was discharged, versus over 16 percent for the hotel as a whole.

Why were 21 employees discharged, and why did 196 resign? According to the figure, 18 different reasons were recorded, plus a miscellaneous category. In addition, 37 employees were unwilling or unable to give any reason for their departure. Since any of the other reasons could have been given without the possibility that anyone would be embarrassed, there is reason to believe a high percentage of these 37 were due to work dissatisfaction, which was given as the specific reason for 32 additional departures. This means that up to 31 percent of all resignations in Housekeeping may be traceable to work dissatisfaction.

Obviously, it is important to pinpoint the reason for each resignation as accurately as possible. In some instances the company can correct conditions that have led to resignations. In other instances the company will be able to retain employees by persuading them that their opportunities for financial advancement and promotion are superior to those offered elsewhere. Where the employee is dissatisfied with his job, his supervisor may be able to recommend him for a position that he will find more interesting.

Too often the employee resigns before the supervisor and department head have had sufficient opportunities to evaluate his skills and potential. Unfortunately, there are many instances where the employee resigns because these managers simply have not taken the time to acquaint him with his real opportunities, or to demonstrate more than a perfunctory interest in his aptitudes, performance, and ambitions.

Any institution that uses lattices of promotion (going across, not up) will find in most cases that it pays. It pays in employee loyalty, for most are very pleased to be offered an opportunity in another department of that same institution.

It is also true in Housekeeping that there are so many entry-level positions to fill and so many different types are hired that often, after an employee has been on the job awhile, we discover hidden abilities. It is not good business practice to transfer the employee to another department too soon because he may be someone who, in a few weeks or months, becomes dissatisfied with any position. But a person who evidences genuine interest in another position should be encouraged to take additional training to learn that new job skill.

## *Why Do They Stay?— Not Why Do They Quit?*

Inordinate amounts of time and literature have been devoted to analyses of employee terminations, both voluntary and involuntary. Repeatedly, it has been determined that a high termination rate is traceable primarily to defects in the selection procedure. Too often it has been assumed from this that it is possible to spot high-risk applicants in the interview and thus reduce the termination rate drastically. To some extent this theory *has* helped weed out applicants seemingly predestined for early termination; but when the employer is so intent on noticing the negatives that cause one problem, he may overlook other weaknesses that will cause the employee to leave for an entirely different set of reasons.

Employers, misled by the idea that you can always spot the "losers," make two major errors: (1) that termination is the result of a defect in the employee rather than, as is often the case, in the institution, its management, department heads, and/or supervisors; and (2) that a sketchy profile of the ideal employee can be compiled from analyses of the departures.

Employees may fail to recognize that an actual profile of the ideal employee already exists in the form of satisfied, long-term personnel who abound in even those departments with turnover rates in excess of 100 percent.

Aspects of the foregoing, which remains valid over the years, were discussed in the October 1972 issue of *The Institute Journal,* a publication of the American Hotel & Motel Association, by Dunnovan L. Sapienza, Ed. D., a noted authority in the field of adult education and expert in foodservice and lodging.

## Company-Sponsored Training Programs

Sapienza sees the honing of skills and preparation for promotion through continuing company-sponsored training programs as the ultimate solution to the turnover crisis. He argues, however, that such programs are only fully effective when it has been possible to identify the present and prospective long-term employees who will participate. They are identifiable because (1) there are employer stimuli to which they are known to respond, and (2) they possess the primary personal factor that triggers the decision, perhaps subconscious, to become or remain a long-term employee.

Company-provided stimuli found to be the most effective in reducing turnover are: (1) employee benefits, (2) supervisor-employee relationships, (3) confidence in management, (4) job security, (5) status and recognition, (6) identification with the company, and (7) opportunity for growth and advancement. Sapienza also found that the longer an employee has been with the institution, the more likely it becomes that he will continue on the job for an indefinite period of time.

As for the primary personal factors responsible for the individual's becoming a long-term employee, Sapienza segregates four main possibilities: (1) a purely intellectual decision in which the employee rationally evaluates the total situation; (2) a favorable decision made because the employee's particular combination of characteristics results in strong feelings of compatibility with the industry, his functions, and his associates; (3) a pragmatic decision to remain on the job because the characteristics of the employment situation are almost uniquely appealing to a particular type of individual; and (4) an almost purely personal decision based on a work situation interpreted by the individual as a pleasurable social situation.

## When and How to Discharge for Misconduct

Despite the high cost of turnover, it is even more expensive to tolerate the low productivity, possible damage to hotel property, and adverse effect on employee morale that results from retention of any employee who, despite repeated warnings and private conferences, persists in misconduct of other than a trifling nature.

Most misconduct cases in Housekeeping fall under one or more of the following.

**Intoxication.** Appearing obviously to be under the influence of alcohol. In some cases these situations would have to be substantiated by a witness.

**Excessive Unexcused Absences or Tardiness.** Every case of unexcused absence and tardiness must be documented. According to the unemployment department, if tardiness and absences are permitted over a period of time, it is assumed that the employer has been willing to accept them in the case of this employee, and therefore he may not be fired unless he is given a certain number of written warnings.

**Dishonesty.** This involves illegal possession of company property or that of fellow employees or guests. Such a claim should be supported by a signed confession but care must be taken that no force or duress is used in obtaining a confession. Some properties ask employees to take polygraph tests. Although these are not admissable in a court of law, if the employee refuses to take the test, or fails the test, it may be reasonable to suspect guilt—especially if there are enough other reasons in his record for termination.

**Conduct Contributing to Moral Delinquency on the Job.** For example, selling narcotics or engaging in prostitution.

**Quarreling.** Loud and abusive language and/or physical fighting in guest areas.

**Refusal to Perform Duties.** Be sure the duties requested are reasonable, that they do not conflict with the union contract, that the employee refused without good cause, and that he did not merely object, delay, or offer a reasonable excuse such as his physical limitations.

**Insubordination.** Refusal to follow instructions or obey orders, excessive talking that interferes with his or other's job performances, horseplay (wasting time unnecessarily), abuse of privileges, and lack of cooperation.

All of these reasons for termination must be documented with a record of each incident in the employee's file. It may be an annoyance to the person administering the department, but it is probably necessary to justify the termination and keep down the company's unemployment insurance payments.

Some things that are *not* considered misconduct include simple negligence or an isolated incident or some types of misconduct. Inefficiency or inability to do assigned work in a satisfactory manner is not considered willfull misconduct. However, if inability to do the job is combined with any of the seven misconduct charges, there is probably sufficient reason to terminate employment, so his records should be checked.

In addition, procedures such as the following are recommended. When hiring a maid she should be told what her quota of rooms is—14, 15, 16—and the length of time she has to become proficient enough to do that number of rooms (anywhere from a week to a month). If she cannot meet her quota, steps should be taken to determine whether additional training will enable her to increase productivity. If not, perhaps she can be given other responsibilities within her capabilities. She should not be allowed to continue as a maid for, let us say, a six-month period during which she almost never reaches her quota, because then the state employment agency would consider that the amount of work she had done had been condoned.

## Performance Records

Performance records on employees should document the following: (1) absenteeism; (2) tardiness; (3) disobedience of house rules (e.g., eating or smoking in guest areas instead of in the employee's cafeterias); (4) using foul language; (5) destroying property; (6) inability to get along with other employees or the supervisor (that could involve verbal language or physical fighting); (7) dishonesty—on the job application, in not reporting an accident, or in reporting an accident that did not happen on the property; (8) not doing the job properly, continual mistakes, very slow performance, improper service, and (9) poor appearance or grooming.

Employee records should also document positive information. This might include good guest relations, compliments from the guests, looking for opportunities to learn more about the job, assuming more responsibility than the job requires, and an excellent attendance record.

Many states require that before being hired the applicant be advised of all house rules such as No smoking in guest areas; in other areas, only plain cigarettes, no cigars, no pipes, no marijuana; no narcotics allowed for use or for sale.

All employees must be told they may not make contacts with guests for purposes of increasing their own income—unless the hotel offers services such as baby-sitting and uses Housekeeping employees.

## Disciplinary Action

Once these rules and regulations are made one is faced with disciplinary action if there is a violation. Disciplinary action usually follows this course:

1. A minor violation results in an oral reprimand with an appropriate review of whatever the rule was that was violated.
2. Major violations or repeated violations will result in a written reprimand by the department head.
3. Three written reprimands will be grounds for immediate dismissal.

Department heads should personally counsel each employee who is given a written reprimand so that he understands why he is receiving it. In large institutions the reprimand (known as a warning slip) is then filed in the personnel office with one copy to any union involved. A third copy goes to the employee who is usually requested to sign it as evidence of having received it. If for any reason he refuses to sign it, Housekeeping should ask his supervisor to explain the seriousness of the incident to the employee, making sure he understands. Refusal to sign the form should be recorded on his records, along with the name of the supervisor involved.

All rules and regulations should always be checked with any union involved. Many unions require that for an employee to be dismissed all three reprimands be given for exactly the same infraction of rules. In some areas there is an arbitration board to see whether the employee may be reinstated or the termination will be enforced.

The administrator of the Housekeeping Department must handle all matters pertaining to employee dismissal so that the institution is not involved with more litigation than is absolutely necessary. It is very bad for the morale of the department and embarrassing to the institution when an employee is absolved by an arbitration board or court of an act that was in fact a serious infringement of house rules.

One other method of discipline can be suspension for a certain number of days for infringement of the rules. (In many cases this system may work better than actual termination.) Unfortunately, sometimes the infringement occurs at a time when the hotel is so busy that the employee cannot conveniently be spared. If the warning and suspension are to be effective, however, action must be taken immediately after the infringement.

Another matter for serious consideration involves the employee who is obviously accident-prone. Always assume the possibility of a major, expensive accident if such an employee is not watched very carefully and restricted to work that presents no major hazards. A good safety program is discussed later in the text, but accidents are significant and must be recorded in the employee's file.

When the employee terminates of his own volition, the employer should make every effort to discover why the employee is leaving, and try to dissuade him if the reasons do not seem sound. That does not mean that one should discourage an employee from leaving to take a job in which he will improve himself, but often the better job he wants may be right there in the same property.

## *Problems of a New Supervisor*

With a change of supervisor, administrator, or director of Housekeeping, long-tenured employees who have become set in their ways are often difficult to win over to a new method of operation. This offers a real challenge to the new department head—that of convincing the old-timers that a new way is better, or at least should be given a try.

An unfortunate fate for a progressive Housekeeper going into a new institution is to inherit a staff that has been there from 15 to 25 years, because new ideas may be difficult to sell. This becomes a problem of motivation. A breaking-in period should be considered under these circumstances, both from the employees' and Housekeeper's points of view. Then the Housekeeper can determine how important it is that the thing she wants done be done exactly one way, or, if given the same results, can she accept its being done by a different method?

With respect to the older employees there is still another difficult situation that will inevitably arise in non-unionized departments. "Our worrisome problem," according to a university housing official discussing his Housekeeping Department, "is the older people here who have not yet reached retirement age, but are not as productive as they once were, primarily because of declining physical strength. These employees have been given small raises over their 20 to 35 years with the department and frankly they are no longer worth what we are paying them by comparison with the younger people who put out more work for considerably less money. However, these loyal old employees cannot humanely be fired or have their salaries cut. It is a dilemma in these days of rising costs."

The dilemma is not confined to Housekeeping Departments. Every personnel director is confronted with the problem of whether to keep reliable, loyal employees rather than more efficient ones whose attendance records may not be so impressive. Perhaps one solution, where unions permit, is less specialization—training employees through the years to be able to handle several jobs so that they can be shifted to different positions as their capabilities or interests change.

## *Evaluation: Performance Appraisal*

For over 25 years the concepts of management by objectives (MBO) and standards of performance (SOP) have been hailed as the miracle ingredients of modern managerial systems.

The MBO might reasonably be described as the logical outgrowth of MBG (management by goals). As defined by many management experts, "goal" is a corporate ideal expressed in vague terms so that it is difficult to determine how long it will take to achieve a corporate goal. It may not even be possible to determine when or if such an end has actually been arrived at. An ambitious lodging company might very well state its corporate goal as the operation of first-class hotels in all major cities of the United States. But there is enough elasticity so that by design, it is never reached.

To move toward its goal, management must decide on appropriate objectives and their order of priority. Unlike goal, there is nothing vague about objective. It has to be specified in terms of such unmistakeable elements as time, space, cost, quantity, and quality; it must also be measurable. A lodging company might set the following as its top-priority objectives: (1) to open Hotel A in city X on April 1, 1983; (2) to open Hotel B in city Y on July 1, 1986.

Management's sweeping objectives are next broken down into subobjectives, which are broken down still further at the department level.

Departmental tasks are defined in terms of key result areas, which represent the most important contribution the individual or team can make to the accomplishment of the department's assigned subobjectives. The method and date by which a task is to be performed plus all additional pertinent, specific data is incorporated into a standard of performance. The SOP also serves as the yardstick by which, at least theoretically, the department head can measure the quality and quantity of every individual's performance. Theoretically, it enables the employee to measure and grade his own performance.

The extreme danger in SOP is that the overzealous department head will become totally immersed in the trappings and paraphernalia of what is, at this level, still impractical for the level of intelligence of workers in most Housekeeping Departments.

When practiced realistically rather than fanatically, however, SOP does at last provide objective criteria for measuring the performance of individuals, teams, and entire departments. It is unquestionably superior to the system previously used, known as the periodic personnel evaluation, which resulted in employees being graded on the basis of accumulated impressions that were largely subjective. The SOP eliminates the popularity contest element from personnel evaluation, neutralizes the personality factor, and congratulates (and often promotes) those who meet or beat their standards of performance consistently.

Ideally, the employee should participate in the development of his own standard of performance. In some cases, he may be able to write it in its entirety. He should always be consulted on the matter both before and during the preparation of his SOP. When completed, it should be reviewed with him as fully as necessary. If for any reason it does not meet with his approval, the most careful consideration should be given to his objections and suggested changes. If an employee is saddled with a standard of performance that he knows he cannot meet, nothing will maintain his morale, and he will begin seeking another job.

## *Self-Designed SOPs*

For the employee who chooses to write his own SOP, the job description discussed in Chapter 6 on personnel will be very beneficial because he needs to know exactly what is expected of him and how it must be done. Standards of performance should cover not only the day-to-day tasks but those that must be done on a periodic basis. Then, when an appraisal of the position is made to the employee, the responsibilities recorded in the job description can be translated into a living part of the job's activities. This also helps the employee to understand those responsibilities, the work goals associated with them, and the degree to which he accomplished the goals. Put into simple language, we can tell him precisely how he is doing. But the objectives should have a time limit, at the end of which an evaluation should be given.

One important aspect of management during the appraisal period has to do with unbiased observation of the kind of work that the person has done. What is he really doing on the job? Could he do a better job with help? Should he be considered for advancement? If he were capable of advancement, what steps should be taken to prepare him for the next assignment?

Another significant point to observe is how the person accepts responsibility. Is he looking for additional responsibility or is he trying to avoid it? In observing the work produced we need to know what the employee emphasizes about his job. Does he work well and quickly? Does he plan his work? Is he enthusiastic? Does he profit from his past mistakes? Does he follow instructions? Does he find better ways to do things? Is he well organized?

These observations should also be discussed with other persons on various levels of management to see if the employee works the same way when other people observe him. Does the employee often consult with others on things that need to be done? Or ways to do a job? In any large department, judging an employee's work performance should be based on several persons' observations because anyone may work or react differently under different conditions with different people.

Theoretically, the employee will refer to his own copy of the standard of performance often enough to be aware at all times of his merits and shortcomings. In actual practice, he may seldom refer to the SOP, and in many cases, may even be unable to locate it on the rare occasions when he might want to look at it. Thus the theory that SOP stimulates continuing self-appraisal for all types and levels of employees is not necessarily true.

### *Frank but Tactful Discussions*

Since the employee will have participated to a substantial degree in the preparation of his own standard of performance, periodic performance reviews will usually find him reasonably aware of the quality of his work. Needless to say, these appraisals should be conducted in complete privacy, with complete candor, and in complete seriousness. Time set aside for an appraisal should be uninterrupted; no telephone calls or visitors should be permitted. The most important thing to that employee is having that conference with his supervisor. If the person in charge is fully familiar with all details of the employee's performance then his strengths can be praised with genuine enthusiasm, and his weaknesses analyzed and prescribed for with understanding and tact.

No matter how well the employee has done his job there are probably at least minor areas in which his performance can be improved still further. If not, the employee must be considered for additional responsibility or promotion. This should be discussed quite frankly with the employee since it will probably be necessary for him to acquire additional skills before he can actually be assigned to a higher position.

If these skills are normally acquired through adult education programs, the employee should be provided with all necessary information concerning such programs and be urged to enroll. If the required skills are normally taught at company-sponsored training programs, he should be advised when the next one will be presented, where, whether it will be held during or after work hours, and whether successful completion of the course will result in immediate promotion. If it will be delayed, he should be told for how long.

One should not help the employee make excuses as to why he did not do something as well as he should have. For example, bringing up an illness in the family only reminds an employee that he has an out for inferior performance. The employee should receive enthusiastic praise for the part of his performance that he has done well. If he has been in the department for some time, this work appraisal should be compared with the previous ones to see if he has improved where he said he would. During any evaluation the employee should be directed toward additional areas of self-improvement.

Supervisors must encourage employees to ask questions, express themselves, and not to repress their frustrations. On the company's table of organization, the supervisor may be the employee's superior, but acting out that role during a performance review is at all times inadvisable.

Finally, all reviews should involve two-way communications. Supervisors must be prepared to do more than listen; they must actually hear and understand.

## Promotion

As a regular policy, the management of the Greenbrier Hotel, White Sulphur Springs, W. Va., always tries to promote from within. When there is to be an opening, it is posted on the bulletin board and anyone who wishes to may apply for that position. In the Housekeeping Department promotions usually go from Linen Room to Uniform Control to Sewing Room to supervisor. Supervisors, however, are non-union positions so the newly promoted person must withdraw from the union. They are then eligible to participate in all management benefits provided by the Greenbrier.

As was pointed out in Chapter 24 in the section on incentives, Marriott Hotels are well known for the generous profit sharing and fringe benefit programs they have devised to keep employees on the job. But the Marriott policy of promotion from within has succeeded so well throughout the chain that it is in itself a powerful inducement to keep employees from looking for greener pastures.

Particularly in organizations in which Housekeeping is a division of the Rooms Department, training and educating the professional Housekeeper naturally prepares him or her for advancement to director of rooms or assistant manager in a hotel.

There is no question that in some institutions, no matter how well educated one is or how well trained, there is little chance for advancement, even in those companies that promote from within. In that case, management runs a strong risk that the individual will defect to another organization in a higher capacity at a higher salary, since the lodging industry is notoriously short of experienced middle management people.

Those companies that are continually accused of pirating people because they do not do enough training of their own are not always villains; often they provide impatient competent people with opportunities to more upward both to greater status and financial rewards.

While there is a strong fraternal feeling among hotel and campus housing people, they nevertheless play musical chairs with jobs. The mobility in the hotel/motel business is, however, generally far greater than in the college housing field where the hours and wages are more predictable and the pace more often can be set by the individual.

## Bibliography

Educational Institute of the American Hotel & Motel Assn. *Supervisory Housekeeping Planning and Operating Motels and Motor Hotels,* Rochelle Park, N.J.: Hayden Book Co., Inc.

Mager, Robert F., and Pipe, Peter. *Analyzing Performance Problems or You Really Oughta Wanna.* Belmont, Calif.: Fearon Pub., 1970.

Vallen, Jerome J. *The Art & Science of Modern Innkeeping.* New York: Ahrens Publishing Co., Inc.

# Chapter 26

# Housekeeping Relationships with Guests and Other Departments

**OVERVIEW** How harmoniously Housekeeping works with other departments is important if people are to feel any sense of accomplishment. Here are problems or challenges that arise when dealing with Engineering, Maintenance, Food and Beverage, Sales, Front Office, Personnel, and Security. Code of conduct with guests is highly pertinent. How room status systems work and their benefits are examined.

No department in the entire lodging industry is as involved with both the occupants and employees as is Housekeeping, which comes into contact with practically all other departments daily. Good working relationships are therefore vital.

In the hotel industry, continuous contact with the Front Office is essential. Housekeeping prepares the most important product a hotel has to sell—the guest room—and the very reputation of the hotel depends on the speed and quality of that preparation. It is urgent that rooms be ready when new guests arrive.

As Hanne Dittler, housekeeping expert for the Westin chain points out: "Since no hotel wants to room a guest in an unmade room, Housekeeping must provide the Front Desk with assurance that rooms will be ready to meet the anticipated arrivals for the day. With the changing patterns of travel, this becomes no small challenge. And as hotels experience more and more large group check-ins and check-outs, the coordination between the Front Desk and Housekeeping becomes critically important."

## Engineering and Maintenance

Cooperation with Engineering and Maintenance is especially critical since Housekeeping depends on them to keep so many things in order. If a towel rack is half off the wall or a baseboard has come loose, a housekeeping inspectress will not be able to report that the room is ready to rent (Figures 26-1 and 26-2). If, however, a carpenter takes care of those details immediately, there will be little delay in readying the room. If a wet-dry vacuum is broken, the sooner the electrician gets to the houseman who has the problem and repairs the machine, the sooner Housekeeping can finish its job. Painters, too, should feel cooperative toward Housekeeping, as it is essential to get rooms painted and back in order quickly.

## Food and Beverage

Refurbishing public areas is an activity for which the Housekeeping Department is responsible (Figure 26-3). This requires that many members of the various departments will necessarily become involved with Housekeeping personnel on every level (Figures 26-4 and 26-5), from the restaurant communicating with the executive Housekeeper or catering manager on

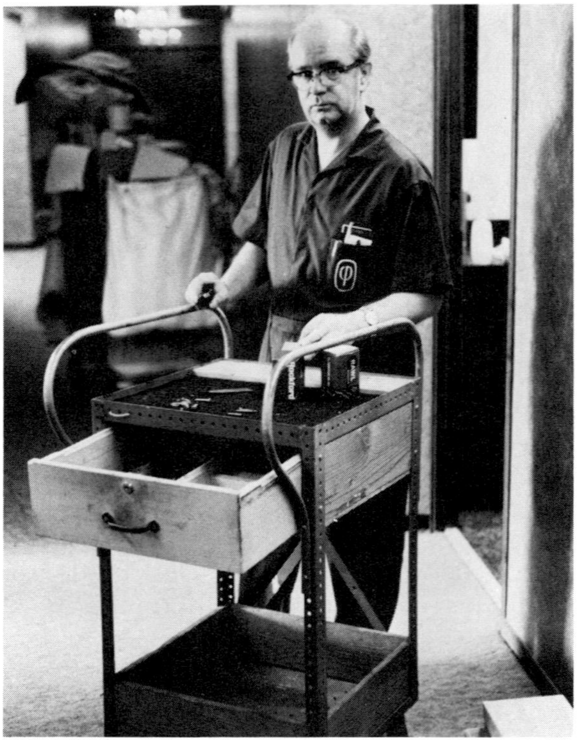

**FIGURES 26-1 and 26-2** *Carpenters in hotels must hurry with repair jobs when there is a shortage of rooms to rent. If a towel rack comes loose, a quick replaster job may have to be done with instant-setting compound. At the Century Plaza Hotel, carpenters use a mobile work bench they fashioned themselves so that work could be done up on the floors rather than having to go back and forth to the carpentry shop.*

REPAIR ORDER

Florida State University

FSU Form 83
1 Sept. 54

No. ____________

TO: ______________________________ JOB NO. ______________

FROM: ____________________________ DATE ________________

Request the following repairs be made:

Requested by ______________

NOTICE OF JOB COMPLETION No. ____________

Building ______________________ Job No. ______________

Date Requested ________________ Date Completed __________

Repaired By ______________________________________

Shop ____________________________________________

Time Required ____________________________________

Comments:

**FIGURE 26-3** *When Housekeeping requests repairs at Florida State University, quadruplicate form is used to emphasize feedback. Office that originates work order expects to have report back on when and by whom work was done. Form is not marked with department names so as to be flexible.* (*It has been in use since 1954!*)

new fabrics for draperies in the dining area, to the head busman who may work with the Housekeeping handyman to hang them.

The professional Housekeeper will keep an eye on the condition of all the furnishings in the various areas and occasionally will question one of the supervisory personnel about any items that seem in imminent need of replacement.

Housekeeping in some properties will be closely associated with the Catering Department if its housemen set up the rooms where the meetings, luncheons, dinners, and banquets are held.

Housekeeping personnel sometimes store catering equipment—linens, tables, chairs, and such items needed for meetings as lecterns and felt table covers.

If Housekeeping cleans the kitchen, there is

certainly going to be interplay with members of the kitchen staff—from the chefs to the stewards. Occasionally there may be complaints from the kitchen that the job was not done properly. At such times the professional Housekeeper must be a diplomat (one of his or her most frequent roles).

Even in properties where the Housekeeping Department does not store, set up, dismantle, or clean up after any catering function or is not in charge of cleaning the kitchens, there still are reasons for the relationship to be cordial since Housekeeping's Sewing Room will no doubt make up the linen discards for use in the kitchens and will repair uniforms for the Food and Beverage and Catering Departments' personnel. Ultimately, Housekeeping handles all linens and uniforms, while maids are constantly asked by guests to remove soiled dishes on trays or tables brought by Room Service. (Some hotels keep multi-shelved carts in the back halls of guest floors for the use of maids needing space for trays. Others keep one busman continually cruising the halls to bring back Room Service materials.)

Housekeeping should be aware that the Sales Department places considerable emphasis on the property's being immaculate so there will be no problem selling it to the public. Some hotels even have Sales spot-check several rooms daily or weekly and report their findings to the Housekeeper.

Good communication with the Purchasing Department is also a necessity, especially if it does all the buying for Housekeeping. It is essential that specifications be accurate, that records be correct, that the right products be purchased at the right price with the right frequency.

Above all, in a large institution there must be the closest possible cooperation among Security, Personnel, and Housekeeping. Because Housekeeping employees work directly within the guest room areas they can be most helpful in reporting suspicious characters or situations that might be irregular and lead to serious trouble. Employees are instructed to report minor irregularities to their supervisors but if a situation seems urgent—such as a gun or large sums of money found in a drawer, or any trace of

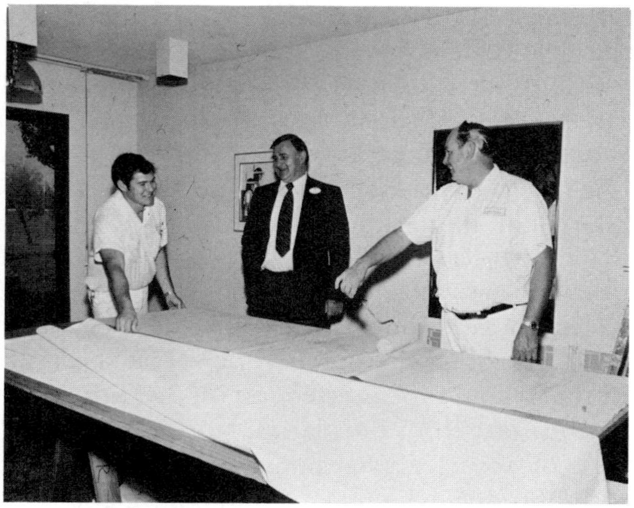

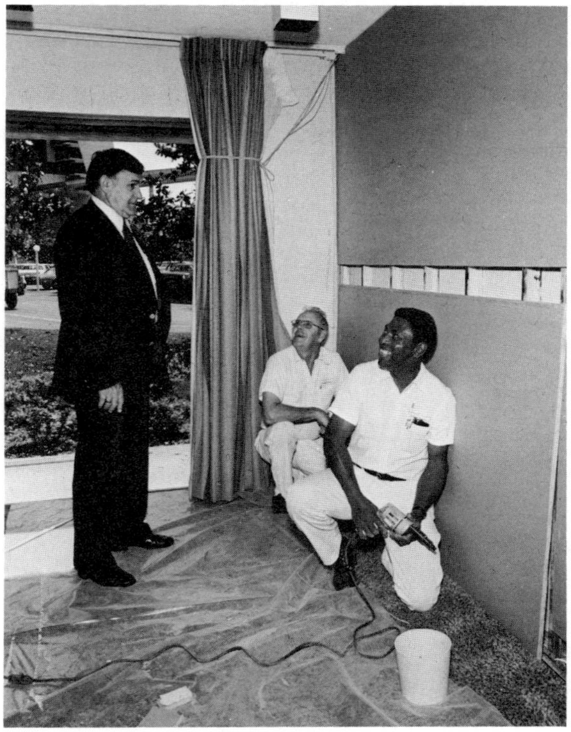

**FIGURES 26-4 and 26-5** *At the Contemporary Hotel, Walt Disney World, Orlando, Fla., guest rooms can be completely redone in a day because of the well-organized mini-rehabilitation team. Bob Chesko, manager of Maintenance, oversees the painting, wallpapering, electrical, and plumbing experts. Carpets are laid, furnishings changed, rooms are magically redecorated in time to be rented that night.*

violence—Security should be called directly on the nearest telephone.

The head of Security should meet with Housekeeping supervisors at least every few months to let them know how they can help. At one meeting the subject might be how to identify "skips"; at another, how to deal with burglars or drug peddlers. Certainly Security will want to discuss life-saving precautions.

Employees should never talk to a newspaper reporter about any hotel crimes. The sheer size of thefts could focus attention on the industry as a prime target for burglaries. Nor should the nature of security precautions ever be discussed with outsiders who may very well have good reason to seek out inside information.

An extremely high percentage of thefts takes place because of carelessness, stupidity, or penny-pinching. For safety's sake, even during the energy crisis, the back of the establishment should be brightly illuminated. Many lighting fixtures are installed so low to the ground that intruders can simply unscrew the bulbs and go about their business.

Many petty thefts are committed by employees especially where the employer, as is often the case, accepts this as natural. Most employees are honest but a single thief may convert the entire organization into a gang of petty thieves. When petty theft begins, the offender should be discovered and dismissed as fast as possible. To prevent this problem from becoming widespread it is imperative that management make it absolutely clear to the staff that theft of any sort shall be considered cause for immediate dismissal, and that whether the value of the stolen item is 50¢ or $50 is absolutely immaterial. Dismissal shall be immediate.

While most crime manifests itself in the motel industry mainly in the theft of merchandise (the TV set being one of the most expensive of such merchandise), within the hotel industry, crime is evident on a broader scale.

Because so many hotels are located in downtown areas, they may be close to a neighborhood where there is a sizable criminal element. Sometimes the busy hotel is a temptation to those unoccupied burglars nearby. The sheer size of a hotel, and its resemblance to a labyrinth, also reduces the chances of quick detection of the criminal. The criminal usually has little difficulty in identifying the house detective. To circumvent this problem some lodging establishments are now hiring women as house detectives.

Furthermore, the large number of people employed in many hotels makes the hotel itself a community, and wherever there is a community, there is crime.

## *Guest Thefts*

Petty theft is so common in the hotel business, and so ineradicable, that the main line of defense is simply to factor the losses into the rate structure. One bit of advice: Do not monogram anything. Most guests are not interested in stealing, but in taking home a free souvenir. If a hotel's towels are not monogrammed they lose their souvenir value. It is advisable to tempt guests with a less costly souvenir, for example, an attractive little ashtray, sewing kit, or even a plastic laundry bag.

Bulky items such as scales, ice buckets, and carafes, are much less likely to be appropriated from a city hotel where those objects would have to be carried out in a suitcase, than from a motor hotel or motel, where guests can easily remove merchandise to cars.

Larger, more valuable items can most easily be safeguarded by keeping the number of entrances and exits to a minimum. If there are side or back doors through which a guest might easily haul out a television set or, for that matter, his own luggage, without being noticed, those doors should be kept either locked or guarded. During the late night hours in particular, everything possible should be done to limit the number of entrances and exits, but be sure that no doors that are fire exits are locked, including the inner doors to the fire doors.

Outsiders breaking into guests' rooms is a less frequent but nonetheless troublesome problem. To eliminate the problem, managers suggest that cashiers always request that the guest turn in his room key when checking out. Hotel rooms ought to be fitted with dead-bolt locks, rather than with spring-latch locks that can be opened faster with a credit card than with the proper key. A room-security system is now available that enables the Front Office to change the lock on each room by punching a central computer after every check-out. Rodeway Inns is experimenting with Digi-key, a four-number combination lock changed for each new guest.

No maids, supervisors, or assistant housekeepers should ever let anyone into a room with their pass keys. Guests of the hotel who forgot their keys will not be insulted if the maid explains she has been instructed not to let anyone in a room.

Housekeeping must protect the institution's property from guests who help themselves when they see storerooms filled with towels, dishes, tableware, draperies, linens, spreads, and large inventories of guest room supplies, such as soaps, instant coffee packs, and stationery.

Those little expendable items have proved very alluring to guests who often grab a handful as they pass the maid's cart. To prevent this, carts sometimes have a custom-made compartmented container with a cover that locks so that supplies left unwatched in the guest room corridors will be secure while maids are inside guest rooms and bathrooms. (Remember, in Chapter 8 we learned the proper procedure is for maids to leave carts in front of open guest room doors, with the carts positioned so that linens face toward rooms for convenience, as well as to remove temptation from passersby.)

As for protecting the more costly furnishings and accessories in guest rooms, that is most effectively accomplished by intelligent initial selection, or by a specific type of installation. Lamps, for example, should be large enough *not* to fit in a guest's suitcase, and should be bolted onto the tables. Pictures should be large too, but more important, anchored with the new types of thief-proof devices available today. Any objets d'art—sculptures, tapestries, book-ends—should be bolted or glued, even in the special suites. This, however, should be done under the supervision of a competent interior designer and a knowledgeable housekeeping executive so that once anchored, there will be no need to relocate them.

Sometimes books are provided in luxury suites and no effort should be made to keep them on the premises; paperbacks are the only practical items to include under such circumstances, but are hardly elegant enough for a $300-a-day suite. Perhaps impressive looking second-hand volumes can be located for the bookshelves of those special suites. It is sometimes even necessary to risk including an expensive set of handsome books if a suite is being rented to truly important people.

First-class hotels are often expected to provide little extras for guests who have forgotten something. Items such as ironing boards and irons, razors, electric shavers, hair dryers, and heating pads are frequently loaned to guests. It is generally Housekeeping's responsibility to store, deliver, and retrieve these items. Guest thievery is surprisingly high on these loans, and management must decide just how this problem should be handled. There are two ways to approach it:

1. Deliver item to guest room, have guest sign a loan slip for it; slip is sent to the Front Desk; item is charged to guest's account; when the article is returned Front Desk is notified and charge is removed from guest's bill.
2. Deliver requested item to guest's room. Even if he is not there, as is often the case, leave the item in the room. Unless it is in short supply and needed by another guest it may be left until guest checks out, and then returned to Housekeeping. There will, of course, be a record of which room the item was sent to, so it is possible to bill the guest after he checks out if he takes the shaver or the hair dryer along with him. This may, however, prove to be extremely poor customer relations.

In some hotels Housekeeping sends a notice to the Credit Department that a specific guest removed hotel property, and leaves it up to the Sales Department to determine whether that guest should be billed or not. If the executive secretary of an association books an $18,000 convention in the hotel, and has pampered a wife who just couldn't get along without the electric iron loaned to her, one can be sure the Sales Department will opt *not* to make that man pay $20 for the iron!

## *Employee Thefts*

Although the Food and Beverage Department is usually the proper office to deal with the security of liquor and foods from storerooms, freezers, and refrigerators, sometimes Housekeeping is responsible for cleaning those areas and it then is necessary for managers to concern themselves with preventing thefts of meat and saleable items such as expensive liqueurs and whiskey.

Techniques used by determined employees are sometimes highly imaginative. One houseman was successful in getting away with dozens of steaks by strapping them inside his thighs. Newspapers in stacks or rolls make effective camouflage for liquor bottles. Uniforms concealed beneath street clothing can be easily removed from the premises. Rarely is a fine and fancy nightgown turned in to Lost and Found because filmy lingerie items are easily smuggled out under the employee's clothing. Even sheets and pillow slips disappear that way. One of the standing hotel jokes is that some new maids "come in a size 8 and go out a size 40" for layers of the bedding have been ingeniously wound round and round and tidily covered with a loose garment.

To eliminate employees making off with company property, any number of protective devices or systems can be used. A secured employee entrance and exit is an obvious answer in a large institution. The guards on duty ask for package passes from employees carrying anything out of the building and have the right to examine bundles or parcels brought into or taken out by employees.

Department heads and supervisory personnel should be authorized to write passes and steps should be taken to ensure that employees cannot forge package passes. Some type of dating or numbering system should be used so that passes cannot easily be duplicated or transferred.

Identification cards with pictures might be required of all employees. Sometimes the name card that the employee wears during the work day will have his picture on it—just to ensure easy identification in a large institution where there is continual turnover and where Housekeeping employees have access to all guest and service areas.

Many stories are told about how large thefts have been committed only because someone in uniform had an official looking form in his hand, and not only acted as if he had the authority to move something, but actually did move it—right out the front door. This should be a lesson that identification beyond a uniform is needed where expensive stores of supplies, equipment, foods, and beverages are being handled.

## *Cordiality Is a Must*

Maids and bellmen should be as cordial to each other as to members of any other department. On the other hand, overdoing it may be an annoyance to guests, for example, employees chatting together in guest room corridors, or maids congregating in the floor linen rooms. When all employees in all departments in a hotel work together, it is possible to lighten the load for everybody. No housekeeper can inspect every item in every room in every department; he or she needs the eyes and ears of all employees to help maintain the establishment in first-class condition.

Employees in one department should feel confident of the complete cooperation of every other department. Any unpleasantness or friction between them must be erased immediately, usually through a conference. Bickering and buck-passing should *not* be tolerated because the one who usually suffers in such cases is the guest. If the guest suffers enough, he may decide to stay somewhere else!

It is important, for example, that the people in the Linen Room understand the pressures under which the Front Desk is working and certainly the Front Desk should understand the pressures on Housekeeping, particularly on days with heavy check-outs and check-ins. There should be some cross-training with Housekeeping supervisors and Front Desk personnel so that each knows what the other is confronted with relative to pleasing the guest. The way employees in both departments will make the utmost effort to cooperate so that all of the guests are taken care of quickly and hospitably.

### *Room Status*

What most often creates friction between the Front Office and Housekeeping is the status of rooms (whether they are occupied, unmade, or cleaned and ready to rent, or out of order). Many electronic systems have been devised for notifying the Front Desk which rooms are ready to rent (Figures 26-6 through 26-9). Any listing of these systems will probably become obsolete within six months to a year because electronic firms are constantly improving present systems or inventing new ones.

Called room status systems, these devices are of great help to Front Office and Housekeeping *if* they actually work. Many of them do not function because they do not represent the proper application for that property, or because the intricate workings of the equipment failed and foolproof repairs have not been possible due to faulty and premature engineering claims.

In a typical system, as soon as a guest room has been cleaned and inspected, a light flashes on a panel at the Front Desk to notify the room clerks that the room whose number is illuminated is ready to sell. Some systems have a special light that indicates the maid has finished cleaning a room but that it has not yet been inspected. The Front Desk can then sell the room with the stipulation that it has not yet been thoroughly checked, or can at least count on that room to be available soon if guests call for reservations.

Some electronic systems serve as a memory, keeping an inventory of the placement of guests, room loan items, and special requests such as bed boards and cribs.

The investment in a status system involves an exceedingly large amount of money. Therefore the unit chosen should be workable for its particular application, by available personnel, and should above all be easily, quickly, and economically repairable.

The majority of systems have three terminals: (1) at the Front Desk, (2) on the floors, and (3) in the Housekeeping Department. The one in the Housekeeping Department cannot be changed from there; it simply shows the status of the rooms, revealing the progress of the maids in each area.

On the floors there is often a board that in-

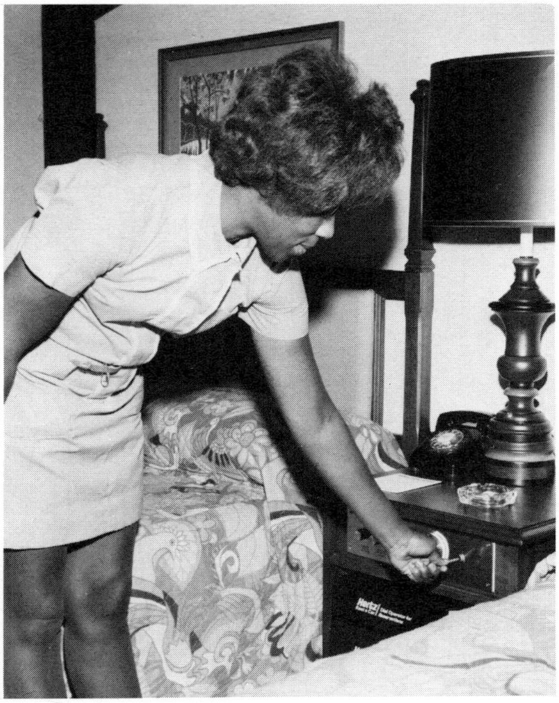

**FIGURE 26-6** *At Mariott Hotel in Los Angeles (as at all the chain's major properties), a computerized room status system helps to get information to Front Desk informing attendants there which rooms are clean and ready to rent. A section housekeeper inserts a jack into a receiver in bedside table.*

**FIGURE 26-7** *A terminal at the Front Desk indicates that the room has been inspected and can be sold.*

**FIGURES 26-8 and 26-9** *In central Linen Room, Housekeeping Department computer terminal has ready all information fed in by section housekeepers as to which rooms are ready. Marriott's director of services explains the capabilities to linen room attendant who will be able to find out condition of all rooms at all times by pressing proper buttons. Housekeeping supervisor at satellite terminal in floor linen room reports ready rooms.*

spectresses or supervisors can view, showing which rooms are check-outs that rooms maids have finished, thus permitting the inspectress to indicate to the Front Desk that these rooms have been inspected and are ready for occupancy. Some systems enable maids to turn a special key in an opening in the radio-TV controls, and thus light up that room number at the Front Desk and on the supervisor's panel to indicate which rooms are ready for inspection. This way room clerks will be aware that certain rooms are soon to be released. Other systems will light up ready rooms only after inspection.

At the Front Desk, personnel are usually able to punch a button and show a check-out or a registration. With some systems it is impossible for a Front Desk clerk to register a guest into a room that is not ready for occupancy. This capability is one of the more salable points of some of the new systems. (Those that operate entirely on lights do not have that facility.)

The electronic system used by the giant, glittering Bonaventure Hotel in downtown Los Angeles has been modified and improved—having been first tried at the Peachtree Plaza in Atlanta—so that it gives three very useful reports:

1. *Occupancy report.* This tells which guest rooms are ready to rent (cleaned and inspected), or vacated and ready to be cleaned.
2. *Floor status report.* This shows which rooms are waiting to be cleaned or are being cleaned, which rooms are out of service for repairs, and which ones are occupied but clean.
3. *Productivity report.* Each time a room attendant completes a room she calls Housekeeping headquarters with her special identification number, reports with a code the room status, and indicates which room she has just cleaned. The computer keeps track of the completed rooms and determines how long it took attendants to clean them.

Although the computerized room status system hasn't eliminated any people on Bonaventure's 275-person Housekeeping staff, it has helped to identify problem areas.

When a room must be put out of order, for

example, to shampoo a carpet or paint a wall, the director of Housekeeping can reserve that room putting one of nine numbers in a code indicating why she is doing so. If the repair is minor (a spot on the wall or a damaged clock) the room number in the computer has a code number or letter and the Front Desk will know what is wrong and can rent the room if necessary.

Extras such as a crib, headboard, or special pillow are also coded on the room number so Housekeeping can keep track of where those items are.

With this complicated, sophisticated equipment has come the ability to do a great many other things. Often the room status is only a small part of a hotel's computer system, which also has the capability of handling reservations and registration, posting guest charges automatically, figuring payroll, doing accounting, scheduling staff, and providing a memory bank that can show a variety of information beneficial both to Housekeeping and the Front Office.

Systems do not always replace employees but they often save time and contribute to the ease and speed of serving the guests. Certainly all lodging units of any size will in the future have some type of status system, and the very large properties will have an exceedingly sophisticated system that will be able to handle scheduling, history, attendance records, payroll, production, and location of employees. Repair requests can also be built into it.

Lacking the installation of an electronic room status system, it is always possible to improvise room control using telephones, pocket pagers, two-way radios, or other devices that enable written communiques such as the Housekeeper's report or house count to be sent two or three times daily to the Front Desk so that all who need to know are aware of room status prevailing at the time.

In some hotels, maids report their ready rooms to Housekeeping headquarters by telephone. There they are charted and reported from that master sheet to the Front Office. In other properties there is a pneumatic tube between Housekeeping and the Front Desk that is used for reporting check-out rooms, ready rooms, rushes, and so forth.

A TeleAutograph system between Housekeeping, Front Office, Engineering and all floors can also be very effective. What is written in one of those transmitting stations is simultaneously written at all of them. The motions of the pencil are actually reproduced, so whoever is doing the writing will have to write large and legibly.

A status system is one of the first steps in automation in Housekeeping, and it is hoped that it will be the forerunner of many other devices that will be developed to reduce labor costs and make for greater efficiency in a department that requires a tremendous amount of manpower.

## *Energy Conservation—Everyone's Responsibility*

Housekeeping is more involved in the business of saving utility costs than any other department, according to one veteran hotelman. "Energy conservation is a major consideration," says Stuart Cross, general manager of Hotel Utah. "Our original structure is 69 years old, therefore there are many challenges in this area. Window frames have become loosened and are cracking and we foresee an extensive restructuring in that area. Meanwhile, we are presently using very heavy blackout drapes which keep the cold and heat from being transmitted through the glass to the rooms. (The glass in this part of the building is single glass and not thermal pane.)

"We are presently experimenting," Cross continues, "with the new fluorescent room lamps which are 40-watt globes but produce the same amount of light as a 100-watt incandescent globe. We are told they guarantee 7500 hours as opposed to the normal 750 hours. This is being done on a trial basis presently on one floor only in the hotel. Preliminary figures indicate the present purchase price to be too high for an optimum payout, but this cost should drop and make these more attractive in the future."

It has always been a policy in most hotels to turn the lights off in public rooms when not in use. Hotel Utah is now policing this action, not only by Convention and Services Department employees, but with security personnel.

Explains Cross, "Although our hotel is generally illuminated by incandescent lights as originally designed and as our decorator redesigned our hotel about eight years ago, we have now replaced all hallway lighting with fluorescent lights. This not only gives consistently good light in those areas where it is needed but it is a more efficient way of lighting the halls.

"In connection with lighting, one of our first energy-saving acts was to turn off and remove the neon lighting atop Hotel Utah. This was a point of controversy among many local citizens; however, the saving of energy offered a valid reason."

Hotel Utah's Maintenance Department is removing the "instant-on" device on room televisions. Although the savings per unit is small, when multiplied by approximately 540 sets it is considerable over a year's time.

Energy conservation projects conducted by Hotel Utah include:

1. Relocation of heat exchanger and addition of another to supply adequate heating capacity with chiller out of the system.
2. Installation of bypass valves and connections to put cooling tower water directly to the chiller condenser during summer operation.
3. Installation of bypass valves and connection to take chillers off heating lines during winter months to minimize heat losses.
4. Investigation into conversion of cooling meeting rooms and restaurant operations at the Temple Square Hotel from the central chiller system to using individual self-contained refrigeration units.
5. Attempt to replace the presently disabled central chiller at the Temple Square Hotel property with a new, more energy-conscious unit. We hope to be able to operate the cooling capacity in stages as needed, and run no excess power not necessary for performance.
6. Preparations to install a heat exchanger in the culinary hot water supply line, to pre-heat the cold water supply to the system. Therefore removing lost heat from steam condensate before it leaves our building.
7. Investigation of more adequate controls to further the use of outside air for cooling, thereby reducing the need for a chiller load.
8. Installation of time clocks on hallway air handlers and exhaust fans to place them on a limited operation schedule.
9. Shutting off all possible fans at night, and leaving them off in all unoccupied meeting rooms and offices when possible.
10. Discontinuation of extended service light bulbs in entire operation where at all possible.
11. Change-over to watt-miser fluorescent light tubes and energy-reducing flood-lights.
12. Installation of hot-water booster coil for closer control of kitchen water, with lower temperature on the remainder of the building.

# Chapter 27

# Safety, Fire Prevention, Health

**OVERVIEW** In these days of active consumerism, OSHA has made many rules to keep workers safe, and Housekeeping has heavy responsibility to follow them and practice preventive measures to cut fire and accident rates. Management of the department must inform, enforce, and train, train, train. Analysis of accidents indicates about 98% are human error and preventable.

The Occupational Safety and Health Act (OSHA) covers approximately 60 million employees in 5 million companies. The act makes it mandatory for every employer to keep his place of business "free from the hazards that are likely to cause death or serious harm to his employees."

Inspectors may make legal entry either to conduct a complete examination of the premises or in response to an individual complaint made by any employee or union. The law specifically forbids giving the employer any advance notice of a scheduled inspection, prohibits the employer from retaliating in any way against an individual complainant, and provides for a fine of up to $10,000 for a violation, plus up to $1000 a day for each day it remains uncorrected, and for jail sentences in extreme instances. Detailed records are required by OSHA of every accident and every illness.

Since only 2 percent of on-the-job accidents result from natural catastrophes (floods, hurricanes, etc.), the other 98 percent are at least theoretically avoidable (Figures 27-1 and 27-2). One study shows that 93 percent of accidents could be eliminated by a combination of proper training, adequate supervision, and employee vigilance. Substantiating this conclusion is the finding that an overwhelming 78 percent of all accidents are caused by the individual's failure to note or to heed a reasonably obvious indication of accident possibility. This may, in turn, point to inadequate training, as in the case of the employee who has never been taught not to throw water on an electrical fire. More often it points to ordinary carelessness.

## OSHA—Legal Aspects

Sound health and safety practices assure more than mere legal compliance with OSHA regulations. Such practices minimize insurance premiums, maximize productivity, and keep up the morale of work teams by keeping the absentee rate down.

The following comprises the nucleus of a sound health and accident program:

1. A Safety Committee with ample representation from the ranks, but with sufficient executive participation to assure the workers of management's interest and support. Limiting committee membership to six months or a year permits broad involvement and constant input of fresh ideas. An ideal member might be an accident-prone worker who is in other respects a valued employee. In a large institution it is desirable for Housekeeping to have its own committee since maids, housemen, and seamstresses have little in common with fry cooks and bellmen.
2. Training and up-date programs that emphasize simple ways of avoiding the avoid-

| Department | Number of Employees | Miscellaneous | Fall at Task Site | Fall-Travelling | Lifting | Pushing-Pulling | Foreign Body in Eye | Struck By | Struck Against | Caught Between | Dermatitis | Falling Object | Lacerations | Burns | Totals |
|---|---|---|---|---|---|---|---|---|---|---|---|---|---|---|---|
| HOUSEKEEPING | 162 | 2 | 1/2 | 1/3 | 2/2 | 3 | | 1/2 | 1/4 | | | | 4 | | 6/22 |
| DINING ROOM & BAR | 200 | 1 | | 1/4 | 2 | | 1 | 1/1 | | | | 3 | 2 | | 2/14 |
| KITCHEN | 89 | 4 | 1 | 1 | 1 | | 2 | 1 | | | | 1/2 | 4 | 3 | 1/19 |
| PARKING | 61 | | 1 | | 1 | | | | 3 | 2/2 | | | | | 2/7 |
| PROPERTY ENG. | 24 | | 1 | 1 | 1 | | | | | | | | | | 3 |
| PROPERTY MAIN-TENANCE | 43 | 1 | | | 1/1 | 1 | 2 | 1 | 2 | | 1/2 | 1 | | | 2/11 |
| LAUNDRY | 37 | 1 | 1/1 | 1/1 | | | | | 1 | | | | | | 2/4 |
| GEN. OFFICE | 191 | | | 1 | | | | | | | | | | | 1 |
| STEWARD | 105 | | 2 | | 3 | 1/1 | 1 | | 1 | | | 1 | 6 | | 1/15 |
| TOTALS | 912 | 9 | 2/8 | 3/11 | 3/11 | 1/5 | 6 | 2/5 | 1/11 | 2/2 | 1/2 | 1/7 | 16 | 3 | 16/96 |

**FIGURE 27-1** *Analysis of Accidents in a Hotel with 1500 Employees During One Year*

Number system: The top number indicates a lost time injury, the bottom number is total injuries. Thus ½ indicates one compensable injury and a total of two injuries requiring outside medical attention. Single numbers are injuries involving no lost time.

BASIC ACCIDENT CAUSES

**10%**
**PHYSICAL**

I Physical Hazards
- a. Mechanical
- b. Electrical
- c. Steam
- d. Chemical

II. Poor Housekeeping
- a. Improperly piled
- b. Congestion

III. Defective Equipment
- a. Machines
- b. Tools
- c. Equipment

IV. Unsafe Building Conditions
- a. Fire protection
- b. Exits
- c. Floors
- d. Openings

V. Improper Working Conditions
- a. Ventilation
- b. Sanitation
- C. Light

VI. Improper Planning
- a. Layout of operations
- b. Layout of machinery
- c. Unsafe practices

VIII. Improper Dress
- a. No goggles—gloves—masks
- b. Unsuitable—long sleeves—high heels

**88%**
**SUPERVISORY**

I. Faulty Instructions
- a. None
- b. Not enforced
- c. Incomplete
- d. Erroneous

II. Inability of Employee
- a. Inexperience
- b. Unskilled
- c. Ignorant
- d. Poor judgment

III. Poor Discipline
- a. Disobedience of rules
- b. Interference by others
- c. Fooling

IV. Unsafe Practice
- a. Chance-taking
- b. Short cuts
- c. Haste

V. Lack of concentration
- a. Attention distracted
- b. Inattention

VI. Mentally Unfit
- a. Fatigue
- b. Excitable
- c. Temper

VII. Physically Unfit
- a. Defective
- b. Weak
- c. Fatigue

**FIGURE 27-2** *98% of all accidents can be avoided, 2% are unavoidable.*

able accident. When people work in close quarters, sweaters snag, loop earrings are snared, long hair can get caught, and rings become hooked; avoid them. Barefooted maids may literally wind up on needles and pins; shoes are mandatory safety equipment. So is cleansing tissue—to protect the hand when reaching in where used razor blades lurk.

3. Eliminating the open invitation to an accident—the slippery walk, loose handles, cracked glass.

4. Enforcement of existing safety rules. If the No Smoking rule does not mean what it says, it should be repealed. Carts, not arms, are the only recommended carrier for quantities of supplies.

5. Continuous analysis of accidents by person, place, occasion. A minority of workers are often responsible for a majority of accidents. Train them, caution them, supervise them, and if necessary, replace them. If the same accident keeps happening at the same place, ferret out its cause and remove it or remodel it.
6. Management sets the pace by example, by enthusiasm, and by openly campaigning for all-out support for the institution's safety program. New workers should be thoroughly indoctrinated in the program, which must also be reviewed regularly with all workers.
7. Making sure everyone knows exactly what to do in case of accident. All plans require that the accident be reported immediately to the injured person's supervisor or department head, who in very large institutions, reports the accident to the Medical Department or the Security Department. In smaller institutions, a common practice is for the supervisor to accompany the injured person to a doctor or hospital.

## *Fire Prevention*

Bringing the institution into compliance with OSHA federal fire regulations is a job for management and specialists, including the community fire department. It is up to management, of course, to make sure all department heads are briefed on specific details of fire prevention plans, location, and usage of equipment, and above all on the precise role of each person or group in the event of fire (Figure 27-3).

**FIGURE 27-3** *At Holiday Inn University, Olive Branch, Miss. (near the company's Memphis headquarters), a fire-fighting class is held as part of regular training for the innkeepers and department heads. Learning to use emergency equipment enables trainees to cope with small fires or help before fire department arrives.*

Department heads should be given appropriate materials (signs, posters, etc.) (Figure 27-4) for use in communicating this plan to all personnel.

The person discovering the fire should immediately make that fact known to the hotel's switchboard operator in the fastest possible manner—either by house phone or by breaking the fire-alarm glass.

Proper housekeeping plays a major part in fire prevention. Cleanliness is fire's natural enemy since it provides nothing for fire to feed upon. Not allowing storage areas to become fire hazards, as well as use of as many noncombustible materials as possible, is also important. Proper equipment maintenance is vital as exemplified by the dangers obviously inherent in frayed electrical cords, sparking motors and the like (Figure 27-5). Maids should be reminded that rooms must always be checked for such common hazards as the travel iron the guest forgot to disconnect or a can of combustible cleaning fluid accidentally left uncapped.

An effective fire prevention program demands daily attention from the Housekeeping Department. At least one day a month, however, the head of Housekeeping should request all personnel to search for, note, and report all observed violations of the institution's fire safety program. Points to be checked may include some or all of the following, depending on the extent of the Housekeeping Department's responsibility:

1. Are regular and emergency exitways (corridors, stairs, doors, fire escapes) kept unobstructed (Figure 27-6)?
2. Are exit lights clean, on, and properly bulbed?
3. Are solid, self-closing metal fire doors designed to deter fire from spreading through them? Check regularly to make sure they are always working properly.
4. Does everyone in Housekeeping know where alarm boxes, extinguishers, and other fire-fighting components are located?
5. Are all of these components located where they should be; in proper working order; unobstructed?

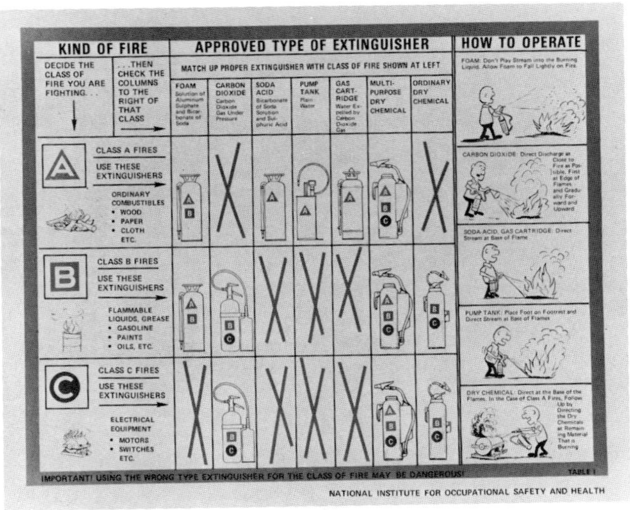


**FIGURE 27-4** *A world of information is contained in this poster put out by the National Institute for Occupational Safety and Health. If employees learn to use the right fire extinguishers chances of disastrous fires are considerably lessened.*

6. If the system includes early-warning detectors and overhead sprinklers (both of which often repay their cost in lower insurance premiums), make sure they are always in proper working order. Allow at least 18 in. clearance between sprinkler heads and possible obstructions.
7. Take care not to overload electrical lines; to replace worn cords and plugs; to avoid overly long lamp cords; to avoid use of defective electric equipment; to repair or replace broken switchplates and other elements of the institution's electrical system.
8. Are all ducts for heating, air conditioning, and ventilation—and especially the exhaust ducts over kitchen ranges—kept clean (Figure 27-7)?
9. If these ducts contain the type of dampers that close automatically when their sensors detect fire, make sure dampers are always in working order.
10. Where No Smoking signs are posted, is the order enforced; in other areas are there enough sand ash urns, and are they easily visible?
11. Are hidden areas (bottoms of elevator shafts, stairwells, and so on) cleaned regularly?

**FIGURE 27-5** *Fire-fighting equipment is available at the Kona Village on the big island of Hawaii where guests rent individual houses made of wood and thatched roofs. Equipment has been disguised as much as possible by being covered with black lava rocks, the dominant element in the landscape.*

12. Are trash collection areas cleaned as often as feasible, and are special precautions (e.g., no smoking) applied in these areas?
13. Have even the plastic plants, and everything else possible, been coated for fire retardancy? Caution: these coatings wash off with frequent cleaning and must therefore be renewed regularly.

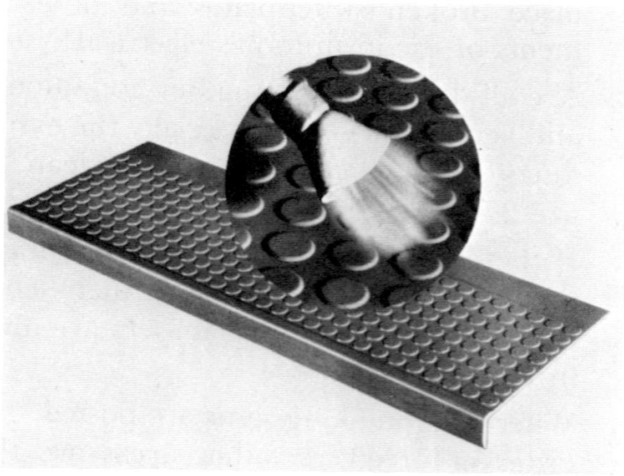

**FIGURE 27-6** *Stair treads are designed for safety underfoot and for fire-resistance. Raised circular discs on surface are easy to clean. Inset shows test for rating flame spread. This rubber tile comes in 24-in square also. Credit: R. C. Musson Rubber Co.*

14. When buying new or replacement items, check to see if the product is available in noncombustible form (e.g., metal wastebaskets). Consider nonflammable blankets and spreads, and the newer fire-retardant mattress covers, all of which reduce the fire hazards posed by persons who insist on smoking in bed.
15. Federal fire regulations mandated by OSHA are both a guide to and a demand for stronger fire prevention measures. Study them carefully; review them regularly.
16. Be sure to seek the counsel of fire department inspectors in determining the efficiency of your present fire prevention program and take advantage of their specialized knowledge to find ways to upgrade your program.

In its Housekeeping Training Manual, Hyatt Corporation has included a helpful section on

**FIGURE 27-7** *Most fires in institutions start in the kitchen in the hood above the range where a grease build-up can quickly ignite. Keeping filters clean is an important part of a proper maintenance program to prevent fire. A filter in a remote location such as shown here can be lifted out with a special mounting bar. (Inset shows enlargement of filter, which in this case was situated farther than usual above cooking surface.) Credit: Flame Guard*

Safety and Fire Prevention, as part of its Hyattrain series. An excerpt follows:

*Fire Prevention*

Cigarettes and matches are the major cause of hotel fires. Employees should smoke in designated areas *only,* and should dispose of ashes and butts safely.

The second most common cause of hotel fires is faulty electric equipment. Housekeeping employees should be alert to this danger and should report frayed cords and loose connections immediately.

Waste accumulation, particularly at the bottom of stairwells and elevator shafts, is an open invitation to fire. Metal receptacles for waste disposal should be conveniently placed throughout the building and emptied with regularity. Oily rags—this includes oil-sprayed dust rags—should be placed in tightly covered metal cans to prevent spontaneous combustion. Solvents and inflammable liquids should be stored in small quantities in metal containers away from heat and flame.

There are three types of fire, and each must be fought in a different way.

## *The "H" in OSHA Stands for Health*

OSHA's concern for employee health is being enforced not only by federal inspectors, but by city, county, and/or state health inspectors as well.

Obviously good health cannot be legislated, nor can employers provide it as a fringe benefit. What they can and must provide, however, is a healthy working environment. This includes reasonable temperatures, adequate ventilation, minimal dampness. It includes provision of a germ-free atmosphere throughout the institution, which means proper sanitation procedures should always be in effect (Figure 27-8).

Most health inspectors are qualified, professional germ fighters. They will, indeed, inspect what the eye cannot see: the bacteria count in laundries and on supposedly sanitized glasses. They will grade the quality of sanitary practices in public and private restrooms, and even in employee locker and shower rooms.

The showers—used with particular frequency by cooks and parking attendants exposed (at least in summer) to high temperatures—are a favorite habitat of germs because of the heat and dampness that combine to produce a potentially swamp-like situation that requires special attention.

While absenteeism is expensive for both employer and employee, it is equally expensive to have a sick person continue working or return to work before he is fully recovered. A sick worker endangers his fellow workers. He also endangers his own safety because illness delays his response to unsafe conditions. Thus a sick worker should be sent home to remain there until completely recovered. In the case of prolonged absence, it is advisable for the supervisor or Personnel De-

**FIGURE 27-8** *Flies are attracted by black light tube in unit (also by aroma of their fallen brothers electrocuted by current and left in tray). Units are ideally installed at level of three or four feet in spots where air currents carry flies or near doors where insects enter. Picture shows typical installation in trash holding room.*

partment to maintain contact to determine when (and if) the individual will actually return to work.

If accurate records are kept and analyzed, it soon becomes apparent that a great deal of absenteeism reflects emotional disturbances that are in no way job-related. Quite often workers susceptible to such problems can be helped by the friendly counsel of supervisors or even by the mere establishment of good rapport with the supervisors.

Where the supervisor suspects real mental illness it should immediately be reported to the department head, who will in turn report the matter to Personnel or the Medical Department. Most workers who are discharged for reasons of emotional problems are capable of handling their job requirements, but cannot cope with themselves, their fellow workers, or associates. It is worth noting, however, that a significant number of cures have been effected by restructuring the worker's job to fit his emotional capability or by reassigning him to a new team, a new supervisor, a new area, or a new function.

While no amateur should assume the role of a psychotherapist, every supervisor should attempt to make his own contribution to solving difficulties of problem workers, initially by conferring privately with the individual and, if subsequently necessary, by requesting a joint meeting with the department head.

Management can further contribute to employee health by insisting on personal cleanliness, by making nutritionally adequate food available (or alternately, a congenial place for employees to bring their own meals), by helping to create a reasonably tension-free working atmosphere, by designating a suitable lounge area for use during coffee breaks, and even by stimulating exercise programs in the guise of a company-sponsored bowling league, softball team, or whatever is feasible in that particular location.

## Health and Safety Guidelines for Housekeeping

1. All appliances such as vacuums and polishers must have grounded connections.
2. Service carts should be equipped with large or wide wheels to make them easier to push, especially over deep-piled hallway carpets.
3. Any objects such as chairs and beds that are heavy or awkward should be lifted or moved by more than one person.
4. Employees should be told never to mix ammonia and bleach to avoid being exposed to a dangerous gas the two substances create.
5. When floors are scrubbed, the area should be identified as slippery by using signs or roping off the area.

## Exits and Exit Markings

1. Every exit must have the word EXIT in plain legible letters not less than 6 inches high with strokes of letters not less than ¾ inch wide.

2. Doors, passageways, or stairways that are neither exits nor ways to exits, but that may be mistaken for exits, must be clearly marked NOT AN EXIT or must be marked by a sign indicating their actual use—STORAGE ROOM or TO BASEMENT.
3. When the direction to the nearest exit may not be apparent to an occupant, a sign with an arrow indicating the direction must be used.
4. Exit access must be arranged so that it is unnecessary to travel toward any area of high hazard potential to reach the nearest exit (unless the path of travel is effectively shielded by suitable partitions or other physical barriers).
5. Nothing may impair the visibility of the exit sign, such as decorations, furnishings, or other signs.
6. A door from a room to an exit or to a way of exit access must be of the side-hinged swinging type. It must swing out in the direction of travel if:
    a. 50 or more persons occupy the room, or
    b. the exit is for an area of high hazard potential.
7. Areas around exit doors and passageways leading to and from the exit must be free of obstructions. The exit route must lead to a public way.
8. If occupancy is permitted at night, or if normal lighting levels are reduced at times during working hours, exit signs must be suitably illuminated.
9. No lock or fastening may be used to prevent escape from inside the building.
10. Where occupants may be endangered by the blocking of any single exit due to fire or smoke, there must be at least two means of exit remote from each other.

### *Power Failures*

Fire and accident are not the only emergencies that hotels and residence halls must be ready to cope with in today's complex world. Brownouts, blackouts, and power failures can be disastrous unless management has a plan. Perfect timing accompanied the San Diego Hilton's fine advanced planning.

Just two weeks after new procedures for dealing with blackouts and power failures was published by Hilton's Engineering Department, the facility in San Diego was struck by a blackout that covered much of Southern California. Although it lasted only a few hours, management was pleased it had met on the subject and was well organized.

The new guidelines for management contained several suggestions that had already been instituted in San Diego. Primary among these was the establishment of a command center, institution of a written action plan, and purchase of certain emergency equipment.

According to Tony Rubino, general manager of the San Diego Hilton, the preplanning was worth the time put into it by the department heads. "When we received the guidelines from the corporate engineering department, I immediately called a department heads meeting to review what we had been sent. We then divided the plan into sections and each of the operating departments, Security, Engineering, Housekeeping, and Food and Beverage were assigned specific jobs to accomplish in case of an emergency. When the lights went out just a few days later there was no panic and everyone knew exactly what was expected of him during the emergency. We had even ordered and received the fluorescent lanterns, AM-FM portable radio, and guest flashlights required by the guidelines, and had these in their assigned places. My office became the command center and we were ready."

After the blackout was over many guests came to the front desk to compliment the staff on the hotel's reaction to the emergency.

From Hyattrain booklet:

**General Safety Rules.** Report every injury on the job to your supervisor immediately.

**Do Not Delay.** Always report any injury, no matter how slight.

Be sure that you understand your job thoroughly and how to do your job safely. If you are in doubt, see your supervisor or your department head.

Report any defective or broken equipment or machines to your supervisor at once. Report any hazards that you may see, such as broken glass, slippery floors, slippery stairs, tripping hazards, burned out electric lights, defective ladders, or defective tools.

Help our new employees become better employees by showing them how to do a job correctly and safely. If you observe hazardous practices report them at once. Prevention may put a stop to an accident.

Whenever possible, use a ladder for a high-reaching job–never use a substitute, such as a chair, box, or a table. Never stand on the edge of a tub or a wash bowl.

Wear safe, comfortable, and sensible shoes when you are at work. Worn out shoes are hazardous to wear.

In the event of an emergency, such as guest or employee injury, illness, fire, or burglary, notify an Inspectress/Inspector or any member of hotel management *immediately!*

## *Occupational Health and Environmental Control*

According to OSHA, persons working in lodging establishments may be exposed to a variety of gases, dusts, vapors, fumes, mists, noise, heat, and cold. The most common exposures are to the following:

1. **Ammonia.** It may be used as a cleaning agent or as a refrigeration gas. Direct skin contact with concentrated liquid ammonia can cause severe burns. When handling liquid ammonia, the individual should wear rubber gloves and goggles or a face shield.

2. **Carbon Monoxide.** Carbon monoxide may be found where ovens or dryers are improperly vented. Doormen may also be exposed where cars are driven into or through the building for loading and unloading passengers or baggage.

3. **Chlorine.** This is used in swimming pools. It may be purchased in the liquid form, as a powder or granules, or as a compressed gas. Respiratory protection should be available where compressed chlorine gas is used or stored.

4. **Drain Cleaners.** These cause skin burns and are harmful when splashed in the eyes. When using drain cleaners, rubber gloves and goggles and/or face shield should be worn. In many operations no Housekeeping employees are ever allowed to use drain cleaners. That is a function for the plumber.

5. **Paints and Adhesives.** They contain a wide variety of solvents and compounds and should be used where there is good ventilation. If this is not possible, respirators for use with organic vapors should be worn. Skin contact with epoxy paints and adhesives should be avoided; however, if this does occur they should be washed immediately.

6. **Perchlorethylene.** The chemical is used in dry cleaning machines found in some hotels. Avoid breathing the vapors and skin contact. A respirator should be available for use when spills occur or when cleaning out the residue. Perchlorethylene is often found in spray-cans and is used as spot remover or furniture cleaner.

7. **Soaps and Detergents.** Contact with these products may cause dermatitis or skin rashes. This often occurs from using too strong a solution. Inhalation of soap dust often occurs when putting material in washing machines, and may cause throat irritation. Disposable paper respirators can be worn during the operation.

8. **Caustic Solutions.** Strong caustic solutions are often used for cleaning reusable filters of range, grill, and broiler exhaust hoods. Controlled procedures are necessary for such cleaning, and protective clothing and equipment must be provided and used.

## *Noise Exposure*

Excessive noise exposure can occur from bands or combos playing in the lounges or bars of hotels and motels. The bartenders, cocktail waitresses, and waiters working in these areas have the greatest risk. Exposure can be reduced by turning down the volume and/or instituting administrative controls. This is one of the most commonly violated standards, however, and can cause permanent hearing damage. To protect employees it is management's responsibility to make sure they are not exposed to levels in excess of the standards. The current standard is 90 decibels A-weighted (dBA) for eight hours. Even at this level hearing damage can be expected in some individuals. It may soon be a requirement, and it is considered good practice, to have hearing checked (audiometric testing) on an annual basis, for all employees exposed to 85 to 90 dBA for eight hours daily. If no hearing loss is observed ear protection is not required.

# Chapter 28

# Laundry

**OVERVIEW** The decision as to whether to buy your own linens and set up your own laundry or use a rental service that provides clean linens according to a contract is difficult, and many factors must be considered. Examples of laundries in different institutions are given, with information about dry cleaning facilities.

An entire textbook could be written on the subject, and, indeed, several books have been written by well-known laundry consultants. The decision on how it is to be handled is complex. In the hospitality/lodging industry there are two possible solutions: one, an on-premise facility, or two, sending laundry out to a commercial operation.

Whether there should be a laundry on the premises in a property of 500 or more rooms is a decision that management must make long before the building is built or is extensively remodeled. Only occasionally is it possible to fit a complete facility into a property after construction is completed.

The decision is made as a result of careful comparisons that take into consideration not just the fairly substantial cost of equipment but also of labor and supplies. Other factors are such unseen costs as: employee taxes and benefits; cleaning and maintaining the laundry; uniforms and printed forms; related expenses incurred by employees working in such other departments as Front Office, Bookkeeping, Receiving, Personnel; the inevitable other expenses including provision for doubtful accounts, insurance, accountant's fees, as well as literally about 18 other items that must figure into a thorough study. Then one must add utility costs (fairly high in a laundry, especially for operating dryers) and sometimes rental of space.

Estimated linen usage (Figure 28-2) must be totaled because if one has the laundry done off-premise, linen is generally rented from the commercial laundry. In most cities this does not mean that linens available will be of inferior quality. On the contrary, most services will buy just what the institution requests, which will eliminate the capital outlay otherwise required. (If the type of linens ordered is unusually unorthodox, the supplier may ask for a substantial deposit.)

Linens washed in an in-house laundry last a great deal longer than those that are sent out. The wear history of linens washed on-premise usually is: sheets 400 to 800 washings, pillow slips about 450 washings, and toweling 300 washings. But many factors affect all of these figures, and washing is not the only problem.

Among Food and Beverage linens (either linen or momie cloth), napkins can be washed approximately 132 times and the table covers about 500. With table covers, usability can be prolonged by mending with patching machines that use heat and pressure to patch in the same color. It is authoritatively stated that linen replacements will run 15 percent less with an on-premise laundry than when it is sent out, but this varies from one area of the country to another

## THE 10 MOST COMMON STAINS AND HOW TO REMOVE THEM FROM WASHABLE FABRICS

Test all fabrics for colorfastness and durability before applying any chemicals; either apply the chemical to an inconspicuous part of the fabric or to a similar piece of fabric. Never use water hotter than recommended in fabric-care chart for ordinary laundering; use suggested bleach.

**Alcoholic Beverages** Soak in cold water before stain has a chance to set; then sponge with liquid detergent and rinse, putting a few drops of vinegar in rinse water. Launder. If stain persists, sponge with solution of one part of alcohol to two of water; if that doesn't work, use alcohol full strength.

**Blood** Soak in cold water; then wash with detergent in warm water. If stain persists, soak in lukewarm water to which ammonia has been added in the proportion of three tablespoons of ammonia to a gallon of water, for at least half an hour. Then wash with detergent again.

**Cosmetics** Sponge with heavy detergent suds. Repeat until all sign of the stain has gone. On some fabrics, letting the material dry between applications of detergent may produce better results. If cosmetics are greasy, follow directions for removing grease stains.

**Egg** Be careful never to use hot water, or stain may become permanently set. If stain has dried, scrape off as much as possible, then sponge with cold water. Work detergent into stain; then launder as usual, using bleach if recommended. Fabrics with special finishes may need further treatment with a spray stain remover for washable fabrics (see Grease). Treat stains from meat juice the same way.

**Fruit** Sponge with cold water first. If stain persists and boiling water will not damage fabric, stretch cloth over a bowl and pour boiling water through the cloth. If stain remains stubborn, sponge with hydrogen peroxide or a lemon-juice solution. If boiling water can't be used, soak in cold water, then rub stain with glycerine and let it remain on fabric for at least three hours. Rinse thoroughly in water to which several drops of white vinegar have been added, then launder.

**Grass** Work detergent into stain and launder, using bleach recommended for that fabric. If stain persists, let fabric dry and sponge with rubbing alcohol, using one to two parts of water on colored fabrics. (Be sure to test solution on scrap first.)

**Grease** Scrape off as much as possible, sponge with liquid detergent, then launder as usual. Non-iron fabrics may also be treated with a solution of two tablespoons of washing soda to a cup of warm water. Rub it into dampened spot and allow to remain for not more than five minutes. If stain persists after either method, sponge with cleaning fluid, let dry and repeat until stain is gone. Another method is to use one of the new spray stain removers. Spray it on stain and launder according to directions on package. This is particularly useful for specially-finished fabrics, which, although stain-resistant, are very hard to get clean once a stain has penetrated the finish.

**Ink, Ballpoint** Ordinary laundering (with bleach, if recommended) will wash out some types of ballpoint ink; others need special treatment and may be set by washing. To determine which type the stain is, mark a similar piece of fabric with the same ink and wash. If laundering does not remove mark, pour denatured alcohol through fabric (testing first on scrap) or rub white petroleum jelly into stain and soak in detergent solution; then launder. You can also spray with hair spray until soaking wet, then scrub by hand, using a thick solution of liquid detergent. (If fabric is not likely to be harmed by a small amount of abrasion, rubbing the stain with an eraser specially designed to remove ballpoint ink *before* you try any of the other methods may do the trick.)

**Ink, Regular** Run cold water through stain until it runs clear; then rub in thick detergent suds. Let stand; then launder, using bleach, if recommended. If stain persists, sponge with chlorine bleach or hydrogen peroxide, or spread a paste of hydrogen peroxide and baking soda on stain and steam; however, all of these methods may affect the color. You might also try a solution of one part of alcohol to two parts of water, then full-strength alcohol. Some ink is permanent and can't be removed.

**Paint** Stains must be treated before they're dry or they will be almost impossible to remove. If they're from an oil-based paint, sponge with turpentine (testing fabric first); then launder, using bleach, if recommended. If stains are from a water-based paint, sponge with liquid detergent and launder. In either case, if traces remain, soak overnight in liquid detergent and launder again.

**Perspiration** Sponge fresh stains with ammonia, older ones with vinegar. Rinse and wash in water as hot as the fabric can take, using bleach, if recommended. If there is any residue of the stain, sponge with cleaning fluid. Since perspiration may weaken a fabric and change the color, it is not always possible to remove its effects completely.

*Reprinted with permission, Family Circle Magazine, November 1972, p. 210.*

**FIGURE 28-1**

*Although many scientifically worded stain-removal charts are available from detergent and laundry machine manufacturers, this page of instructions is explicit, easy to follow, and works.*

| | NO.OF BEDS | REQUIRED 1 TURN | NO.OF TURNS | TOTAL REQUIRED | WEIGHT EACH (POUNDS) | DAILY LBS. OF CLEAN-ING (1 TURN X WEIGHT) |
|---|---|---|---|---|---|---|
| **ROOM LINEN** | | | | | | |
| **BED LINEN** | | | | | | |
| Kings | | | | | | |
| Queens | | | | | | |
| Doubles | | | | | | |
| Twins | | | | | | |
| Pillow Slips | | | | | | |
| | | | TOTAL BED LINEN | | | |
| **TERRY** | NO.OF ROOMS | | | | | |
| Bath Mats | | | | | | |
| Bath Towels | | | | | | |
| Face Towels | | | | | | |
| Wash Cloths | | | | | | |
| | | | TOTAL TERRY | | | |
| | | | TOTAL ROOM LINEN | | | |
| **RESTAURANT LINEN** | NO.OF TABLES | | | | | |
| **TABLE COVERS** | | | | | | |
| 54 x 54 | | | | | | |
| 64 x 64 | | | | | | |
| 82 x 82 | | | | | | |
| Banquet | | | | | | |
| **NAPERY** | | | | | | |
| 20 x 20 | | | | | | |
| 22 x 22 | | | | | | |
| Banquet | | | | | | |
| | | | TOTAL RESTAURANT LINEN | | | |
| **KITCHEN LINEN** | NO.OF ITEMS | | | | | |
| Kitchen Towels | | | | | | |
| Bar Mops | | | | | | |
| Grill Cloths | | | | | | |
| Chef Aprons | | | | | | |
| Chef Caps | | | | | | |
| Chef Coats | | | | | | |
| Chef Pants | | | | | | |
| Uniforms | | | | | | |
| Bartender Coats | | | | | | |
| | | | TOTAL KITCHEN LINEN | | | |
| | | | RESTAURANT AND KITCHEN LINEN | | | |
| | | | TOTAL DAILY USAGE (POUNDS) | | | |
| | | | SUB-TOTAL | | | |
| | | | ADD:_____% OF RECYCLING | | | |
| | | | SUB-TOTAL | | | X 365 |
| | | | TOTAL ANNUAL USAGE (POUNDS) (To Schedule 2) | | | |

**FIGURE 28-2** *Estimated Linen Usage*

*If management wants to study the advisability of having its own laundry, this form will help the Housekeeper figure weights of linens used and compare on-premise performance with rates that commercial laundries charge for equal amounts. To figure costs of linens, use same form, changing the "weight each" column to "price each." Add to total sales tax and freight for total cost of inventory.*

and, of course, is also affected by the amount of linen that "walks" (guests are particularly fond of washcloths for souvenirs).

## *Linen Contracts*

Outside laundries for large and medium-sized operations have wear factors of: sheets 250 to 300 washings, pillow slips 150, toweling 150. Thus it is readily seen that the life of linens is reduced by 50 percent. In Food and Beverage the same lower life expectancy is true: napkins 75 washings and table covers 100 to 150 washings. This also will vary in different parts of the country.

There are, in addition, extra costs on outside laundry contracts: sorting, special folding, separation of sheets and pillowcases, separation of stained and repaired linen, and extra delivery services. In considering any outside laundry contract, either the Housekeeper should be intimately involved in the details of this contract if the manager sets it up, or before settling on a contract, the Housekeeper should check with the manager. If possible, more than one bid should be requested; three are customary.

Certain factors should be considered for that contract:

1. The delivery service—the number of days and the number of deliveries per day; also the time of day
2. How they will handle the guests' work
3. How much mending an outside laundry will do, and whether they remove stains or whether the Housekeeper must set up the stain-removal program
4. How to ensure that all supplies used by the laundry shall be of the best quality
5. Specification for the fold required on towels and sheets
6. The need for uniforms to be returned on hangers
7. Assuming that there is insurance on the laundry, where should coverage begin; what adjustments will be made on mechanical damages; the settling of claims and shortages
8. How the billing is to be made—weekly, monthly; what sort of emergency service is available for Saturdays and holidays
9. How is it possible to establish an accurate count of linens both in and out of the laundry
10. What sort of equipment is used for pick-up and delivery

In establishing a contract, the first requirement is to select a reliable firm with a reputation for integrity. (Later in this chapter a contract based on hotel occupancy is described.)

The executive Housekeeper must set up an inventory system. For example, when clean laundry is received, each basket or container should be clearly marked with the items, date, time of delivery, and the itemized accounting. In the case of weight cost for hotel linens, weight should be clearly marked on the basket. The hotel should check periodically the weight and cost of all baskets and containers. Claims and shortages also often occur on guest laundry, and it should be determined in advance how this should be handled.

There are many advantages for a small or medium-sized operation having its own laundry facilities (Figures 28-3 and 28-4). For instance, it may cut down considerably on the linen inventory necessary for food and beverage operations because tablecloths and napkins can be washed after each meal. If they do not require ironing, they can immediately be reused. This certainly will reduce the par and it may help to eliminate the problem of cooks using table linens (destined for an outside laundry) to wipe up kitchen floors or tables. It certainly means a quicker change from soiled to clean linen, since it requires only an hour or less for a load to be washed, dried, and folded, and there is no transportation delay.

Additional advantages include no need for pick-up and sorting, as a busboy or a houseman can deliver the load. When linen is processed on the premises it is put right back on the shelf and there is no possibility of its becoming confused with that of another organization.

Before deciding on an on-premise laundry, there are negative factors that should be considered—water supply (Figure 28-5), and whether fuel and labor usage can be inexpensive

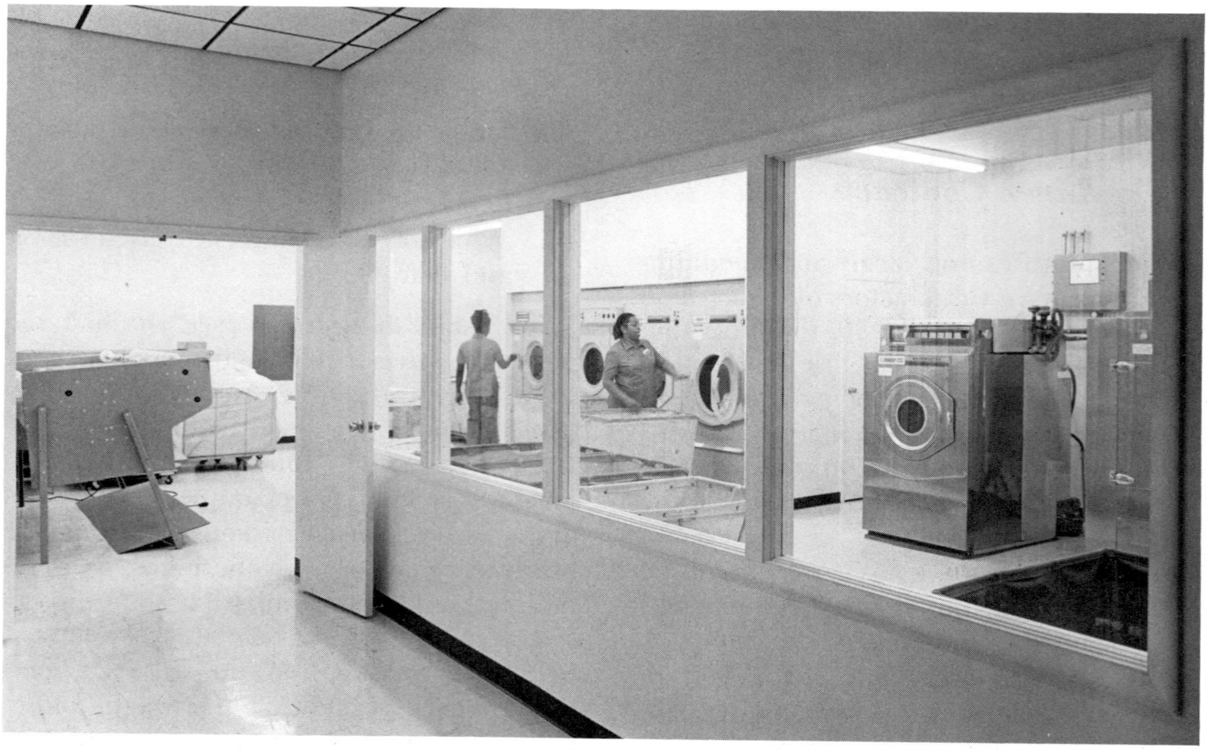

**FIGURE 28-3** *Housekeeping Department at Holiday Inn University has completely glassed-in laundry. The ultramodern facility, complete with sheet and towel folders, is a laboratory, classroom, and in-house laundry.*

**FIGURE 28-4** *Holiday Inn University executive Housekeeper Mrs. Mattie Bowman (left) inspects a stack of linens with one of the university's Housekeeping staff. Towels were folded with automatic folder. Holiday Inn management teams who attend the school want to learn more about such operations as laundry.*

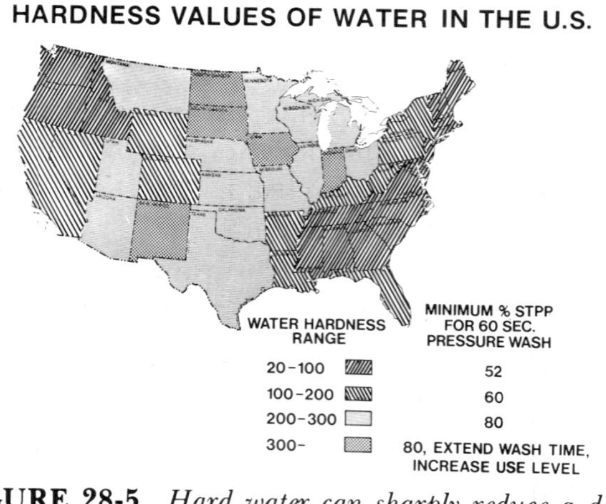


**FIGURE 28-5** *Hard water can sharply reduce a detergent's cleaning power, so a formulation that works in a soft-water state like Georgia may be ineffective in New Mexico, a hard-water state. Researchers find that careful proportioning of polyphosphates and other additives solves the hard water problem.*

enough to make the operation worthwhile (i.e., offering a lower cost than an outside laundry firm).

At best, and only when optimum conditions prevail, actual savings run from between 8 and 10 percent, and then only when certain conditions are present, such as:

1. Consistent, near-capacity occupancy in the hotel or motel
2. Superb laundry personnel management
3. Favorable energy cost rates to heat water and dry the linen
4. Efficient laundry layout designed with proper equipment and no mechanical breakdown
5. Low-cost investment and internal costs
6. Excellent control of linen inventory with few losses

Let just one factor be negative and these small cost savings change overnight to a severe profit drain.

The Linen Supply Association of America will provide material, formulas, and advisors to help Housekeepers make the decision that will work best in each facility.

If it *is* decided to have an on-premise laundry, there are several decisions that will affect the Housekeeping Department. Most important is whether the laundry is under the direct supervision of the Housekeeper, or run by a laundry manager as an entirely separate department. In very large properties it will be a separate department with its own manager. The unit operated under the housekeeper's jurisdiction may be of regular size, or if it only handles no-iron bed linens it may be just a mini-laundry.

A department that is an entirely separate operation will not be considered in this text other than to state that in such cases it is imperative that the Housekeeper and the laundry manager be on most cooperative terms. The manager, of course, not only wishes to control costs but also to fulfill the needs of the property. In the hospitality and lodging industry there is often a crisis when a specific article (or 500 of them!) that was not anticipated being in demand is suddenly needed for a banquet or a guest room. There are endless surprises that can crop up; therefore a cooperative spirit must exist between the two departments.

The greatest revolution in the laundry and linen world occurred in the late 1960s and early 1970s with the introduction of polyester-cotton sheets and pillowcases. Poly-cottons are here to stay and lodging facilities that do not have laundries may wish to consider installing mini-laundries for these bed linens. Given normal usage, it is estimated that poly-cottons outwear the traditional percales by being able to withstand at least 50 percent more washings.

For institutions using poly-cottons and having an on-premise laundry, there are very specialized ways of handling these bed linens that should be strictly adhered to.

Beginning in 1973, it was almost impossible to procure percale because most of the cotton material was being put into poly-cotton. Although a few percales returned to the market in 1978 they are very expensive. To date, poly-cotton has not been fully translated into the terrys for towels and bathmats, since many housekeepers feel they are not sufficiently absorbent to please guests in the luxury hotels.

## *Equipment for Various Sizes of Laundries*

For the medium or small property, if space is available, a small washer-extractor that takes 50 to 75 pounds of laundry could be considered in conjunction with a 50- to 100-pound dryer. In some areas, where union jurisdiction permits, the Linen Room workers could be instructed in the use of the machines, thus saving the expense of additional employees.

In considering a small washer-extractor, a high-speed model with vibration mount should be considered. On a ground-level cement floor, the cement base is practical. On other floors vibration can be eliminated by the mount. Location will be determined by the availability of water and drains, proximity to the linen chutes, the labor available, and whether local unions will permit Housekeeping employees to operate laundry machines (Figure 28-6).

The washer must be capable of handling all types of fabrics. Poly-cotton fabrics need gradual

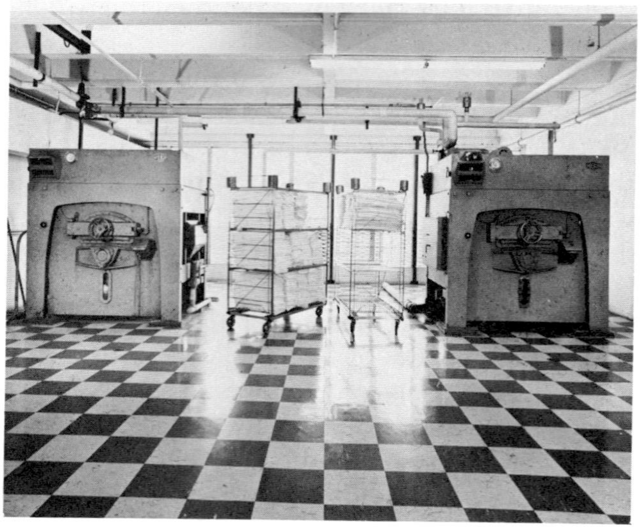

**FIGURE 28-6** *This 300-room downtown Howard Johnson motor hotel found an ideal location for its linen and laundry department even though there was no space available on the ground floor. A parking garage occupies floors 2 through 7, and the guest rooms are located on the next 10 stories. Six parking places on the seventh floor were taken for the Laundry and Linen Departments, so the hotel's linens can be chuted down from the guest room floors above. The washer-extractors are equipped with hydro-cushion suspension, which helps prevent vibration on upper floor installations such as this one. According to Otto K. Felkner, the manager, there have been no problems with noise or vibration since the machines were installed in 1968. "No guest has ever complained about the machines and guests have never been conscious that we had a laundry operating on the floor below their rooms."*

cool-down rinse and a gentle final spin to prevent wrinkling. So in choosing equipment be sure that it can perform under all circumstances so that there is no yellowing or graying in the material.

## *Mini-Laundries*

There are new liquid detergents that make mini-laundries much simpler (Figure 28-7). The old-fashioned system used powders formulated for cotton, not for blends. Liquid detergents made it possible for the leading chemical firms to mesh dispensing units with machine cycles electronically. Now laundry workers can add the detergent directly from shipping containers into the dispensers. This reduces a great deal of handling. Remember that originally a laundry needed washers, extractors, dryers, and ironers. But permanent press, or poly-cotton, eliminated ironing except where state law still requires it. (Originally, ironing was considered a sanitary process by which the extreme heat would kill any germs.)

This type of mini-laundry does not require the space or the number of operators that were formerly needed. The operator can be doing many other things during the wash and dry cycles. In fact, the manufacturers of the washer-extractors will help plan and determine equipment location, and often give guidance on production analysis, cost analysis, equipment recommendation, and lay-out. One of the major factors to consider is cost.

For the average hospitality institution, laundry costs are nearly 4 percent of sales which represents a significant cost area. Any major reduction in laundry costs can lead to substantial improvement in the establishment's profit.

One of the ways in which the no-iron fabrics save money is that it only takes 15 to 20 minutes instead of 40 to 60 minutes to wash them; drying time is even shorter, however, someone will be needed for folding unless an automatic roller folder is purchased (Figure 28-8).

Even if the mini-laundries are able to process only the terry goods, they will achieve a large cost savings, terry being the heaviest of the laundry items. Thus it can represent a large saving to the hotel. Terry includes bath towels, hand towels, washcloths, bathmats, and in some cases, bedspreads. The savings in washcloths alone would be tremendous. If percale or muslin sheets have to be sent out so they can be ironed, there would still be a reduction of the amount of laundry leaving the premises.

Some small hotels and motels take their laundry to the local laundromat. Others try to persuade laundry equipment firms to install a laundromat on their property so that they can pay for it over a period of time. This can be done either by using coin slots to pay by the load or by making a fixed monthly payment. Laundromats are potential money-makers for certain types of institutions, for example, resorts where guests, because they will stay more than a day or two, would also be likely to use the equipment.

All equipment for use with no-iron fabrics

**FIGURE 28-7** *Washing machines in this mini-laundry use automatic dispensing system with liquid detergents. Operators load machine and press buttons marked "towels," "uniforms," "sheets," etc., and programmed washing (based on water and soil in laundry) starts. Prescribed amounts of chemicals for suds, bleach, and sour/soft cycles are fed into machine from containers at base of machine. Operators feel secure that machines are being fed as they are able to see level of liquids. Dryers have permanent press controls, reset button, indicator lights to show cycles, and buzzer that sounds at end of cooling-down cycle. Credit: E. L. Fraser*

has been dramatically improved in recent years. The fear that laundry left in a damp or wet bundle for any period of time might become permanently wrinkled has been largely dispelled. The only thing that will really cause permanent wrinkles is a quick change of temperature. It is still recommended that after being laundered, no-iron bed linens be allowed to rest an extra 24 hours on the shelf before they are reused.

Except in very large properties the trend is certainly toward in-house mini-laundries. In the large property it is still considered more efficient to have a full laundry, but such a facility is changing rapidly and requires far less valuable square footage since no-iron linens require fewer pieces of equipment.

The mini-laundry rarely can be useful for guest laundry because that will be a mix of different fabrics. It will have to be sent out and a contract with a laundry will be necessary if this service is offered. But the no-iron fabric definitely affected the guest laundry situation—for many guests, even in luxury properties, do some of their own laundry. Therefore many hotels now provide a small clothes line and some type of washing detergent in the bathrooms. The blend fabric has also had an impact on valet services because wrinkles in most no-iron clothes drop out easily and thereby eliminate much of the pressing service that was once necessary.

Regardless of the system used, the important factor concerning laundries is that linens be available when needed. The efficiency of maids and housemen is greatly reduced if they are not.

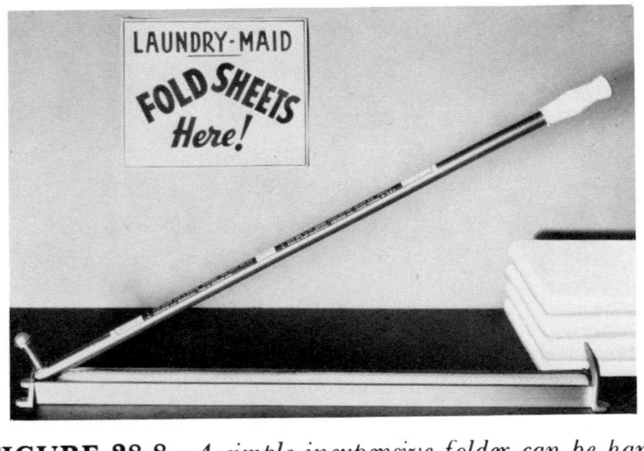


**FIGURE 28-8** *A simple inexpensive folder can be handled by one operator who can put out about 70 sheets per hour (as compared with two operators, folding by hand, putting out 55 to 60 per hour). Two operators at this folder can do 120 sheets per hour.*

Some system should be worked out so that Linen Rooms are stocked before people come to work. One solution would be to have an ample inventory so that there would always be linens in the floor closets ready for maids to use when they make up the rooms.

If an outside laundry is used, delivery times and dates should be set up so that the linens are available at the times needed. There is never any excuse for a maid's making up a room and leaving the spread turned back and the pillows with no slips because the laundry has not been delivered. A person paying the rates charged in the hospitality industry today should never go into a room that is not completely made up, nor should housekeepers ever have to turn down a request for towels or washcloths or other linens.

If the mini-laundry is the best way to assure an adequate linen supply, then it should certainly be discussed with management and the installation encouraged. Perhaps leasing equipment may be the answer to the situation. When a mini-laundry is decided on, it should be installed as close to the Linen Room as possible so that Housekeeping personnel will not have to make extra trips to see if the machines require attention. The most efficient location will be recommended by the laundry equipment supplier at the time a suggested layout and estimate are given.

**FIGURE 28-9** *Linen is transported in special electric carts, complete with thatched roof, in Kona Village, Hawaii. Each guest unit is a separate Polynesian-styled house located several hundred feet away from its neighbor on a gravel path. Soiled laundry is put in bags on trucks; clean linen packs are carried out by maids who enjoy informal uniforms—aloha shirts, shorts, and thong sandals.*

In sprawling properties, it may seem logical to have more than one so that linens do not have to be transported so far from the laundry to guest rooms (Figure 28-9). In the long run, however, it would take more employees, more keys and controls, more inventory, more maintenance on machines. It is far less expensive and easier to buy a few more carts and pay a houseman for the few hours a day required to wheel clean laundry to the remote guest areas.

### *Training Laundry Personnel*

As soon as the laundry is installed, training workers should take top priority. Assistance in this will be given by the equipment supplier. In this competitive age, having the machines perform properly is important, so it is in the suppliers' best interest to see that operators do not damage equipment or linens, but instead get excellent results (Figure 28-13). The wrong procedures may result not only in stained or discolored linens (through mixing types of laundry, e.g., heavily soiled rags or uniforms with lightly soiled linens), but the life of the linens can be considerably shortened.

Says Nena Holder, former executive Housekeeper of the Holiday Inn chain, "Laundry workers should be carefully trained in the care of uniforms and bed, bath, table and other linens. The secret of care control is to take remedial action on any problem, no matter how small, as soon as possible. Whether it is a rip in a drapery, a stain in the sheet, or a spill on the bedspread, the sooner it is given attention the smaller the problem will be, and the more likely that permanent damage will not occur. Nowhere is control more needed than in the laundry. Training employees to presort, pretreat, and use correct formulas for each type of textile is a control which can lengthen the life of any item."

Conversely, some high-rise motor hotels and

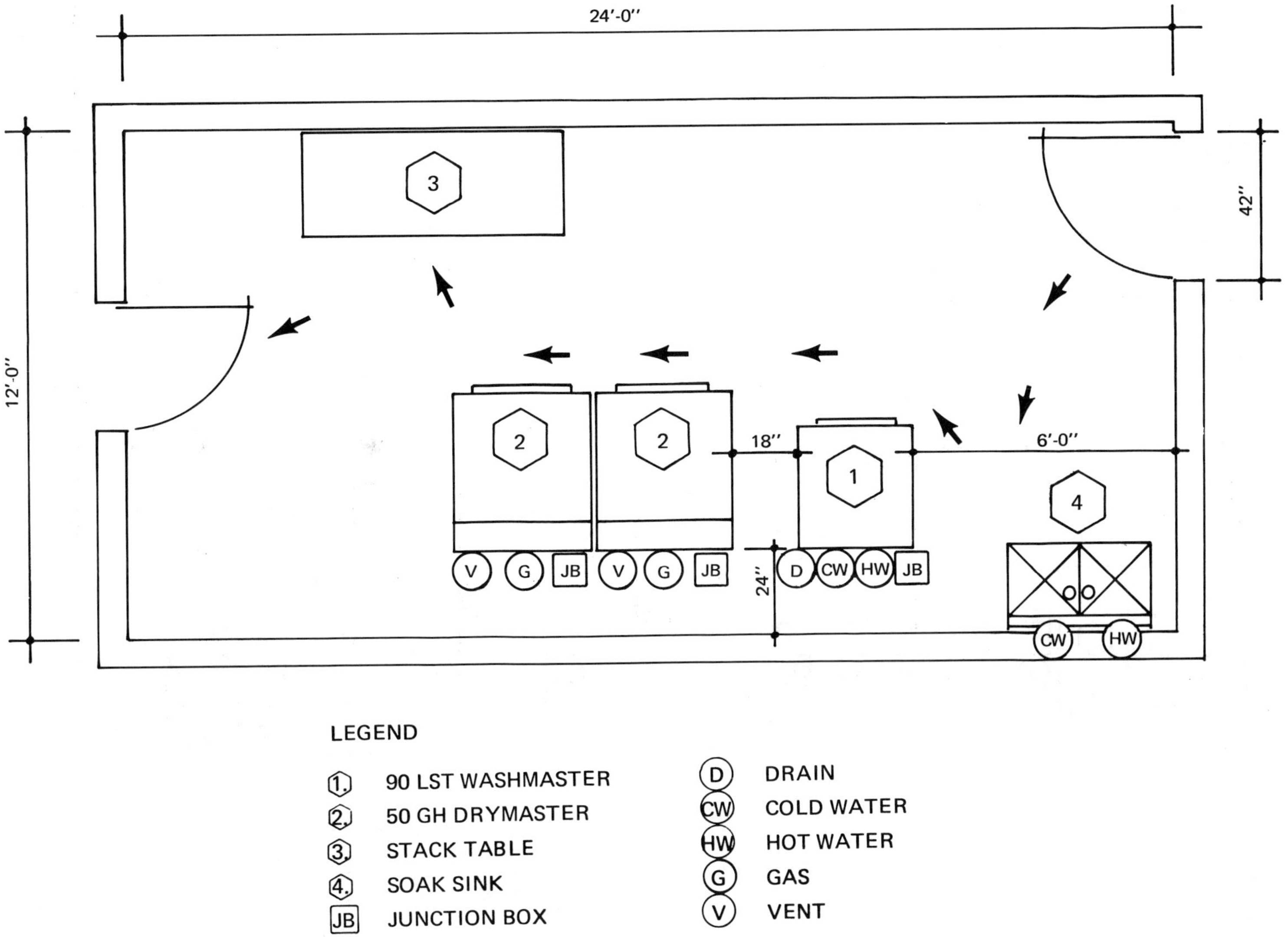


**FIGURE 28-10** *Laundry Layout For an 80 Room Motel or Hotel (Using One Employee)*

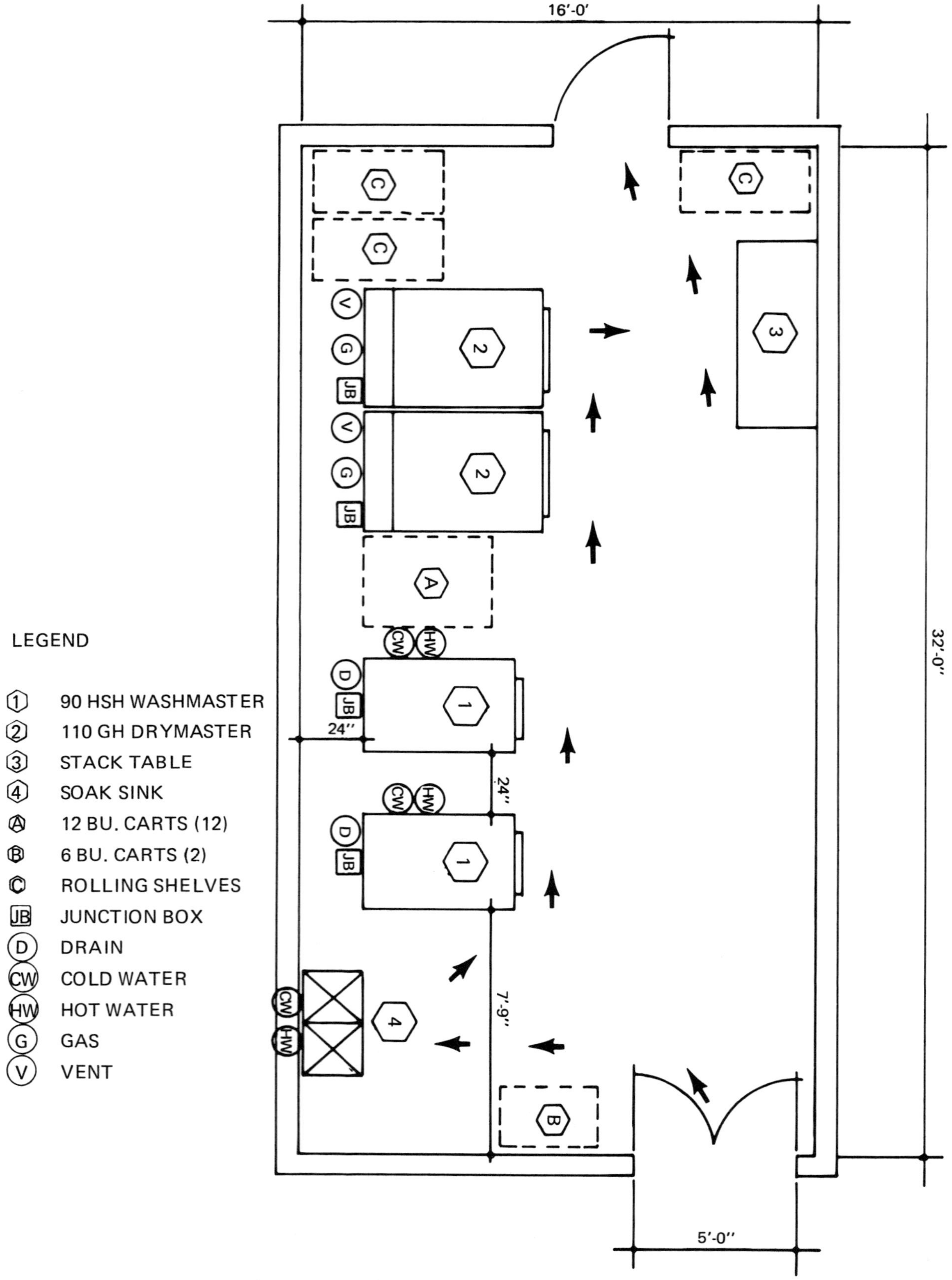


**FIGURE 28-11** *Laundry Layout for a 175 Room Motel or Hotel (Using* Three *Employees)*

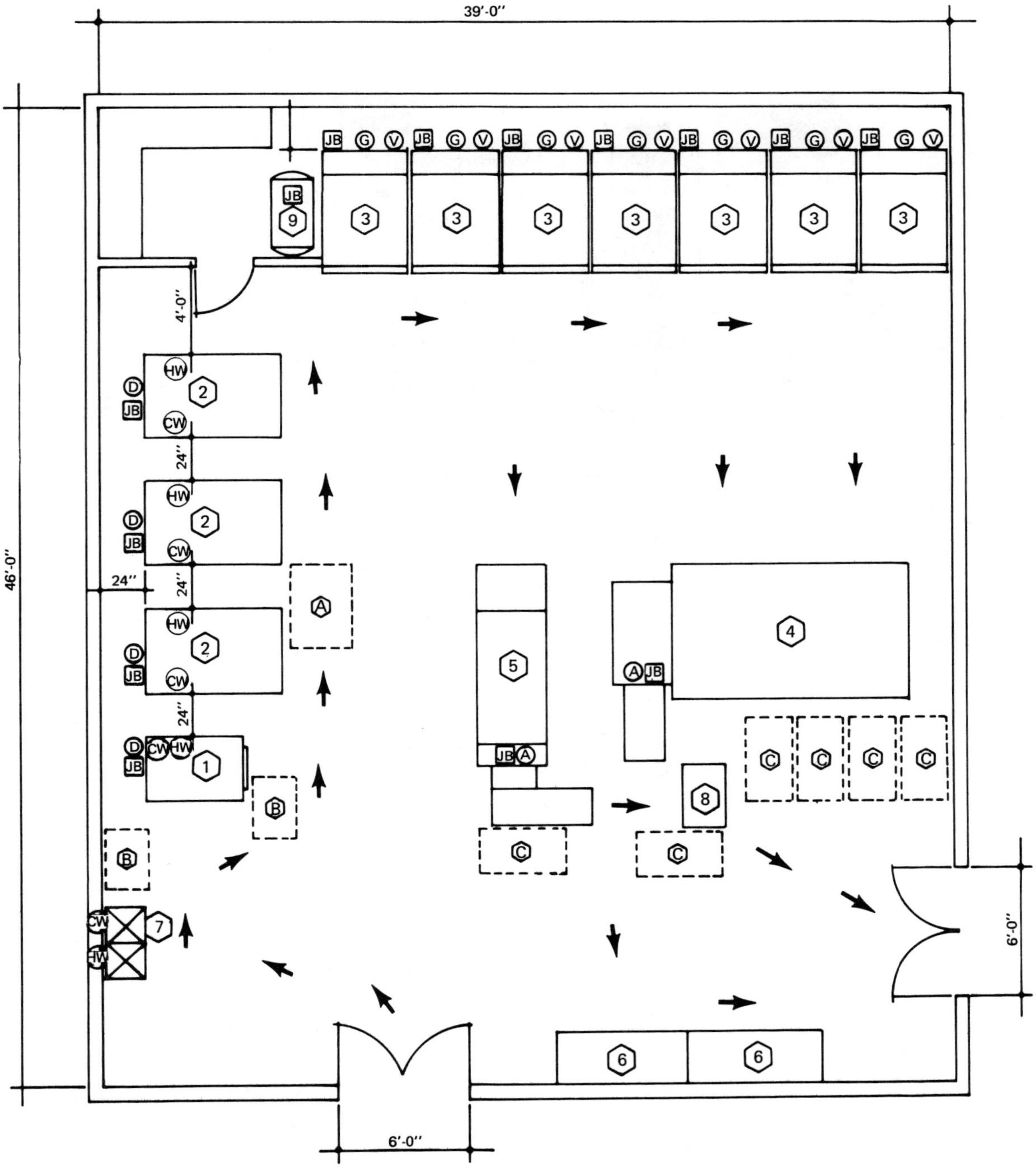


LEGEND

| | | | | | |
|---|---|---|---|---|---|
| 1 | 90 HSH WASHMASTER | 8 | TYING MACHINE | CW | COLD WATER |
| 2 | 225 HSH WASHMASTER | 9 | 305 TM 5 HP AIR COMPRESSOR | HW | HOT WATER |
| 3 | 110 GH DRYMASTER | A | 12 BU CARTS (24) | G | GAS |
| 4 | GALAXIE SHEET FOLDER | B | 6 BU CARTS (4) | V | VENT |
| 5 | WISHBONE SMALL PIECE FOLDER | C | ROLLING SHELVES | A | AIR |
| 6 | STACK TABLE | JB | JUNCTION BOX | | |
| 7 | SOAK SINK | D | DRAIN | | |

**FIGURE 28-12** *Laundry Layout for a 650 Room Motel or Hotel (Using* Five *Employees)*

**FIGURE 28-13** *Laundry at Marriott Hotel Tucson, Arizona, processes its bed and table linens at half the cost of contract linen service. For 312 rooms, laundry has three washers—one large, two medium—and four dryers. Most of the laundry crew are deaf-mutes. Company that makes machines does cost study; plans laundry; lays it out; obtains all supplies and equipment including carts, chutes, etc.; and trains employees. Interior furnishings of guest rooms were based on hotel having an on-premise laundry; bedspreads are washable, match sheets.*

more particularly, residence halls, have had great success with a mini-laundry on each floor. One laundry station can handle from 30 to 100 rooms, says one expert. One operator can handle linen from 60 to 70 guest rooms comfortably in a day (Figure 28–10).

## *Dry Cleaning Equipment*

The business of cleaning and pressing guest clothing, employees' uniforms, and draperies in a hotel can be exceedingly expensive. To set up an in-house facility will require skilled people to take care of it. A new system has been set up for institutional dry cleaning so that the unit and cleaner can be housed right in the building and operated by hotel or college personnel at relatively low cost.

The Century Plaza Hotel has an impressive quantity of cleaning and pressing—175,000 pounds annually. It was cleaning the draperies that caused Jim Johnson, the original laundry-valet manager, and his staff the greatest concern. "They're made of conventional curtain fabrics," he explains, "but lined with a special blackout material. The lining is a supported vinyl that is extremely sensitive to high temperatures, and difficult to dry-clean without causing blisters."

The hotel's conventional dry cleaning equipment was unable to provide safe treatment for the draperies. "Our original machine used perchlorethylene, which is still the most widely used dry cleaning solvent," said Johnson. "But perc was too aggressive for certain fabrics, including our draperies. Our total, annual dry cleaning bill was more than $10,000."

The contract cleaning arrangement was

costly in another way. To achieve the greatest economy it was necessary to send out all the draperies on a floor at one time, and they would be gone for at least two days. Naturally, the rooms could not be rented without curtains, so the hotel stood to lose something like $65,000 a year in potential revenue.

These combined cost factors prompted Johnson and his associates to look for a faster, less expensive way of handling the drapery problem. "It took us four or five years of intensive investigation and evaluation to find what we wanted. In that time we evaluated five or six different dry cleaning systems, and none of them met our needs. Our chief complaint was that they caused bubbling in the vinyl material. Finally we looked into a system based on a fluorinated hydrocarbon solvent—Du Pont's 'Valclene' dry cleaning fluid, as a possible answer."

After preliminary tests convinced Century Plaza management that Valclene was the answer, a 40-pound capacity machine designed especially for use with that fluid was installed, and it paid for itself in less than two years.

"We completely eliminated the need to close a whole floor of rooms while the draperies were being cleaned. With a 27-minute dry cleaning cycle required for the draperies, we have them back and hanging in the rooms within two hours."

There is no lingering odor when draperies are back in place as their would be with other solvents. Nor is there noticeable color change.

The installation has helped the hotel improve its dry cleaning service for guests, too. Conventional perchlorethylene dry cleaning is too rough for such things as bonded fabrics, knits, suedes, leathers, and garments with fur or plastic ornamentation. Now the hotel valet service can handle just about anything without having to sort garments according to fabric, and without having to remove ornamentation and replace it after cleaning. Other hand-labor operations, such as spotting and finishing, are also minimized. Guests get three- to four-hour service. It is also completely nonflammable, an important consideration in a big hotel. Of course, proper ventilation and care in using the fluid are mandatory.

## *New Type of Linen Supply Contract*

The Frontier Hotel and American Linen Service, both of Las Vegas, agreed to a new form of linen supply contract designed to take the bookkeeping out of Housekeeping. Since studies indicate the hotel's monthly occupancy rate is almost directly reflected in corresponding changes in linen supply consumption, the novel contract calls for a flat monthly charge based on 100 percent occupancy. This guaranteed maximum charge is multiplied by the hotel's actual occupancy rate to determine the exact charge for linen supply services at the end of each month.

Innovators of the plan analyzed that less than a 5 percent fluctuation in linen consumption occurred in Housekeeping accounts during periods of peak occupancy. Further analysis indicated that any drop in the occupancy rate was matched by an almost identical drop in the amounts of linen used.

Both the linen supplier and hotel have achieved substantial savings by eliminating time-consuming, traditional linen-counting in favor of a flat monthly charge based solely on the 600-room hotel's monthly occupancy rate.

Terms of the contract have relieved the Housekeeping Department of virtually all of the burdensome details associated with counting, sorting, and record keeping, as well as the mechanics and occasional friction related to the exchange of damaged linens. The hotel also reports a savings through simplification of accounting procedures.

## *Laundry Safety*

With general OSHA standards there is a section addressed specifically to laundries. (Dry cleaning operations are exempt from this section, but are subject to the requirements of the general OSHA standards.) The following are from the special section on Laundry Machinery and Operations.

1. Power-marking machines must be equipped or guarded to prevent injury to fingers.

2. Washing machines and drying tumblers must be equipped with interlocking devices that will prevent the inside cylinder from moving when the outer door on the case or shell is open, and will also prevent the door from being opened while the inside cylinder is in motion.
3. Extractors must be equipped with an interlocking device that will prevent the cover from being opened while the basket is in motion, and will also prevent the power operation of the basket while the cover is not fully closed and secured. This device should not prevent the movement of the basket by hand to ensure even loading.
4. Power wringers must be equipped with a safety bar or other guard across the entire front of the feed or first rolls.
5. Steam pipes that are within seven feet of the floor or working platform, and with which the worker may come into contact, must be insulated, covered with a heat-resistive material, or otherwise properly guarded.
6. Each power-driven machine must be provided with means for disconnecting it from the source of power. Starting and stopping devices for machines must be located so as to be operable from the front.

## *Sources of Information*

Linen Supply Assn of America
P.O. Box 1283
Hallandale, Fla., 33009

Pellerin Milnor Corp.
P.O. Box 400
Kenner, La. 70063

E. L. Fraser (Econ Systems)
Division Economics Laboratory
4 Corporate Park Drive
White Plains, N.Y. 10604

Washex Machinery Corp.
5000 Central Freeway
Wichita Falls, Texas 76306

McCabe Spreader Feeder
1318 Grinnel
Lubbock, Texas 79403

UniMac Company, Inc.
Marianna, Fla. 32446

## *Bibliography*

Bradley, L. A. The No-Iron Laundry Manual. *Cornell Hotel & Restaurant Administration Quarterly. 1970.*

Nolte, William. It's Time to Update No-Irons. *The Executive Housekeeper.* March, 1974.

# *Chapter 29*

# *Record Keeping—Personnel, Purchasing, and Control*

**OVERVIEW** Professional Housekeepers' offices must have adequate storage and scrupulously kept files, as there are many forms to be provided to employees and many records to be retained: payroll; attendance; room history showing repairs and replacements; purchasing records with details of suppliers; warranties; key control; and an endless number of supply lists for various workers' specific tasks. The larger the property the more likely that management will require Housekeeping to maintain numerous sets of detailed records. Let us first consider payroll, the biggest single item in the budget.

## *Payroll*

The number of hours each employee works must be accurately recorded. This is usually done by the employee himself, either by inserting a card bearing his name into a time clock that stamps his arrival and departure times, or by signing in and out on a daily time sheet. In a small department, one employee may be assigned to record this information in a monthly or annual diary. Whatever the system, it is as important to insist on prompt departure as prompt arrival, since most union contracts call for overtime pay.

The Housekeeper must be familiar with union, city, county, state, and federal regulations concerning wage minimums, overtime rates and regulations, whether tips and meals are part of remuneration, and fringe benefits. In smaller institutions, the Housekeeper may even have to calculate deductions for taxes, union initiation, and dues.

Many states require that payroll records be kept for at least three years. If the staff is large, this means ample storage space must be provided for a great many records, all of which must be readily available to government and other authorized inspectors.

Records of specific daily work assignments should also be maintained and kept as a permanent part of each employee's personnel file to simplify determining who was in a given area on any given day. These records should be maintained and released as directed by the Security Department or management representative.

Employee files should also include the Personnel Department's evaluation data (see Chapters 6 and 25) as well as a record of dates on which earned vacation time was taken, and those on which other absences were approved, with reasons for granting approval.

Also helpful are records showing daily productivity of individuals, teams, and combined staff, which provide an accurate measurement of departmental strengths and weaknesses.

Most housekeepers keep detailed room refurbishment records that show dates, prices, sources of supply, and descriptions of all furnishings and accessories, including amounts of extra fabrics, sample swatches, and descriptions of color schemes. Comparable records are kept on refurbishment of public areas. Analysis of these records shows life and upkeep of each purchase, and thus its true cost per day, month, or year. From such data the desirability of changing brands can be determined. This record keeping should be coordinated with the Purchasing Department to avoid duplication of work.

The institution's room-assignment records indicate which rooms have had the most usage and must be frequently checked to determine refurbishment requirements. In a hotel, this often applies to rooms on lower levels, which—being bargain-priced because of obtrusive street noise, absence of desirable view, and so forth—are favored by family groups. The best rooms—connecting and corner rooms and those with exceptional views—are even more frequently occupied because room clerks tend to assign them so often to VIPs and regular customers.

If guest rooms are badly damaged by guests who may have had a wild party or a fight, the executive Housekeeper should immediately be notified so that the damage can be examined and assessed. Then the Housekeeper notifies the credit manager posthaste that such charges should be put on the bill for that particular room.

## *Damage and Replacement*

Housekeeping will keep a book that records the date, room number, damage done, and costs to repair or replace. The professional Housekeeper will keep a list of costs of furnishings for guest rooms and for public areas so that if he or she is not on duty the assistant housekeeper will have access to it and can give that information to the credit manager. Very often in cases of this sort the assistant manager will want to accompany the Director of Housekeeping to view the actual scene of the damage, or, if it is simply one item, it may be sent to the assistant manager or manager's office.

Whether or not the charges are actually put onto the destructive guest's bill is a matter for the Sales Department to decide, based on how good a customer the guest or his company is and what the circumstances were. If the amount of business brought in by that convention is sizable enough, the Sales Department may deem it well worth the price of the $75 lamp that was broken at the boisterous cocktail party.

Sometimes, if an inordinate amount of soiling should occur, there may have to be extra charges for clean-up, but rarely, as this is to be expected in hospitality suites. In large commercial properties, certain suites are designated for that purpose, and kept separate from those very elegant ones used for the individual room guests.

Whether those charges reported by Housekeeping actually do go onto the bill or ever get paid should never be known by *anyone* in the Housekeeping Department, because it might affect their attitudes toward that same group—association or company—in the future; if it were not paid, the service might be diminished. That should never be allowed to happen. All guests should be treated equally—and that means splendidly.

Needless to say, all rooms and public areas should be systematically inspected by supervisors at regular intervals. Records should show dates of inspection, inspector's name, and recommendation, plus notation of action taken or scheduled. In residence halls, men's areas almost always sustain more abuse than women's, thus requiring considerably more repair and refurbishment.

Major housekeeping projects (e.g., scrubbing and painting patios and balconies) should coincide with periods of low occupancy. A record of past scheduling may be helpful in setting future dates for these projects (which, incidentally, should *never* be postponed if there is heavy soil or damage).

Record keeping can be extremely time-consuming; therefore it should be limited to projects that develop information more valuable than the cost of recording it and should never duplicate the work of another department (e.g., room-painting records are probably kept by engineering, maintenance, or the painting department).

## *Purchasing*

If Housekeeping does its own purchasing in a medium-sized or large institution (over 50 rooms), records should be arranged so a secretary or clerk can prepare purchase orders for the Housekeeper's signature.

In one frequently used system, files are arranged by product type, for example, soap. Individual sheets identify each soap by brand name, size, purpose, monthly consumption, quantity to be ordered, and frequency of ordering (e.g., quarterly), along with name of sales representative, address and phone number, plus most recent price. Records state which manufacturers offer cut-price promotions and when (e.g., each spring or fall).

Housekeeping purchases hundreds of items. Here are some hints on how-to-do-it from the experts.

Concentrate your business with a few suppliers. The greater your purchasing power, the greater your bargaining power—not only in terms of quantity discounts, but in terms of better service. This policy also cuts time spent with salesmen and reduces paperwork for Housekeeping and Accounting.

Do not forget, though, it is from salesmen that one learns what is new on the market and what other colleges, hotels, and other facilities are doing to increase efficiency. Setting aside certain hours or days to see only those salesmen who have new products to present (and then by appointment only) enables Housekeeping to minimize untimely interruptions.

Before ordering a product in quantity, enough samples should be obtained to find out if employees will use it properly; if proper usage requires complicated instruction; and if it really does the job under actual work conditions (i.e., when used by employees who are not overly concerned with making it work).

A suave salesman may make a disposable folding dustcloth seem indispensable. If maids will not fold it in such a way as to keep exposing new surfaces and thus assure maximum mileage, it is just another over-priced product that will not do what it is supposed to do because the maids will not make it perform properly.

Quantity purchasing should be limited to products that will not become obsolete. Almost any toilet tissue or soap will stay in style for years, while a three-year supply of furniture polish might soon seem antiquated as newer polishes reach the market.

Buying cleaning supplies in bulk (even by the barrel) may be economical if the person who decants liquids or scoops out powders into small containers is employed at a low enough wage to keep this added labor cost from cancelling the initial savings.

If Housekeeping has only limited storage space, it is necessary to resist over-stocking bulky items at bargain prices. Pity the poor housekeeper who saved $3.00 each on down pillows but has no place to store them. The boxes can also be a fire hazard and may take up space needed to store frequently used floor machines. In some instances, however, it is possible to purchase in quantity and have the supplier store the material in his warehouse until needed. In others it is wise to purchase before the effective date of a price increase and specify that shipment be made later.

## *Reevaluating Products*

At least once a year, every product purchased by Housekeeping should be reevaluated for performance, durability and the like, and to determine whether there is something new and better on the market. It is also important to exchange product information with other departments and to check prices on housekeeping supplies available from their major suppliers.

In multi-unit institutions—hotel chains, motel franchises, and colleges with more than one campus—purchasing is often done centrally, with individual properties requisitioning what they need at a much lower price than they could obtain individually. Units will almost always find a savings by purchasing through headquarters those items with the corporate identification—soap, hangers, bath linen—that all properties need.

Hotel and motel chains often own subsidiaries that supply all their needs, from furniture and wall coverings to facial tissues and room keys. Hyatt Corporation owns Rosemont Purchasing; Westin Hotels set up Westin Services and Supply; Holiday Inn, Innkeepers Supply; Sheraton, Standard Wholesale Supply Corp. Quite often these supply companies design interiors and provide furnishing and supplies to other institutions, including competitive hotels.

According to V. E. Newberry, purchasing director of Lourdes Hospital in Paducah, Ky., specifications (called specs) are "detailed descriptions of parts or a whole—a statement of particulars as to size, color, quality, and performance. Thus, as we examine a need, we begin to develop specifications. In the example, we found a chair to be too low. We may have determined also that a stool was too high. So we need something in between; we need something adjustable. This is one part of the specifications. Other parts could be color, casters, etc. Specifications should also include materials of which an item may be made, such as fabric or plastic, metal or wood. Specifications should be used as a guide—not an absolute—and they should be as flexible as possible so a substitute can be made. They should be written so you will not become confused as you look at more and more items."

If a specific item is desired and the professional Housekeeper is sure that one particular manufacturer does a far better job than his competitors, it is customary to put "A. J. Smith Co. or equal" on the specifications that go to the Purchasing Department.

Here are terms used in purchasing:

**Purchase Requisition.** Formal written request that Purchasing uses as a basis for issuing a purchase order. This should not be confused with the stores or supply requisition, which is a request for delivery from the institution's central store room.

**Purchase Order.** Offer to buy, until accepted by vendor (supplier) without exception. When vendor accepts the order, it becomes a contract and must be considered as such. The purchase order should contain all pertinent information needed by the vendor to perform delivery and invoice in accordance with buyer's wishes. Purchase orders are usually written in multiple copies.

**Acknowledge (Acknowledgment).** Communication from vendor to buyer accepting purchase order.

**Delivery Receipt.** This may be a copy of invoice or a special form used for delivery only. Usually there are two copies, one of which is signed and returned to vendor; the second is retained for receiving files.

**Invoice or Billing.** Formal written request by vendor for payment for goods or services.

**Invoice Terms.** Conditions set forth in invoice governing delivery of and payment for goods or services. These should be identical to purchase order terms.

**Claim.** Claim should be filed either with the shipper or carrier when loss or damage occurs in transit. Party against whom a claim is filed is determined by FOB (free on board) point—the point at which title to goods passes from vendor (shipper) to buyer.

**Vendor.** Supplier—shipper or consignor—from whom a purchase is made.

**Expedite.** Effort to improve a promised delivery situation.

**Inventory Control.** Process of limiting amount of money invested in inventory while ensuring sufficient supplies or goods to continue serving the institution.

**Inventory Turn (Turnover).** Number of times in a given period inventory is depleted and replenished.

**FOB Point.** Point at which title to goods passes.

**Cash Discount.** Percentage buyer may deduct from amount of invoice—excepting freight charges—if he fulfills certain conditions when paying for purchases. Sometimes these terms are written 1-10, net 30, 2-10 EOM, for example.

## *Watch That Warranty!*

When a prospective purchase has been narrowed down to two or three brands, it is time to check each manufacturer's printed warranty.

Most warranties cover replacement of defective parts. Ideally, however, it should state that if the product is defective when delivered, it will be replaced rather than repaired. Parts should be guaranteed for at least one year.

If a warranty specifies repairs, they should either be performed on-site, or the product should be removed for repair at the seller's expense. Consider the plight of a Dubuque, Iowa motel operator who might buy a John J. Jones Mark IV heater for the motel swimming pool, only to learn when warranty-authorized repairs are required that he must ship either the defective parts or complete (plumbed in!) heater to the manufacturer's plant in North Hollywood, Calif. At his own expense!

Many warranties do not include labor, which is almost always far greater than the cost of parts. The warranty that provides free labor for up to two years provides far greater value than one that covers labor for 30 days or not at all.

How long will parts be available? Some manufacturers can provide parts for models discontinued 50 years ago; others cannot provide parts for models discontinued last year. The most convincing proof of the manufacturer's policy is his current parts catalog; the sales representative should always be able to provide a copy of it.

The salesman should also be asked for sample copies of the manufacturer's operating and maintenance instructions, as well as service manuals that should be available in the workers' native language as well as in English. All manuals should be checked for clarity and completeness by the persons who will actually use them.

Once the decision has been made to purchase, the salesman should be asked to provide assurances that the following demonstrations will be conducted on-site for the manager, the housekeeper, supervisor, and employees, at dates agreed upon before the order is placed: (1) how to use the product efficiently, effectively, and safely; (2) how to maintain the product in proper working order; and (3) how to service it after the warranty has expired. Supervisory people who will attend these demonstrations should be asked to study pertinent literature in advance, which the salesman should, of course, be asked to provide in suitable quantities.

There should be ample time set aside for questions to be answered and for the actual operators to practice while the demonstrator is there to help them learn.

## Controls

Almost every department head of every institution is continuously looking for "controls" as if they were the true panacea. The reader may have noted how many activities are referred to using that term—Linen Control, Uniform Control, Beverage Control, Cost Control, and on and on.

No matter how well written the manuals with rules and regulations covering employee behavior and dress may be, or how explicit the instruction manuals are, there is no insurance that these rules and these instructions will be followed. Controls are needed to implement them.

And the controls—the inspectors who inspect the inspectors—all must have written procedures and checklists. That means that quantities of the printed word abound throughout most of the lodging industry today. The writings are voluminous, and regrettably, too often unused.

The supervisor or manager must measure the results of an employee's work—casually by observation, carefully by scrutinizing it closely, or by testing and using a check sheet to be sure every point is covered.

Take the simple matter of attendance. The manual says the maids and housemen are supposed to arrive at 8 AM. To prove that they did arrive, they sign in on the time sheet (they may also punch the time clock); then they pick up their floor or area assignments for the day. In a very large Housekeeping Department, however, this does not guarantee that maids are actually in place and productive immediately. Thus floor supervisors keep another attendance record to verify arrival time of employees in their areas.

## Key Control

As important to control as attendance are the keys given maids and housemen. When they are issued master keys for their sections it is evidence that they are being trusted with the welfare of many guests and their possessions. Of course, instructions have been given these employees on how careful to be with keys, but it takes another set of forms and more supervisors to double-check that keys are being handled responsibly.

It is interesting that some hotels do not instruct maids to check the television or radio to be sure it is working, the reason given being that they will then be tempted to watch it or use that as an excuse for having it on while they work. Regardless of the rules, it is almost inevitable that some maids will watch television anyway for they can always claim that the guest had left it on.

Because maids, supervisors, and inspectresses might not be able to see just what works in the room without trying all the appliances—lights, plumbing fixtures—one prestigious hotel, the Del Coronado in San Diego, has initiated an effective policy. The hotel provides room cards in all rooms so that guests, should they locate an item that requires repair, can notify Housekeeping, with the card, of the pending problem and have it taken care of immediately by the appropriate department.

Throughout this book we have been emphasizing controls continuously, from key control (Chapters 6 and 25) to avoiding aerosol sprays, to checking for holes in linens. In fact, almost every activity that has been described has included a way of saving costs while still assuring fine service. No guest should ever feel that his room cleaning has been rationed or his supply of soap decreased.

But cutting down on waste is the major control, and it can be practiced if all employees have an interest in helping the hotel or residence hall to conserve. They can best be motivated to exercise controls if they are kept informed about how the business is doing and are made to feel part of the total operation.

## *Forms Used Daily*

These are examples of forms from several different institutions, each of which will handle these controls in various ways.

**1. Housekeeper's Room Check Reports** (Figures 29-1 and 29-2). Usually this is filled out by the maid, called Housekeeper in this hotel. It is first filled in before rooms are made up so supervisors can count rooms needing maid service. When maid's day is complete she turns the form in again with all rooms made up and ready to rent. As five of her rooms were unoccupied and ready, she did five rooms in another section. Many hotels no longer do an AM room check as it disturbs guests. At 7 o'clock in the mornings, night staff sends Housekeeping a report of occupancy and the number of expected check-outs.

**2. Housekeeper's Report** (Figure 29-3). Made out by room control person; sent to Front Office at least twice daily as double-check on exact room status. Information on which rooms

HOUSEKEEPERS
ROOM CHECK REPORT

SECTION 36

Darlene Cooper — Housekeeper

7/12 — Date

| ROOM | READY to RENT | Occupied | CO | DND | COTS | REMARKS |
|---|---|---|---|---|---|---|
| 516 | | X1 | | | | |
| 517 | X | | | | | |
| 518 | X | | | | | tile broke |
| 519 | | X2 | | 1 | | |
| 520 | | | X | | | |
| 521 | | | X | | | |
| 522 | | X2 | | X | | |
| 523 | | ?X | | | | Sleep out |
| 524 | | 1X | | 1 | | |
| 525 | | X2 | | | | |
| 526 | | | X | | | |
| 527 | X | | | | | |
| 528 | X | | | | | |
| 529 | X | | | | | |
| 530 | | X3 | | | 1 | |
| 531 | | X2 | | X | | |

**FIGURE 29-1** *A maid's early morning room check.*

HOUSEKEEPERS
ROOM CHECK REPORT

SECTION 36

Darlene Cooper — Housekeeper

7/12 — Date

| ROOM | READY to RENT | Occupied | CO | DND | COTS | REMARKS |
|---|---|---|---|---|---|---|
| 516 | X | X1 | | | | |
| 517 | X | | | | | |
| 518 | X | | | | | tile broke |
| 519 | | X2 | | 1 | | |
| 520 | X | | X | | | |
| 521 | X | | X | | | |
| 522 | | X2 | | X | | |
| 523 | ? | ?X | | | | sleep out |
| 524 | | 1X | | | | |
| 525 | | X2 | | | | toilet stuck |
| 526 | X | | X | | | |
| 527 | X | | | | | |
| 528 | X | | | | | |
| 529 | X | | | | | |
| 530 | X | X3 | | | 1 | Light bulb |
| 531 | | X2 | | X | | |

**FIGURE 29-2** *Her same report at the end of day.*

are vacant and ready to rent is called in to Front Desk as they are prepared. Rooms designated 000 are out of order. Where both vacant and occupied are checked, or room numbers are circled, room had been made up since report was first started.

**3. Club's Daily Room Report** (Figure 29-4). Combines several forms used in larger hotels.

**4. Inspectress Report** (Figure 29-5). A Please Do is given to maid who goes back to room and corrects situation as necessary.

**5. Request for Repairs by Maintenance or Engineering** (Figure 29-6). Waldorf-Astoria's is made out in triplicate by inspectress or assistant housekeepers; Century Plaza's is called repair schedule (see Chapter 26).

**6. Housekeeping Repair Checklist** (Figure 29-7). This form is used for many different jobs, so that the repairman who goes up to the floors will know which rooms need attention. It is made out by assistant housekeeper based on the repair schedule turned in by inspectresses or floor supervisors.

**7. Attendance Card.** Kept in payroll file, it has 365 dates on it. Payroll clerk marks date employee started and then checks sign-in sheet daily and marks cards for each hourly Housekeeping employee.

DATE *February 3*

| ROOM | Occupied | Vacant | Baggage | ROOM | Occupied | Vacant | Baggage | ROOM | Occupied | Vacant | Baggage | ROOM | Occupied | Vacant | Baggage | ROOM | Occupied | Vacant | Baggage |
|---|---|---|---|---|---|---|---|---|---|---|---|---|---|---|---|---|---|---|---|
| 100 | ✓ | | ✓ | 153 | ✓ | | ✓ | 200 | O | O | O | 251 | | X | | B-1 | ✓ | | ✓ |
| 102 | ✓ | | ✓ | 154 | ✓ | | ✓ | 201 | O | O | O | 252 | | X | | B-2 | ✓ | | ✓ |
| 103 | | X | | (155) | ✓ | X | ✓ | 202 | O | O | O | 253 | | X | | B-3 | ✓ | | ✓ |
| 104 | | X | | 156 | | X | | (203) | ✓ | X | ✓ | 254 | | X | | B-4 | ✓ | | ✓ |
| 105 | | X | | 157 | | X | | 204 | | X | | 255 | | X | | B-5 | ✓ | | ✓ |
| 106 | | X | | 158 | | X | | 205 | | X | | 256 | | X | | B-6 | ✓ | | ✓ |
| (107) | ✓ | X | ✓ | 159 | ✓ | | ✓ | 206 | | X | | 257 | | X | | B-7 | | X | |
| 108 | ✓ | | ✓ | 160 | | X | | 207 | | X | | 258 | ✓ | | ✓ | B-8 | | X | |
| 109 | ✓ | | O | 162 | ✓ | | ✓ | 208 | | X | | 259 | ✓ | | ✓ | B-11 | ✓ | | ✓ |
| 110 | ✓ | | ✓ | 163 | X | | X | 209 | | X | | (260) | ✓ | X | ✓ | | | | |
| 111 | ✓ | | ✓ | 164 | | X | | 210 | | X | | 263 | ✓ | | ? | 328 | X | | X |
| 114 | ? | X | ✓ | 165 | ✓ | | ✓ | 211 | | X | | 264 | ✓ | | ? | 330 | ✓ | | ✓ |
| 115 | | X | | 166 | ✓ | | ✓ | 214 | | X | | 265 | | X | | 331 | | X | |
| 116 | | X | | 167 | ✓ | | O | 215 | | X | | 266 | | X | | 332 | | X | |
| 117 | | X | | 168 | | X | | 216 | | X | | 267 | | X | | 333 | ✓ | | ✓ |
| 118 | X | | ✓ | 169 | | X | | 217 | | X | | 268 | | X | | 334 | ✓ | | ✓ |
| 119 | X | | ✓ | 170 | | X | | 218 | | X | | 269 | | X | | 335 | ✓ | | ✓ |
| 120 | X | | ✓ | 171 | | X | | 219 | | X | | 270 | ✓ | | ✓ | (336) | ✓ | X | ✓ |
| 121 | X | | ✓ | 172 | | X | | 220 | | X | | 271 | ✓ | | ✓ | 337 | | X | |
| 122 | | X | | 173 | | X | | (221) | ✓ | X | ✓ | 272 | ✓ | | ✓ | 338 | | X | |
| 123 | | X | | 174 | | X | | 222 | ✓ | X | ✓ | 273 | ✓ | X | ✓ | 339 | ✓ | | ✓ |
| 124 | | X | | 175 | | X | | 223 | ✓ | | ✓ | 274 | | X | | 340 | ✓ | | ✓ |
| 125 | | X | | 176 | | X | | 224 | | X | | 275 | | X | | 341 | | X | |
| 126 | | X | | 177 | | X | | 225 | ✓ | | ✓ | 276 | | X | | 342 | ✓ | | ✓ |
| 127 | | X | | 178 | | X | | (227) | ✓ | X | ✓ | 277 | | X | | 343 | | X | |
| 128 | O | O | O | 179 | | X | | 228 | O | O | O | 278 | | X | | 344 | ✓ | X | ✓ |
| 130 | O | O | O | 180 | | X | | 230 | O | O | O | 279 | | X | | (345) | ✓ | X | ✓ |
| 131 | O | O | O | 181 | | X | | 231 | O | O | O | 280 | | X | | 347 | | X | |
| 132 | O | O | O | 182 | | X | | 232 | ✓ | | ✓ | 281 | | X | | 348 | | X | |
| 133 | | X | | 183 | | X | | 233 | | X | | 282 | | X | | 349 | | X | |
| 134 | | X | | 184 | ✓ | | ✓ | 234 | | X | | 283 | | X | | 350 | | X | |
| 135 | | X | | 185 | ✓ | | ✓ | 235 | | X | | 284 | | X | | 351 | | X | |
| 136 | | X | | 186 | ✓ | | ✓ | 236 | | X | | 285 | | X | | 352 | | X | |
| 137 | | X | | 187 | | X | | 237 | | X | | 286 | | X | | 353 | | X | |
| (138) | ✓ | | ✓ | 188 | | X | | 238 | | X | | 287 | | X | | 354 | | X | |
| 139 | ✓ | | ✓ | 189 | ✓ | | ✓ | 239 | | X | | 288 | | X | | 355 | | X | |

**FIGURE 29-3** *Housekeeper's Report*

11–05
REV. 12/70

# L. A. A. C.

**DAILY ROOM REPORT**

"HOUSEKEEPER/FRONT DESK" –
In "Guests" column use:
T.–to indicate Transient
P.–to indicate Permanent
O.O.–to indicate Out of Order
EMP.–Employee
V.–Vacant

Day__________________ Date__________________ Made by______________

| RECAP. | | Quantities | Room | Guests | Charge | Room | Guests | Charge | Room | Guests | Charge |
|---|---|---|---|---|---|---|---|---|---|---|---|
| 1 Total Available Rooms | Forward | | | | | | | | | | |
| | TODAY | | | | | | | | | | |
| | To Date | | | | | | | | | | |
| 2 Used by Employees | Forward | | | | | | | | | | |
| | TODAY | | | | | | | | | | |
| | To Date | | 1002 | | | 1102 | | | 1202 | | |
| 3 Out of Order | Forward | | 1003 | | | 1103 | | | 1203 | | |
| | TODAY | | 1004 | | | 1104 | | | 1204 | | |
| | To Date | | 1005 | | | 1105 | | | 1205 | | |
| 4 Available for Rental 1—2—3 | Forward | | 1006 | | | 1106 | | | 1206 | | |
| | TODAY | | 1007 | | | 1107 | | | 1207 | | |
| | To Date | | 1008 | | | 1108 | | | 1208 | | |
| 5 Permanent Rooms Rented | Forward | | 1009 | | | 1109 | | | 1209 | | |
| | TODAY | | 1010 | | | 1110 | | | 1210 | | |
| | To Date | | 1011 | | | 1111 | | | 1211 | | |
| 6 Transient Rooms Rented | Forward | | 1012 | | | 1112 | | | 1212 | | |
| | TODAY | | 1014 | | | 1114 | | | 1214 | | |
| | To Date | | 1015 | | | 1115 | | | 1215 | | |
| 7 Total Rooms Rented 5 and 6 | Forward | | 1016 | | | 1116 | | | 1216 | | |
| | TODAY | | 1017 | | | 1117 | | | 1217 | | |
| | To Date | | 1018 | | | 1118 | | | 1218 | | |
| 8 Vacant Rooms 4—7 | Forward | | 1019 | | | 1119 | | | 1219 | | |
| | TODAY | | 1020 | | | 1120 | | | 1220 | Library | × × |
| | To Date | | 1021 | | | 1121 | | | 1221 | | |

**FIGURE 29-4** *Club's Daily Room Report*

**8. Floor Attendance Sheet** (Figure 29-8). This is filled out by floor supervisor who has maids sign in both morning and after lunch.

**9. Key Sign-out Slips.** These have places for key numbers, time-in, time-out, inspectress's or room attendant's name. Each is filled in by floor supervisor who retains the record.

**10. Key Inventory** (Figure 29-9). This is master key control from linen room—can be set up with any number of keys, any codes.

**11. Night Housekeeper's Work Sheets.** Each room is listed and marked for work to be done by night crew; gives status of rooms.

**12. Night Supervisor's Report** (Figure 29-10). Each maid is given a list of these rooms for turndown or other services.

## *In-Office Records and Forms Used Periodically*

1. *Public areas furnishing records.* Master budget made out by director of Housekeeping will show items that need repair and which subsequently were replaced.
2. *Room furnishing record* (Figure 29-11). If major rejuvenation program is in progress, a large board with rooms all charted may also be necessary.

CENTURY PLAZA HOTEL
CENTURY CITY / LOS ANGELES

PLEASE DO

ROOM 823 TIME 11:05 DATE 4/17

TO Jane FROM RDR DONE BY 1:45

**REFRIGERATOR:** inside, outside, drip tray, ice trays, defrost
**ICE BUCKET:** clean, replace, lid, Room Service Menu
**GLASS SHELF:** clean, dust: top
**GLASSES:** clean, replace, need______, position
**BATH:** supply shelf, clean, soap, beads, Trav., cap, sewing kit, basin, mirror, light panel, drain, soap dish, faucets, marble, Kleenex dispenser, drain stopper, overflow valve, ash tray, matches, aerator
**TOILET:** lid, seat, lower outside, inside, TP holders, toilet paper, chrome Scales, waste basket, striped rug, floor crooked
**TUB:** shower doors, frames, tracks, drain, chrome, tile, aerator, ceiling, bath mat, rubber mat, top of doors, bottom of tub, grouting, shower floor, remove soap
**SUPPLIES:** wash cloths, bath towels, hand towels, refold, shoe mitt
**BATHROOM:** floor, walls - wash near______
**MIRRORS:** clean, dust
**VANITY SHELF:** polish, dust lamp, bench
**ROOM AIR CONDITIONER:** 72°, off
**BEDS:** remake, krinkle, dust ruffle even, dust headboard, corners
**BEDSPREAD:** position, soiled
**PICTURES:** dust, straighten
**PILLOWS:** position, ends out, gold on top
**NIGHT STAND:** radio, clock, blanket control on 5, lamp, ash tray
Look under bed, dresser, night stand, telephone table
**LUGGAGE BENCH:** position, wipe
**GAME TABLE:** polish, writing folio, magazine, clean drawer paper, ash tray, matches
**LANAI:** clean furniture, position, rail, floor, lights, diffuser
**CHAIRS:** dust, brush out, position, turn cushion
**CORNER TABLE:** phone, pen, pad, lamp, Directory of Services, ash tray, matches, position
**WASTE BASKET:** empty, clean liner, position
**T. V.:** dust, screen, T.V. tents
**DRESSER:** polish, lamp, paper, phone books, utility bag, Bible
**CLOSET:** wooden hangers, plastic hangers, laundry bags, dust the rod and shelf
**DO NOT DISTURB SIGN:** new card
Close curtains, draperies
**RUGS:** vacuum, spotted
**VENTS:** dust bathroom, room
Lampshades, pictures, mirrors - straight, dust
**FURNITURE:** line up correctly, too much polish
**COFFEE TABLE:** dust, position
**FLOOR LAMP:** dust, shade
Lights out
**COMMENTS:**

**FIGURE 29-5** *Inspectress Report or "Please Do"*

3. *Room physical history* (Figure 29-12). Can be combined with furnishings record; includes repairs on all items. Recorded on this particular form that fits into a notebook are the executive Housekeeper's or manager's inspections. In a property with deluxe standards repairs are done immediately, so items marked C (in poor condition) are attended to right away. Note dates of inspection are same dates of work orders. Although form has no place for recording date work was completed, one can assume it was done as soon as possible and that the executive Housekeeper or assistant checked with the engineers or Maintenance Department or electricians daily until work was completed and costs recorded.
4. *Warranties and guarantees.* On equipment usually; to be kept with instructions for each machine.

21-2

**ORDER TO BUILDING SUPERINTENDENT** 02851

MAKE A SEPARATE ORDER FOR EACH JOB

*The Waldorf-Astoria*

FROM ____________ DATE ____________ 19 ____

LOCATION ____________

NATURE OF WORK ____________

APPROVED DEPT. HEAD ____________

| STARTED | COMPLETED | TOTAL TIME | | | |
|---|---|---|---|---|---|
| | | | MATERIAL $ | | |
| | | | LABOR $ | | |
| | | | TOTAL $ | | |

ORIGINAL and DUPLICATE TO BUILDING SUPT.
PINK COPY REMAINS IN BOOK

Mechanic ____________

TIME STAMP ALL COPIES

CONTROL DATA CORP. KALAMAZOO, MI 49002 309,125-2

**FIGURE 29-6** *Request for Repairs*

CENTURY PLAZA HOTEL
CENTURY CITY / LOS ANGELES

HOUSEKEEPING REPAIR CHECK LIST

| 5 | 6 | 7 | 8 | 9 | 10 | 11 |
|---|---|---|---|---|---|---|
| 513 | | | | | | |
| 515 | | | | | | |
| 512 | 604 | 743 | | | | |
| 514 | 638 | 748 | | | | |
| 516 | | | | | | |
| 518 | | | | | | |
| 519 | | | | | | |
| 534 | | | | | | |
| 536 | | | | | | |
| 538 | | | | | | |
| 537 | | | | | | |
| 533 | | | | | | |

| 12 | 14 | 15 | 16 | 17 | 18 | PH |
|---|---|---|---|---|---|---|

Door Seals — (Gaskets)
MR.

**FIGURE 29-7** *For Repair Man—Rooms*

__________ FLOOR

__________ DATE

__________ DAY

__________ FLOOR SUPERVISOR

| | TIME IN | NAME | POSITION | FLOOR & CART NO. | LUNCH OUT | LUNCH IN |
|---|---|---|---|---|---|---|
| 1 | | | | | | |
| 2 | | | | | | |
| 3 | | | | | | |
| 4 | | | | | | |
| 5 | | | | | | |
| 6 | | | | | | |
| 7 | | | | | | |
| 8 | | | | | | |
| 9 | | | | | | |
| 10 | | | | | | |
| 11 | | | | | | |
| 12 | | | | | | |

**FIGURE 29-8** *Floor Attendance Sheet*

5. *Damage book.* Indicates which rooms or public areas were damaged by guests, with dates.
6. *Request for repairs (on items going out of building).* Receipts from companies doing repairs may be sufficient.
7. *Form for employees' packages.* OK'd by department head; given to guard as employees exit.
8. *Form for employees' uniform request.* Given by personnel to new employees.
9. *Purchase requisition.*
10. *Log book.* This has records of all telephone requests from guests or other departments.
11. *Lost and found log book* (Figures 29-12 and 29-13). This includes dates, descriptions of articles, and names of persons who found them.
12. *Damage reports and charges* (Figures 29-14 and 29-15). Interdepartmental communication goes to accounting office so that student can be charged for repairs. Students must sign form agreeing to pay for their share of damages.
13. *Guest loan slips* (Figure 29-16). These records are sent to Front Office cashier to put on guest account until returned.

## Lists Kept by Director of Housekeeping

1. Cost of furnishings repairs
2. Cost of room furnishings
3. List of guest request items and costs (this information is kept handy in case items are damaged and guests must be charged)
4. List of rooms in which mattresses have been replaced

## CENTURY PLAZA HOTEL

CENTURY CITY / LOS ANGELES

**KEY INVENTORY**

Morning Check__________ Date__________

| 5 | 6 | 7 | 8 | 9 | 10 | 11 | 12 | 14 | 15 | 16 | 17 | 18 | 19 | |
|---|---|---|---|---|---|---|---|---|---|---|---|---|---|---|
| 2 | 2 | 2 | 2 | 3 | 4 | 2 | 2 | 2 | 4 | 3 | 6 | 10 | 4 | |
| 4 | 7 | 5 | 6 | 7 | 8 | 4 | 5 | 3 | 6 | 4 | 7 | 11 | 6 | |
| 6 | 5 | 6 | 7 | 9 | 9 | 6 | 6 | 7 | 8 | 8 | 8 | 13 | 11 | |
| 8 | 6 | 13 | 11 | 12 | 10 | 9 | 11 | 13 | 13 | 10 | 10 | 14 | 13 | |
| 9 | 8 | 14 | 14 | 14 | 11 | 12 | 12 | 14 | 16 | 15 | 12 | 18 | 14 | __________ |

Afternoon Check__________

| 5 | 6 | 7 | 8 | 9 | 10 | 11 | 12 | 14 | 15 | 16 | 17 | 18 | 19 | |
|---|---|---|---|---|---|---|---|---|---|---|---|---|---|---|
| 2 | 2 | 2 | 2 | 3 | 4 | 2 | 2 | 2 | 4 | 3 | 6 | 10 | 4 | |
| 4 | 7 | 5 | 6 | 7 | 8 | 4 | 5 | 3 | 6 | 4 | 7 | 11 | 6 | |
| 6 | 5 | 6 | 7 | 9 | 9 | 6 | 6 | 7 | 8 | 8 | 8 | 13 | 11 | |
| 8 | 6 | 13 | 11 | 12 | 10 | 9 | 11 | 13 | 13 | 10 | 10 | 14 | 13 | |
| 9 | 8 | 14 | 14 | 14 | 11 | 12 | 12 | 14 | 16 | 15 | 12 | 18 | 14 | __________ |

Evening Check__________

| | | | | | | | |
|---|---|---|---|---|---|---|---|
| 12A5 | 12A7 | 5A9 | 14A11 | 12A14 | 7A16 | 9A18 | |
| 10A6 | 13A8 | 3A10 | 9A12 | 9A15 | 9A17 | 5A19 | __________ |

Night Check__________

| 5 | 6 | 7 | 8 | 9 | 10 | 11 | 12 | 14 | 15 | 16 | 17 | 18 | 19 | |
|---|---|---|---|---|---|---|---|---|---|---|---|---|---|---|
| 2 | 2 | 2 | 2 | 3 | 4 | 2 | 2 | 2 | 4 | 3 | 6 | 10 | 4 | |
| 4 | 7 | 5 | 6 | 7 | 8 | 4 | 5 | 3 | 6 | 4 | 7 | 11 | 6 | |
| 6 | 5 | 6 | 7 | 9 | 9 | 6 | 6 | 7 | 8 | 8 | 8 | 13 | 11 | |
| 8 | 6 | 13 | 11 | 12 | 10 | 9 | 11 | 13 | 13 | 10 | 10 | 14 | 13 | |
| 9 | 8 | 14 | 14 | 14 | 11 | 12 | 12 | 14 | 16 | 15 | 12 | 18 | 14 | __________ |

Master Key Check__________

| | | | | | | | |
|---|---|---|---|---|---|---|---|
| A-18 | A-32 | A-45 | A-94 | A-98 | A-131 | D.R. | |
| A-19 | A-33 | A-77 | 4-95 | A-128 | A-132 | | |
| A-31 | A-35 | A-93 | A-97 | A-130 | A-100 | | __________ |

Reserve Key Check__________

| | | | | | | | |
|---|---|---|---|---|---|---|---|
| 5A5 | 8A7 | 13A9 | 3A11 | 5A14 | 5A16 | 17A18 | |
| 4A6 | 8A8 | 2A10 | 10A12 | A15 | 11A17 | 12A19 | __________ |

**FIGURE 29-9** *Key Inventory*

**NIGHT SUPERVISORS REPORT**

DATE

| PRE REGS | VIPS & SPEC. ATTN. | NEW BAGGAGE & RUSHES |
|---|---|---|
| | | |
| SOFA BEDS | RECHECKS | |
| | | |

**FIGURE 29-10**

| | | CHANGE | CHANGE |
|---|---|---|---|
| Pt. - Paper | | | |
| Ceiling | | | |
| Carpet | | | |
| Draperies | | | |
| Beds | | | |
| Chests | | | |
| Desks | | | |
| Nite Stands | | | |
| Mirrors | | | |
| Easies | | | |
| St. Chairs | | | |
| Lamps | | | |
| Pictures | | | |
| Sofas | | | |
| Coffee Tables | | | |
| End Tables | | | |
| Bath | | | |

**FIGURE 29-11**

**FIGURE 29-11** *Room Furnishing Record Greenbrier's Housekeeping Department keeps furnishings records on a 5-in. by 8-in. card for each guest room. Every item is described so that Mrs. Kappa will be able to plan refurbishing for the room without even having to go into it. She keeps fabric inventory and record of which rooms have each fabric. Upholsterer updates fabric inventory once a month so department can order ahead if supplies run low.*

**FIGURE 29-12** *At check-out time, counter in Linen Room is busy place as maids return their keys, turn in security sheets, and any lost and found items (e.g., vase on counter) not already given to supervisors. Assistant housekeepers and management trainees, left, note found items and name of maid who located them. Maids check work sheet before leaving to be sure they are scheduled for next day.*

ROOM NO ________

FORM 85

## ROOM MAINTENANCE RECORD

| PAINTING AND WALL WASHING | | | |
|---|---|---|---|
| | PAINTING (DATE) | PAINT COLOR | WALL WASHING (DATE) |
| ENTIRE ROOM | | | |
| BATHROOM | | | |
| CEILING | | | |
| WALLS | | | |
| | | | |
| CLOSET | | | |
| RADIATOR | | | |
| | | | |
| | | | |

| CARPET |
|---|
| NEW (MONTH AND YEAR LAID) |
| SHAMPOO (DATE) |
| |
| |

| INVENTORY OF FURNISHINGS | | |
|---|---|---|
| ITEM | NO. ITEMS | DESCRIPTION |
| CARPET | | |
| DRAPERIES | | |
| BEDSPREAD | | |
| BED (S) | | |
| DRESSER | | |
| DESK | | |
| DESK-DRESSER | | |
| DESK CHAIR | | |
| OCCASIONAL CHAIR | | |
| CLUB CHAIR | | |
| SOFA | | |
| PICTURES | | |
| LUGGAGE RACK | | |
| MIRROR | | |
| NIGHT STAND | | |
| COFFEE TABLE | | |
| END TABLE | | |
| WASTE BASKET | | |
| VALANCE BOARD | | |
| LAMP TABLE | | |
| | | |
| | | |

**FIGURE 29-13** *Room Maintenance Record*

THE UNIVERSITY RESERVES THE RIGHT TO INSPECT THE LIVING QUARTERS OF ANY STUDENT RESIDING ON UNIVERSITY PROPERTY

CHECK IN

Cost of repairing damages or replacing losses will be charged to the occupants of the room at the time of discovery, unless the party directly responsible is known. Damage to the general area, such as hallways and toilets, will be charged proportionately to each occupant of the floor, section, or dormitory as prescribed by the Director of Student Housing, unless the party directly responsible is known.

After you sign below, you will be issued a key. You will not be allowed to move to another room or receive another key to your room until you have returned the key or paid one dollar, as directed by the Resident Advisor, for the key issued you.

If you leave the campus without returning the key or paying for it, you will be charged for the key. After a key charge has been sent to the Bursar, no credit will be given for the key if it is returned to the office.

Athletics of any sort or manner, or on any scale, in the vicinity of the dormitories is forbidden. This regulation is for your protection, physically and financially.

EXCERPT FROM MISSISSIPPI CODE ANNOTATED, SECTION 2085, 1945 .....

A student of any college, university or school, who shall carry, bring, receive, own, or have on the campus, college or school grounds, or within two miles thereof, any weapon the carrying of which concealed is prohibited, shall be guilty of a misdemeanour, and, on conviction, be fined not exceeding three hundred dollars or imprisoned in the county jail not exceeding three months, or both .........

To insure us that you are willing to be responsible for the proper care of your room, and its furnishings, sign below. At the time of clearance you should request this sheet.

I, hereby sign in receipt for the dormitory room and furnishings therein, in the condition indicated (form on the reverse side). I agree to pay a proportionate share of the cost of repairing or replacing any damages to or losses of these items and that of the general area that occur during my occupancy of this dormitory.

Key, Hook Number 1476 — HARD LUCK (Manager: Print occupants full name)

Issued on 8/27/79 (Date) — [signature] Signature of Occupant

- - - - - - - - - - - - - - - - - - - - - - - - - - - - - - - -

Dormitory West Tower Section 8 Room 804

Key WAS ______ Returned Date 4/15/80 Busy Bee Resident Advisor's Signature

**FIGURE 29-14** *Replacement Form, University of Mississippi*

UNIVERSITY OF MISSISSIPPI
UNIVERSITY, MISSISSIPPI

INTERDEPARTMENTAL COMMUNICATION

TO: • Mr. Doyle L. Russell, Director of Accounting — DATE 7-31-80

FROM: • Binford T. Nash, Director of Student Housing

SUBJECT: • Dormitory Damage Charge
Due and payable at the Bursar's Office

LUCK, Hard

West Towers 804 Inspect. report dated 7-31-80

8 1/2" x 11" burned spot on door

1/2 charge = $10.00

Info: Roommate (No Luck) charged remainder by separate memo.

**FIGURE 29-15** *Interdepartmental Communication*

**RADIO LOAN RECORD**

Room:_______ Date:_______ Time:____

| Guest's Name | Radio No. | Returned (Date & By Whom |
|---|---|---|
| | | |
| | | |
| | | |
| | | |
| | | |

On day of departure please call Ext. 7215 for radio to be returned.

Guest please sign here______________

**FIGURE 29-16** *Forms for items loaned to guests are made up in duplicate so that guest will have one copy to remind him to return the borrowed item. Most places will charge for the item if it is not officially returned. There is a separate form made out for each—hair dryers, irons, and so on.*

| Spotted Walls | Torn Wallpaper | Loose Soap Dish | Draperies & Curtains Pins/Rods | Bed Frames | Lamps | Draperies & Curtains Soiled/Torn | Dust Ruffles Soiled/Torn |
|---|---|---|---|---|---|---|---|
| 512<br>514<br>519<br>538<br>537<br>741<br>743 | 635<br>604<br>608<br>741 | | | | 514<br>604<br>608<br>743 | 519 (Torn)<br>743<br>745<br>747<br>749 } Soil | 512<br>514<br>516<br>518<br>743 |

| Tile Repair | Chair Knobs | Dresser Handle | Chair or Sofa Soiled | Lamp Shades Soiled | Carpet Repair Burn or Stretch | Carpet Spotted | Furniture Repair: Room | Furniture Repair: Item |
|---|---|---|---|---|---|---|---|---|
| | | | 533 | 512<br>514<br>516<br>518<br>519 | 533 | 533 | 533 | Sofa arm chipped |

| Foyer Lights | Clock | Rubber Bumper | Screw Cap | Decals | Other |
|---|---|---|---|---|---|
| 741 | | | | 635<br>604<br>608 | Door Seals<br>512 513<br>514 515<br>516<br>518<br>519<br>534<br>536<br>538<br>537<br>533 |

JRS.

**FIGURE 29-17** *Repair Schedule*

5. Miscellaneous lists of floors or rooms which have had periodic repairs done—decals replaced, door hardware polished (Figure 29-17).
6. For contents of maids' carts, list is kept in linen room on guest room floor.
7. For housemen's carts—each function has different list of supplies. Lists kept where carts are stored or supplies issued. These may include such functions as rest room cleaning, relamping, or night clean-up inside and out public areas.

# Chapter 30

# Restoration and Rehabilitation

**OVERVIEW** Preservation of historical landmarks and restoring old buildings pose new challenges to the professional Housekeeper who must learn to restore, clean, and maintain perishable old furnishings, and keep a lovely patina on them. Such majestic old places as the Hotel Utah have done a thorough job of restoration, as have the Grand Hotel in Mackinac, Mich., and the St. Regis in New York City.

"We're going to see more and better restoration. We're just beginning to learn how to do it," according to Phil Pistilli, president of the Alameda Plaza Hotel and past president of AH&MA. "What we did nine years ago at the Raphael in San Francisco was unique at the time, now it's getting to be the norm. Many people are searching for old buildings. That's why it's difficult to find them—and nobody wants to sell them—they all want to restore them."

About the interiors of restored hotels, Pistilli says, "Use as much as you can of the old existing furnishings, but you can't really use that many antiques because of wear and tear. But I think we should study how to keep architectural elements like those magnificent arches," he points to in his Chicago Raphael property. According to Pistilli there's not much economy rehabilitating an old place. Sometimes you just break even.

Fortunately, today upkeep of the beautiful old buildings of the past is being sponsored increasingly by interested groups who are dedicated to keeping America's heritage alive. Never has the National Preservation Society had more support and never have more old hotels been the center of attention as today when few new structures can be built and rehabilitation is, at last, considered the patriotic and profitable way to expand the lodging business.

## The Hotel Utah

Let us consider an exemplary case, the Hotel Utah built in 1911, and relatively unchanged since then.

This hotel is one of the country's most elegant urban properties. Under managing director Stuart Cross's guidance, it is more beautiful and functional than ever after three years of extensive restoration. In addition to refurbishing the original crystal-and-velvet-era appointments, an addition was built bringing the number of guest rooms and suites to 560.

The hotel was able to locate and use the company that built the structure in 1911 to the specifications of a successful, luxury-oriented founder of a local mining company (Figures 30-1, 30-2, and 30-3).

The designers and architect duplicated everything from the shape of the bricks to the cornices and decorative sculpture, while installing the modern conveniences necessary to a forward-looking luxury hotel. Thus the Hotel Utah has retained both its traditional elegance and reputation for fine service.

"We were very conscious of keeping as close to the original design as possible," says Cross. "The Hotel Utah is the social center of town. Local business is very important to us. In fact, a lot of out-of-town business is based on local enthusiasm."

The Mormon Church, which has controlling interest in the hotel, made a large financial commitment to the project of $40,000 per room, a total investment of $22 million. Facing the famed Mormon Tabernacle, the Hotel Utah, too, is a centerpiece of the city and another symbol of the Church's seminal influence.

Therefore owners, operators, architect, and interior designers were guided by a joint commitment to renovate and restore rather than apply superficial decoration. The designers, Heinz Janders Interior Consultants, San Francisco, and architect Robert Fowler, Salt Lake City, worked closely to maintain the continuity between the building shell and the space it encompasses between the almost 70-year-old structure and the new extension.

"We tried to incorporate the new with the old," says Lilianna Liu, Heinz Janders' associate and asssistant on the project. "Because materials were just not available, we were forced to do an interpretation of the original concept. I suppose you could call it 'Gold Rush Era' " (Figures 30-4 through 30-7).

While the hotel is opulent in places, as in the lobby, Janders designed for a restrained formality, working mainly with custom-made items and the original furnishings—remnants of the 1911 design and the Dorothy Draper redesign in the 1950s.

**FIGURE 30-1** *This is a duplicate of the original four-sided settee in the lobby of Hotel Utah. It is covered in a deep red velour and the center contains a fresh flower bouquet, changed daily.*

**FIGURE 30-2** *Close-up of the drapery fabric—a small gold satin print on a cream-colored background. Typical window treatment: sheer curtains and an overdrape with a heavy blackout lining. (These rooms are located in the newer north wings of the hotel.)*

The scope of the task was enormous. In addition to the old and new guest rooms, Janders restored the huge public areas that include 25 meeting rooms and three restaurants.

"I came here after hearing about the renovation," says Cross. "I thought it would be a chal-

**FIGURE 30-3** *This pilaster chair was originally used in the lobby around the columns; now 16 of the same chairs are placed around the mezzanine area. The wood has been restored and they have been reupholstered in a tapestry fabric, a muted pattern in shades of ruby and beige with green and gold accent stripes.*

**FIGURE 30-4** *Only a few of the original Chinese Chippendale chairs remain on the premises. The two pictured are in the reception area of the Catering/Convention Services Department. Wood has been restored and fabric is a soft rose velvet.*

**FIGURE 30-5** *An outside window of the Hotel Utah's Empire Room. The drapery includes a sheer, a separate blackout curtain, and a draw drapery, plus a gold velvet side drapery with a brocade trim.*

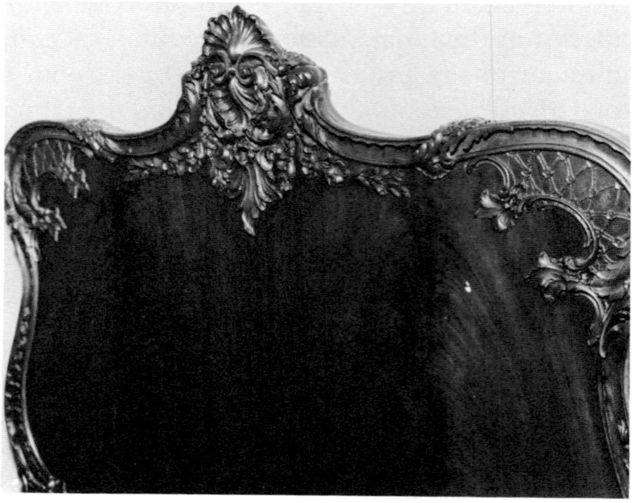

**FIGURE 30-6** *Detail of an original headboard. The hotel is limited as to the original bedroom furnishings, all of which are contained in one small area on the second floor called The Heritage Rooms.*

lenge to bring the service up to the quality of the property."

A newcomer to both Salt Lake City and the Hotel Utah, Cross displays the same love for the property as employees who have been with the hotel through many of its seven directors.

Since approximately 1960, the Hotel Utah has been upgrading its facilities and until the major remodeling and restoration project of some seven or eight years ago, the plumbing equipment, which was taken out of the original

**FIGURE 30-7** *A close view of an original desk and chair in one of the Heritage Room furnishings.*

rooms was saved. Therefore the hotel has its own source of supply for repairing and improving some of the equipment it is presently using. When it is necessary to call on a craftsman for repair work, such as matching tile and woodwork, the work must be contracted out. The craftsmen are limited but so far they have found those needed.

It will be no surprise that the hotel maintenance manager says the newer products don't hold up as well as some of the original products used in this building, and he attributes this to better workmanship and materials in older fittings.

### *New York's St. Regis*

One of New York City's most massive restoration projects has recently been started at the famed St. Regis Sheraton Hotel. According to a spokesman for the organization responsible for architectural restoration, this project, called Renaissance, one of the largest and costliest of its kind, will leave the venerable hotel in perhaps even finer structural condition than when it opened more than 75 years ago.

Involved in this project is total cleaning and waterproofing of the entire building. For cleaning the masonry surface, the firm hired to restore the property will use a high-pressure water treatment that has been proved effective in removing heavily embedded grime and dirt, while in no way damaging the stone facade (Figure 30-8). As part of this exterior renovation, all windows—more than one thousand—will be recaulked, all broken or cracked panes will be replaced, and all the ornate copper hardware on the casement windows repaired and refinished.

"An unusual aspect of this portion of the St. Regis project," notes the contractor, "is in the decision made by the building's owners to restore the patina of the ornate copper facade of the building."

This patina, which is a result of exposure to the elements, is often valued aesthetically for its color. In the case of the St. Regis, certain areas of the building have received more exposure than others with the result that some portions of the facade have built up greater discoloration. The

**FIGURE 30-8** *Work in progress at New York City's St. Regis Hotel. Darkened portion shows building prior to masonry cleaning. Following masonry cleaning, the entire building will be waterproofed.*

decision involved a choice of restoring the copper to its original condition or artificially aging it with chemicals. The St. Regis chose the latter. Although this was purely an aesthetic decision, it underscores the attention to detail that has characterized this project. The company developed a chemical formulation specifically for this project that took into account not only the age of the copper and its present condition, but the various types of exposure that it would receive.

In an energy-conscious time, a new awareness has also marked the restoration of the St. Regis. Not only has this work been designed to bring the property back to the highest standards in terms of aesthetics, but much of the activity is focusing on reducing energy costs. To that end, the entire building is being weatherproofed, with

**FIGURE 30-9** *Workman completes hand-polishing of ornate bronze railing in Waldorf Astoria lobby. All marble and metal elements in this area have been hand-stripped, cleaned, and repolished. This is the first time in many years a renovation project of this scope has been undertaken at the hotel.*

**FIGURE 30-10** *Workman puts finishing touches to gold leaf on the torchieres fronting the Plaza entrance. The renovation of these famed symbols of New York's Grand Hotel has been a costly hand-executed task, but indicative of the type of maintenance and restoration work now being demanded by New York hotels.*

all masonry surfaces sealed and repointed. All windows and doors are being fully weather-stripped, and its distinctive slate mansard roof is being completely restored with all broken or missing slate replaced.

Figures 30-9 and 30-10 are examples of other renovations in progress in New York hotels.

## *The Peabody Hotel*

The Peabody Hotel, a downtown Memphis, Tenn. landmark for more than 50 years, is undergoing renovation. Built in 1925, the Peabody has been closed since 1975.

The Belz Investment Co., the present owner, obtained a loan guarantee for more than $11 million from the U.S. Economic Development Administration and $1 million additional assistance from a U.S. Department of Housing and Urban Development grant.

Renovation of the Peabody is one of several signs of new life in downtown Memphis (Figure 30-11). Long-blighted Beale Street, known as the birthplace of the blues, is being redeveloped for entertainment, retail, and office use.

Figure 30-12, however, shows how a restoration can go wrong.

## *The Grand Hotel*

Built in 1887, the Grand Hotel opens in mid-May each year and its doors do not close until November, when the hordes of Midwestern vacationers and conventioneers from around the country seek warmer climes.

The Grand Hotel dominates Mackinac Island. Its grand rows of ionic columns and fluttering flags stand out as symbols of the hotel's social and economic prominence.

In keeping with the modern demands of a brisk convention business in addition to the regular tourists who come to sample the island's throwback atmosphere and renowned fudge, the hotel is undergoing a complete redesign of its vast interiors. Called on to head the project was Carleton Varney, president of Dorothy Draper & Co., Inc., a New York design firm.

**FIGURE 30-11** *Even birdlife may be part of the professional Housekeeper's domain, as were the famous ducks in the Peabody Hotel. They delighted visitors and brightened bellmen's days too, but they added problems to floor maintenance crew.*

During the course of the five-year redesign, Varney and his staff have implemented a scheme that reflects both aspects of the hotel's business and remains faithful to the luxurious, comfort-oriented program that has been the resort's hallmark.

"We hope the redesign will be a continuous program," says R. D. Musser, president, owner, and hands-on operator of the property. Musser purchased the hotel from his uncle recently.

The Grand Hotel has 262 rooms serviced by 400 employees. Since the property is just shy of 100 years old, the guest rooms come in a variety of shapes and sizes—who ever heard of portion control in the nineteenth century? Yet in every case, Varney succeeded in incorporating his unique, elegant, summer-like mixture of floral and geometric patterns.

In both public areas and guest rooms Varney selected natural fabrics. Even the lamp shades are done in washable fabric. Carpets in the expansive lobby and attached Audubon Bar are 70% wool, 30% nylon.

The lobby is long and, by Grand Hotel standards, narrow, with the hotel's trademark geranium theme woven in the carpet. Furnishings convey a subdued, refined feeling through plush material and rich colors.

Ornaments include hand-painted murals at each end of the room, highly polished wood tables, and a gleaming breakfront that serves as a backdrop to afternoon tea.

The idea was to make the design look as if it had been there forever. Everything was custom-made, as few older items were salvaged.

The Audubon Bar features authentic Audubon Society prints and a few touches of local color, including turtle-shaped ash trays (Mackinac is the American Indian word for turtle).

The dining room runs about half the length of the hotel's famous porch—said to be the longest in the world—and seats 1100, using 60-

**FIGURE 30-12** *"A renovation that just missed" might be appropriate comment for this lobby. It was once very grand, but has been tampered with by designers who have not helped maximize existing attributes. In an effort to make the ceilings seem lower, paneling and painting have been stopped partway up the wall. This succeeded in curtailing any grandeur, at the same time failing to give a guest anything in a size to relate to. Painted ceilings only detract from grace of chandeliers, which are hung too high to be admired.*

year-old chairs recovered with tufted red oilcloth. In the Dorothy Draper tradition, the ceiling is painted light blue to give an added dimension to the airy room.

Varney's own designs transferred to material grace both windows and hanging light fixtures—fabric has been wound around the supporting chains.

In the dining rooms and throughout, the hotel the sprinkler system, which hangs from the sky-like ceilings, is purposely painted to let people know the protection is there.

## *Caring for the Treasures of the World*

Many older establishments are at least partially furnished with valuable and precious antiques. These items require special care for preservation. More and more citizens are donating beautiful furnishings and art works for historical preservation. In European countries the gift may be the complete estate, including all home furnishings and art. In America our history is of fewer centuries but the objects are just as precious to us. In some instances (Hearts's castle) there is a combination of imported items with American construction and styles.

The wear and tear exacted by tours and tourists must be carefully repaired each year. Special care is given to floors, doors, carpets, draperies, and chandeliers.

Old formulas and systems may be the answer to some problems, as a modern tool or chemical is often destructive to delicate work.

The lovely antique glass or painted metal wall sconce will be better cleaned by hand with a small soft brush to remove dust than by trying to use a vacuum cleaner attachment.

Locating the personnel for such tasks is perhaps difficult, but the older generation is interested in historical preservation, and by special arrangements and short shifts can often be interested in doing the work. These persons are more readily persuaded to spend time and energy in repair and restoration of treasures than younger workers.

## *Woods*

Several firms have developed runners of carpet that can be laid over carpet or floors for protection during tours. Groups follow set patterns in viewing rooms and areas, but if possible, alternate routes should be devised to equalize the wear on floors and or carpets. Stanchions should always outline the tour route to prevent people from wandering into restricted areas and handling fragile items.

Constant dusting is also essential to maintain the patina on furniture and wood surfaces. Wooden panels, doors, and window ledges should be oiled and rubbed at least yearly with good light oil that will assist in preservation. One modern item is lemon oil, which helps the wood retain color with the slight bleaching quality of lemon as well as leaving a slight trace of fresh lemon odor that offsets the mustiness of old wood. If expense of cleaning supplies is a major budget factor, lemon oil can be made by combining 1 gallon of good oil (some firms use mineral oil) and 2 ounces of lemon extract. Where possible, on stained and unpainted wood apply the oil with 0000 steel wool. This will combine removal of air pollutants with feeding the wood.

Painted surfaces can be cleaned professionally with the use of a special dough. This method was used on the walls and ceiling of the Sistine Chapel at the Vatican. It removes built-up soil from painted surfaces without injury to the paint.

For paintings and framed canvases it may be possible to enclose them in additional glass cases. Post notices Do Not Touch, as the oil and soil of many hands can do irreparable harm to a surface. Tradition has it that one contributing factor to the beauty and patina of the rosewood and mahogany furniture of our deep South was hand-rubbing of the wood by the house slaves.

The greatest danger to treasures (with the exception of china and porcelain) is moisture or water. Great care must be taken that roofs do not leak, windows are rain-proof, and damage from broken water pipes is immediately repaired. Water marks on floors and panels may often be removed with a slight solution of Clorox and water. *If possible* place the item in the sun or expose it to as much natural light as available.

Books and leather items are particularly sensitive to moisture. Temperature controls are extremely importanr to pictures (oils, watercolors, etc.) books, and wood. Depending upon areas and climate, a determination should be made about ideal room temperature and moisture where such treasures are involved. Because of moisture in the air, air conditioning must be rigidly controlled if used at all. By and large, cooler temperatures are better for preservation of most items.

## *Wall Coverings*

In nineteenth century America an expensive and exotic treat that conveyed a special sense of hospitality, the pineapple, came to be incorporated into formal wallcovering designs as a sign of warm conviviality, and thus the motif passed into the American heritage. It also has become part of a Colonial traditional hand-print wall covering collection styled by Bill Giardiello.

Other patterns feature such tradition-inspired motifs as Colonial floral stripes, Greek keys, damasks, Oriental toiles, fleurs de lis, and formal florals.

Although one supplier responsible for the Dwoskin Heritage collection presents a historically accurate palette that includes such period hues as cranberry, bayberry green, hearth brown, off-white, cream, and mustard, patterns may be custom colored to suit any decor needs.

The hand-prints are sturdy, scrubbable paper-backed vinyls selling in a suggested retail price range of about $25.00 per yard.

An enormously helpful book/catalog, *Fabrics for Historic Buildings,* was published by the National Trust for Historic Preservation in 1977. Written by Jane C. Nylander, curator of textiles and ceramics at Old Sturbridge Village, Mass., printed by Preservation Press, it shows examples of which fabric weaves and designs are appropriate for which styles and periods, with pictures, descriptions, and sources of products.

Nylander warns against using original old fabrics, explaining that light weakens them, so their display should be limited. Reproductions, she maintains, are so authentic today that they are highly acceptable and certainly more practical where the draperies, bedspreads, and upholsteries will be subjected to hard use by the public. This book is instructive, interesting, and indispensible to any Housekeeper who will be selecting fabrics for a restored building.

According to David L. Owen, Director of Housekeeping/Design at Denver's Brown Palace Hotel (see Chapter 1), "the availability of remodeling supplies is probably the greatest problem. Lumber, lighting fixtures, fabrics, furniture must usually be ordered far in advance of project start-up dates to ensure having the right quality and quantity on hand so projects can stay on schedule. American-made items have always been erratic in delivery and quality control where European-made items have been unusually fine in all ways. Now our overseas suppliers seem to be trying to emulate the U.S. in this area.

"I have been guaranteed quantity on almost all items used in our remodeling projects up to one year in advance to ensure our supply as well as freezing prices. All indications show no ceiling on prices and we have saved by doing this."

## *Fabrics*

Fabrics such as tapestry covers and wall hangings should be kept as dust-free as possible with frequent vacuuming. Repair of these is possible with the use of invisible thread and a backing fabric carefully applied. Museums have perfected this technique to display fragments many hundreds of years old. This painstaking work is best done by an experienced seamstress with patience and an appreciation of fine things.

In some cases reweaving is possible, depending on the location of the wear. Attempts should be made to use the same product (linen, wool, or cotton hand-dyed to match) for mending since synthetic threads may do harm to an old fabric.

To freshen colors in upholstery and carpets use a mild solution of water and household ammonia (1 tablespoon to 1 quart of warm water), wringing out the cloth as nearly dry as possible to remove dust and some grime and brighten colors without damaging. Care must be used not to have a strong ammonia solution, as it will remove colors.

## *Old Care Techniques*

Keep a record of special cleaning and products used. Prepare a list of experts in the area or on staff that can be called in for consultation on problems.

The following is a list of historical remedies to try for restoring treasures, before using the harsher new chemicals.

**Copper.**

1. Half a lemon dipped in salt. Rub well, rinse in warm water, and polish with soft cloth.
2. Weak ammonia solution will remove the greenish deposit on copper.
3. Wash in warm water then rub on a paste of salt, fine sand, and vinegar with a piece of flannel; wash again in warm water. After cleaning rub the item with a little oil to retard tarnish.

**Brass.**

1. Equal parts of flour and salt moistened with vinegar.
2. Mixture of 1 oz. oxalic acid to a pint of soda water will remove tarnish. Use a cloth wrung out of the solution and rub brass well; wash off the solution and polish with chamois.
3. Worcestershire sauce can be rubbed on and washed off.

**Bronze.**

1. Wash well, then apply with a cloth a solution of 1 part muriatic acid to 2 parts water.
2. Let dry, polish with little oil.

**Crystal or Glass.**

1. Make a paste of dry mustard and oil and apply to ring in a vase.
2. Construction sand and water solution to remove discoloration in bottle or jar.
3. Crushed eggshells and soap and water; shake vase or bottle well and rinse.
4. Clean with solution of salt and vinegar (a dessert spoon of salt moistened with vinegar and shaken in the glass or decanter).

**Ivory.**

1. A paste of lemon juice and salt or whiting (dry bleach). Leave on for one or two hours if badly discolored.
2. Keep ivory in the sun, as dark areas will discolor.
3. Fine sandpaper may be used on dark areas but exercise great care not to mar the finish. Polish with chamois after use.

**Marble.**

1. Toothpaste will remove rings.
2. Wipe with light solution of household bleach and place in the sun; repeat if necessary.
3. Equal parts of soap quicklime and caustic potash. Apply with brush and allow to stand several days. Wash or wipe off.

**To Polish Marble.**

1. Make a paste by sifting together ½ pound soda, ¼ pound powdered pumice, and ¼ pound chalk. Add water, rub well into marble. Let dry, wash off with soap and water, dry.
2. Furniture polish.
3. Milk.

**Marble Bust.**

1. Dust very well and then wash in a very weak solution of hydrochloric acid (soap injures the color of marble).

**Mildew.**

1. Make a mixture of soap, starch, and salt; cover spots and expose to sun.
2. Soak in milk overnight and then place in sun to dry.
3. For drastic action make a diluted solution of chloride of lime and soak for ½ hour, rinse well.

**Mother-of-Pearl.**

1. Wash with powdered bleach and cold water. *Do not* use warm water and soap.

**Pewter.**

1. Wash in warm soapy water and dry by rubbing round and round, never up and down.
2. If in poor condition, dip a cloth in equal parts of turpentine and linseed oil, then in rottenstone. Rub round and round, wash in warm soapy water.

**Silver.**

1. Silver long stored and with heavy tarnish can be soaked in potato water two hours.
2. Flat silver may be cleaned by boiling in a solution of 1 teaspoon salt, 1 teaspoon soda to each quart of water in an aluminum pan. Boil one to two minutes, then wipe dry.
3. Chloride of lime will remove dried ink from silver ink wells.
4. A piece of alum in the silver drawer will retard tarnish.

**China.**

1. Baking soda rubbed on with a damp cloth will remove spots.
2. Mending broken china may be done with a paste of powdered quicklime and white of an egg beaten well and warmed slightly.

**Iron Rust.**

1. Rub on lemon juice and salt. If stubborn let stand one hour or apply a second time, then rinse in water and little ammonia.

2. Oxalic acid or salts of sorrel.
3. As last resort saturate stain with diluted hydrochloric acid and rinse at once with warm water and little ammonia; rinse in clear water.

**Papier mache.**

1. Sponge in lukewarm water using a little soap if spotted or greasy. Dry thoroughly and polish with a soft cloth. Never use boiling or hot water.

**Lacquered Goods.**

1. Dip in warm water containing lemon juice or sour milk, rub with piece of flannel, polish with chamois.

**Leather.**

1. To revive luster in Morocco or other leather apply white of egg with a sponge.
2. Clean leather with a brush dipped in a weak solution of soda water and then rinsed and dried.
3. If the leather is very old it may be softened with a little oil and then rubbed with a sponge dipped in a little turpentine.

**Tortoise Shell.**

1. Brighten with rottenstone fine powder mixed with oil, then silver polish and then chamois. *Do not* ever let stand in water.

**Veils or Lace.**

1. Wash gently in lukewarm water, do not wring out, but agitate up and down by hand, rinse, and squeeze. Then dip in diluted sugar-water (1 pint warm water and 3 lumps sugar), squeeze gently, and pin on a cloth to dry. If still limp, press between pages of waxed paper with warm iron.
2. To preserve sick lace, soak in hot milk and borox to prevent yellowing. Store in blue tissue paper in dark area, as light will discolor.
3. To clean, steep overnight in water in which borox has been dissolved. Then wash in gentle soap and rinse several times. Then dip in warm sugar water and pin on a cloth to dry.
4. To wash, use a bottle with the solution as this will prevent the lace tearing or twisting.
5. Really rare lace may be folded with dry powdered magnesia sprinkled between the layers and left for several days. The magnesia will have absorbed much of the dirt and dust.

**Tar.**

1. Remove with cloth saturated in turpentine, benzine, or gasoline in well-ventilated area.

**Velvet** (to raise pile).

1. Hold pile out over a hot iron covered with a wet cloth; the steam will raise the pile and remove creases.

**Wallpaper.**

1. Can be cleaned with a cloth and uncooked oatmeal. Be sure to cover entire surface and have dropcloths to catch residue.

**Zinc.**

1. Wash thoroughly in warm soapy water, dry, and rub with cloth dipped in turpentine.

**To Remove Bruises from Furniture.** Wet the part with warm water. Fold a piece of brown paper several times, soak in warm water, and lay on the bruise. Then apply a warm—*not hot*—iron until the moisture has evaporated. If the dent is not raised to the surface, repeat the process.

**Furniture Scratches.**

1. Use a mixture of equal parts of linseed oil and turpentine.
2. Rub with a scratch-removing polish or with a walnut half.

**Furniture Worms.**

1. Send piece to an expert who will fumigate with a special gas.
2. Home treatment is to rub the unfinished areas with benzine and inject liquid into holes. Do this in May and end of July for several years.

**Furniture Marks.**

1. Use 0000 steel wool dipped in oil and then cigarette ash; rub gently.
2. Rub with linseed oil.

**Coral.**

1. To clean, use tissue paper to polish.
2. Boil in a solution of mild soap and soda, then dry and polish with chamois.

**To Clean Oil Paintings.**

1. Rub a slice of freshly cut potato dipped in cold water over the picture, wipe off with a sponge and then with lukewarm water. Dry and polish with a piece of old silk. Then rub surface with flannel dipped in a *very little* linseed oil, which will preserve the paint but not attract dust. Maintain a temperature about 70°.

**Matting on Paintings.**

1. To clean, use damp cloth being careful to avoid excessive moisture.
2. Dust often and occasionally wash with salt water.
3. Bran water may be used in place of salt water (1 gallon water, 2 handsful bran).

**Ink Removal from Carpets.**

1. Soak with milk and keep applying and soaking up ink stain. When completely removed wash with soapy water and then clear water. Soak up water, let dry.
2. Gentle soap, lemon juice, and starch in solution.

**Paint.**

1. To remove from fragile fabrics, use chloroform on a cloth and rub gently, being sure to be in a well-ventilated area.
2. Equal parts of ammonia and turpentine will remove from fabric. Saturate the spot two or three times and then apply soapy water if possible.

**Iron Rust.**

1. Remove with lemon juice and salt.
2. Rub with a cork dipped in olive oil.

**Chandeliers and Mirrors of Glass or Crystal.**

1. Wipe with cloth wrung out of a solution of water and alcohol.
2. Polish with newspaper.
3. Air-blasting will clean a chandelier by blowing dust away. Camera shops have products called OMIT and Super Blast (compressed air spray for cleaning cameras).
4. Drip-dry washing can be done with cleaner in aerosol cans. Floor or furniture under the fixture should be covered to catch the drip.

**Leather Books.**

1. Use a soft dust cloth or the round dusting brush of the vacuum cleaner. To hand-dust hold the book with the headcap down. (The headcap is the covering leather that has been shaped over the headbands at the head and foot of the backbone of a book.) Slant the spine up toward you and stroke the tops of the pages away from you with the dust-cloth or brush. This prevents dust from seeping down between leaves.
2. There are special oil treatments for old leather bindings, but they must be correctly mixed and applied.
3. Books react to climate in a room. They should have humidity of 50 to 65 percent and a constant temperature of 60 to 70 degrees. Good circulation is essential.
4. If books are attacked by mildew wipe them with soft clean rag. Take books out of the room. Wipe and brush, sun and air them before returning to shelves.

**Cane.**

1. During the hot season cane should be dampened once a month to counteract dryness. With a sponge lightly moistened in warm water go over the entire canework area. Let dry before using.

2. To remove soil use a sponge wrung out nearly dry from mild soapy warm water. Rinse off soap with clear warm water and the sponge.

**Wicker.**

1. Natural unpainted wicker may be washed. To clean and bring up the color, mix a cup each of peroxide and vinegar into a pail of water, and apply to furniture with a sponge. Rinse with clear water and dry thoroughly.
2. Painted wicker should be dusted regularly with a damp rag and should be cleaned twice a year with clear water and a sponge.
3. Pieces of furniture both painted and unpainted can be wet down once a year with the garden hose or in the shower (cushions removed), which will help prevent drying out. Clean before repainting; also, file off fuzzy fibers with very fine (000) sandpaper following the lines of the grain. Wicker may be repainted many times, as the paint helps the looks and preservation.

**Equipment Essential for Caring for These Treasures.**

1. Sturdy high and low ladders
2. Soft-bristled brushes for dusting delicate fixtures
3. Stiff brushes for cleaning construction dirt, cement, or bricks
4. Long-handled soft brushes for dusting high places
5. Sponges, preferably real
6. Heaters and fans for temperature control
7. Brooms, dusters
8. Stanchions and cords for crowd control
9. Tour carpet
10. Vacuum sweepers with all attachments
11. Back-pack vacuums for high work
12. Floor polishers, buffs, waxers, and strippers
13. Custodian carts for transporting equipment
14. Large quantities of natural-fiber cloth, cotton, wool, silk (synthetic fabrics are not as usable, do not absorb soil)
15. Gentle soap
16. Lemon oil is good light oil for rubbing wood. Fresh lemons and baking soda are helpful in bleaching.

Keep a handy supply of products quickly available for cleaning and spotting so that instant repair can be made.

## *Products for Use*

**I. Abrasives**

They should only be used when all else has failed. They must be clearly understood, as a harsh product can do irreparable harm.

1. Sandpaper and glasspaper may be used in sheets or as discs for power tools. They are used for stripping down and smoothing wooden surfaces. With paint remover they may also be used to remove stubborn old paint and varnish.
2. Wet and dry emery paper in various grades can be used dry or soaked in water as a lubricant without dissolving into pulp. This helps to give an almost mirror finish to an object to be painted or varnished. An object is rubbed down with wet and dry paper between coats until the surface feels perfectly smooth.
3. Steel wool in various grades is also very useful for refinishing. It may even be used with metal polish on metal items to remove stubborn stains. Can be used with rust removal products for good results. Do *not* use on a glossy finish. Do use in conjunction with wood finishing materials. Remember, the finer the steel wool the finer the work.
4. Other abrasive products for special uses
    a. Emery powder
    b. Tripoli powder
    c. Rottenstone (a mineral from England used mixed with oil)
    d. Jeweler's rouge (powdered iron oxide)
    e. Pumice powder
    f. Crocus powder
    g. Carborundum
    h. Whiting

All may be used moistened on soft rags, swabs, or with small brushes such as old toothbrushes.

To join the following materials after degreasing, before abrasion and after:

Brass—abrade with emery

Ceramics and porcelains—abrade with carborundum and water

Copper—abrade with emery

Glass—abrade with carborundum

Gold—abrade with fine emery or crocus paper

Lead, tin, and solder—abrade with fine emery

Leather—degrease with great care and abrade with glasspaper

Silver—abrade with fine emery

Steel and iron—abrade with emery

Stone—abrade with a wire brush

Wood—abrade with glasspaper or fine steel wool

**II. Acids**

They should be used with great care and with rubber gloves. A solution of bicarbonate of soda will neutralize the acid if an accident occurs. When diluting acid always add the acid to the water, never vice versa.

1. Hydrochloric acid (spirits of salts) is never used on stainless or other metals or with nylon or other man-made fibers.
2. Acetic acid (vinegar)
3. Oxalic acid is poisonous, but has many uses as stain remover. A 5 percent solution will remove ink stains. 2 teaspoonfuls to 2 pints cold water will remove blood stains. In a saturated solution it will remove black water marks on wood.
4. Citric acid (lemon juice is a natural citric acid and good gentle bleach). A 10 percent solution is often used for stain removal.
5. Nitric acid is a *strong* fuming acid—use with great care. In a very dilute solution may be used for cleaning gilt and gilding. It will darken soft soldering.
6. Muriatic acid is hydrochloric acid. In dilute solution often used for cleaning tiles after fire damage.
7. Accumulator acid is diluted sulphuric acid, used in making chlorine gas for bleaching.

**III. Adhesives**

These have become so strong and heat- and water-resistant that even ceramics and china can be invisibly mended. Thoroughly clean the surfaces before adding adhesive. To clean the old surface the following may help.

1. Resinous cements are dissolved by alcohol.
2. Gelatine or animal glues may be removed by warm water.
3. Celluloid cements remove with amyl acetate or acetone.
4. Dissolvex will remove modern epoxy glues.

To repair after cleaning

1. Scotch glue for woods. Be careful in preparing as this glue must *never* boil. It is used warm. Items should be clamped or weighted. Do not use in damp areas as the glue will grow fungus. Heat also affects this and causes release of hold.
2. Durofix is a celluloid cement.
3. Aerolite a strong resin adhesive.
4. 2-tube adhesives—many makes.

The enemies of treasures are myriad—commonplace and unusual. Time itself has inescapable effects. Atmosphere causes some items to disintegrate. Insects can cause damage and too high a temperature causes cracks and warps. So be constant with care and attention. Be sure that any repairs and maintenance are correctly done and only the proper products used. Proceed with caution at all times, know when you can do it, and even more important, when to seek expert advice.

## *Bibliography*

*Mrs. Beeton's Household Management.* London: Ward Loch & Co.

*The White House Cookbook.* 1899.

Epsie Kinara. *The Care and Keeping of Antiques.* Hawthorn Books, 1971.

Suzanne Beedell. *Restoring Junk.* David McKay Co. 1970.

# Chapter 31

# What Ultimately Makes a Housekeeper "Professional"?

**OVERVIEW** The ability to make up a budget (which will be the guide for activities all year) and understanding hotel and institutional accounting sets the professional Housekeeper above others in supervisory positions in the department, and makes him or her material for promotion to management. Forecasting labor needs, room repairs, and replacements that will prove necessary are vital to running the department well. Here we examine opportunities in the field, salaries on several levels, how the chains train, successful community involvement, sources of seminars, importance of NEHA, and the need for education.

Not only effectively administering the department, but understanding the financial aspects of the entire institution will be essential if Housekeepers are to be called professionals and be considered qualified management personnel.

## *Budget*

Definitions for budget vary widely but one that is easily understood is "an estimate of the proposed expenditures for one or two years." It should be carefully prepared and be used to govern a department's spending. It can be considered the framework within which the Housekeeping Department operates.

### *Guide to Running the Department*

It is the guide as to which things need repair or replacement. It helps to determine what costly pieces of equipment should be purchased, and helps to pinpoint the areas where emphasis will be placed for the coming year. (Perhaps, for example, the dormitory may be adding a snack bar. This means labor for clean-up in public areas undoubtedly will have to be increased.)

The budget is an instrument used by management for controlling and directing activities, especially the purchasing activities. One of the advantages of budgeting is that it makes a department head carefully plan the operation for the ensuing period of time.

This is made easier if proper records have been kept so that one knows what items, based on age, are going to need replacement. Conferences with management are a helpful way to keep informed of contemplated changes in the institution's standards. The department head must know what the approximate sales (volume of business) will be for the upcoming period, as any large increase or decrease in occupancy or banquet business would certainly affect many things

Housekeeping is expected to provide. For this reason, management does a forecast for the entire year and the various departments base their proposed expenditures on that.

## *Capital Improvements*

In the hospitality field, the housekeeper's main involvement in the preparation of the budget deals with *capital improvement* which includes purchasing new equipment and furnishings, renovation, and building new facilities that would improve service. Capital expenditures are for assets and improvements with a useful life of more than one year (e.g., furniture and equipment); these are assets for which tax deduction is spread over the number of years that represent their approximate life expectancy. Direct expenses are for material with a life expectancy of less than one year (e.g., cleaning compounds, light bulbs, guest stationery).

A capital improvement budget must be carefully documented. Management tends to request that equipment and furniture be repaired rather than replaced. It is up to the housekeeper to indicate the age of materials scheduled for replacement and to present facts and figures showing that repairs are impractical, or will be required so frequently as to be uneconomical. If new equipment on the market will more than repay its cost through savings effected (e.g., by radically reducing the number of man-hours required to perform a housekeeping function), the housekeeper may be able to obtain approval for a machine costing in excess of $2500. Otherwise, it may prove almost impossible, even if the machine costs only $250.

The capital expenditures budget must usually be accompanied by forms on which the housekeeper summarizes the justification for each purchase.

In connection with the capital improvement, there will also be an expense column, that is, labor for installation or fabricating—as when the Sewing Room makes draperies, or carpenters construct shelving (Figures 31-1 and 31-2).

A uniform system of accounting is used in many hotels and by many different departments for which separate statements and budgets are set up. There are the Rooms Department; Food and Beverages; Administrative and General (A & G); Advertising and Business Promotion (A & BP); Parking and Garage; Laundry and Valet; Maintenance and Utility (M & U). Housekeeping is part of the Rooms Department so the housekeeper's figures are turned into the resident manager who makes out the complete Rooms Department budget as well as the monthly and annual operating statements.

Operating statements list expenses that occur on a monthly basis, including labor, uniforms, laundry, linens and blankets, guest room supplies, and cleaning supplies. These items all appear on the Rooms Department Operating Statement (Figure 31-3), plus other direct expenses (over which Housekeeping has little control) such as telephone, postage, telegraph, reservations, guest history, travel agents' commissions, and printing and stationery.

Assuming it has been determined that suppliers' costs continue to be competitive and that the amount of usable inventory on hand has been verified, direct expenses represent the least controversial part of the Housekeeping budget. Management will almost never request that torn or worn linens be used, that soap be eliminated from guest rooms, or that the department switch to cheaper and inferior cleaning compounds.

That part of the budget covering payroll will almost always be approved if it includes no increase in manpower and if projected wage increases are only those agreed upon in union contracts. Reasonable merit increases for valued workers, especially for supervisors who would be difficult to replace, tend to be readily approved except during periods of financial downturn. Any request for additional staff members, however, will be carefully scrutinized, and must be supported with evidence that it will benefit the entire institution and not merely the Housekeeping Department.

## *Priorities for Replacement*

When preparing a budget all items should appear on a priority basis:

*1st priority.* Those items that are absolutely essential in the coming year.

| ROOMS | CAPITAL | EXPENSE | TOTAL |
|---|---|---|---|
| 1st Priority | | | |
| 1) Carpet | | | |
| 4 hospitality suites in penthouse | | | 7,500.00 |
| 2) Mattresses, springs and frames | | | |
| 100 Twin mattresses & springs | 120.00 ea. | | 12,000.00 |
| 25 Queen mattresses & springs | 146.00 ea. | | 3,750.00 |
| 100 Twin frames | 12.50 ea. | | 1,250.00 |
| 24 Queen frames | 19.00 ea. | | 456.00 |
| 3) Glass curtains | | | |
| 4000 yards Alabaster Guest Rooms | | | 14,000.00 |
| 4) Redo carpets, draperies, upholstery in 130 rooms—01-02-60-26-06-12-21-29-37-47- on each floor | 1,600.00 ea. | | 195,000.00 |
| 5) Lamps | | | |
| 25 Guest Room Variety | 19.00 ea. | | 843.00 |
| 400 Shades | 19.00 ea. | | 7,500.00 |
| 6) Bedspreads—500 | | | 20,000.00 |
| Gold Double—200 | | | |
| Gold Twin—150 | | | |
| Green Twin—150 | | | |
| 7) 12 Upright vacuums | | | 750.00 |
| 8) Back-Pak vacuum for high places | | | 94.00 |
| 9) 1 Dry foam shampoo machine | | | 625.00 |
| 2nd Priority | | | |
| 1) Draperies | | | |
| Penthouse—5 suites | 625.00 ea. | | 3,125.00 |
| Labor | | 940.00 | |
| 2) Carpet | | | |
| 1914 | | | 5,000.00 |
| 1915 | | | 5,000.00 |
| Corridors—7 carpets ordered | | | 52,500.00 |
| 3) New plastic plants—penthouse, corridors, suites | | | 3,125.00 |
| 4) 5 Housemen,carts | | | 625.00 |
| 3rd Priority | | | |
| 1) 25 Balcony chairs | | | 958.00 |
| 2) 8 Love seats for 02s | | | 3,000.00 |
| 3) 2 Carousel hangers for costume control | | | 375.00 |
| 4) Beepers, two-way | | | 750.00 |

**FIGURE 31-1** *Capital Budget for Rooms Housekeeping (Rooms Department)*

*2nd priority.* New items that would add a touch of luxury or give a fresh look to the operation.
*3rd priority.* Items that are certain to be needed within the next two years (included in case a mass purchase of like furnishings would reduce prices appreciably).

In a hotel, club, residence hall, or resort that has been open for three or more years, it becomes necessary to renovate, refurbish, or redecorate certain areas. Replacement items involve major expense—carpets, draperies, case goods, furniture, beds, springs and mattresses, lamps. These will be for the guest rooms as well as the public areas. One instrument that makes it much easier for the housekeeper to plan the yearly capital budget is a record showing when things were purchased and how long they were intended to wear. A projected budget should always cover at least ten years in the future, then one can quickly pick up whatever needs replacement or particular attention. Naturally, in the

| PUBLIC AREAS | CAPITAL | EXPENSE | TOTAL |
|---|---|---|---|
| 1st Priority | | | |
| 1) South Mezzanine—Draperies | | | 6,250.00 |
| 4 rooms—Directors' Room, Governor's Room and 2 Dining Rooms—8 windows | | | |
| 2) Lobby | | | |
| Austrian draperies | | | 25,375.00 |
| 3) Lamps | | | |
| 3 extra | | | 281.25 |
| 24 shades | | | 500.00 |
| 4) Garden Court | | | |
| Carpet | | | 6,875.00 |
| Blinds | | | 12,500.00 |
| 5) Ballroom | | | |
| Carpet per yard $22.50 | | | 78,750.00 |
| Labor | | 7,625.00 | |
| Walls | | | 31,325.00 |
| Labor | | 6,250.00 | |
| 2nd Priority | | | |
| 1) Fiesta Room | | | |
| Carpet—center dining room and bar | | | 12,500.00 |
| 2) Swimmers': elevator entrances | | | |
| Carpet—indoor-outdoor | | | 375.00 |
| 3rd Priority | | | |
| 1) Patio Plants | | | 1,875.00 |
| 2) Las Vegas Room | | | |
| Draperies | | | 1,250.00 |
| Carpet | | | 12,500.00 |
| 3) Latin Room | | | |
| Labor to install carpet | | 3,125.00 | |
| Carpet | | | 15,000.00 |
| 4) Convention Area (4 dining rooms) | | | |
| Redo wall decor and carpet | | | 10,000.00 |
| 5) Employees' Dining Area Carpet | | | 1,875.00 |

**FIGURE 31-2** *Capital Budget for Public Areas*

| | | CURRENT MONTH | | YEAR-TO-DATE | | |
|---|---|---|---|---|---|---|
| | | THIS YEAR | LAST YEAR | THIS YEAR | BUDGET | LAST YEAR |
| 1 | GROSS SALES | | | | | |
| 2 | Permanent Rooms | | | | | |
| 3 | Transient Rooms | | | | | |
| 4 | | | | | | |
| 5 | TOTAL GROSS SALES | 243,879 | 195,817 | 1,751,997 | 1,525,000 | 1,308,444 |
| 6 | | | | | | |
| 7 | DIRECT EXPENSES | | | | | |
| 8 | Total Labor Costs (Line 47) | 48,841 | 41,093 | 382,292 | 345,800 | 302,394 |
| 9 | Laundry and Dry Cleaning | 6,018 | 4,556 | 36,875 | 28,700 | 22,438 |
| 10 | Uniforms | 1,158 | 2,961 | 4,418 | 4,400 | 3,404 |
| 11 | Linens and Blankets | 2,289 | 1,472 | 14,494 | 15,200 | 9,920 |
| 12 | Cleaning and Cleaning Supplies | 1,448 | 2,622 | 7,778 | 7,300 | 6,812 |
| 13 | Guest Room Supplies | 4,946 | 12,277 | 35,425 | 32,900 | 29,850 |
| 14 | Printing and Stationery | 724 | 1,246 | 1,787 | 1,600 | 1,936 |
| 15 | Telephone, Telegrams and Postage | 134 | 180 | 1,070 | 4,000 | 1,572 |
| 16 | Travel Agents' Commissions | 13,306 | 18,448 | 77,784 | 70,000 | 113,635 |
| 17 | Reservations | 4,888 | 768 | 30,850 | 26,600 | 5,193 |
| 18 | | | | | | |
| 19 | | | | | | |
| 20 | | | | | | |
| 21 | | | | | | |
| 22 | Miscellaneous | 2,171 | 4,328 | 10,508 | 9,700 | 7,676 |
| 23 | TOTAL DIRECT EXPENSES | 85,923 | 89,951 | 603,281 | 546,200 | 504,830 |
| 24 | ROOMS DEPT. GROSS PROFIT | 157,956 | 105,866 | 1,148,716 | 978,800 | 803,614 |
| 25 | % of Gross Profit to Gross Sales | 64.8 | 54.1 | 65.6 | 64.2 | 61.4 |
| 26 | STATISTICS | | | | | |
| 27 | % of Occupancy | 95.4 | 77.5 | 87.2 | 77.8 | 68.0 |
| 28 | % of Double Occupancy | | | | | |
| 29 | Average Transient Rate | 16.68 | 16.39 | 16.68 | 16.24 | 15.86 |
| 30 | Average House Rate | | | | | |
| 31 | LABOR ANALYSIS | | | | | |
| 32 | Assistant Manager & Front Office | 5,098 | 3,737 | 35,112 | 35,652 | 25,627 |
| 33 | Reservations | 2,128 | 2,038 | 12,768 | 14,855 | 15,376 |
| 34 | Housekeeping | 24,609 | 22,428 | 207,484 | 190,144 | 161,452 |
| 35 | Service | 7,645 | 5,770 | 63,483 | 56,449 | 53,818 |
| 36 | | | | | | |
| 37 | | | | | | |
| 38 | | | | | | |
| 39 | | | | | | |
| 40 | Miscellaneous | | | | | |
| 41 | Vacations | 2,540 | 1,836 | 13,625 | 12,600 | 8,915 |
| 42 | Recovered Salaries | | | | | |
| 43 | Total Salaries and Wages | 42,020 | 35,809 | 332,472 | 309,700 | 265,188 |
| 44 | Payroll Taxes & Employee Relations | 6,753 | 5,231 | 49,360 | 35,400 | 36,774 |
| 45 | Employee Meals and Rooms | 68 | 53 | 460 | 700 | 432 |
| 46 | Contract Labor | | | | | |
| 47 | TOTAL LABOR COSTS | 48,841 | 41,093 | 382,292 | 345,800 | 302,394 |
| 48 | % Total Salaries & Wage to Sales | | | | | |
| 49 | % of Total Labor Cost to Sales | | | | | |
| | ROOMS OCCUPIED | 14,476 | 11,948 | 105,116 | 93,900 | 82,489 |

HOTEL ______ XYZ ______ 8 MONTHS ENDED ______ August ______

**FIGURE 31-3** *Rooms Department Monthly Operating Statement*

*On the Monthly Operating Statement in Item No. 34, Housekeeping Labor Analysis–all Housekeeping activities are figured in, including fringe benefits, Sewing Room labor and expense (though much of its activity may have been for Food and Beverage), Uniform Control (though uniforms were distributed to all departments). Reason for Housekeeping's being concerned with this figure and No. 9, 10, 11, 12 and 13 is to compare costs for each year. If there is a notable discrepancy, that area can be quickly analyzed.*

inspections that the director of housekeeping or the assistants do all during the year, other items will be discovered that are not wearing so well as they should, or are wearing better than anticipated.

In addition to maintaining a complete physical inventory of all materials by age, condition, purchase price, and established cost to repair or replace, some Housekeepers who have a wide variety of styles and qualities in guest rooms may recommend changes that are a matter of opinion, depending on the taste of the guest. Thus the housekeeper's budget presentation will be strongly reinforced by detailed room-classification records. Class A rooms require no capital expenditure. Class B rooms need partial refurbishing—perhaps bedspreads, lamps, chairs, carpets, or draperies. Class C rooms need major refurbishing.

Management can sample rooms in each class and very quickly determine the validity of the housekeeper's judgment and recommendation. This determination will be further influenced by the effectiveness of her presentation of proposed color schemes, changes in furniture and furnishings as indicated by sample pieces, photos, fabric swatches, together with competitive bids wherever possible.

## *Requests for Monies*

One of the first considerations in preparing the budget is to be absolutely honest in the requests for monies. If management ever gets the idea that twice as much as was really needed was budgeted because only 50 percent of the amount was expected, then that is exactly what will happen in the future. This is a very unprofessional approach.

If one uses the priority system as shown in the sample budget, then there is no need to *overbudget.* Management has been shown the exact situation.

In most institutions, Housekeeping is responsible for furnishings throughout the entire facility. The housekeeper therefore must (1) maintain liaison with heads of other departments; (2) make a physical inspection of all facilities and equipment for which Housekeeping is responsible; (3) confer with other department heads to obtain agreement on proposed expenditures to be incorporated in the Housekeeping or another department's budget; (4) keep abreast of design and equipment developments pertinent to the needs of these other department heads; and (5) establish contact with suppliers.

Housekeeping must above all be able to maintain regular liaison with Engineering and Maintenance Departments concerning matters of mutual interest. These would include all repairs from new tile, equipment and caulking for bathrooms, and painting, to replacement of wallpaper, refinishing wood surfaces, and so on.

Certain support positions are traditionally included in the Housekeeping Department: (1) the Sewing Room, which does mending and sewing for all departments; (2) Linen Control, which in a hotel provides substantial services to restaurant and catering departments; (3) Uniform Control, which issues and cares for all uniforms, regardless of department; and (4) Property Maintenance, which includes not only public areas and rooms, but back-of-the-house areas, too.

Once the proposed annual budget for Direct Expenses and Payroll has been approved, it is a primary function of the head of the department to make sure Housekeeping operates within its limitations. The budget must be reviewed with management at least semiannually, almost invariably no less than quarterly, and quite often monthly (Figure 31-4).

Since perfection in forecasting is impossible, it might very well be construed as only a minor matter if as many as half of the accounts have been overspent at any given time—*if* the other half of the accounts have been underspent by an equal amount. If the total spent exceeds the total budgeted by only a small margin, management will generally prove tolerant during prosperous times, or when the department head provides satisfactory evidence that the deficit will be wiped out before the end of the year. Otherwise management may request that the budget be revised downward to assure elimination of the deficit.

When this happens an immediate conference should be scheduled with all assistants reporting directly to the head Housekeeper, as well as with

| | | CURRENT MONTH | | YEAR-TO-DATE | | |
|---|---|---|---|---|---|---|
| | | THIS YEAR 1981 | LAST YEAR 1980 | THIS YEAR 1981 | BUDGET | LAST YEAR 1980 |
| 1 | GROSS SALES | | | | | |
| 2 | Permanent Rooms | | | | | |
| 3 | Transient Rooms | | | | | |
| 4 | | | | | | |
| 5 | TOTAL GROSS SALES | | | | | |
| 6 | | | | | | |
| 7 | DIRECT EXPENSES | | | | | |
| 8 | Total Labor Costs (Line 47) | 5.50 | 5.16 | 5.46 | 5.52 | 5.50 |
| 9 | Laundry and Dry Cleaning | .63 | .57 | .52 | .45 | .40 |
| 10 | Uniforms | .12 | .37 | .06 | .06 | .06 |
| 11 | Linens and Blankets | .24 | .18 | .21 | .24 | .18 |
| 12 | Cleaning and Cleaning Supplies | .15 | .33 | .10 | .12 | .12 |
| 13 | Guest Room Supplies | .51 | 1.54 | .51 | .52 | .54 |
| 14 | Printing and Stationery | .07 | .15 | .03 | .02 | .03 |
| 15 | Telephone, Telegrams and Postage | .02 | .03 | .02 | .06 | .03 |
| 16 | Travel Agents' Commissions | 1.38 | 2.21 | 1.11 | 1.11 | 2.07 |
| 17 | Reservation Service | .51 | .09 | .43 | .42 | .11 |
| 18 | | | | | | |
| 19 | | | | | | |
| 20 | | | | | | |
| 21 | | | | | | |
| 22 | Miscellaneous | .23 | .52 | .15 | .15 | .15 |
| 23 | TOTAL DIRECT EXPENSES | 8.91 | 11.15 | 8.60 | 8.67 | 9.19 |
| 24 | ROOMS DEPT. GROSS PROFIT | 16.36 | 13.29 | 16.34 | 15.63 | 14.61 |
| 25 | % of Gross Profit to Gross Sales | | | | | |
| 26 | STATISTICS | | | | | |
| 27 | % of Occupancy | | | | | |
| 28 | % of Double Occupancy | | | | | |
| 29 | Average Transient Rate | | | | | |
| 30 | Average House Rate | | | | | |
| 31 | LABOR ANALYSIS | | | | | |
| 32 | Assistant Manager & Front Office | .52 | .46 | .49 | .57 | .46 |
| 33 | Reservations | .22 | .25 | .18 | .24 | .27 |
| 34 | Housekeeping | 2.55 | 2.82 | 2.97 | 3.03 | 2.92 |
| 35 | Service | .79 | .72 | .90 | .90 | .97 |
| 36 | Property Maintenance | | | | | |
| 37 | | | | | | |
| 38 | | | | | | |
| 39 | | | | | | |
| 40 | Miscellaneous | .26 | .24 | .19 | .19 | .15 |
| 41 | Vacations | .26 | .24 | .19 | .19 | .15 |
| 42 | Recovered Salaries | | | | | |
| 43 | Total Salaries and Wages | 4.35 | 4.50 | 4.74 | 4.93 | 4.81 |
| 44 | Payroll Taxes & Employee Relations | .70 | .66 | .70 | .49 | .67 |
| 45 | Employee Meals and Rooms | .00 | .00 | .02 | .02 | .02 |
| 46 | Contract Labor | | | | | |
| 47 | TOTAL LABOR COSTS | 5.05 | 5.16 | 5.46 | 5.54 | 5.50 |
| 48 | % of Total Salaries & Wages to Sales | | | | | |
| 49 | % of Total Labor Cost to Sales | | | | | |
| | ROOMS OCCUPIED | 14,476 | 11,948 | 105,116 | 93,900 | 82,489 |

HOTEL ____XYZ____ ____8 MONTHS ENDED ____August____ 1981

**FIGURE 31-4** *Rooms Department Cost Per Occupied Room*

*Costs per occupied room enables professional Housekeeper to see areas where great discrepancies exist from month to month and to pinpoint problems where economy may be possible.*

all other supervisors. (Such a conference should, of course, be called semiannually even when all budget targets are being met.) The primary purpose of this meeting is to find realistic ways of eliminating the deficits on overspent accounts. The secondary purpose is to explore the possibility of achieving still greater efficiency where the controlling account is on target or even underspent.

## *Labor and Cost Analysis*

The first question to ask is whether every worker is performing up to par. If not, is the problem due to poor morale in the department, or being caused by a small group of individuals? Can it be traced to poor training or inadequate supervision? Or are the sub-par workers incompetent as a result of poor selection procedures?

Is the Housekeeping Department performing only necessary tasks, or could some of the tasks be eliminated altogether or reduced in frequency? Is the department overstaffed, resulting in each worker's job involving a slightly lower workload than the employee is capable of performing? Is the marketplace checked regularly for improved supplies and equipment to enable the workers to increase their efficiency? Does every worker know exactly what his job is, and does that job really represent a full day's work? Can one or more workers be replaced by machines?

At each of these semiannual meetings, every job and every employee should be carefully evaluated. In some cases, perhaps, jobs can be restructured to achieve greater compatibility between the work requirement and the employee's skills and temperament. Conversely, instead of revising the job to fit the worker, perhaps the worker can be shifted to another more suitable job.

Consideration should also be given to the possibility that certain assignments can be accomplished with no loss of effectiveness by less skilled and therefore lower-paid employees. At the same time, the possibility should not be overlooked that the most economical way to accomplish an assignment may actually call for employment of more highly skilled and more highly paid employees.

All such meetings should, of course, include a thorough review of the wage structure in general, as well as of compensation of certain individual employees. None of this, obviously, can be accomplished intelligently unless the head of the department is thoroughly familiar with the terms of applicable union contracts. He or she should know when a lesser-paid employee can be hired instead of one more highly paid.

Occasionally, Housekeeping may be able to achieve a major economy by completely abolishing a traditional function of the department. Hilton Hotels, for example, found that guests no longer expect maids to turn their beds down at night, and thus eliminated this service unless specifically requested. The exception is in the case of studio rooms, where the chain finds it safer and more economical to have the maids convert the sofas into beds rather than expose them to undue wear and tear that might result from exasperated guests using excessive force to accomplish the mystifying conversion.

Larger institutions often have specialized personnel capable of helping to modernize the facility at very modest expense. Staff painters can make an important contribution by painting interiors and furniture, as well as public areas, in colors more appropriate to the late twentieth century than to the nineteenth. An upholsterer can often modernize furniture by replacing venerable coverings with fresh and modern fabrics. The staff carpenter can easily remove antiquated legs from chests of drawers and dressers, reframe pictures, build new lamp bases. Worn tops can be covered or replaced by plastic laminates that are immune to liquor stains and cigarette burns, never need to be refinished, and reduce cleaning and polishing time almost to zero.

The semiannual meeting should not, however, be looked upon as nothing more than an occasion for cutting costs and services. It should, above all, be approached as an opportunity to find new ways to maintain and improve housekeeping services within realistic budget limitations of constantly upgrading the image of the department and the institutions by a never-ending search for increased budgetary efficiencies.

## *Forecasting*

In addition to the annual budget, the housekeeper will often be asked to prepare a long-range forecast, which is in effect an estimated budget for as little as two years or as long as ten years. Since it is reasonable to assume only minor long-range variations in consumption of supplies or the department's manpower requirement, the forecast is primarily intended to determine when large sums of money will have to be spent for furniture, furnishings, and equipment. This can be gauged from a careful analysis of (1) the age and price of those current items as shown on purchase orders or special records, and (2) the life expectancy of each component as indicated by replacement records. In the case of newer institutions, the forecast will be based in part on information gathered from furnishings suppliers, which will in turn be based both on the replacement records of other institutions, and on the past experience of the department head.

While the long-range forecast is primarily a tool of management, its importance to Housekeeping is great since it constantly rivets the attention of the department on a program of permanent and continuous upgrading (Figure 31-5). Without regular attention, gradual deterioration in the facility may be overlooked until shabby gentility becomes glaringly apparent. At that point Housekeeping often has to launch a crash remodeling program of such proportions that normal operation of the facility has to be seriously interrupted over an extended period of time.

Most high-rise facilities repeat the same floor plan on all guest room levels so that each vertical tier (e.g., rooms 1206, 1306, 1406) will be comprised of identical rooms, uniformly decorated and furnished. Thus ideally, refurbishing should be approached as a vertical, not horizontal project.

This offers important advantages. It is obviously more economical to purchase identical furniture, furnishings, and carpet for an entire tier of rooms than for a single room. But in a hotel, Front Office personnel can make room assignments more intelligently when they know exactly what type of room they are assigning to each guest. This becomes virtually impossible if each room in a property is individually rather than uniformly decorated and furnished. It is therefore vital that the long-range forecast indicate which tiers of rooms will be refurbished in each of the several years covered by the projection.

In a small operation the matter of a budget may be very simple, especially if there are no large food and beverage areas and the labor force is small. However, in order to have efficient management it is essential that some type of budget be prepared every year, so the owner or general manager will have an idea of what the expenses of the Housekeeping Department will be.

## *Opportunities in the Housekeeping Profession*

Since capable housekeepers are in such short supply, pirating is inevitable—one hotel luring the housekeeper away from another with promises of higher pay and more prestige. Pirating is also responsible for the high turnover among management pirates themselves. Professional Housekeepers turnover almost as frequently as hourly employees, and will probably continue to do so as long as the demand far exceeds the supply.

But pirating need not be thought of as a villainous activity. Actually, in a very real sense it provides advancement opportunities to industry personnel who, despite outstanding qualifications, might otherwise be deprived of such opportunities because their present organizations may only rarely have openings in their middle and top management ranks.

According to the AH&MA Educational Institute, there is a very definite need "for lodging supervisors and managers, qualified not only by experience but also by education—a professional knowledge of hotel/motel operations must include not only an awareness of the technical mechanics of operations, but a training exposure to communications, human relations, teaching others, and leading people.

"Educated estimates indicate a conservative need for 100,000 employees new to the industry,

ROOM ______________ INSPECTOR ______________ DATE ______________

| | OK | NEEDS ATTENTION | MISSING | REMARKS |
|---|---|---|---|---|
| **ALCOVE HEADER & LIGHT** | | | | |
| Diffuser glass | | | | |
| Door wood & framing finish | | | | |
| **ENTRY** | | | | |
| Marble threshold | | | | |
| Door closure | | | | |
| Lock/Night latch | | | | |
| Door seal gasket | | | | |
| Door sweep | | | | |
| **BAR UNIT** | | | | |
| Finish on bar case | | | | |
| Motor noise | | | | |
| Finish on shelves over unit | | | | |
| **CEILING FIXTURE** | | | | |
| Appearance | | | | |
| Wall switch (lighted) | | | | |
| **BATHROOM** | | | | |
| Door finish | | | | |
| Wall switch (lighted) | | | | |
| Aluminum wall unit | | | | |
| Soap dish | | | | |
| Bottle opener | | | | |
| Panel under sink | | | | |
| Lav. pressure & drain flow | | | | |
| Towel shelf and bar | | | | |
| Sliding panels on bath | | | | |
| Clothesline unit | | | | |
| Wallpaper | | | | |
| Tile | | | | |
| Ceiling (above bath) | | | | |
| Faucet strainer | | | | |
| **BEDROOM** | | | | |
| Dresser drawer operation | | | | |
| Bed frame noise | | | | |
| Radio operation | | | | |
| Electric clock | | | | |
| Message light operation | | | | |
| Elec. blanket control (lighted) | | | | |
| TV operation | | | | |
| Club/or love seat chair arms | | | | |
| Game table | | | | |
| Drapery operation | | | | |

**FIGURE 31-5** *Guest Room Inspection Record*

*Guest Room Inspection Record is used twice a year by the executive Housekeeper, or an assistant to ascertain just what repair budget is going to be needed the next year.*

each year, just to satisfy requirements of existing properties in the United States alone.

"Add to this number, the needs of the huge numbers of new properties and new rooms under construction now—some 1100 establishments averaging over 200 rooms each. This new construction represents more than 200,000 new jobs.

"Perhaps 10 percent—some 30,000—of the total requirement is for people with supervisory ability and a knowledge of lodging operations," go AH&MA predictions.

Opportunities abound—the profession needs more competent people and will continue for many years to have openings for supervisory personnel. There is little reason, however, to suppose that service departments such as Housekeeping and Maintenance will always be so difficult to staff with satisfied employees, since the growth of the United States economy is increasingly due to expansion of service industries rather than manufacturing.

Trammell Crow, a Texas millionaire who has invested heavily in a new chain of hotels, holds out rosy hope for the future of service-oriented industries. "Right now," he explains "we can only hope to motivate employees a little at a time, change their attitudes gradually and convince them that it is noble to serve others. There won't be any sudden breakthrough, but 20 years from now there will be a need to find new ways to fill one's life because *man will have few occupations available to him.* Top plateaus of production will have been reached and man will find that *service* can be an honorable profession."

Many people are Housekeeping Department heads today despite having had no special training. They perhaps worked in a hotel or college in some other capacity, had astute business sense, and an interest in the property. Managers often would rather hire someone they know is reliable and who will learn housekeeping, than take a chance on a stranger. As Housekeeping becomes increasingly more professional and qualified persons become more plentiful, managers will doubtless show a shift in this attitude. The following information appeared in *Executive Housekeeper* magazine, July 1978. The salaries have been increased 12 percent and the room rates about 20 percent (Table 31-1).

Economy rate hotels and motels are those with best single-occupancy rooms renting for less than $30 per night in or near a city, and for less than $25 per night on the road.

Moderate rate hotels and motels are those with best single-occupancy rooms renting from $30 to $50 per night in or near a city, and from $25 to $35 per night on the road.

Luxury rate hotels and motels are those with best single-occupancy rooms renting for over

**TABLE 31-1 Executive Housekeepers in the Lodging Field**

**A. National averages**

| | **Low** | **Average ($) Salary** | **High** | **Number of Respondents** |
|---|---|---|---|---|
| Men and women | 6940 | 13,730 | 28,000 | 32 |

Note: There were 3 men and 29 women respondents.

**B. Guest room classification and salary ranges**

| | **Low** | **Average ($) Salary** | **High** | **Number of Respondents** | **Percentage of Respondents** |
|---|---|---|---|---|---|
| Economy rate | 8400 | 10,060 | 11,870 | 5 | 15.63 |
| Moderate rate | 6940 | 12,620 | 24,920 | 15 | 46.88 |
| Luxury rate | 9320 | 16,920 | 28,000 | 12 | 37.50 |

Note: Figures are combined for men and women.

$50 per night in or near a city, and over $35 per night on the road.

These price demarcations were established solely for the purposes of this survey and are by no means a rigid definition of economy, moderate, and luxury establishments.

## *Experience*

Miss Florence Cunningham, former corporate director of housekeeping of Sheraton Corp., feels that management must immediately begin to understand Housekeeping and plan for its future. The basis for her concern was the fact that about 90 percent of the professionals in the field today are in their fifties; consequently, the number of available replacements at any given time will be far from adequate, unless remedial action is begun now.

"Experience is the best means of training," she explains, "and I would like to see more cross-pollination of ideas, more exposure to different housekeeping situations, more crossing over between residence halls, small motels, large clubs, and giant hotels in order to give the housekeeper of tomorrow an idea of how various maintenance and cleaning situations are handled, and how one can best motivate employees."

## *Realistic Training Programs Do Happen*

What actually has been and is being done today to help prepare employees for that promotion from within? This is a situation every organization would implement if only there could be adequate replacements available when needed in their own institutions.

An excellent example of what can be accomplished was revealed in Denver when a special training program called Project CHRIE (Career Housekeeping and Related Institutional Education) took on the specific problem of developing shy, insecure, and/or uneducated employees in Housekeeping and Maintenance Departments. Those who showed an interest in furthering their careers, who wanted to take on more responsibility, and were capable enough to be leaders but needed a gentle push or a chance to practice supervisory skills before being thrown into a leadership situation, were invited to become students. Of the 40 who started, 34 graduated from the three-month course in how to be a supervisor and felt ready for promotion.

The institutions for which they worked paid them while they attended classes (during working hours) and paid additional wages when other employees filled in for them. The program was taught by instructors highly skilled in building student confidence and bringing out latent leadership qualities. After completion of the program, changes in student attitudes were obvious—the graduates expressed and demonstrated an increased sense of worth and ability. The experiment resulted in all 34 graduates being promoted. Seven of them enrolled in additional vocational courses at Denver City College because of increased confidence in their ability to climb the management ladder.

Project CHRIE was funded by the federal government. It is not to be confused with CHRIE, a national group, the Council of Hotel, Restaurant and Institutional Education, which wholeheartedly endorsed the smaller project. The course was directed by Don Hinderleider, a housekeeping consultant who did considerable research before setting it up. Hinderleider's studies revealed some interesting, although contradictory, information. Most housekeepers questioned across the country said that their employees did *not* want a promotion to supervisor. Employees questioned, on the other hand, said they would like to be promoted. (Actually, the housekeepers did not feel the employees were capable of being supervisors and had not explored the subject with them at any length.)

The key to the dilemma was found when housekeepers replied to a questionnaire about what specific training programs employees needed in order to become supervisors. Three areas in which formal training was urgently required were mentioned by almost every respondent: communications, human relations, and how to train others. It was then decided that these subjects would comprise the basic curriculum, and that emphasis on such functional matters as cleaning techniques would be inappropriate and ineffective.

This original project is now being repeated at community colleges around the country, and institutions are willing to pay to have their potential Housekeeping supervisors trained this way.

## *Terminology Needs Improvement*

We in the hospitality industry are aware of the negative connotation of the term Housekeeping and hope that a new name for the Housekeeping Department, such as environmental services or guest services (depending on the nature of the establishment) will eventually find broad acceptance. Titles for housekeepers in the college and university housing field include:

- Assistant housing manager
- Housekeeping supervisor
- Custodial supervisor
- Supervisor
- Supervisor—housekeeping services
- Supervising housekeeper-women
- Janitor supervisor-men
- Coordinator of housing facilities

Often, executive Housekeepers are considered by management as less than fully capable administrators of their departments. Where this is the case, it is usually because the person who became the department head was given very little management training to prepare for the responsibility. Continuous institutional management training and education at all levels is imperative.

In addition to knowing how to make out a budget and read a profit and loss statement, the housekeeper must always continue to increase her understanding of people—the most important resource of the department. Some Housekeeping managers will become so involved in business, so much the administrator, that overseeing the employees is delegated to the assistant housekeepers.

One should never isolate oneself from the mainstream of the Department—the employees and the guests. In addition to being on the floors frequently and conducting meetings, simply answering the phone in the Linen Room several times every day (at different hours when certain types of problems arise) will keep the housekeeper posted on the real world as it affects Housekeeping.

In an effort to be a good administrator, a housekeeper should not become so involved with job specifications, work flow patterns, work simplification procedures, standards of performance, and management by objectives as to forget that these matters are important only insofar as they affect the performance of individuals. It is necessary to use the tools of management—to set up staffing charts, write job specifications, and predict productivity for the staff before the institution opens. However, based on the realities of human performance, these documents, procedures, and policies may all have to be altered when one finds out what really works and what does not.

Sheraton Hotels in North America, in conjunction with its many new openings in the early 1980s, instituted a new amenity program. The upgrade included larger soaps, oversized and luxurious bath linens, alarm clocks, flowers, shampoo, pulsating showerheads, shoe-shine machines, valet stands, and more. This was brought about because studies indicated items that are important to occasional and frequent guests. Instrumental in this survey were their executive Housekeepers, who fine-tuned the program to fit each hotel from a practical standpoint. Yearly reviews of the program take place and are part of one of the corporate Housekeeping seminars.

The Corporation has identified the executive Housekeeping position as one that requires not only technical expertise, but strong people skills, training skills, and dedication. We are putting people through training programs and then into the housekeeping area with the intention of moving them into further management responsibility within two years. A position in Housekeeping is no longer a dead end.

Training has been reemphasized, especially in the area of advanced budget and control skills and personnel skills. The company is averaging two to three North American sessions per year and at least one for each international division.

Quality Audit has taken a new look at Sheraton, according to its assistant director of rooms operations, Jeff McIntyre. "Along with the in-

**FIGURE 31-6** *Sheraton's advanced Housekeeping seminar of December 1979 shows a group of people who are highly motivated and likely to be in the Corporation for some years to come. Many of these people will go through housekeeping areas for two years and then move into further management responsibilities.*

spection process," he reports, "we have a quality Audit session in our training classes that teach how to objectively quality-audit your own hotel. This process is recommended to be used for all professional Housekeepers" (Figure 31-6).

## *Plums for the Achievers*

If one achieves the status of department manager or an even higher position, there are many rewards in store in addition to job satisfaction. According to *Planning and Operating Motels and Motor Hotels,* by G. Podd and J. Lesure, Hayden Book Co., Inc., these plans may include:

**1. Annuities.** These may be paid for entirely by the company, or may be of the mutually contributory type. Payments almost invariably begin at retirement, and are treated for tax purposes as regular income.

**2. Life Insurance.** This may be provided solely by the company or again, be mutually contributory. Usually a term insurance policy that cannot be converted to cash, the benefit is seldom taxed.

**3. Tax-free Compensation.** Here the company agrees either to reimburse the executive for all income taxes paid by him, or it may provide a fixed amount in net income each year. (Valid except in New York state.)

**4. Deferred Income**

a. *The adequate level plan.* Here the executive is paid an adequate salary, but less than his real worth. Payment of the difference, with interest, is usually deferred to at least age 55, and usually to retirement, when the individual's reduced income will assure its being taxed at a lower rate.

b. *Consulting or part-time basis plan.* Under this plan, executives reaching a certain age (usually retirement) receive a percentage (sometimes as much as one-half) of the highest salary level they have achieved in return for their consulting services for a fixed period, which may vary from one year to life.

c. *Profit-sharing plans.* Under these government-approved and government-regulated plans, the company invests a share of annual profits in proportion to the individual's salary, and sometimes seniority, and pays it out upon retirement in a lump sum or in installments. In the event of death, it is usually payable to the individual's estate. This type of plan is rare in the lodging industry.

**5. Other Insurance.** This usually covers travel and hospitalization; other types of insurance, except term life insurance, are taxed as ordinary income if provided. A type used by some organizations is known as "key man" insurance, which actually makes the company the beneficiary, but the company in turn may make a part of the benefit available, either in other insurance or cash payment, to an executive's estate. The individual may pay for these benefits in part. Less frequently, they are paid for in full by the company.

**6. Club and Association Memberships.** A good many companies allow reasonable memberships that are helpful in business to executives.

**7. Health Examinations.** Most companies provide some form of physical checkup (usually annually). This, of course, benefits both company and individual; costs are paid by the company.

**8. Company Transportation.** Some establishments provide company cars for their executives; in the case of the chief executive, perhaps even a chauffeured limousine. When management personnel are transferred, moving costs are paid.

**9. Entertainment and Expense Allowances.** Entertainment and other incidental expenses that benefit business are generally covered. Some companies even go so far as to stipulate a budgeted amount that may be spent. One hotel company allows its executives below rank of managers $150 a month for entertainment as long as such expenditures are for sound business reasons.

**10. Housing and Maintenance.** Complete housing and maintenance are usually limited to certain industries where it is to the company's advantage. Hotels, primarily resorts, sometimes provide complete room and board, either in the hotel proper or in separate buildings.

**11. Vacation-leaves-travel.** There is now a tendency among hotels to provide longer split vacations to executives. Various types of special leaves for periods of time with full or part-time pay, are used by some organizations where such leaves can be granted without affecting the operation.

**12. Stock Options.** Large, publicly held companies sometimes offer stock options to their highest-ranking executives. For example, these grant the individual five years to buy a specified number of shares of company stock at the price prevailing when the agreement is made.

**13. Miscellaneous.** Trade and purchase discounts, executive dining privileges, paid magazine subscriptions, company scholarships, and paid educational programs (either all or part paid by the company) are other benefits.

Now that women are being recognized as effective executives, there will be many opportunities for the housekeepers of tomorrow to use their positions as a springboard to management.

Another bright prospect for the housekeeping world is that the basic idea that giving service is desirable in itself is gaining more acceptance.

The role of Housekeeping manager (and indeed, of room attendant, custodian, or porter) may take on much greater significance as we enter a new era, not of conspicuous consumption, but of conserving resources and recognizing service as a uniquely honorable profession (Figure 31-7).

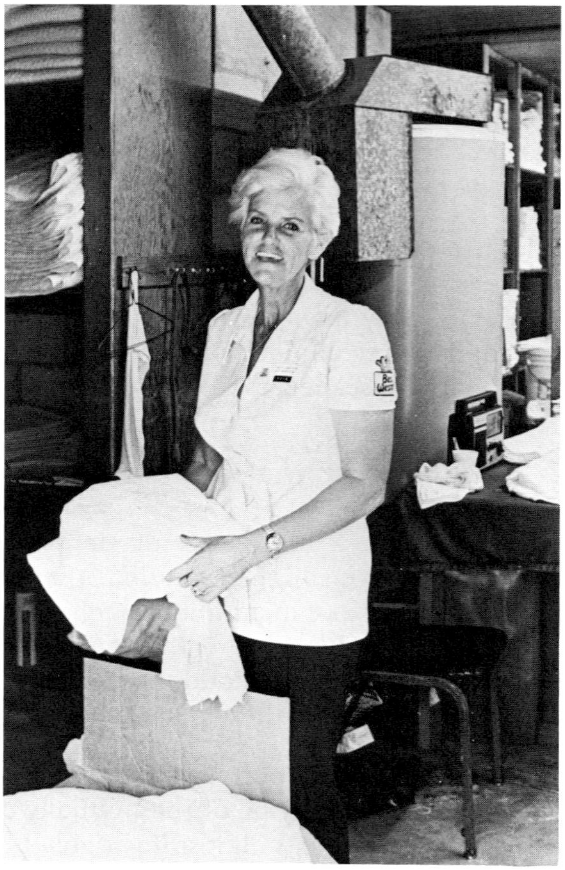

**FIGURE 31-7** *Becoming involved in the community is especially helpful to enhancing the image of Housekeeping. Fay Klein, Housekeeper of the Best Western Key Ambassador Hotel in Key West, Florida, responded to a call to help victims of Hurricane David in the Dominican Republic. She contacted hotels and motels in the southern-most city to ask for contributions of linens, bedding, and personal clothing that was to be discarded. She said the response was overwhelming. Every business she called had something to offer. She delivered an entire truckload of items at the end of the week to the coordinator of the countywide effort in Marathon, Florida.*

## *Moves to Improve the Image*

Mrs. Brian Walker, when director of Housekeeping at the Houston (Texas) Oaks Hotel, suggested the following activities for improving the status and image of Housekeeping and thus attracting more qualified people:

1. Stress professionalism in all education for housekeeping supervisory staffs. Because the role of housekeeper requires a business-like understanding of the operation of the institution—hotel, college, club, or retirement residence—adequate preparation for the position involves several years of formal study and related experience.
2. Set up on-going educational programs in many different departments of the hotel or residence hall to give exposure to all facets of the business.
3. Have personnel extend knowledge voluntarily to be more valuable to management—for example, learn interior design or accounting.
4. Take an active part in the community—in business or cultural activities—as upgrading the housekeeper's image can be quite a boon when recruiting new employees.
5. Create a multi-company pool of replacement personnel. For example, a group of hotels, clubs, and colleges jointly contribute toward a fund for hiring extra trainees, then when an opening appears a qualified person could be available. Many of the trainees would, of course, be present employees who would become qualified for increasingly important positions. Such enlightened self-interest on the part of management would repay its cost in increased efficiency and morale.

Susan Picht, executive Housekeeper of the Westin Hotels, believes that housekeeping is most definitely a young woman's field. Active in the National Executive Housekeeper's Association, she puts great stress on the fact that age should not be a deterrent in welcoming young women to the field. She had two young trainees, just out of college, whose work performance was highly satisfactory, and there was no problem with the staff's accepting these youthful supervisors.

Mrs. Picht, however, raises the question of "What do we have to offer young women after they have become executive housekeepers and have been there for some time? They are not going to be content to be in the same position for the next 30 years. Where are they going in the future? To larger properties? In what capacities?"

Many of them are obviously interested in a future in management, Front Office, Sales, Food and Beverage, Reservations, Resident Manager, Manager. Is it realistic to hope for such opportunities?

On this point Florence Cunningham, now at the Boston Sheraton, makes some pertinent observations. "Male housekeepers," she believes, "seem to learn so much so fast and soon get ambitious, then want to move on. Women find it a diversified and interesting job because it is not confining . . . it requires communication with all other departments and offers the opportunity to perfect a property or even take on the highly responsible job of rehabilitation, involving interior design and the careful selection of all replacement furnishings."

One who speaks up about the questionable image of Housekeeping while doing a superb job is Louise Wheeler, manager of Housekeeping at the Contemporary Hotel in Disney World (Figure 31-8).

"I have very definite ideas on what I think must be done if Housekeeping is to survive as something more than a hodge-podge of the lame, the halt, and the blind. The hotel industry could do several things to raise Housekeeping from the last place on earth where people want to work to reasonably desirable status as a means of making a livelihood. Unless attitudes change, the best of hotels can only dwindle into mediocrity while top management runs around wringing their hands and wondering what happened.

"NEHA can prattle on forever on the matter of certification, but all the certified Housekeepers in kingdom come won't matter at all if they can't get their staffs out from under the everlast-

**FIGURE 31-8** *Louise Wheeler, center, stops by the hotel's logo (at left) with two of her five supervisors (left to right) Dorothy Vann and Maxine Bell. Their hotel has 100 percent occupancy and averages 3.8 people per night, making a heavy burden for Housekeeping. Ms. Wheeler came to open the hotel on a temporary basis. Over five years later she's still enjoying it. She speaks boldly on the need to improve the image of Housekeeping, and bring it to the front of the house.*

ing 'below stairs' image that eats away at any normal person's desire to be a part of it.

"Why, where, and when was Housekeeping banished to this back-of-the-house category when it is they who make or break *all* areas of the house? This has always been a mystery to me.

"I sometimes feel that those members of my staff who are really doing excellent work are almost an 'endangered species' and we must do whatever it takes to keep them and find more like them."

There are other careers besides being an executive Housekeeper that may evolve from experience in that position. Consultant is one. A housekeeping consulting firm could be extremely helpful to administrators and managers who might be building or remodeling but do not have qualified authorities in these areas on their staffs. Architects and interior designers could realize the same benefit when working on projects for which a professional housekeeper has not yet been hired.

Housekeeping consultants could also render invaluable service to manufacturers of textiles, uniforms, laundry equipment, or any other product whose purchase is specified or influenced by the professional housekeeper. A highly qualified consultant not only knows what will work and what will not, but also knows what will appeal to housekeepers. This person would also know what is likely to be of very little interest to housekeepers because of specialized conditions. These reactions would probably rarely be anticipated by the manufacturer who is about to go into mass production of a housekeeping item.

## *Education*

As discussed in Chapter 5, there is little agreement about what type and level of education an executive Housekeeper needs. It depends considerably, of course, on the caliber and size of the particular institution, the extent to which the individual seeks to gain or avoid responsibility, and quite obviously, the individual's ultimate ambitions.

## *Seminars on Housekeeping*

Educational Institute of American
Hotel and Motel Assn.
Attn: Director of Industry Relations
1407 S. Harrison Rd.
East Lansing, Mich. 48823
(517) 353-5500

Science Engineering Assn. Inc.
3630 Peachtree Rd.
Atlanta, Ga. 30319
(404) 261-2050

University of Denver
Beaumont Center for Management
Development
Denver, Colo. 80208
(303) 753-2927

Department of Engineering
University of Wisconsin-Extension
432 North Lake St.
Madison, Wisc. 53706

Housekeeping supervisors on large campuses, which may house from 5000 to 20,000 students, are usually expected to have a bachelor's degree. This requirement does not reflect the complexity of campus housekeeping, but rather the level of education needed to take advantage of the numerous chances for advancement. The housekeeping supervisor may have an opportunity to move up to residence hall manager or to housing director; the more education an individual has, the more likely the promotion is to materialize.

Needless to say, the individual will also feel more comfortable in the college community if he has an education equal to those in the management positions around him. Many persons strongly prefer to work on a campus because the opportunity for more education is not only at hand, but, in some cases, is available at a discount.

As explained in Chapter 5, the National Executive Housekeepers Association (NEHA) has set up educational standards, and anyone wishing to receive NEHA certification must comply with them. Belonging to NEHA and having a certificate is, of course, helpful to the professional housekeeper. Because that association is comprised primarily of hospital housekeepers, however, membership is not essential to those who choose to make their careers in other areas.

## *NEHA-Approved Courses*

Over 100 colleges and universities throughout the United States are now offering courses of study approved by NEHA leading to certified membership (CEH).

Some courses are structured to lead to a bachelor or associate degree, or students may take the 320-hour program that does not necessarily lead to a degree.

Essentially, the position of institutional housekeeping manager is closely related to that of business manager. Thus the requirements for a course in business management, with some subject modifications, will also satisfy NEHA educational requirements. The affiliation of institutions offering approved courses is the result of long-time effort by the association to improve educational opportunities for persons interested in institutional or professional housekeeping as a career.

The association distributes funds, through the Institutional Housekeepers Educational Trust, Inc., to schools for scholarship assistance. The funds are administered according to regulations of the schools.

Additional information about this program may be obtained from National Executive Housekeepers Association, Inc., 414 Second Avenue, Gallipolis, Ohio 45631.

Wherever there is a problem, there is also a solution. If one housekeeping department finds it difficult to deal with something, obviously others do, too. Often when a group of people get together for serious discussions, they find new ways to solve their problems—reason enough for housekeepers to establish organizations that encourage a frequent exchange of ideas.

## *Future—Space-Age in Housekeeping*

As expenses for labor and equipment rise, it is quite possible that lodging establishments—especially those that cater to family trade—may come to a do-it-yourself housekeeping program. This would mean that housekeeping services would be optional at additional cost. Clean linen would be given to guests as they checked in and daily thereafter, so that they could change sheets or towels as they wished; perhaps even electric sweepers would be provided.

Modern technology may be developed to the point where rooms will be heated and cooled so accurately that no blanket is needed, with washable spreads doubling as blankets for guests who insist on them.

For bathroom cleaning, an electronic device may simply "de-charge" the dirt particles and remove them from all tile and toilets.

There surely will be some sort of instant cleaning system for guest rooms. Just as super-high temperatures clean an oven, perhaps developments in microwave or laser will provide a product that melts soil but is otherwise harmless. Or there may be dirt-repellant finishes, just as there are electronic precipitators today. (These are very costly units that attract dust electronically to keep a room dust-free.)

There may be such innovations as single-service sheets on a roll that can be pulled down from head to foot and torn off along perforations. These disposable sheets then would go into an incinerator that would heat water for the institution.

Perforated, single-service pillow slips are another distinct possibility. Sheets and pillowcases as soft as a sumptuous paper dinner napkin would feel luxurious indeed.

We urgently need to cut the amount of labor needed for current cleaning techniques, because Americans are increasingly unwilling to perform hard or tedious work. The lodging industry is lagging behind many others in making use of electronics and streamlined procedures.

Ambitious, intelligent people with business, engineering, psychology, or home economics backgrounds have a chance to move up rapidly in Housekeeping and make a reputation for themselves. There is an urgent need for builders of better mousetraps and for people-oriented managers who are highly skilled in human relations techniques—especially those related to motivation. As academically trained specialists enter the field to take charge of highly sophisticated new housekeeping systems, it is inevitable that Housekeeping will earn recognition as an indispensable, respected profession within the growing, gregarious lodging industry.

## *Suggested Reading*

From the Educational
Institute of American
Hotel/Motel Assn.

*Check In-Check Out,* by Jerome J. Vallen, Dean, College of Hotel Administration, University of Nevada, Las Vegas (1974).

*The Laws of Innkeepers,* by John H. Sherry, Professor of Law, Cornell University, Ithaca, N.Y. (1972).

*Modern Hotel-Motel Management* (3rd Edition), by Gerald W. Lattin, Dean, School of Hotel, Food and Travel Services, Florida International University (1964).

*The Art & Science of Managing Hotels, Restaurants-Institutions,* by Jerome Vallen (1978).

*Readings on Managing Hotels-Restaurants-Institutions,* by Dunnovan L. Sapienza, James R. Abbey, Jerome J. Vallen; College of Hotel Administration, University of Nevada, Las Vegas (1977).

From CBI Publishing

*Science of Housekeeping,* by Georgina Tucker and Madelin Schneider, (1973), third printing (1977).

From Whitney Library of Design,
1515 Broadway New York 10036

*Specifications for Commercial Interiors . . . Professional Liabilities, Regulations and Performance Criteria,* by S. C. Rezinoff

# *Glossary*

**Adjoining Rooms:** Two or more rooms side by side with or without a connecting door between them. In other words, rooms can be adjoining without being connecting.

**AH&MA:** American Hotel and Motel Association. In 1980 it was 70 years old and had a membership of over one million rooms.

**A la Carte:** Meal where each item on the menu is priced separately (run of the menu).

**AP (American Plan):** Rate includes three full meals and room; (full board or full pension).

**Back-of-the-House:** Service areas not exposed to the public.

**Back-to-Back:** Heavy check-out and check-in on the same day, so that as soon as a room is made up a new guest checks in it.

**Banquet Houseman:** Does set-ups and cleaning.

**Bath Attendant:** Maid or houseman who cleans only bathrooms.

**Beepers—One-Way:** One-way communication device—message can be given but not answered.

**Booking:** Reservations that are made for rooms or parties.

**Bridge the Beds:** Put two twin mattresses crosswise over the box springs to form a king-sized bed.

**Brush Up—BUP:** Tidying a room after a guest has checked out (beds having been done earlier) and replacing bathroom supplies.

**Buckets:** Where the Front Desk Cashier puts guest folios.

**Building Superintendent (sometimes called Chief Engineer):** Person in charge of the maintenance of the building; he may or may not belong to any particular department.

**Busboy or Busman:** Person who clears the table in a restaurant after each course.

**Cabana:** A room adjacent to pool area, with or without sleeping facilities, usually separate from hotel's main building.

**Case Goods:** Furniture made of wood with storage areas; term now includes most furniture except beds.

**Cash Bar:** Private room bar set-up where guests pay for drinks.

**Census:** Term used for hospital occupancy.

**Check-In:** Hotel day starts at 6:00 AM; however, occupancy by arriving guests may not be possible until after the established check-out time (usually 1:00 PM).

**Check Out:** (verb) To vacate a hotel room, (taking luggage), turn in key, and pay bill; (noun, hyphenated) a room that a guest has officially vacated. Check-out time is usually 1:00 PM.

**Chief Engineer:** Person in charge of the Engineering Department; he maintains a crew of plumbers, electricians, and other specialized personnel.

**Closed Dates:** Dates on which no rooms can be rented because of a full house.

**Commercial Rate:** Rate agreed on by company and hotel for all individual room reservations.

**Comp:** Complimentary, no charges for room or service(s).

**Confirmed Reservation:** An oral or written confirmation by hotel that a reservation has been accepted (written confirmations are preferred).

There is usually a 6:00 pm (local time) check-in deadline. If guest arrives after 6:00 pm and hotel is filled, assistant manager will make every effort to secure accommodations in another hotel. (Does not apply to guests with confirmed reservations where late arrival has been specified.)

**Connecting Rooms:** Two or more rooms with private connecting doors permitting access between them without going into the corridor.

**Continental Breakfast:** Consists of juice, toast, roll or sweet roll; coffee (tea or milk). (In some countries: coffee and roll only.)

**Continental Plan:** Rate includes breakfast and room. Commonly called bed and breakfast.

**Contract Furnishings:** Furnishings designed for institutional use as opposed to home use; purchased on a contract that includes specifications, delivery time, and so on.

**Covers:** Number of persons served at a food function.

**Crinkle Sheet:** Distinctively woven sheet (resembling seersucker) used to cover and protect blanket.

**Cut-Off Date:** Designated day when buyer (upon request) must release or add to function room or bedroom commitment. On certain types of groups, rooming lists should be sent to the hotel at least two weeks prior to arrival.

**Day Rate:** Usually one-half regular rate of room, for use by guest during a given day up to 5:00 PM. Sometimes called a use rate.

**Deluxe:** Finest type hotel (private bath and full service).

**Demi-Pension:** Rate includes breakfast and lunch or dinner and room.

**Deposit Reservation:** A reservation for which hotel has received cash payment for at least first night's lodging in advance and is obligated to hold the room regardless of guest's arrival time. Hotel should preregister this type of guest. Cancellation procedure: this type of reservation should be cancelled as early as possible but a minimum of 48 hours prior to scheduled date of arrival in a commercial hotel. For resort hotels, customer should verify cancellation policy at time of making reservation.

**DNP:** Do Not Post (usually found on events sheet) means do not show on event board.

**Double:** Room with a double or queen-size bed.

**Double-Double:** Room with two double beds.

**Duplex:** A two-story suite (parlor and bedroom[s] connected by a stairway).

**Efficiency:** An accommodation containing some type of kitchen facility.

**Entry-Level Jobs:** Jobs one can start without having prior experience, and for which the least amount of training and instruction is needed.

**EP (European Plan):** No meals included in room rate.

**Executive Chef:** Skilled and experienced cook who manages all kitchen employees and activities; *chef de cuisine* is European term for this position.

**Extend the Beds:** Add extensions to make beds longer.

**Farm-Out:** Sending guests who have reservations that cannot be honored to other hotels with vacancies. This is done only when there are no rooms available even though guests have reservations.

**Felts:** Table pads; mattress covers.

**First Class:** Medium range (usually has private or semiprivate bath).

**Flat Rate:** Specific room rate for group, agreed on by hotel and group in advance.

**Flush Valve:** Handle and hardware used to flush toilet.

**Food and Beverage Cashier:** Person usually stationed in kitchen near door to dining area, who adds up restaurant checks, accepts money, and makes change for waiters and waitresses.

**French Service:** Each food item individually served on a plate at table by waiter, as opposed to serving a plate that has been completely set up in the kitchen.

**"Front":** Word used by Front Desk personnel to summon a bellman.

**Front Desk:** Area where guest checks into hotel, keys are kept, and mail is distributed, and from which information is dispensed.

**Front Office:** Area where information regarding guests is kept; also assistant manager's offices.

**Front Office Cashier:** Person who adds up all charges made to a room and collects money upon departure of guests who occupied it.

**Front Office Clerk:** Person who checks in guests and keeps track of rooms available.

**Front-of-the-House:** Entire public area.

**Full Bar Set-Up:** Completely stocked bar with liquor, various mixers, and garnishes for drinks.

**Full Comp:** No charges made for room, meals taken in hotel, telephone, valet, or any items.

**General Clean:** Indicates thorough cleaning of guest room and bath; done on a periodic basis.

**Guarantee:** Figure given by function planner to hotel, at least 24 hours prior to function, stating number of persons to be served. Most hotels are prepared to serve at least 5 percent over the guaranteed figure. Payment is made on a basis of the guaranteed number of covers or total number served, whichever is greater.

**Guaranteed Payment Reservation:** Room set aside by hotel at request of the customer in advance of guest's arrival. Payment for room is guaranteed and will be paid by company or organization even though guest may not arrive—unless cancellation procedure in effect in each property is adhered to.

**Guaranteed Reservation:** A confirmed reservation with a promise to accommodate, or if unable, to pay for a room elsewhere, including transportation there and back to that hotel next day. Guest guarantees to pay if a no-show.

**Guest Charge:** Anything put on guest's bill—purchases, room service, telephone, valet, or whatever.

**Held Luggage:** Guest's property held in lieu of payment for accommodations.

**Hospitality:** A room used for entertaining; usually a function room or parlor.

**Hospitality Suite:** A parlor with connecting bedroom(s) to be used for entertaining.

**Host Bar:** Private room bar set-up where drinks are prepaid by a sponsor (known also as a sponsored or open bar).

**Hotel Garni:** A hotel without dining facilities (except breakfast).

**House Count:** Number of rooms rented on a particular date.

**Houseman:** Employee who works for Housekeeping or Maintenance Departments; does heavy cleaning jobs, lifting, and moving.

**Housing Bureau:** The local convention bureau will (for certain convention groups) act as a housing bureau and assign rooms in various participating hotels in city or area. Sometimes a hotel clearing house is a semiofficial organization that assigns rooms in hotels for individuals and/or groups that are seeking accommodations in a certain area.

**IFB (Including Fringe Benefits):** Meals, insurance, vacation pay, pension fund, uniforms, and so on.

**Inspected:** Room has been thoroughly checked by an inspectress, supervisor or housekeeper.

**Junior Suite:** A large room with a partition separating the bedroom furnishings from the sitting area.

**King:** Largest size bed available from manufacturers; may be 80 inches by 80 inches or 72 inches by 72 inches; may be formed by putting two twin mattresses crosswise on twin box springs.

**Lanai:** A room overlooking water or a garden, with a balcony or patio. (Found in resort hotels mainly.)

**Make Up:** Change linen on beds, clean room and bathroom while guest is registered in room.

**MAP (Modified American Plan):** Rate includes breakfast, dinner, and room.

**MIP:** Most important person.

**MTD:** Month to date—these are the accounting totals showing the revenues and expenditures for a specific month as of a specific date.

**Napery:** Table linens.

**NEHA:** National Executive Housekeeper's Association.

**Night Auditor:** Person who balances hotel accounts and posts all guest charges on bills.

**Night Spread:** Lightweight spread used on bed at night to protect blanket and give each guest a clean cover.

**No-Show Employees:** Personnel who do not come to work on a scheduled work day and do not call in to explain their absence.

**No-Show Guests:** Guest with confirmed reservation who does not check into hotel or call to cancel.

**Occupancy:** Number of rooms actually in use.

**Open:** Rooms available for renting.

**Operation:** Functioning of a hotel, especially activities, dealing directly with serving guests.

**Out of Order:** Status of a guest room not rentable because it is being repaired or redecorated.

**Oversold (also Overbooked):** Reservations have been accepted beyond a hotel's capacity to provide rooms.

**Paid Bar:** Private room bar set-up where all drinks are prepaid. Tickets for drinks are sometimes used.

**Par:** Number of sets of linen needed per bed or sets of towels per guest in use, in laundry in linen room.

**Parlor:** A living or sitting room not used as a bedroom (called a salon in some parts of Europe).

**Parlor Maid:** Maid who services suites and public areas.

**Plant:** The entire hotel operation.

**Pocket Pager—1 Way:** A communication device that permits one to receive messages but not answer them.

**Powder Room Maid:** Woman who cleans public area ladies' rooms.

**Preregistered:** No delay check-in, usually provided guests who have stayed in hotel previously; often room assignments based on guests' previous preferences.

**Processing:** Filling out various insurance and tax forms—done immediately after an employee is hired.

**Property:** A hotel's building, land, and all facilities connected with it.

**Property Maintenance:** Department in very large hotels, often a division of Housekeeping, that does heavy cleaning in front and back of house, usually includes night cleaning crew; may be in charge of keeping up exteriors and grounds.

**Queen:** Middle-sized bed—larger than a double, smaller than a king; dimensions 60 inches by 80 inches or 60 inches by 72 inches.

**Rack Rate:** Current rate charged for each accommodation as established by hotel's management.

**Rebate:** Part or all of rental refunded to guest.

**Reception Hostess:** Woman who greets and registers guests.

**Rehabilitation:** Retraining a person to do things correctly; helping him to adjust to society, or providing work opportunity to a person with a prison record.

**Roll In:** Put rollaway bed in guest room.

**Roll Out:** Take rollaway bed out of guest room.

**Rooming List:** List of names submitted in advance by buyer to occupy previously reserved accommodations.

**Run-of-the-House Rate:** An agreed-upon rate generally priced at an average figure between minimum and maximum for group accommodations for all available rooms except suites. Room assignments usually made on a best-available basis.

**Runner:** Person who supplies restaurants with food from storage areas.

**Sample:** A display room used for showing merchandise. It may or may not be provided with sleeping facilities.

**Security:** The department in charge of protecting both employees and guests from thefts and vandalism.

**Single:** Room with one bed for one person.

**Skips:** People who leave hotel without paying bill.

**Sleep Out:** Guest who rented room but did not sleep there.

**Sommelier:** Wine waiter.

**Sous-Chef:** Cook who is second in command of kitchen; in large hotels there may be several, each in charge of a specific restaurant.

**Stewards:** In kitchens: dishwashers, often in charge of storing and transporting dishes, glassware, and tableware; banquet stewards set up and dismantle furniture in banquet and meeting rooms.

**Studio:** A one-room parlor set-up having one or two couches that convert to a bed (sometimes called an executive room).

**Suite:** A parlor connected to one or more bedrooms. When requesting a suite, guests should always designate the number of bedrooms needed in a suite, since several can connect.

**Support Positions:** Those jobs that directly assist a department head.

**Table d'Hote:** Full course meal with limited choice as opposed to a la carte.

**Tel Autograph:** Used to produce a facsimile telegram, reproducing graphic matter by means of a transmitter in which the motions of a pencil are reproduced by a receiving pen controlled by electromagnetic devices.

**Third Sheet:** A night spread or cover used to protect blanket.

**Tidy-Up:** To straighten and clean a room after guest's departure when full service has been given earlier.

**Toilet Tissue:** Toilet paper—not to be confused with facial tissue—placed for convenience of guests in bathrooms or public lavatories.

**Tourist or Economy:** Commercial-type hotel (usually without private bath).

**Trainee:** Usually a person in training for further management responsibility.

**Turndown:** Evening service—removing bedspread and turning down bed, straightening room, and replenishing used supplies and linen.

**Twin:** A room with twin beds.

**Twin Double:** A room with two double beds for two, three, or four persons; sometimes called a family room or double-double.

**Vacant and Ready:** A room that is unoccupied, cleaned, and ready for renting.

**VIP:** Very important person.

**Walkie Talkie—2-Way:** Communication system where messages can be both received and answered.

**Walk-Ins:** A guest who appears at hotel in person and requests room when not holding a reservation.

**Water Closet:** Toilet.

**"We Walked One Or Two":** A few people with reservations could not be accommodated so they were taken (gratis) to another hotel where rooms had been procured and paid for.

# *Appendix*

## *Metrics—Measurements of the Future*

Every phase of a lodging operation will be affected by the switch to the metric system. The new wing about to be added to your building will be located on a hectare (about 2.5 acres); laundry will be weighed in kilograms instead of pounds; carpeting will be measured in meters or square meters instead of feet, inches, and square yards; and detergents may come in 4-liter containers instead of gallons.

Although it may be up to ten years before the United States switches entirely to metrics, the complexity and expense of the changeover will require careful planning. Management should take the planned approach rather than casually drifting into the new system or instigating a crash program that will upset employees. Having to replace tools or containers, recalibrate equipment, translate documents, re-mark worksheets, and train all employees, *will* be costly but can be simplified (and may even be interesting) if procedures are studied well in advance.

Formally, the metric system is known as the International System of Units (abbreviated SI), and its adoption will ultimately make all calculations in the Housekeeping Department easier, whether they are being done by a seamstress measuring draperies or by the specifier of cleaning supplies.

"Despite Congressional stalling that has delayed passage of a metric act for the last 196 years," says Dean A. C. Ingersoll, director of Continuing Education in Engineering and Mathematics at the University of California, at Los Angeles, "metrication is inevitable, as evidenced by the near-record turnout for the UCLA metric conference; it drew the largest number of registrants since our post-Sputnik seminar in 1957. Before long, the English units that the U.S. now uses—ounces, yards, etc.—will sound as archaic as terms such as stones, furlongs and barleycorns."

The metric system is a single, simplified language built on units of tens, hundreds, and thousands. The American currency system is a perfect example of metrics, with ten pennies to a dime, ten dimes to a dollar. One can quickly convert $100.00 into anything from $50.00 to $1.00 bills, or even into nickels or dimes.

When measuring length in metrics, ten millimeters equal one centimeter; ten centimeters equal one decimeter; ten decimeters equal one meter, and so on. Area and volume are measured in square or cubic meters; capacity (or volume) in liters; mass and weight in grams. The logic of the metric system is again illustrated by the fact that these are all based on the power of 10.

Training materials—slides, publications, cassettes—need not necessarily be custom-designed, but it is absolutely imperative that they meet the highly specific needs of the employees who are to learn from them. Employees should be taught only what they need to know to perform the functions of their specific jobs. A houseman would not need to know about angular accelerations or kinematic viscosity.

A helpful current text is *Thinking Metric,* by Thomas and Marilyn Gilbert.[1] An easy reference and self-teaching guide, it stresses the need to relate metric measurements to everyday things so that people can remember easily relative sizes and weights rather than memorizing SI equivalents of the English system. For example, once one realizes that a little fingernail is 10 millimeters wide and recalls how wide a 35-mm slide is, then one can begin to think metric.

---

1. Thomas and Marilyn Gilbert, *Thinking Metric,* (New York: John Wiley & Sons, Inc., 1973).

## METRIC CONVERSION CHART

**Approximate Conversions TO Metric Measures**

| WHEN YOU KNOW | MULTIPLY BY | TO FIND |
|---|---|---|
| | *LENGTH* | |
| inches | 2.5 | cm |
| feet | 30 | cm |
| yards | 0.9 | m |
| miles | 1.6 | km |
| | *AREA* | |
| sq. inches | 6.5 | sq. cm |
| sq. feet | 0.09 | sq. m |
| sq. yards | 0.8 | sq. m |
| sq. miles | 2.6 | sq. km |
| acres | 0.4 | hectares |
| | *MASS (Weight)* | |
| ounces | 28 | g |
| pounds | 0.45 | kg |
| short tons (2,000 lb) | 0.9 | tonnes |
| | *VOLUME* | |
| teaspoons | 5 | ml |
| tablespoons | 15 | ml |
| fluid ounces | 30 | ml |
| cups | 0.24 | liters |
| pints | 0.47 | liters |
| quarts | 0.95 | liters |
| gallons | 3.8 | liters |
| cubic feet | 0.03 | cubic meters |
| cubic yards | 0.76 | cubic meters |
| | *TEMPERATURE (exact)* | |
| Fahrenheit temperature | 5/9 (after subtracting 32) | Celsius temp. |

**Approximate Conversions FROM Metric Measures**

| WHEN YOU KNOW | MULTIPLY BY | TO FIND |
|---|---|---|
| | *LENGTH* | |
| millimeters | 0.04 | inches |
| centimeters | 0.4 | inches |
| meters | 3.3 | feet |
| kilometers | 0.6 | miles |
| | *AREA* | |
| sq. centimeters | 0.16 | sq. inches |
| sq. meters | 1.2 | sq. yards |
| sq. kilometers | 0.4 | sq. miles |
| hectares (10000 $m^2$) | 2.5 | acres |
| | *MASS (Weight)* | |
| grams | 0.035 | ounces |
| kilograms | 2.2 | pounds |
| tonnes (1,000 kg) | 1.1 | short tons |
| | *VOLUME* | |
| milliliters | 0.03 | fluid ounces |
| liters | 2.1 | pints |
| liters | 1.06 | quarts |
| liters | 0.26 | gallons |
| cubic meters | 35 | cubic feet |
| cubic meters | 1.3 | cubic yards |
| | *TEMPERATURE (exact)* | |
| Celsius temperature | 9/5 (then add 32) | Fahrenheit temp. |

### STYLE

| **Basic Units** | **Abbrev.** |
|---|---|
| meter | m |
| liter | l |
| gram | g |
| kilojoule | kj |

| **Prefixes** | **Abbrev.** | **Number** |
|---|---|---|
| milli | m | .001 |
| centi | c | .01 |
| deci | d | .1 |
| deka | da | 10 |
| hecto | h | 100 |
| kilo | k | 1000 |

Prefixes are always used with units.
  Ex: mm = millimeter, cm = centimeter
No periods are used after abbreviations.
  Ex: cm not cm.
No plurals are used with abbreviations.
  Ex: 3mm not 3mms

Spaces separate numbers, not commas.
  Ex: 439 268.32 not 439,268.32
Decimals are used, not fractions.
  Ex: 1.5 not 1-1/2

**Length.** Common metric units of length are: *millimeter* for small dimensions; *centimeter* for daily practical use; *meter* for expressing dimensions of larger objects (and short distances), and *kilometer* for longer distances. The centimeter is about four-tenths of an inch. Meter is about 40 in., and kilometer about six-tenths of a mile. When drawing to metric scale, engineering and product dimensions are in millimeters; architectural drawings can be in millimeters or centimeters. On land surveys, meter is the unit; for maps kilometer is used.

**Area.** Small areas are usually measured in square centimeters ($c^2$). In building and construction, square meter ($m^2$) is used and is about 20% larger than a square yard. The hectare, used for land surveys, is about 2.5 acres.

**Volume.** For volume, most convenient unit is cubic decimeter ($dm^3$), commonly referred to as the liter. A liter is slightly larger than the U.S. liquid quart, but smaller than the U.S. dry quart and British Imperial quart.

**Mass.** There are slightly less than 30 grams in one avoirdupois ounce. For most other uses, kilogram is convenient and is approximately 2.2 pounds. The metric ton, 1000 kg, is used for large shipments. A liter of water at standard temperature and pressure has a mass of one kilogram.

**Temperature.** All countries using the Metric System, use the Celsius (°C) scale (formerly centigrade) for ordinary measurement of temperature. On the Celsius scale, pure water at standard atmospheric pressure freezes at 0°; boils at 100°. Normal human body temperature is 37°; comfortable room temperature about 22°.

## Bibliography


American National Standards Institute. *Standard Metric Practice Guide* (ASTM E 380-70). 1430 Broadway, New York, N.Y. 10018.

Brown and Sharpe, Industrial Products. *Metric Planning.* North Kingston, R.I. 02852.

Chisholm, L. J. *Units of Weight and Measure: International (Metric) and U.S. Customary; Definitions and Tables of Equivalents.* National Bureau of Standards. NBS Misc. Pub. 286.

U.S. Government Printing Office. *The Modernized Metric System* (wall chart). Superintendent of Documents, Washington, D.C. 20402. NBS Special Publication 304.


## Sources of Information

American National Standards Institute
1430 Broadway
New York, N.Y. 10018

Library Filmstrip Center, 3033 Aloma, Wichita, Kansas, 67211.

Swani Publishing Co., P. O. Box 248, Roscoe, Ill. 61073.

Global Engineering Documentation Services, Inc., P. O. Box 2060, Newport Beach, Calif. 92660

Metric Assn., Inc., Sugarloaf Star Route, Boulder, Colo. 80302.

# *About the Authors*

**Georgina Tucker** is herself truly a professional Housekeeper. Now a consultant who advises hoteliers in all parts of the world, Mrs. Tucker has been involved with executive foodservice or housekeeping in hotels throughout her career. She was Director of Housekeeping for Los Angeles' famed Century Plaza Hotel for over 10 years, having opened it as one of Western International Hotel's flagship properties. She began her association with WIH after her graduation from Washington State University in the 1930s, and most of her career was with WIH.

Her assignments as a consultant have been with hotels from coast to coast, in Hawaii, Alaska, Japan, Canada and Greece. She helped to open WIH properties in Japan and the Far East, as well as Mexico City and the United States.

Starting in 1966 when she assumed her position as Century Plaza's Director of Housekeeping, she was active in the National Executive Housekeepers Association of which she was a board member, and the Los Angeles Hotel Housekeepers Association, which she helped found. She has been an editorial board member of *The Executive Housekeeper Magazine;* and was a member of the American Home Economics Association, and Home Economists in Business, having been chairman of the Washington State Chapter.

One of the few educators in her field, she has taught Executive Housekeeping at Pepperdine University (Los Angeles); lectured at numerous colleges and universities, and provided internships for students seeking degrees in Institutional Housekeeping. She is helping to establish housekeeping courses at La Canada Community College on the San Francisco Bay peninsula, and instructing in Housekeeping Management at Golden Gate University in San Francisco. Mrs. Tucker is listed in *Who's Who of American Women,* and is the author of another book, *The Science of Housekeeping.* In 1978 she was awarded the Home Economics Distinguished Alumnus Award by Washington State University.

**Madelin Schneider,** Editor of *Hotels & Restaurants International* magazine, (formerly *Service World International*) Mrs. Schneider served as Executive Editor of *Restaurants & Institutions Magazine* for three years and has been affiliated with the publication since 1958. She began as Furnishings-Decorating-Housekeeping Editor, then served as Design Trends Editor in the magazine's Chicago headquarters. From 1963 through 1968, while writing part-time for the magazine, she earned her teaching credentials at San Francisco State University and taught school in California at both the high school and college levels. Her subjects were Journalism, English, Design, and Arts and Crafts.

In 1968, she set up the West Coast Editorial Office for *Institutions Magazine* in San Francisco, then in 1972 located in Los Angeles, where she was responsible for writing and researching articles on Maintenance, Product Design, Interior Design, Management, Training, and Education in the Foodservice and Lodging Industry.

Mrs. Schneider has lectured at hotel, restaurant, and institutional management courses, and taught institutional interior design. Because she is continually interviewing knowledgeable management people in the lodging industry, she has become thoroughly familiar with both the problems and the solutions that are presented in *The Professional Housekeeper.* Mrs. Schneider also served as editor of *The Science of Housekeeping.*

Now in Cahners Publishing's Chicago headquarters, Mrs. Schneider supervises an editorial staff, and continues to interview, lecture, and write.

# Index